Contents

KU-524-682

Oxford Handbook of
Clinical Medicine

Oxford Handbook of Clinical Medicine

THIRD EDITION

R. A. HOPE
J. M. LONGMORE
T. J. HODGETTS
and
P. S. RAMRAKHA

Oxford   New York   Tokyo

OXFORD UNIVERSITY PRESS

Oxford University Press, Great Clarendon Street, Oxford OX2 6DP

Oxford New York
Athens Auckland Bangkok Bogota Bombay Buenos Aires
Calcutta Cape Town Dar es Salaam Delhi Florence Hong Kong
Istanbul Karachi Kuala Lumpur Madras Madrid Melbourne
Mexico City Nairobi Paris Singapore Taipei Tokyo Toronto
and associated companies in
Berlin Ibadan

Oxford is a trade mark of Oxford University Press

Published in the United States
by Oxford University Press Inc., New York

© R. A. Hope and J. M. Longmore 1985;
R. A. Hope, J. M. Longmore, P. A. H. Moss, and A. N. Warrens 1989;
R. A. Hope, J. M. Longmore, T. J. Hodgetts, and P. S. Ramrakha 1993

First published 1985
Second edition 1989
Indian impression 1985
German translation 1988
French translation 1989
Spanish translation 1991
Third edition 1993
Reprinted with corrections and updates 1993, 1994, 1996, 1997
Romanian translation 1995
Reprinted with further corrections and updates 1996

A catalogue record for this book is available from the British Library

Library of Congress Cataloging-in-Publication Data available

ISBN 0 19 262115 7

Printed in China

Preface to the third edition

When we embarked on this third edition we were expecting to do no more than add a few updates to each section. But medical practice as well as medical science has changed enormously in only three years. There is hardly a page which has not been substantially rewritten. Even an institution as solid and permanent as the British Home Office has altered its telephone number.

Dr Tim Hodgetts and Dr Punit Ramrakha have rewritten and reworked the text to ensure that it is not only up-to-date but also exactly relevant to today's clinicians and clinical students. Their knowledge and perspectives have immensely enhanced the book, and working with them has been the greatest pleasure in preparing this new edition.

The main difficulty in writing the third edition has been to keep it down to size. The editions, like our children, have a tendency to grow. Although we have been keen to curb its growth, behind the *Handbook* there is a ghost which is growing and changing almost every day. This ghost, like Descartes's conception of the human mind, is in a machine. It is the updatable electronic version. The version of this work which you hold in your hand is a snapshot of this ghost. As is the way with even the best family snapshots, some members are inevitably caught misbehaving—thrusting themselves forward, or hiding behind each other, as they compete for the photographer's attention. This ceaseless ferment is not to be deprecated: it is the essence of late twentieth-century medicine. (You have the possibility of improving this snapshot in a number of ways with your own additions to the blank pages so that the book becomes a personal record of your developing view of medicine.) The OUP/*BMJ* update boxes, the updating fax service, and the electronic version (p16) are all now available to help in this process.

New chapters in this edition include *Radiology* and *Oncology*. Other new topics include the major disaster; burns; blast injury, and a selection of emergency procedures. There are new pages on essential epidemiology; health economics and QALYs; experts and expert-systems; impairment, disability and handicap; stroke, falls and hypotension; kinder interventions in geriatric medicine; travel advice; needle-stick injuries; TB-with-AIDS; atrial septal and ventriculoseptal defects; the treatment of acute renal failure; vascular disease and renal failure; haemochromatosis; hyperviscosity syndromes; and thyroid eye disease.

There are also some thumb-nail descriptions of past medical personalities.

1993

J.M.L.
R.A.H.

From the preface to the first edition

This book, written by junior doctors, is intended principally for medical students and house officers. The student becomes, imperceptibly, the house officer. For him we wrote this book not because we know so much, but because we know we remember so little. For the student the problem is not simply the quantity of information, but the diversity of places from which it is dispensed. Trailing eagerly behind the surgeon, the student is admonished never to forget alcohol withdrawal as a cause of post-operative confusion. The scrap of paper on which this is written spends a month in the white coat pocket before being lost for ever in the laundry. At different times, and in inconvenient places, a number of other causes may be presented to the student. Not only are these causes and aphorisms never brought together, but when, as a surgical house officer, the former student faces a confused patient, none is to hand.

This book is intended for use on the ward, in the lecture-theatre, library, and at home. Each subject only occupies one page, and opposite is a blank page for additions, updatings, and refinements. If they face the relevant page of print, they will be automatically indexed.

Clinical medicine has a habit of partly hiding as well as partly revealing the patient and his problems. We aim to encourage the doctor to enjoy his patients: in doing so we believe he will prosper in the practice of medicine. For a long time now, house officers have been encouraged to adopt monstrous proportions in order to straddle simultaneously the diverse pinnacles of clinical science and clinical experience. We hope that this book will make this endeavour a little easier by moving a cumulative memory burden from the mind into the pocket, and by removing some of the fears that are naturally felt when starting a career in medicine, thereby freely allowing the doctor's clinical acumen to grow by the slow accretion of many, many days and nights.

R.A.H.
J.M.L.

Drugs

While every effort has been made to check drug dosages in this book, it is still possible that errors have been missed. Dosage schedules are continually being revised and new side-effects recognized. Oxford University Press makes no representation, express or implied, that drug dosages in this book are correct. For these reasons, the reader is strongly urged to consult the *Update* section of the electronic form of this text (see p17–18), the *British National Formulary*, or the drug company's printed instructions before administering any of the drugs recommended in this book.

▶ Except where otherwise stated, drug doses and recommendations are for the non-pregnant adult who is not breast feeding.

Acknowledgements

In preparing this third edition we have had a great deal of help from a number of people. We would like to thank, in particular, those who advised on specific sections of the book, giving us the benefit of their experience and knowledge: Dr C Alcock, Dr N Beeching, Ms J Boorman, Dr J Brierley, Dr D Chapman, Dr A Edwards, Ms R James, Ms A Jennings, Dr J Kay, Dr T Littlewood, Dr G Luzzi, Dr W McCrea, Ms S Nicholls, Dr G Rocker, Dr J Satsangi, Dr D Sprigings, Mr P Stanton, Dr D Walters, Dr J Wee, Ms E Weekes, Dr S Winner, Dr A Zeman.

We thank Mr E Buckley and Mrs Anne Joshua for their great help in harmonizing the electronic and paper versions of this text. Their unstinting help has enabled one of our favourite typographic maxims to be approached in this work: *to be perfect is to have changed many times.*

This book can only keep abreast of readers' needs through the enormous amount of feedback which readers send us. We have tried to write to all who have been kind enough to write to us and apologize to anyone we have missed. In some instances comments were sent but either no name was given or we were unable to decipher the writing. We thank these anonymous people as well as the following for their valuable help in developing this handbook: R Adamson, X Airton, L Alan, A Aldridge, M Ali, A Alizai, S Al-Motairi, C Antonetti, M Anwar, P Atack, B Badgruddin, A Baig, N Balaswiya, M Beirne, M Beranek, E Berinou, I Bleehen, K Boddu, B Bourke, A Broadbent, J Brooks, J Brown, K Burn, P Buttery, E Cameron, P Camosa, N Caporaso, O Carr, R Casson, R Chaira, W Chicken, T Choudhry, G Chowdharay-Best, J Collen, E Collier, M Collins, B Colvin, R Coull, J Cox, A Das-Gupta, Y de Boer, U Desai, A de Silva, J Devine, E Dickinson, A Ditri, J Dobbie, N Duthie, T Eden, V Eden, N Elkan, D Ellis, A Ezigweth, M Faragher, C Finfa, P Fisichella, E Fitzpatrick, M Gaba, M Gagan, I Ganans, J Germain, S Ghaem, J Gibson, J Gilbody, J Gotz, H Gray, D Habart, E Hammond, C Hanna, R Haughton, R Haynes, W Hell, T Hennigan, D Herold, M Hewitt, P Hollinshead, J Hopkinson, P Hrincko, I Ibrahim, A Iqbal, A Jennings, G Johnson, N Joshi, A Joshua, Z Kango, A Kelly, A Khalid, Y Kontrobarsky, H Kuralco, J Lagercranbe, G Lane, F Larkin, R Lawson, S Lee, J Lewis, J Linder, G Lomax, M Lowenthal, G Lyons, P Mackey, C Martin, T Matteveci, A McCafferty, D McCandless, H McIntosh, B Melrad, W Mgaya, K Mitchell, S Monella, M Morgan, B Morton, D Moskopp, S Moultrie, K Muthe, E Neoh, M Newell, B Pait, B Parkin, N Paros, F Phraner, F Pilsczek, D Poppe, M Porter, E Pringle, M Procopiou, J Pryse, B Pÿnenburg, A Ramachandran, N Richmond, M Robert, J Robinson, A Rosenthal, J Rudd, S-U Safer, P Saunders, J Schmitt, S Schwarz, P Seeley, F Sellers, V Sepe, Shahnaz-Hayat, I Silva, E Smeland, R Smit, K Spreadbury, D Stead, J Stern, A Struber, A Taimoor, H Talat, A Tavakkolizadeh, D Taylor, N Taylor, M Tengoe, L Thomson, W Thomson, T Toma, D Turner, S Vaidya, C van der Worp, J Verghese, L Walker, S Walker, T Walton, L Watson, H Weerasooriya, W Westall, T Whitehouse, T Wiggin, J Williamsen, M Wong, J Wood, R Zajdel.

The British Lending Library, and staff at the Cairns Library, Oxford, and at the Worthing Postgraduate Library have been most helpful in tracing references and books.

Finally we would like to thank the staff of OUP for their help and support.

Symbols and abbreviations

►	this is important
►►	don't dawdle! Prompt action saves lives
♂/♀	male to female ratio
~	approximately
−ve	negative
+ve	positive
↓	decreased
↑	increased
↔	normal
°	degrees
A₂	aortic component of 2nd heart sound
Ab	antibody
Ag	antigen
ABPA	allergic bronchopulmonary aspergillosis
ac	*ante cibum* (before food)
ACE	angiotensin-converting enzyme
ACTH	adrenocorticotrophic hormone
ADH	antidiuretic hormone
ADL	activities of daily living
AF	atrial fibrillation
AFB	acid-fast bacillus
AFP	(and α-FP) alpha-fetoprotein
AIDS	acquired immunodeficiency syndrome
Al	aluminium
ALL	acute lymphoblastic leukaemia
Alk phos	alkaline phosphatase (also ALP)
AMP	adenosine monophosphate
ANA	antinuclear antibody
ANAC	anti-neutrophil cytoplasmic antibody
APTT	activated partial thromboplastin time
ARDS	adult respiratory distress syndrome
ASD	atrial septal defect
ASO	antistreptolysin O
AST	asparate transaminase
ATN	acute tubular necrosis
AV	atrioventricular
AVM	arteriovenous malformation(s)
AXR	abdominal x-ray (plain)
AZT	zidovudine
Ba	barium
BAL	bronchoalveolar lavage
BCR	British comparative ratio (≈INR)
bd	*bis die* (twice a day)
BKA	below-knee amputation
BMJ	*British Medical Journal*
BNF	*British National Formulary*
BP	blood pressure
Ca	carcinoma
CABG	coronary artery bypass graft
cAMP	cyclic AMP
CAPD	chronic ambulatory peritoneal dialysis

CBD	common bile duct
CC	creatinine clearance
CCF	congestive cardiac failure (ie RVF with LVF)
CHB	complete heart block
CI	contraindications
CK	creatine (phospho)kinase (also CPK)
CLL	chronic lymphocytic leukaemia
CML	chronic myeloid leukaemia
CMV	cytomegalovirus
CNS	central nervous system
COAD	chronic obstructive airways disease
COC	combined oral contraceptive (oestrogen+progesterone)
CPAP	continuous positive airways pressure
CPR	cardio-pulmonary resuscitation
CRP	C-reactive protein
CRF	chronic renal failure
CSF	cerebrospinal fluid
CT	computerized tomography
CVP	central venous pressure
CVS	cardiovascular system
CXR	chest x-ray
d	day(s) (also expressed as x/7)
DIC	disseminated intravascular coagulation
DIP	distal interphalangeal
dl	decilitre
DoH	Department of Health
DM	diabetes mellitus
DU	duodenal ulcer
D&V	diarrhoea and vomiting
DVT	deep venous thrombosis
DXT	deep radiotherapy
EBV	Epstein–Barr virus
ECG	electrocardiogram
Echo	echocardiogram
EDTA	ethylene diamine tetraacetic acid (full blood count bottle)
EEG	electroencephalogram
EM	electron microscope
EMG	electromyogram
ENT	ear, nose, and throat
ESR	erythrocyte sedimentation rate
EUA	examination under anaesthesia
FB	foreign body
FBC	full blood count
FDP	fibrin degradation products
FEV_1	forced expiratory volume in 1st second
FFP	fresh frozen plasma
F_IO_2	partial pressure of O_2 in inspired air
FROM	full range of movements
FSH	follicle-stimulating hormone
FVC	forced vital capacity
g	gram
GA	general anaesthetic
GB	gall bladder
GC	gonococcus
GFR	glomerular filtration rate
GGT	gamma glutamyl transpeptidase

GH	growth hormone
GI	gastrointestinal
GP	general practitioner
G6PD	glucose 6-phosphate dehydrogenase
GTT	glucose tolerance test (also OGTT: oral GTT)
GU	genitourinary
h	hour
HAV	hepatitis A virus
Hb	haemoglobin
HBsAg	hepatitis B surface antigen
HBV	hepatitis B virus
Hct	haematocrit
HCV	hepatitis C virus
HDV	hepatitis D virus
HHT	hereditary haemorrhagic telangiectasia
HIDA	hepatic immunodiacetic acid
HIV	human immunodeficiency virus
HOCM	hypertrophic obstructive cardiomyopathy
HONK	hyperosmolar non-ketotic (diabetic coma)
HRT	hormone replacement therapy
HSV	herpes simplex virus
ICP	intracranial pressure
IDA	iron-deficiency anaemia
IDDM	insulin-dependent diabetes mellitus (compared with NIDDM: non-IDDM)
Ig	immunoglobulin
IHD	ischaemic heart disease
IM	intramuscular
IFN-α	alpha interferon
INR	international normalized ratio (prothrombin ratio)
IPPV	intermittent positive pressure ventilation
ITP	idiopathic thrombocytopenic purpura
ITU	intensive therapy unit
iu/IU	international unit
IV	intravenous
IVI	intravenous infusion
IVC	inferior vena cava
IVU	intravenous urography
JAMA	*Journal of the American Medical Association*
JVP	jugular venous pressure
KCCT	kaolin cephalin clotting time
kg	kilogram
kPa	kiloPascal
L	left
l	litre
LBBB	left bundle branch block
LDH	lactate dehydrogenase
LFT	liver function test
LH	luteinizing hormone
LIF	left iliac fossa
LKKS	liver, kidney (R), kidney (L), spleen
LMN	lower motor neurone
LP	lumbar puncture
LUQ	left upper quadrant
LVF	left ventricular failure
LVH	left ventricular hypertrophy

μg	microgram
MCV	mean cell volume
MDMA	3,4-methylene dioxymethamphetamine
ME	myalgic encephalomyelitis
MET	maximal exercise test
mg	milligram
MI	myocardial infarction
min(s)	minute(s)
ml	millilitre
mmHg	millimetres of mercury
MND	motor neurone disease
MRI	magnetic resonance imaging
MS	multiple sclerosis (do not confuse with mitral stenosis)
MSU	midstream urine
NAD	nothing abnormal detected
NBM	nil by mouth
ND	notifiable disease
NEJM	*New England Journal of Medicine*
NG(T)	nasogastric (tube)
NIDDM	non-insulin-dependent diabetes mellitus
NMDA	N-methyl-D-aspartate
NR	normal range (=reference interval)
NSAID	non-steroidal anti-inflammatory drugs
N&V	nausea and/or vomiting
OCP	oral contraceptive pill
od	*omni die* (once daily)
OD	overdose
OGD	oesophagogastroduodenoscopy
OGTT	oral glucose tolerance test
OGS	oxogenic steroids
OHCS	*Oxford Handbook of Clinical Specialties* (4th ed OUP)
om	*omni mane* (in the morning)
on	*omni nocte* (at night)
OPD	out-patients department
ORh−	blood group O, rhesus negative
OT	occupational therapist
OTM	*Oxford Textbook of Medicine* (OUP 1987)
P₂	pulmonary component of 2nd heart sound
PaCO₂	partial pressure of CO_2 in arterial blood
PaO₂	partial pressure of O_2 in arterial blood
PAN	polyarteritis nodosa
PBC	primary biliary cirrhosis
PCR	polymerase chain reaction
PCV	packed cell volume
PE	pulmonary embolism
PEEP	positive end-expiratory pressure
PEFR	peak expiratory flow rate
PERLA	pupils equal and reactive to light and accommodation
PID	pelvic inflammatory disease
PIP	proximal interphalangeal
PL	prolactin
PND	paroxysmal nocturnal dyspnoea
PO	*per os* (by mouth)
PR	*per rectum* (by the rectum)
PPF	purified plasma fraction (albumin)
PRL	prolactin

PRN	*pro re nata* (as required)
PRV	polycythaemia rubra vera
PTH	parathyroid hormone
PTT	prothrombin time
PUO	pyrexia of unknown origin
PV	*per vaginam* (by the vagina)
qid	*quater in die* (4 times a day)
qds	*quater die sumendus* (to be taken 4 times a day)
qqh	*quarta quaque hora* (every 4 hours)
R	right
RA	rheumatoid arthritis
RBBB	right bundle branch block
RBC	red blood cell
RFT	respiratory function tests
Rh	rhesus
RhF	rheumatic fever
RIF	right iliac fossa
RUQ	right upper quadrant
RVF	right ventricular failure
RVH	right ventricular hypertrophy
R_x	*Recipe* (treat with)
s/sec	second(s)
S_1, S_2	first and second heart sounds
SBE	subacute bacterial endocarditis (more correctly IE = infective endocarditis)
SC	subcutaneous
SD	standard deviation
SE	side-effect(s)
SL	sublingual
SLE	systemic lupus erythematosus
SOB	short of breath (SOBOE: on exercise)
SR	slow release
stat	*statim* (immediately; as initial dose)
SVC	superior vena cava
SXR	skull x-ray
T°	temperature
$t_{1/2}$	biological half-life
T_3	triiodothyronine
T_4	thyroxine
TB	tuberculosis
TIA	transient ischaemic attack
tid	*ter in die* (3 times a day)
tds	*ter die sumendus* (to be taken 3 times a day)
TRH	thyroid-releasing hormone
TPR	temperature, pulse, and respirations count
TSH	thyroid-stimulating hormone
u/U	units
UC	ulcerative colitis
U&E	urea and electrolytes
UMN	upper motor neurone
URT	upper respiratory tract
URTI	upper respiratory tract infection
USS	ultrasound scan
UTI	urinary tract infection
VDRL	venereal diseases research laboratory
VE	ventricular extrasystole

VF	ventricular fibrillation
VMA	vanilyl mandelic acid (HMMA)
V/Q	ventilation:perfusion ratio
VSD	ventriculo-septal defect
WBC	white blood cell
WCC	white cell count
wk(s)	week(s)
WR	Wasserman reaction
yr(s)	year(s)
ZN	Ziehl–Neelsen

Note: other abbreviations are given in full on the pages where they occur.

Topics adapted since the last printing

This page documents principal updatings from 1993.07.02 to 1997.01.04
► See the electronic version (p16–17) for more recent updates, and a description of what each update entails.

The *greater than* signs (>) indicate how important the update is; one is the minimum (least important); five is the maximum (most important). ►► indicates emergency topics.

Topics adapted since the last printing (cont.)

Contents

Note: the content of individual chapters is detailed on each chapter's first page.

Love in the time of cholera[1]

'[Dr. Juvenal Urbino] tried to impose the latest ideas at Misericordia Hospital, but this was not as easy as it had seemed in his youthful enthusiasm, for the antiquated house of health was stubborn in its attachment to atavistic superstitions, such as standing beds in pots of water to prevent disease from climbing up the legs, or requiring evening wear and chamois gloves in the operating room because it was taken for granted that elegance was an essential condition for asepsis. They could not tolerate the young newcomer's tasting a patient's urine to determine the presence of sugar, quoting Charcot and Trousseau as if they were his roommates . . . while maintaining a suspicious faith in the recent invention of suppositories.

Cholera became an obsession for him. He did not know much more about it than he had learned in a routine manner in some marginal course, when he had found it difficult to believe that only thirty years before, it had been responsible for more than one hundred forty thousand deaths in France, including Paris. But after the death of his father . . . he studied with the most outstanding epidemiologist of his time, and the creator of the *cordons sanitaires*, Professor Adrien Proust, father of the great novelist . . .

In less than a year his students at Misericordia Hospital asked for his help in treating a charity patient with a strange blue coloration all over his body. Dr. Juvenal Urbino had only to see him from the doorway to recognize the enemy . . . Dr. Juvenal Urbino alerted his colleagues and had the authorities warn the neighbouring ports . . . and he had to restrain the military commander of the city who wanted to declare martial law and initiate the therapeutic strategy of firing the cannon every quarter hour [to purify the air] . . .

The patient died in four days, choked by a grainy white vomit, but in the following weeks no other case was discovered despite constant vigilance. A short while later *The Commercial Daily* published the news that two children had died of cholera . . . Her parents and three brothers were separated and placed under individual quarantine, and the entire neighborhood was subjected to strict medical supervision. One of the children contracted cholera but recovered very soon, and the entire family returned home when the danger was over. Eleven more cases were reported in the next three months, and in the fifth there was an alarming outbreak, but by the end of the year it was believed that the danger of an epidemic had been averted. No one doubted that the sanitary rigor of Dr. Juvenal Urbino, more than the efficacy of his pronouncements, had made the miracle possible. From that time on, and well into this century, cholera was endemic not only in the city but along most of the Caribbean coast and the valley of the Magdalena, but it never again flared into an epidemic. The crisis meant that Dr. Juvenal Urbino's warnings were heard with greater seriousness by public officials. They established an obligatory Chair of Cholera and Yellow Fever in the Medical School, and realized the urgency of closing up the sewers and building a market far from the garbage dump. By that time, however, Dr. Urbino was not concerned with proclaiming victory, nor was he moved to persevere in his social mission, for at that moment one of his wings was broken, he was distracted and in disarray and ready to forget everything else in life, because he had been struck by the lightning of his love for Fermina Daza.'

1 Gabriel García Márquez 1989 *Love in the Time of Cholera*, trans. E. Grossman, Penguin Books. Reproduced with kind permission of the original publishers Jonathan Cape. Copyright © 1988 by Alfred A. Knopf, Inc. Reprinted by permission of the publisher.

Thinking about medicine 1

Ideals

2   Decision and intervention are the essence of action: reflection and conjecture are the essence of thought: the essence of medicine is the combining of these realms of action and thought in the service of others. We offer the following ideals to stimulate both thought and action: these ideals are hard to reach—like the stars—but they can serve for navigation during the night.

- Do not blame the sick for being sick.

- If the patient's wishes are known, comply with them.

- Work for your patients, not your consultant.

- Use ward rounds to boost the patients' morale, not your own.

- Treat the whole patient, not the disease—or the ward sister.

- Admit people, not 'strokes', 'infarcts' or 'crumble'.

- Spend time with the bereaved; you can help them shed tears.

- Question your conscience—however strongly it tells you to act.

- The ward sister is usually right; respect her opinion.

- Be kind to yourself—you are not an inexhaustible resource.

How to advise a patient

It is an interesting fact that many patients do not take our advice: fortunately, as we rarely check compliance, it is a fact of which we remain ignorant. As our advice can be quite good, it is wise to try to increase the chance that it is taken.

Avoid jargon 'Jaundice' meant 'generally sick and unwell with yellow vomit' to 10% of patients in one study.

Avoid vague terms These will mean one thing to the doctor and another to the patient; studies show that the patient's interpretation will be more pessimistic than the doctor's. For example, 'you might possibly need a colostomy' meant that the chance was <50% to the surgeon, but was interpreted by the patient to mean that the chance was >60%. Use figures such as '1-in-10', and make sure the patient understands them.

Help patients to remember Give the most important instruction first. This will be remembered best. Expect half of all subsequent advice to be forgotten.[1] Two 'tricks' help here:

Explicit categorization: 'I will tell you the treatment first, then what is wrong, and then the outcome. First the treatment . . .'

Specific advice: 'Lose 1 lb each week', not 'lose weight'.

Give written information This increases compliance.[2] Use short sentences with short words—more like the *Sun* than the *Daily Telegraph*. Flesch's formula[2] quantifies this: 'reading ease' $= 207 - 0.85\,W - S$ (W is the mean number of syllables per 100 words; S is the mean number of words per sentence). Most prose falls between 0 and 100 (100 is very easy). Aim for >70.

NB: not all patients can read, and many non-readers will be shy about revealing this fact. One tactful way to find out if your patient can read is to use the chart on p37. If he cannot read *Blood* in large type but correctly reports that the octopus in the series of pictures is holding a small bucket, then reading *is* a problem, and this fact is ascertained via a non-threatening eye test.

If a video machine is available, this may be helpful (not just to non-readers) in showing how to use an inhaler in asthma etc. Videos are available (*What you really want to know about . . .*) for topics such as asthma, bronchitis, colostomies, Crohn's disease, ulcerative colitis, depression, and glue ear—starring well-known comedians.[3] (Advice may be more readily acted on if it comes from a well-known, well-loved source.)

Be persuasive The satisfied patient takes your advice.[3] Be friendly; have pleasant surroundings where possible; keep waiting times short. Find out what the patient expects from you.

Five questions when prescribing off the ward
- How many daily doses are there? (Three is better than four.)
- Are many other drugs to be taken? Can they be reduced?
- The bottle: can he read the instructions—and can he open it?
- How will you measure how effective your treatment is, eg home blood glucose tests, weekly weight or peak flow measurements?
- Enlist the spouse's help in ensuring compliance. How will this be checked? Eg count the remaining pills at the next visit.

1 P Ley 1976 in *Communications Between Doctors and Patients*, ed A Bennett, OUP   2 R Flesch 1948 *J Appl Psych* 32 221   3 Available from Videos For Patients, 68a Delancey St, London, UK NW1 7RY   4 V Francis 1969 *NEJM* 280 535

The bedside manner

4 Our bedside manner matters because it indicates to patients whether they can *trust* us. Where there is no trust there can be little healing. A good bedside manner is not static: it develops in the light of patients' needs—but it is always grounded in those timeless clinical virtues of honesty, humour, and humility in the presence of human weakness and human suffering.

The following are a few examples from an endless variety of phenomena which may arise whenever doctors meet patients. One of the great skills (and pleasures) in clinical medicine is to learn how our actions and attitudes influence patients, and how to take this knowledge into account when assessing the validity and significance of the signs and symptoms we elicit. The information we receive from our patients is not 'hard evidence' but a much more plastic commodity, moulded as much by the doctor's attitude and the hospital environment as by the patient's own hopes and fears. It is our job to adjust our attitudes and environment so that these hidden hopes and fears become manifest and the channels of communication are always open.

Anxiety reduction or intensification Simple explanation of what you are going to do often defuses what can be a highly charged affair. With children, more subtle techniques are required such as examining the abdomen using the child's own hands, or examining his teddy bear first.

Pain reduction or intensification Compare: 'I'm going to press your stomach. If it hurts, just cry out' with 'I'm going to touch your stomach. Let me know what you feel' and 'Now I'll lay a hand on your stomach. Sing out if you feel anything'. The examination can be made to sound frightening, neutral, or joyful, and the patient will relax or tense up accordingly.

The tactful or clumsy invasion of personal space During ophthalmoscopy, for example, we must get much nearer to the patient than is acceptable in normal social intercourse. Both doctor and patient may end up holding their breath, which neither helps the patient keep his eyes perfectly still, nor the doctor to carry out a full examination. Simply explain 'I need to get very close to your eyes for this'. (Not 'We need to get very close for this'—one of the authors was kissed repeatedly while conducting ophthalmoscopy—by a patient with frontal lobe signs. There are dangers in suggesting too much familiarity.)

The induction of a trance-like state Watch a skilful practitioner at work palpating the abdomen: the right hand rests idly on the abdomen, far away from the part which hurts. He meets the patient's gaze: 'Have you ever been to the sea-side?' (His hand caresses rather than penetrates) '. . . Imagine you are back on the beach now, perfectly at ease, gazing up at the blue, blue sky (he presses as hard as he needs) . . . 'Tell me now, where were you born and bred?' If the patient stops talking and frowns only when the doctor's hand is over the the the right iliac fossa, he will already have found out something useful.

Asking questions

This is an activity in which we all invest a good deal of time—and careful attention as to what makes questions effective or ineffective will be handsomely repaid.

Leading questions On seeing a patient's blood-stained handkerchief you ask: 'How long have you been coughing up blood for?' 'Six weeks, doctor.' So you assume that he has been coughing up blood (haemoptysis) for 6 weeks. In fact, the stain could be due to an infected finger, or to epistaxis (nose bleeds). On finding this out at a later date (and perhaps after expensive and unpleasant investigations) you will be annoyed with your patient for misleading you—whereas he was only trying to be polite by giving you the sort of answer you were obviously expecting. With such leading questions as these, the patient is not given an opportunity to deny your assumptions.

Questions which suggest the answer 'Was the vomit red, yellow or black—like coffee grounds?'—the classical description of haematemesis (the vomiting of blood). 'Yes, black, like coffee grounds, doctor.' The doctor's expectations and excessive hurry to get the evidence he needs into the format of a traditional mould have so tarnished this patient's story as to make it useless.

Open questions The most open of all questions is 'How are you?'. The question suggests no particular kind of answer, so the direction the patient chooses to take offers very valuable information. Other examples include gentle imperatives such as 'Tell me about the vomit' 'It was dark.' 'How dark?' 'Dark bits in it.' 'Like . . . ?' 'Like bits of soil in it.' This information is pure gold, although it is not cast in the form of 'coffee grounds'.

Reticular questions These are particularly useful in revealing if symptoms are caused or perpetuated by psychological mechanisms. They probe the network of causes and enabling conditions which allow nebulous symptoms to flourish in family life. Until this sort of question is asked these illnesses may be refractory to treatment. Eg: 'Who is present when your headache starts? Who notices it first—you or your wife? Who worries about it most (or least)? What does your wife do when (or before) you get it?' Think to yourself: 'Who is his headache?'

Echoes The doctor repeats the last few words the patient has said, prompting new intimacies, otherwise inaccessible, as the doctor fades into the background and the patient soliloquizes 'I've always been suspicious of my wife.' 'Wife . . .' 'My wife . . . and her father together' 'Together . . .' 'I've never trusted them together' 'Trusted them together . . .' 'No, well, I've always felt I've known who my son's real father was . . . I can never trust those two together.' Without any questions you may unearth the unexpected, important clue which throws a new light on the history. ▶ If you only ask questions you will only receive answers in reply.

Death: diagnosis and management

Diagnosing death The patient is pulseless, apnoeic, with fixed pupils, and no heart sounds. If on a ventilator brain death may be diagnosed even if the heart is still beating.

Brain death—British criteria[1] Brain death is death of the brainstem, which is recognized by establishing:
- Deep coma with absent respirations (hence on a ventilator).
- The absence of drug intoxication and hypothermia (<35°C).
- The absence of hypoglycaemia, acidosis, and U&E imbalance.

Tests: All brainstem reflexes should be absent.
- Fixed, unreacting pupils. Absent corneal response.
- No vestibulo-ocular reflexes, ie no eye movement occurs after or during slow injection of 20ml of ice-cold water into each external auditory meatus in turn. Visualize the tympanic membrane first to eliminate false-negative tests.
- No motor response within the cranial nerve distribution should be elicited by adequate stimulation.
- No gag reflex or response to bronchial stimulation.
- No respiratory effort on stopping the ventilator and allowing $PaCO_2$ to rise to 6.7kPa.

Other considerations: Repeat the test after a suitable interval—usually 24 hours. Spinal reflexes are not relevant to the diagnosis of brain death. An EEG recording is not required, nor is a neurologist. The doctor diagnosing brain death must be a consultant (or his deputy registered for >5 years). The opinion of one other doctor (any) should also be sought.

The USA criteria for brain death[2] are slightly different, eg an EEG is required to confirm the absence of cerebral activity if brain death is to be diagnosed within 6 hours of apparent cessation of brain activity. Diagnosis of brain death is allowed in cases of intoxication if isotope angiography shows absent cerebral circulation, or if the intoxicant has been metabolized.

The donation of organs: The point of diagnosing brain death is partly that this allows organs (such as kidney, liver, or heart and lungs) to be donated and removed with as little damage from hypoxia as possible. Do not avoid the topic with relatives. Many are glad to give consent, and to think that some good can come after the death of their relative, and that some part of the relative will go on living, giving a new life to another person.

After death Inform the GP and consultant. See the relatives. Provide a death certificate promptly. If the cause of death is in doubt, or if it is due to violence, injury, neglect, surgery or anaesthesia, alcohol, suicide, or poisoning, inform the Coroner or, in Scotland, the Procurator Fiscal.

1 Medical Royal Colleges 1976 *BMJ* ii 1187   2 Medical Consultants 1981 *JAMA* **246** 2184

Facing death

On most wards death is a daily occurrence, and as helping patients face death is one of our prime functions, this also ought to be a daily occurrence—so if you have not engaged in this activity recently, something, somewhere, has gone wrong.

Whenever you find yourself thinking: 'It is better for the patient not to know', suspect that you mean: 'It is more comfortable for me not to tell'. We find it difficult to tell for many reasons: it distresses the patient; it may hold up the ward round; we do not like to acknowledge our own impotence to alter the course of disease; telling reminds us of our own mortality; and it may unlock our previous griefs. We use many tricks to avoid or minimize the pain: *rationalization* ('The patient would not want to know'); *intellectualization* ('Research shows that 37% of people at stage 3 survive 2 years . . .'); *brusque honesty* ('You are unlikely to survive 6 months.'—and, so saying, the doctor rushes off to more vital things); *inappropriate delegation* ('Sister will explain it all to you when you are calmer').

Why it may help the patient to be told
- He already half knows—but everyone shies away so he cannot discuss his fears (of pain, or that his family will not cope).
- There may be many affairs for the patient to put in order.
- To enable him to judge if unpleasant therapy is worthwhile.

▶ *Most patients are told less than they would like to know.*[1]

How and what to tell
- Put yourself in the patient's place. What are his worries likely to be (about himself; his family)?
- Allow the patient to control the information. Give sufficient information, and then the opportunity to ask for more.
- Be sensitive to hints that he may be ready to learn more. 'I'm worried about my son.' 'What is worrying you most?' 'Well, it will be a difficult time for him, (pause) starting school next year.' Silence, broken by the doctor 'I get the impression there are other things worrying you'. The patient now has the opportunity of proceeding further, or of stopping there.
- Ensure that the GP and the nurses know what you have and have not said. Also make sure that this is written in the notes.

Stages of acceptance
Accepting imminent death often takes time, and may involve passing through 'stages' on a path. It helps to know where your patient is on this path (but progress is rarely orderly, and need not always be forwards: the same ground may need to be covered many times). At first there is *shock* and *numbness*, then *denial* (which reduces anxiety), and then *anger* (which may lead you to dislike him), then *grief*, and then *acceptance*. Finally there may be an intense longing for death as the patient moves beyond the reach of worldly cares.[2]

Recognizing depression
which will benefit from antidepressants is difficult in those who are physically ill because symptoms like early morning waking and loss of appetite may be due to the physical illness and not the depression. Be guided by morbid thoughts, such as undue guilt and low self-esteem and the inability to feel pleasure (anhedonia).

1 A Stedeford 1984 *Facing Death*, Heinemann   **2** J S Bach 1727 *Ich habe genug*, Cantata number 82 on the Feast of the Purification. See also R Buckman 1992 *How to Break Bad News* Macmillan, London

Is this new treatment any good?

This question frequently arises after reading original papers of clinical interest. Should a new treatment actually be tried on my patient today—or is this treatment, which appears to offer real advantage, really something needing further substantiation? Not only writers of textbooks but also all medical practitioners have to decide what new treatments to recommend, and which to ignore. In assessing the use of research papers, ask the following:

1 Does the research address a significant problem?

2 Does it give a clear answer? Is the answer clinically significant, as well as being statistically significant?

3 Are the patients analysed in the paper similar to my patients?

4 Does the journal employ 'peer review' (in which so-called 'experts' vet the paper before its publication)? This system is far from perfect, but it is better than no system at all.

5 Are the statistical analyses valid? Much must be taken on trust as many analyses depend on sophisticated computing. Few papers, unfortunately, present 'raw' data. Nevertheless, it is important to look out for obvious faults, by asking questions such as:

- Is the sample large enough to detect a clinically important difference, say a 20% drop in deaths from disease X? If the sample is small, the chance of missing such a difference is high. In order to reduce this chance to less than 5%, and if disease X has a mortality of 10%, >10,000 patients would need to be randomized. If a small trial which lacks power (the ability to detect true differences) *does* give 'positive' results, the size of the difference between the groups is likely to be exaggerated. (This is known as a type I error; a type II error applies to results which indicate that there is no effect, when in fact there is.) ▶ So beware even quite big trials which purport to show that a new treatment is equally effective as an established treatment.
- Were the compared groups chosen randomly, and did the result produce two groups that were well matched?
- Were the two treatments being compared carried out by practitioners equally skilled in each treatment?

6 Was the study 'double-blind' (both patient and doctor are unaware of which treatment a patient is having?). Could either have told which was given—eg by the metabolic effects of the drug?

7 Was the study placebo-controlled? Note: valuable research can go on outside the realm of double-blind, randomized controlled trials—but conclusions need even more care in interpreting.

8 Has time been allowed for criticism of the research to appear in the correspondence columns of the journal in question?

9 If I were the patient, would I want the new treatment?

A single well-planned clinical trial may be worth centuries of uncritical medical practice; but a week's experience on the wards may be more valuable than a year spent reading journals. This is the central paradox in medical education. How can we trust our own experiences knowing that they are all anecdotal; how can we be open to novel ideas but avoid being merely fashionable? An attitude of wary open-mindedness may serve us best.

Essential epidemiology

Epidemiology is the study of the distribution of clinical phenomena in populations. Its chief measures are prevalence and incidence.

The prevalence of a disease is the number of cases, at any time during the study period, divided by the population at risk. If the population at risk is unclear, then the population must be specified—for example, the prevalence of uterine cancer varies widely, depending on whether you specify the general population (men, women, boys, and girls) or only women, or women who have not already had a hysterectomy.

The incidence of a disease is the number of new cases within the study period which must be specified, eg annual incidence. Point prevalence is the prevalence at a point in time. The lifetime prevalence of hiccups is ~100%; the incidence is millions/yr—but the point prevalence may be 0 at 3 am today if no one is actually having hiccups.

Association Epidemiological research is often concerned with comparing the rates of disease in different populations. For example, the rate of bronchial carcinoma in a population of men who smoke, compared with men who do not smoke. A difference in rates points to an association (or dissociation) between the disease and factors which distinguish the populations (in this case, smoking or not). If the rates are equal, association is still possible—with a confounding variable (eg both groups share the same very smoky environment).

Ways of accounting for associations: A causes B; B causes A; a third agent, P, causes A and B; it may be a chance finding.

There are 2 types of study which explore causal connections:

1 *Case–control (retrospective) studies:* The study group consists of those with the disease (eg lung cancer); the control group consists of those without the disease. The previous occurrence of the putative cause (eg smoking) is compared between each group. Case–control studies are retrospective in that they start after the onset of the disease (although cases may be collected prospectively).

2 *Cohort (prospective) studies:* The study group consists of subjects exposed to the putative causal factor (eg smoking); and the control group consists of subjects not so exposed. The incidence of the disease is compared between the groups. ► A cohort study generates incidence data, whereas a case–control study does not.

Matching An association between A and B may be due to another factor P. To eliminate this possibility match study and control groups for P. One powerful but unreliable way to do this in cohort studies is for the subjects to be allocated to the groups randomly, but check important 'P's have been distributed evenly between groups.

Overmatching Suppose that unemployment causes low income, and low income causes depression. If you matched study and control groups for income, then you would miss the genuine causal link between unemployment and depression. In general, avoid matching factors which may intervene in the causal chain linking A and B.

Masking If the subject does not know which of two trial treatments she is having, the trial is single-blind. To further reduce risk of bias, the experimenter should also not know (double-blind).
► In a good treatment trial, the blind lead the blind.

Further reading: D Sackett 1991 *Clinical Epidemiology*, 2 ed, Little Brown, Boston.

Six to five against

10 Your surgical consultant asks whether Gobble's disease is more common in men or women. You have no idea, and make a guess. What is the chance of getting it right? Common sense decrees that it is even chances; 'Sod's Law' predicts that whatever you guess, your answer will always be wrong. A less pessimistic view is that the balance is only slightly tipped against you: according to Damon Runyon, 'all life is six to five against'.[1] This might be called *Runyon's Law*.

The doctor as gambler Given any set of information, different diagnoses have different odds. Any new information simply changes these odds.

Does a new symptom suggest a new disease, or is it due to the disease which is already known? Here the gambler scores over the intuitive clinician, as the answer is often counter-intuitive. Suppose that S is quite a rare symptom of Gobble's disease (occurring, let us say, in only 5% of patients). But suppose that S is a very common symptom of disease A (occurring in 90%). If we have a patient whom we already know has Gobble's disease and who goes on to develop symptom S, isn't S more likely to be due to disease A, rather than Gobble's disease? The answer is usually no: *it is generally the case that S is due to a disease which is already known, and does not imply a new disease*.[2]

The reason for this can be seen by considering the 'odds ratio', ie the ratio of the [probability of the symptom, given the known disease] to [the probability of the symptom given the new disease × *the probability of developing the new disease*].

This ratio is, usually, vastly in favour of the symptom being due to the old disease because of the prior odds of the two diseases. For example, a 50-year-old man with known carcinoma of the lung has some transient neurological symptoms and a normal CT scan. Are these symptoms due to secondaries in the brain or to transient ischaemic attacks (TIAs)? The chance of secondaries in the brain which cause transient neurological symptoms is 0.045 given carcinoma of the lung. The chance of such secondaries not showing up on a CT scan is 0.1. Therefore the chance of this cluster of clinical symptoms is 0.0045 (ie 0.045 × 0.1). The chance of a normal CT scan and transient neurological symptoms given a TIA is 0.9. However, the chance of a 50-year-old man developing TIA is 0.0001. Therefore the odds ratio is 0.0045/(0.9 × 0.0001). That is, the odds ratio is ~50 to 1 in favour of secondaries in the brain.

Note: it is only very occasionally that the prior odds of a new disease are so high that the new disease is more likely. For example, someone presenting with anaemia already known to have breast cancer, who lives in an African community where 50% of people have hookworm-induced anaemia, is likely to have hookworm as well as breast carcinoma.

1 D Runyon 'A Nice Price', in *Runyon on Broadway*, 1950, Constable
2 J Bella 1985 *Lancet* **i** 326

Investigations change the odds

Diagnostic medicine is for gamblers (see p10). We treat investigations as though they tell us the diagnosis. But the logic of investigations is rather different. As you are taking a history and examining the patient, you make various wagers with yourself (often barely consciously) as to how likely various diagnoses are. The results of subsequent tests simply affect these odds. A test is, generally, only worthwhile if it alters the diagnostic odds in a way which is clinically important.

The effect of the test on the diagnostic odds

To work this out you need to know the *sensitivity* and *specificity* of the investigation. All investigations have false positive and false negative rates, as summarized below:

		Patients with the condition	Patients with-out condition
TEST RESULT	Subjects appear to have the condition	True +ve (a)	False +ve (b)
	Subjects appear not to have condition	False −ve (c)	True −ve (d)

Sensitivity: how reliably is the test +ve in the disease? a/a+c
Specificity: how reliably is the test −ve in health? d/d+b

Suppose we have a test in which the sensitivity is 0.8 and the specificity is 0.9. The *likelihood ratio* of the disease given a positive result (LR+) is the ratio of the chance of having a positive test if the disease is present to the chance of having a positive test if the disease is absent [0.8/(1−0.9)]; ie 8:1 in the above example. More generally, LR+ = sensitivity/(1−specificity); and LR− (likelihood ratio of the disease given a negative result) = (1−sensitivity)/specificity. (1−0.8/0.9, ie 2:9 in the above example).

Is there any point to this investigation? Work out the 'posterior odds' assuming first a positive and then a negative result to the test. This is done from the equation:

$$posterior\ odds = (prior\ odds) \times (likelihood\ ratio).$$

For example, if your clinical assessment of a man with exercise-induced chest pain is that the odds of this being due to coronary artery disease (CAD) is 4:1 (80%), is it worth his doing an exercise tolerance test (sensitivity 0.72; specificity 0.8)?[1] If the test were positive the odds in favour of CAD would be 4 × 0.72/(1−0.8) = 14:1 (93%). If negative, they would be 4 × (1−0.72)/0.8 = 1.4:1 (58%).

An experienced clinician is likely to have higher prior odds for the most likely diagnosis. The above example shows that with high prior odds, the test has to have a high sensitivity and specificity for a negative result to bring the odds below 50%.

1 D Simel 1985 *Lancet* **i** 329

Quality, QALYs, and the interpretation of dreams

Resource allocation: how to decide who gets what Resource allocation is about cutting the health cake—whose size is *given*. What slice should go to transplants, new joints, and services for dementia? Cynics would say that this depends on how vociferous is each group of doctors. But others try to find a rational basis for allocating resources. Health economists (econocrats) have invented the QALY for this purpose.

Making the cake Focusing on how to cut the cake diverts attention from that central issue: how large should the cake be? The answer may be that more should be spent on some health service, not at the expense of some other health gain, but at the expense of something else.[1]

What is a QALY? 'The essence of a QALY (Quality Adjusted Life Year) is that it takes a year of healthy life expectancy to be worth 1, but regards a year of unhealthy life expectancy to be worth <1. Its precise value is lower the worse the quality of life of the unhealthy person.'[2] If a patient is likely to live for 8 years in perfect health after one form of treatment, he gains 8 QALYs; if another treatment would give him 16 years but at a quality of life which is rated by him at only 25% of the maximum, he would gain only 4 QALYs. The dream of health economists is to buy most QALYs for a given budget: to do this they need to know the cost per QALY of each treatment. They have worked out that the cost per QALY is £194 to correct scoliosis, £592 for a shoulder replacement, and £1412 for a kidney transplant.[3] What truths and wishful thinking lie behind these figures?

The attraction of QALYs QALY analysis is a a better way to ration care than techniques based on raw survival figures. QALYs counter vested interests, and help question assumptions—eg each QALY gained by post-MI thrombolysis costs half as much in those >65 years compared with those <55 (in whom there are fewer death to prevent).[4] If resources are scarce, should we restrict coronary care to the elderly?

Problems with QALYs ● Injustice: it is reasonable for an *individual* to choose the treatment which is most likely to maximize QALYs—but applying QALYs to the allocation of resources is quite different as it involves choosing between the welfare of different people. Those with mental or physical handicaps will have less claim on resources because their handicap means that each year of their life is given <1 QALY. And, in general, the old do badly because palliative intervention is likely to result in fewer QALYs because their life expectancy is less.

● QALYs may not add up. Can one apply valuations made by summing group preferences to individuals who may not share the group's scale of preferences? *Question*: if a vase of flowers is beautiful, are 10 vases of flowers ten times as beautiful?—or might the scent be overpowering?

● Is welfare all important? 1000 people may benefit from constipation treatment—which might give more QALYs for the money it takes to save one person's life. In allocating resources, perhaps the needs of the dying person should come before the benefits to the constipated.

● QALYs do not suggest *who* should choose between treatments of similar QALY value. Doctors do not like deciding (they are put in an invidious position) and polls show that patients do not want to decide—so this leaves only politicians, which neither group may be happy to trust.

1 J Bell 1988 *Philosophy and Medical Welfare*, Cambridge University Press   **2** A Williams 1985 *Health and Social Service Journal* 3   **3** R Klein 1989 *BMJ* ii 275   **4** A Renton 1992 *BMJ* i 182

Psychiatry on medical and surgical wards

▶ Psychological problems are common in your patients, their relatives and your colleagues.

▶ Seek appropriate help for your own problems.

Depression This is common, and commonly ignored. 'I would be depressed in her situation . . .', you think to yourself, and so you do not think of offering treatment. The usual guides (eg early morning waking, loss of appetite and weight) are of little help in diagnosing depression, as they are very common on general wards. Persistent low mood (without moments of happiness), thoughts of guilt with an excessive degree of self-blame, and thoughts of being worthless are, on the other hand, useful signs.

In general, if in doubt, recommend an antidepressant—and see if it helps. In patients less than 65 years old use amitriptyline or imipramine (which is less sedating). Contraindications: glaucoma, severe cardiac disease. In those over 65 years old, use a newer (but less well tried) antidepressant, such as dothiepin. Dose: (the same for all 3 drugs above) increase in steps of 25–50mg over the course of a week until 150mg/24h PO is given (unless prevented by side-effects: dry mouth, blurred vision, retention of urine, drowsiness, arrhythmias, fits). If there is no effect after 3 weeks, increase the dose to 200mg/24h PO. Keep at this dose for 3 more weeks before deciding that the drug is ineffective. If so, tail off over a week. If it is effective keep at the top dose for 2 months and then reduce to 75–100mg/24h PO for a further 4 months before tailing off over 2 weeks.

Alcohol This is a common cause of problems on the ward (both the results of abuse, and the effects of withdrawal). See p682.

The violent patient First ensure your own and others' safety. Do not tackle violent patients until adequate help is to hand (eg hospital porters or police). Common causes: *alcohol intoxication*, *hypoglycaemia*, *acute confusional states* (OHCS p350).

Once adequate help arrives, try to talk with the patient to calm him—and to gain an understanding of his mental state. If this fails it will be necessary to restrain him. English law allows you to do this (p12). Measure the blood glucose, or give IV dextrose straight away (p738). If hypoglycaemia is not the cause, before further investigation is possible, chemical restraint may be needed (eg haloperidol up to 20mg IM stat, then 5mg/h).

The Mental Health Act (1983, England and Wales) Unlike English common law, this is rarely relevant in general hospitals. Section 5 allows detention of a patient already receiving in-patient treatment—if he is suffering from a mental disorder, and is a danger to himself or others. An acute confusional state (p446) is a mental disorder. Application must be made to the hospital managers by the consultant in charge or his deputy. Detention is for <72h. For further details, see OHCS p400.

What is the mechanism?

Like toddlers, we should always have the question 'Why?' on our lips. The reason for this is not that we are always searching for the ultimate cause of phenomena (although there is a place for this): it is so that we can choose the simplest level for intervention. Some simple change early on in a chain of events may be sufficient to bring about a cure, whereas later on in the chain such opportunities may not arise.

For example, it is not sufficient for you to diagnose heart failure in your breathless patient. Ask: 'Why is there heart failure?' If you do not do this, you will be satisfied with giving the patient a diuretic drug—and any side-effects from these such as uraemia or incontinence induced by polyuria will be attributed to an unavoidable consequence of necessary therapy.

If only you had asked 'What is the mechanism of the heart failure?' you might have found an underlying cause, such as anaemia coupled with ischaemic heart disease. You cannot cure the latter, but treating the anaemia may be all that is required to cure the patient's breathlessness. But do not stop there: ask 'What is the mechanism of the anaemia?' Look at a blood film for the answer. You discover that there is hypochromia and a low serum ferritin (p572)—and at last, you might be tempted to say to yourself, I have the prime cause: iron-deficiency anaemia. Here is the end to my chain of reasoning.

Wrong! Put aside the idea of prime causes, and go on asking 'What is the mechanism?' Return to the blood film to decide if the cause is dietary or bleeding (suggested by polychromasia, p576). You decide that the iron deficiency is dietary, and this is confirmed by going over the history again (never think that the process of history taking is over after your first contact with your patient). Why is the patient eating a poor diet? Is he ignorant, or too poor to eat properly? Go back to taking the history again, and perhaps you will find that the patient's wife died a year ago, he is sinking into a depression, and cannot be bothered to eat. He would not care if he died tomorrow. (The social history, so often of vital importance, is all too often diminished to a few questions about alcohol and tobacco.)

You now begin to realize that simply treating the patient's anaemia may not be of much help to him—so go on asking 'Why?': 'Why did you bother to go to the doctor at all, if you are not interested in getting better?' It turns out that he only went to see the doctor to please his daughter. Such a patient is most unlikely to take your treatment, unless you really get to the bottom of what he cares about. In this case, his daughter is what matters, and unless you can enlist her help, all your therapeutic initiatives will fail. Talk with his daughter; offer help for the depression; and teach her about iron-rich foods, and, with luck, your patient's breathlessness may gradually begin to disappear.

On being busy: Corrigan's secret door

Unstoppable demands, increasing expectations as to what medical care should bring, the rising number of elderly patients, coupled with the introduction of new and complex treatments all conspire, it might be thought, to make doctors ever busier. In fact, doctors have always been busy people. Sir James Paget, for example, would regularly see more than 60 patients each day, sometimes travelling many miles to their bedside. Sir Dominic Corrigan was so busy one hundred and thirty years ago that he had to have a secret door made in his consulting room so that he could escape from the ever-growing queue of eager patients.

We are all familiar with the phenomenon of being hopelessly over-stretched—and of wanting Corrigan's secret door. Competing, urgent, and simultaneous demands make carrying out any task all but impossible: the house officer is trying to put up an intravenous infusion on a shocked patient when his 'bleep' sounds. On his way to the telephone a patient is falling out of bed, being held in, apparently, only by his visibly lengthening catheter (which had taken the house officer an hour to insert). He knows he should stop to help, but instead, as he picks up the phone, he starts to tell Sister about 'this man dangling from his catheter' (knowing in his heart that the worst will have already happened). But he is interrupted by a thud coming from the bed of the lady who has just had her varicose veins attended to: however, it is not her, but her visiting husband who has collapsed, and is now having a seizure. At this moment his cardiac arrest 'bleep' goes off, summoning him to some other patient. In despair, he turns to Sister and groans: "There must be some way out of here!" At times like this we all need Corrigan to take us by the shadow of our hand, and walk with us through his metaphorical secret door, into a calm inner world. To enable this to happen, make things as easy as possible for yourself.

Firstly, however lonely you feel, you are not usually alone. Do not pride yourself on not asking for help. If a decision is a hard one, share it with a colleague. Secondly, take any chance you get to sit down and rest. Have a cup of coffee with other members of staff, or with a friendly patient (patients are sources of renewal, not just devourers of your energies). Thirdly, do not miss meals. If there is no time to go to the canteen, ensure that food is put aside for you to eat when you can: hard work and sleeplessness are twice as bad when you are hungry. Fourthly, avoid making work for yourself. It is too easy for junior doctors, trapped in their image of excessive work, and blackmailed by misplaced guilt, to remain on the wards re-clerking patients, re-writing notes, or re-checking results at an hour when the priority should be caring for themselves. Fifthly, when a bad part of the rota is looming, plan a good time for when you are off duty, to look forward to during the long nights.

Finally, remember that however busy the 'on take', your period of duty will end. For you, as for Macbeth:

> Come what come may,
> Time and the hour runs through the roughest day.

The Internet and medicine

This facility for the global sharing of electronic information has been variously dubbed as the biggest time-wasting development in medicine in recent years, or, alternatively, as a priceless asset in finding certain more or less vital pieces of information. There is no doubt that it has the power to completely reformulate and democratize the power-balance in the doctor-patient relationship. The doctor used to know everything, and the patient was the hungry supplicant craving for information. With the Internet it is easy for the patient to know more than his specialist. Indeed it is possibly now the case that *only* patients can keep up-to-date with their ailments—as the professionals have too many commitments on their time to find out all there is to know about the nostril, the cerebellum, or whatever is their specialism.

▶ The Internet underlines the ancient precept that what matters is *judgement*, not *information*. The Internet may tell us what *can* be done for our patients, perhaps some very complex surgery (we may even see this being done on our screens), but wise old physicians will still be best at advising on what *should* be done, and *when*. The slow accretion of experience at the bedside is now all the more valuable for existing in a counterpoint-type of relationship with information technology. The skill is to make this counterpoint an exercise in harmony.

Here are some possibly useful Internet addresses[1] *Journal of the American Medical Association*—http://www.amaassn.org/journals/standing/jamahome.htm. This site is praised as being one of the best examples of medicine on the internet. It includes *American Medical News* and an HIV information centre.
New Scientist—http://www.newscientist.com/
British Medical Journal—http://www.bmj.com/bmj
New England Journal of Medicine—http://www.nejm.org/
The Lancet—http://www.thelancet.com/
Sowerby Unit for Primary Care Informatics—http://www.ncl.ac.uk/~nphcare/
Department of health (UK)—http://www.open.gov.uk/doh/dhhome.htm
CDC centres for disease control and prevention—http://www.cdc.gov/cdc.html
Communicable disease surveillance—http://www.open.gov.uk/cdsc/cdschome.htm
General Practice On-line—http://www.priori.com/journals/GP.htm
The virtual hospital—http://indy.radiology.uiowa.edu/VirtualHospital.html
The visible human project—http://www.nlm.nih.gov/research/visible/
Paranoia drugs information server—http://www.paranoia.com/drugs/

NB: Not all Internet information is peer-reviewed, and issues of quality are very much to the fore. The above have been chosen because of general appeal, but it is not exhaustive. Many rare diseases have their own home page, typically run by national or international patients' groups.

Useful introductory and advanced tests are available in hard copy, which can enable people with a personal computer and a Modem to connect to the Internet, and, perhaps, to use it productively.[2]

Oxford Handbooks on-line Updates are available at the following site, run by Oxford University Press: http://www.oup.co.uk.

This site provides regular updates to this volume, and its sister, the *Oxford Handbook of Clinical Specialities*. We cannot expect to realize fully our aim of detailing all the significant advances in medicine: but the scope of these two volumes, combined with their conveniently placed blank pages, ensures that at least we have the resource to which we can append advances in almost any sphere of clinical medicine. This Internet site will also offer a route into our 'diagnostic' and decision support facility provisionally named *Oxford Clinical Mentor*—see opposite.

1 E. Penman 1996 *Pulse* 56 97 **2** BC McKenzie 1996 *Medicine and the Internet*, OUP

Experts and expert systems

The best way to get help with a difficult diagnosis is to ask an expert. But what should you ask her, and how do you decide what the expert should be an expert in? If you end up sending someone to see a dermatologist when he needed a microbiologist, or to a neurologist when he needs a cardiologist, he may die a preventable death. We use the tools that come to hand when there is work to be done. If you only have a hammer, it is all too easy to turn every problem into a nail, so that you can get to work on it. Dermatologists provide topical solutions, surgeons provide operative solutions, and psychiatrists provide solutions by listening. Most of the time this works out very well because the doctor who first sees the patient knows in which direction (if any) the patient should be pointed. This doctor might do well to ask 'If I were a dermatologist, what would I do? . . . If I were a psychiatrist, what would I do?'—and so on, and then choose what seems to be the best course of action, asking herself: 'If I adopt a surgical solution (if this is what she has chosen), what might I be missing?' And it is here that decision support has its most practical application. Consider a patient who is fed up with his chilblains and headaches. A dermatologist may concentrate on the chilblain, a psychiatrist on the feeling of being fed up, and a neurologist may concentrate on the headache. Each person is doing what he or she thinks is a good job. In fact, what was needed (in this real patient) was either a different specialist—or an electronic system for diagnosis. Headache and chilblains can be fed into an electronic system (eg *Oxford Clinical Mentor*[1]), which, within a few seconds, comes up with the idea that the man could have endocarditis if the chilblain is really a vasculitis, and goes on to suggest looking for splenomegaly, murmurs and haematuria; and, if any tests are needed, to include blood cultures. To work this trick what is needed is a list of keywords, preferred terms and synonyms, and a database of a large number of diseases, along with details about mistakes doctors have made, so that these too can be taken into account.

It may be irritating to hear that this well-looking man with chilblains and headache could have endocarditis ± a mycotic aneurysm—until your eye strays to Sister's note 'about one plus of haematuria', whereupon irritation turns to a shiver as you realize that you may be sitting next to the embodiment of one of nature's most deadly and stealthy diseases.

Mentor-type databases (Iliad and Internist-1/QMR are others) provide a way of connecting apparently unrelated phenomena: eg knee pain and constipation may be explained by pseudogout. The computer postulates hypothyroidism to explain constipation, and hypothyroidism is one of pseudogout's recognized associations (p666).

Another use for these databases is for diagnosing very rare diseases. The database referred to above (*Oxford Clinical Mentor*) includes this book (and the *Oxford Handbook of Clinical Specialties*) as well as *The Oxford Handbook of Clinical Rarities* which only exists electronically.

Keeping up-to-date: '*Do you know, boy, that when they make those maps of the universe, you are looking at the map of something that looked like that six thousand million years ago? You cannot be much more out of date than that, I'll swear.*'[2] This edition may not be quite that out of date by the time you read it—but the same principle applies. Modem-updated electronic databases bring you nearer—not to the big bang of oup creation, but to the tiny tinkerings of their mundane authors, and their latest small, but sometimes not so inconsequential, thoughts, ideas, and updatings.

1 *OXFORD CLINICAL MENTOR*—from OUP: +44 (0) 1865 267979, Electronic Publishing, Great Clarendon Street, Oxford OX2 6DP and EMIS: +44 (0) 13 2591122)   **2** G Green 1957 *Under the Garden*, Penguin

2 | History and physical examination

Relevant pages in other chapters:
Dictionary of symptoms and signs (p38–58); the elderly in hospital (p64); mini-mental test score (p72); pre-operative assessment (p82); the acute abdomen (p112); lumps (p144–52); hernias (p158 & p160); varicose veins (p162); jugular venous pressure (p258); heart sounds (p260); murmurs (p262); points on examination of the chest (p324); urine (p370); testing peripheral nerves (p408); dermatomes (p410); nystagmus (p422).

Relevant pages in the Oxford Handbook of Clinical Specialties:
Vaginal and other gynaecological examinations (*OHCS* p2); abdominal examination in pregnancy (*OHCS* p84); the history and examination in children and neonates (*OHCS* p172–6); examination of the eye (*OHCS* p476); visual acuity (*OHCS* p476); eye movements (*OHCS* p486); ear, nose and throat examination (*OHCS* p528 & p530); facial nerve lesions (*OHCS* p566); skin history and examination (*OHCS* p576); examination of joints—see the contents page to Chapter 9 (*Orthopaedics and trauma, OHCS* p604).

▶ The way to learn how to elicit physical signs is at the bedside, with guidance from an experienced colleague. The pages in this chapter dealing with physical signs are not intended as a substitute for this process: they are simply an *aide-mémoire*.

Taking a history

An accurate history is the biggest step in making the correct diagnosis and will direct further investigations and subsequent management of the patient. Thus, taking a history should precede both examination and treatment in all but the extremely ill. Treatment of a patient begins the moment one reaches the bedside. Try to put the patient at ease: a good rapport may relieve distress on its own. Always introduce yourself to the patient and check that the patient is seated/lying comfortably. Often some general questions (age, occupation, marital status) help break the ice and give some general assessment of his higher mental functions.

Presenting complaint (PC) 'What has been the trouble recently?' Record the patient's own words rather than, eg 'dyspnoea'.

History of presenting complaint (HPC) When did it start? What was the first thing noticed? Progression since then. Try to characterize pain and symptoms roughly as below:
- Site; Radiation; Intensity; Duration; Onset (gradual, sudden);
- Character (sharp, dull, knife-like, colicky);
- Associated features (nausea, vomiting, etc.);
- Exacerbating and alleviating factors;
- 'Ever had it before?'

Direct questioning (DQ) Specific questions about the diagnosis you will already have in mind and review of the relevant system.

Past Medical History (PMH) Ever been in hospital for anything? Illnesses? Operations? Ask specifically about diabetes, asthma/bronchitis, TB, jaundice, rheumatic fever, high BP, heart attacks/angina, stroke, epilepsy, anaesthetic problems.

Medications and allergies Any tablets, injections? Any 'off the shelf' drugs? Ask the features of the allergy; it may not have been one.

Social and family history (SH/FH) Probe without prying. 'Who else is there at home?' Job. Marital status. Spouse's job and health. Housing. Who visits?—relatives, neighbours, GP, nurse. Who does the cooking and shopping? What can the patient not do because of the illness? Age, health and cause of death if known of parents, sibs, children; ask about TB, DM, IHD, and other diseases of relevance.

Tobacco, Alcohol and 'Recreational' drugs How much? How long?

Functional enquiry (p22) To uncover undeclared symptoms. Some of this may already have been incorporated into the history.

Functional enquiry

Just as skilled acrobats are happy to work without safety nets, so older clinicians usually operate without the functional enquiry. But to do this you must be experienced enough to understand all nuances of the presenting complaint.

General questions may be the most significant, eg in TB or cancer:

Weight loss	Night sweats	Any lumps	Fatigue
Appetite	Fevers	Itch	

Cardiorespiratory symptoms Chest pain (p256). Exertional dyspnoea (quantify exercise tolerance eg in number of flights of stairs). Paroxysmal nocturnal dyspnoea. Orthopnoea, ie breathlessness on lying flat (a symptom of left ventricular failure). Ankle oedema. Palpitations (awareness of the heartbeat). Cough. Sputum. Haemoptysis (coughing up blood). Wheeze.

Gastrointestinal symptoms Abdominal pain (nature: constant or colicky, sharp or dull; site; radiation; duration; onset; severity; relationship to eating; alleviating, exacerbating and associated features). Other questions:

Indigestion	
Nausea, vomiting	Stool: colour, consistency, blood, slime
Swallowing	Difficulty flushing away
Bowel frequency	Tenesmus or urgency

Tenesmus is the feeling that there is something in the rectum which cannot be passed (eg due to tumour). Haematemesis is vomiting blood. Melaena is altered (black) blood passed PR (see p494).

Genitourinary symptoms Incontinence (stress or urge, p74). Dysuria (painful micturition). Haematuria (bloody micturition). Nocturia (needing to micturate at night). Frequency (frequent micturition) or polyuria (passing excessive amounts of urine). Hesitancy (difficulty starting micturition). Terminal dribbling.

Vaginal discharge. Menses: frequency, regularity, heavy or light, duration, painful. First day of last period (LMP). Number of pregnancies. Menarche. Menopause. Chance of current pregnancy?

Neurological symptoms Sight, hearing, smell, taste. Seizures, faints, funny turns. Headache. 'Pins and needles' (paraesthesiae). Weakness ('Do your arms and legs work?'), poor balance. Speech problems (p430). Sphincter disturbance. Higher mental function and psychiatric symptoms (p34 and p72). The important thing is to assess function: what can and cannot the patient do at home, work, etc?

Musculoskeletal symptoms Pain, stiffness, swelling of joints. Diurnal variation in symptoms. Functional deficit.

Thyroid symptoms *Hyperthyroidism:* Prefers cold weather, sweaty, diarrhoea, oligomenorrhoea, weight loss, tremor, visual problems. *Hypothyroidism:* Depressed, slow, tired, thin hair, croaky voice, heavy periods, constipation, dry skin.

As set out here, history taking may seem deceptively easy—as if the patient knew the hard facts, and the only problem was extracting them. But what a patient gives you is a mixture of hearsay ('she said I looked very pale'), innuendo ('you know, doctor, down below'), legend ('I suppose I did bite my tongue—it was a real fit, you know'), exaggeration ('I didn't sleep a wink'), and impossibilities ('The Pope put a transmitter in my brain'). The great skill (and pleasure) in taking a good history lies not in ignoring these garbled messages, but in making sense of them.

Physical examination

With a few exceptions (eg BP, breast lumps) physical examination is not a good screening test. Plan your examination to emphasize the areas that the history suggests may be abnormal. A few well-directed, problem-oriented minutes can save you hours of fruitless, though very thorough, physical examination. You will still be expected to examine all 4 systems (p26–32), but with time you will be able to rapidly exclude any major undisclosed pathology. Practice is the key.

Look at the patient as a whole for a few seconds and decide on how sick he seems to be. Is he well or *in extremis* and try to decide why you think so? Is he in pain and does it make him lie still or writhe about? Is his breathing laboured and rapid? Is he obese or cachectic? Is his behaviour appropriate?

Specific diagnoses can often be made from the face and body habitus and these may be missed unless you stop and consider them: eg acromegaly, thyrotoxicosis, myxoedema, Cushing's syndrome or hypopituitarism. Think about Paget's disease, Marfan's, myotonia and Parkinson's syndrome. Look for rashes, eg the malar flush of mitral disease and the butterfly rash of SLE.

Assess the degree of hydration by examining the skin turgor, the mucous membranes, peripheral perfusion and temperature, postural changes in blood pressure and urine output (see p44). Try to distinguish water from saline loss (p628).

Check for cyanosis (central and peripheral) (p42). Is the patient jaundiced? Yellow skin is unreliable and may resemble the lemon tinge of uraemia, pernicious anaemia, carotenemia (sclerae are not yellow) or Ca caecum. The sign of jaundice is yellow sclerae seen in good daylight. Pallor is a non-specific sign and may be racial, familial or cosmetic. Anaemia may be assessed from the skin creases and conjunctivae (pale if Hb <9g/dl). Koilonychia and stomatitis suggest Fe-deficiency. Anaemia with jaundice suggests malignancy or haemolysis. Pathological hyperpigmentation is seen in Addison's, haemachromatosis (slate-grey) and amiodarone, gold, silver and minocycline therapy.

Palpate for lymph nodes: examine for nodes in the neck (from behind), axillae, groins, epitrochlear and in the abdomen (see p50). Look for subcutaneous nodules (p52).

Don't forget to look at the temperature chart and the results of urinalysis where indicated.

The hands

A wealth of information can be gained from shaking hands and rapidly examining the hands of the patient. Are they warm and well perfused? Warm, sweaty hands may signal hyperthyroidism while cold, moist hands may be due to anxiety. Are the rings tight with oedema? Lightly pinch the dorsum of the hand—persistent ridging of the skin suggests loss of tissue turgor. Are there any nicotine stains?

25

Nails The nails are affected by a variety of metabolic disturbances:

Koilonychia (spoon-shaped nails) suggests iron deficiency but may occur in other conditions such as syphilis or ischaemic heart disease. Onycholysis (destruction of nails) is seen with hyperthyroidism, fungal nail infection and psoriasis. Beau's lines are transverse furrows that signify temporary arrest of nail growth and occur with periods of severe illness. As nails grow at roughly 0.1mm/day, by measuring the distance from the cuticle it is often possible to date the stress. Mees' lines are paired white, parallel transverse bands sometimes seen in hypoalbuminaemia. Terry's lines are white nails with normal pink tips seen in cirrhosis. Pitting of the nails is seen in psoriasis and alopecia areata.

Splinter haemorrhages are fine longitudinal haemorrhagic streaks, which in the febrile patient may suggest infective endocarditis. They may be normal—being caused, for example, by gardening, when their subconjunctival correlates will be *absent*. Nail-fold infarcts are characteristically seen in vasculitic disorders.

Clubbing of the nails occurs with a wide variety of disorders (see p42). It is not generally seen in simple emphysema. There is an exaggerated longitudinal curvature, loss of the angle between the nail and the nail bed and the nail feels 'boggy'. Cause unknown but may be due to increased blood flow through multiple AV shunts in the distal phalanges.

Chronic paronychia is a chronic infection of the nail-fold and presents as a painful swollen nail with intermittent discharge. Treatment involves keeping the nail dry and antibiotics eg erythromycin 250mg/6h PO and nystatin ointment.

Changes occur in the hands in many different diseases. Palmar erythema is associated with cirrhosis, pregnancy and polycythaemia. Pallor of the palmar creases suggests anaemia. Pigmentation of the palmar creases is normal in Asians and blacks but is also seen in Addison's. Dupuytren's contracture is fibrosis and contracture of the palmar fascia and is seen in liver disease, trauma, epilepsy and simple ageing.

Swollen proximal interphalangeal (PIP) joints with distal (DIP) joints spared suggests rheumatoid arthritis; swollen DIP joints suggests osteo-arthritis, gout or psoriasis. Look for Heberden's (distal) and Bouchard's (proximal) 'nodes' (osteophytes—bone overgrowth at a joint—seen with osteoarthritis).

The cardiovascular system

History
Age, occupation, ethnic origin

Presenting symptoms	Risk factors for IHD
Chest pain	Smoking
Dyspnoea—exertional?	Hypertension
orthopnoea? PND?	Diabetes mellitus
Ankle swelling	Hypercholesterolaemia
Palpitations	Family history of IHD
Dizziness, blackouts	

Past history
Angina or MI
Rheumatic fever
Intermittent claudication

Previous investigations & interventions
ECG, echocardiogram, angiogram, angioplasty, CABG

Examination
Inspection A general impression of the patient may be formed whilst taking the history (see before). Is the patient in distress or pain? Note the general colour, facies and body habitus. Look out for the features of Marfan's (aortic incompetence). Down's (ASD, VSD) and Turner's (coarctation of the aorta). Rheumatological disorders are associated with cardiac lesions (eg ankylosing spondylitis—aortic incompetence).

The face may provide a clue as to the cardiovascular disease. Mitral stenosis is associated with a characteristic *malar flush*. Examine the eyes for *Argyll Robertson pupils* (p56) which may be seen with syphilitic aortic incompetence. Prosthetic valves produce low grade haemolysis and *jaundice*. *Xanthelasma* and *corneal arcus* suggest hyperlipidaemia. *Proptosis*, *lid lag* and *lid retraction* may alert you to Graves' disease (p542) as the cause of the irregular rhythm and heart failure. Then inspect the mouth for a *high arched palate* (Marfan's syndrome—aortic incompetence). *Mucosal petechiae* may be due to infective endocarditis. Note the state of *dentition*. Look at the tongue and lips for *cyanosis* (p42).

Pick up the patient's hand (p25). Look for the peripheral stigmata of endocarditis: splinter haemorrhages, Osler's nodes (tender lumps in the pulps of fingertips), Janeway lesions (red macules on the wrist and dorsum). Note any tendon xanthomas and nicotine stains.

Next examine the pulse: rate and rhythm at the radial pulse and character and volume at the carotid or brachial pulse (see p54). Feel for the peripheral pulses. *Radio-radial delay* and *radio-femoral delay* (ie asynchrony of the two pulses) is seen in coarctation of the aorta.

Take the blood pressure. The jugular venous pressure (JVP) may be assessed from the external jugular (often unreliable as can be obstructed). Ideally look for the internal jugular (see p258).

History and physical examination

The cardiovascular system

Inspect the precordium for scars (the lateral thoracotomy scar of mitral valvotomy; midline sternotomy scar), deformity and visible pulsations. Remember pacemaker boxes. Palpate for the apex beat. This is the point furthest from the manubrium where the heart can be felt beating. Normally this is in the 5th intercostal space in the mid-clavicular line (5th ICS MCL) (see p40). Palpate for a parasternal heave of right ventricular hypertrophy and for precordial thrills (palpable murmurs).

Auscultation Listen with the bell and the diaphragm at the apex (the mitral area). Identify the first and second heart sounds and decide if they are normal. Then listen for added heart sounds and murmurs. Repeat at lower left sternal edge and then in the aortic and pulmonary areas (right and left of the manubrium). Re-position the patient in the left lateral position: again feel the apex beat (?tapping) and listen specifically for the diastolic rumble of mitral stenosis. Sit the patient up and listen at the lower left sternal edge for the blowing diastolic sound of aortic regurgitation—accentuated at the end of expiration.

The chest Percuss the back to exclude a pleural effusion and listen for inspiratory crepitations due to left ventricular failure. Feel for sacral oedema.

The abdomen Lie the patient flat and feel for hepatomegaly (right ventricular failure) and check if it is pulsatile (tricuspid incompetence). Feel for splenomegaly (endocarditis) and for an abdominal aneurysm. Palpate the femoral arteries and auscultate for any bruits.

The extremities Go on to feel for the peripheral pulses and look for any signs of peripheral vascular disease, oedema, clubbing, xanthomata and varicose veins (if CABG may be a possibility). Listen over the femoral pulses for bruits.

Remember to examine the fundi for hypertensive change (p684) or Roth's spots (endocarditis), the urine (haematuria in endocarditis) and the temperature chart.

The respiratory system

History
Age, occupation, ethnic origin

Presenting symptoms
Cough
Sputum, haemoptysis?
Dyspnoea
Wheeze
Chest pain
Sinusitis
Hoarseness
Night sweats

Past history
Pneumonia, bronchitis
Tuberculosis
Previous CXR abnormalities
Drugs (eg steroids, bronchodilators)
Allergies

Social history
Smoking
Occupation (farming, mining)
Pets
Family history of atopy, TB, emphysema

Examination
Undress the patient to the waist and position him sitting at the edge of the bed.

Inspection Try to assess the general health of the patient—is he distressed? Cachetic? Dyspnoeic patients are sitting forward, with elbows supported, using accessory muscles of respiration and often exhibit nasal flaring and intercostal recession with inspiration. Count the respiratory rate and note the pattern of breathing. Look for chest wall deformities (p40). Inspect the chest for scars of previous operations, chest drains or radiotherapy (skin thickening and tattoos demarcating the field of irradiation). Note the movement of the chest wall—is it symmetrical? If not, pathology is on the restricted side.

Examine the hands for *clubbing* (p42), peripheral cyanosis, nicotine staining and wasting of the intrinsic muscles of the hand—seen in T1 lesions (Pancoast's tumour, p706). Palpate the wrist for tenderness (hypertrophic pulmonary osteoarthropathy, p342). Check for *asterixis* (CO_2 retention flap). Palpate the pulse for obvious *paradox* (weakens in inspiration), take the BP if necessary.

Inspect the face—look out for the ptosis and constricted pupil of a Horner's syndrome and check the tongue and lips for the bluish tinge of central cyanosis.

Feel the trachea in the sternal notch (it should pass slightly to the right). If deviated, concentrate on the upper lobes for pathology. Note the presence of *tracheal tug* (descent of trachea with inspiration, suggesting severe airflow limitation). Palpate for cervical lymphadenopathy. Assess chest expansion using hands or tape measure (normal 5cm). Pathology lies on the side with limited movement. Ask the patient to repeat '99' whilst palpating the chest wall. The differences in vibration or vocal fremitus follow the same pattern as for vocal resonance.

Percussion Percuss all areas including the axillae, clavicles and supra-clavicular areas. Listen and *feel* for the nature and symmetry of the sound. Distinguish between normal, resonant (hyperexpanded chest or pneumothorax), dull (over liver, consolidated lung) and stony dull (pleural effusion).

Auscultation—Ask the patient to breathe in and out through an open mouth—deeply and not too fast. Listen for change in *intensity* (increased or decreased) or change in *nature* (ie bronchial breathing, high pitched or blowing; symmetrical or asymmetrical).

Listen for *added sounds* eg wheeze (poly- or mono-phonic; inspiratory or expiratory), crackles (early vs late/pan-inspiratory, fine vs coarse), rub (suggests pleurisy—eg from pulmonary infarct or pneumonia).

Vocal resonance is the auscultatory equivalent of tactile vocal fremitus and is altered in the same way by pathology—ask the patient to repeat '99'. Muffled over normal lung but increased and clearer over consolidated lung. In some cases whispered speech is heard clearly over consolidated lung (*whispering pectoriloquy*).

If an abnormality is detected, try to localize it to the likely abnormal segment (see figure below).

Look at the **jugular venous pressure** (p258) and examine the heart for signs of **cor pulmonale** (p366). Look at the **temperature chart** and at the sputum. Pink froth suggests pulmonary oedema. Absolutely clear sputum is probably saliva. Large volume purulent sputum is seen in bronchiectasis. Green sputum suggests infection. Haemoptysis (coughing up blood) may be a sign of sinister disease and requires further investigation (see *haemoptysis*, p48 and *sputum*, p56).

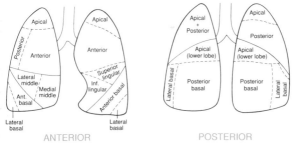

The respiratory segments supplied by the segmental bronchi.

Gastrointestinal history

Presenting symptoms
Abdominal pain
Nausea, vomiting, haematemesis
Dysphagia (p492)
Indigestion (dyspepsia, p490)
Recent change in bowel habit
Rectal bleeding or melaena
Diarrhoea or constipation
Appetite, weight change
Mouth ulcers
Jaundice
Pruritus
Dark urine, pale stools

Social history
Smoking, alcohol
Overseas travel, tropical illnesses
Contact with jaundiced persons
Occupational exposures
Sexual orientation

Past history
Peptic ulcer
Carcinoma
Jaundice, hepatitis
Blood transfusions, tattoos
Previous operations

Treatment
Steroids, the Pill
NSAIDs
Antibiotics
Dietary changes

Family history
Irritable bowel disease
Inflammatory bowel disease
Peptic ulcer
Polyps, cancer
Jaundice

Genitourinary history

Age, occupation, ethnic origin

Presenting symptoms
Fever, loin pain
Frequency, urgency, dysuria
Polyuria
Nocturia
Stream: hesitancy, terminal dribbling
Urine colour, haematuria
Urethral discharge
Sex—any problems?
Menstrual history: menarche, meno-
    pause, length of periods, amount,
    pain? intermenstrual bleeding?
Vaginal discharge
Pain on intercourse (dyspareunia)

Past history
Urinary tract infection
Renal colic
DM, high BP
Gout
Analgesic use
Previous operations

Social history
Smoking
Sexual orientation

The gastrointestinal system

Inspect (and smell) for signs of chronic liver disease:

Hepatic foetor	Gynaecomastia	Clubbing (rare)
Purpura	Scratch marks	Muscle wasting
Spider naevi (fill from the centre)	Jaundice	
Leuconychia (hypoalbuminaemia)	Palmar erythema	
Liver flap (asterixis, a coarse irregular tremor)		

Inspect for signs of malignancy, anaemia, jaundice, hard Virchow's node in left supraclavicular fossa (p142).

Inspect the abdomen Note visible pulsation (aneurysm), peristalsis, scars, striae, masses, distension, genitalia, herniae. If the veins of the abdominal wall look dilated assess the direction of flow. In IVC obstruction, flow below the umbilicus is upwards; in portal hypertension (*caput medusae*), it is downwards.

Palpation Adjust the patient so that his head rests on only one pillow, his arms at his side. Make sure that he and your hands are warm. While palpating, be looking at his face to assess any pain. First palpate gently through each quadrant, starting away from the pain. Note tenderness, guarding (involuntary tensing of abdominal muscles because of pain or fear of it), and rebound tenderness (greater pain on removing hand than on gently depressing abdomen: it is a sign of peritoneal inflammation); Rovsing's sign (p114).

Palpating the liver: Begin >10cm below the right costal margin with the patient breathing deeply. Use the radial border of the index finger to feel the liver edge, moving it up 2cm at a time at each breath. Assess its size, regularity, smoothness, and tenderness. Is it pulsatile? Confirm the lower border and define the upper border by percussion (normal upper limit is in 5th intercostal space): it may be pushed down by emphysema. Listen for an overlying bruit. See p48 for causes of hepatomegaly (enlarged liver).

Palpating the spleen: Start in the RIF and move towards the left upper quadrant with each respiration. Features of spleen which differentiate it from kidney: one cannot get above it, overlying percussion note is dull, it moves with respiration, it may have a palpable notch on its medial side. If you suspect splenomegaly but cannot detect it, assess the patient in the right lateral position with your left hand pulling forwards from behind the rib cage; also percuss in the mid-axillary line in the 10th interspace: it should be resonant.

Palpate the kidneys bimanually with the left hand under the patient able to push up in the renal angle. Attempt to ballot the kidney. It moves only slightly with respiration.

Assess other masses by the scheme on p152.

Percuss for the shifting dullness of ascites: the level of right-sided flank dullness increases by lying on the right.

Auscultation Bowel sounds: absence implies ileus; they are enhanced and tinkling in bowel obstruction. Listen for bruits.

Examine Mouth, tongue, rectum, genitalia, urine, as appropriate.

The neurological system

History of neurological symptoms should be taken from the patient and if possible, from a close friend or relative as well. The patient's memory, perception or speech may be affected by the disorder making the history difficult to obtain. Note the progression of the symptoms and signs: gradual deterioration (eg tumour) vs intermittent exacerbations (eg multiple sclerosis). Ask about the following:

Age, occupation, ethnic origin. Right- or left-handed?

Presenting symptoms
Headache
Problems with:
—Vision (blurring or double vision)
—Hearing (tinnitus, vertigo)
—Smell, taste
—Speech (dysarthria, dysphasia p430)
Pain
Pins and needles (paraesthesiae)
Numbness or abnormal sensations
Weakness
Poor balance
Sphincter disturbance
Abnormal involuntary movements
Fits, faints or 'funny turns'
—frequency, duration
—mode of onset, termination ?warning
—details of typical attack
—tongue biting? incontinence?
—time to recovery; confused after?
Cognitive state (p72)

Past history
Meningitis, encephalitis
Head or spine trauma
Seizures
Previous operations
Risks for cerebrovascular disease (p432) (high BP, DM, smoking, AF)

Medications
Current and previous

Social and family history
What can the patient do? or not do?
Family history of neurological or psychiatric disease
Consanguinity

The neurological examination

Higher mental function Conscious level, orientation in time, place and person, memory (short- and long-term). See p72.

Speech Dysphasia, dysarthria, dysphonia. See p430.

Skull & spine Malformation. Signs of injury. Palpate scalp. If there is any question of acute spinal injury, *do not move the spine*. Carotid/cranial bruits.

Cranial nerves Sit facing patient. Causes of lesions: p458.
I *Smell:* Test ability of each nostril to differentiate smells.
II *Acuity* and its correctability with glasses or pin-hole. Visual fields: Losses; inattention. Tests and sites of lesions are given in full in OHCS p492. Pupils (p54): Size, shape, symmetry, reactions to light (direct and consensual) and to accommodation if reactions to light are poor. *Ophthalmoscopy:* Darken the room. Instil tropicamide 0.5% one drop, if necessary. Select the focusing lens for the best view of the optic disc (pale or swollen?). If the view is obscured, examine the red reflex with your focus on the margin of the pupil. You will get a view of the fovea if you ask the patient to look at the ophthalmoscope's light after drops.

History and physical examination

The neurological system

III IV VI *Eye movements* (OHCS p486), nystagmus p422. (III palsy: ptosis, large pupil, eye down and out; IV palsy: diplopia on looking down or in; VI palsy: diplopia on lateral gaze).

V *Motor palsy:* 'Open your mouth': jaw deviates to side of lesion.

V *Sensory:* Corneal reflex lost first; check all 3 divisions.

VII *Facial* droop and weakness. Lower ²/₃ in UMN lesion; whole of one side of face in LMN lesion: 'raise your eyebrows'; 'show me your teeth'. Test taste with tastes, eg mint.

VIII *Hearing:* Ask patient to repeat a number whispered in an ear, while you block the other. See p424. *Balance:* See p420 & p422.

IX X *Gag reflex:* Palate pulled to normal side on saying 'Ah'.

XI *Trapezii:* 'Lift your shoulders' against resistance. *Sternomastoid:* 'Turn your head to the left/right' against resistance.

XII *Tongue movement:* Deviates to side of a lesion.

Sensation Light touch (cotton wool), pain (pin-prick), vibration (128Hz tuning fork), joint position sense. See dermatomes p410–11.

Motor system *Drift:* Patient sitting, arms stretched out in front, eyes closed. Do arms drift? Then ask patient to touch nose with index finger (coordination). *Abnormal posture. Muscle bulk* (wasting). *Abnormal movements. Fasciculation* (visible muscle twitching that does not move the limb). *Tone:* Look for spasticity (pressure fails to move a joint, until it gives way—like a clasp-knife), rigidity (lead-pipe), rigidity + tremor = cogwheeling, clonus (rhythmic muscle 'beats' in response to sudden stretching: eg gastrocnemius on ankle dorsiflexion). *Power:* Oppose each movement: see root values, p408. *Coordination:* Finger–nose, rapid alternating movements, heel–shin. *Reflexes:* Biceps (C5–6), triceps (C7–8), supinator (C5–6), knee (L3–4±L2), ankle (S1–2), abdominals (lost in upper motor neurone—UMN—lesions), plantars (up-going in UMN lesions). See myotomes p408.

Gait Have patient walk: normally; heel-to-toe; on heels; then on toes. Ask him to stand feet together. If balance is worse on shutting his eyes, Romberg's test is +ve, implying abnormal joint position sense. If he cannot perform this even with eyes open, this may be cerebellar ataxia, but is not Romberg's positive.

Bear in mind textbooks present idealized situations: often one or more sign is equivocal or even contrary to expectation; don't be put off, consider the whole picture, including the history.

The psychiatric assessment

Introduce yourself, ask a few factual questions (precise name, age, marital status, occupation, and who is with them at home). These will help your patient to relax.

Presenting problem Then ask for the main problems which have led to this consultation. Sit back and listen. Don't worry whether the information is in convenient form or not—this is an opportunity for the patient to come out with his or her worries unsullied by your expectations. Ask: 'Are there any other problems?' After 3–5 minutes you should have a list of all the problems (each sketched only briefly). Read them back to the patient and ask if there are any more.

History of presenting problem For each problem obtain details, both current state and history of onset, precipitating factors and effects on life—this may take up to 20 minutes.

Check of major psychiatric symptoms Check those which have not yet been covered: *depression* (low mood, thoughts of worthlessness, hopelessness, suicidal thoughts and plans: 'Have you ever been so low that you thought of harming yourself?', 'What thoughts have you had?', sleep disturbance with early morning waking, loss of weight and appetite); *hallucinations* ('Have you ever heard voices when there has not been anyone there, or seen visions?') and *delusions* ('Have you ever had any thoughts or beliefs which have struck you afterwards as bizarre?'); *anxiety* and avoidance behaviour (eg avoiding shopping because of anxiety or phobias); *alcohol* and *other drugs*.

Family history (health, personality and occupation of parents and siblings) and family medical and psychiatric history.

Background personal history Try to understand the presenting problem. *Biography* (relationships with family and peers as a child, school and work record, sexual relationships and current relationships and family). Previous ways of dealing with stress and whether there have been problems and symptoms similar to the presenting ones. *Previous personality* (mood, character, hobbies, attitudes and standards).

Past medical and psychiatric history

The mental state examination Assess state at interview. Record examples of abnormal thought or speech. Record information under the following headings: *appearance*, *behaviour*, *speech* (both form eg pressure of speech, and content), *mood*, *abnormal experiences* (hallucinations) or *abnormal beliefs* (delusions). Assess *cognitive function*: orientation in time, place and person; short-term memory (give name and address and test recall 5mins later); long-term memory (eg recent political events), concentration (months of the year backwards, digit span).

The psychiatric assessment

Method and order of a routine examination

1 Look at the patient. Is he healthy, unwell, or *in extremis*?
2 Put the thermometer in his mouth.
3 Examine the nails and hands.
4 Work up the arm: pulse rate and rhythm. Keeping your finger on the pulse, count respiratory rate, and specifically consider: Paget's, acromegaly, endocrine disease (thyroid, pituitary, or adrenal hypo- or hyperfunction), body hair, abnormal pigmentation, skin conditions.
5 Blood pressure.
6 Conjunctivae (anaemia) and sclerae (jaundice).
7 Read thermometer.
8 Examine mouth and tongue (cyanosis, smooth, furred, ?beefy)
9 Examine the neck from behind: nodes, goitre.
10 Make sure the patient is at 45° to begin cvs examination in the neck: jvp; feel for character and volume of carotid pulse.
11 The precordium. Look for abnormal pulsations. Feel the apex beat for character and position. Feel for parasternal heave and thrills. Auscultate the apex (bell) in the left lateral position, then the other three areas (p26) and carotids (diaphragm). Sit the patient forward and listen in expiration.
12 Sit the patient forward to look for sacral oedema, and look for ankle oedema.
13 Now begin the respiratory examination with the patient at 90°. Observe respiration and posterior chest wall. Assess expansion, then percuss and auscultate the chest with the bell.
14 Sit the patient back. Feel the trachea. Inspect again. Assess expansion of the anterior chest. Percuss and auscultate again.
15 Examine the breasts and axillary nodes (p138).
16 Lie the patient flat with only one pillow. Inspect, palpate, percuss and auscultate the abdomen.
17 Look at the legs: swellings, perfusion, pulses, oedema.
18 Rapid neurological examination: *Cranial nerves:* Test pupil responses and look in the fundi. Test corneal reflexes. 'Open your mouth; stick your tongue out; screw up your eyes; show me your teeth; raise your eyebrows'.
Peripheral nerves: Look for wasting and fasciculation. Test tone in all limbs. 'Hold your hands out with your palms towards the ceiling and fingers wide. Now shut your eyes.' Watch for pronator drift. 'Keep your eyes shut and touch your nose with each index finger'. 'Lift your leg straight in the air. Keep it there. Put your heel on the opposite knee (eyes shut) and run it up own your shin'. You have now tested power, coordination, and joint position sense. Tuning fork on toes and index fingers to assess sensation.
19 Examine the gait and the speech.
20 Any abnormalities of higher mental function to pursue?
21 Perform rectal and consider vaginal examination.
22 Examine the urine with dipstick and microscope if appropriate.

In general, go into detail where you find something wrong.

Pictures to test naming and recognition

N.5 renaissance

N.8 accession

N.12 examiner

N.18 passion

N.36 blood

The British Faculty of Ophthalmologists' test types
- test each eye separately
- hold text at 30cm from eye
- record smallest type accurately read

3 | Dictionary of symptoms and signs

Symptoms are features which the patient reports to the doctor. Physical signs are elicited on examination by the doctor. Together they constitute the clinical features of the condition in that patient. To discuss any single feature in isolation is artificial as each is interpretable only in the context of the others. Nonetheless, it is still valuable to have an understanding of the pathophysiology and significance of individual features. In this chapter, we discuss some of the more important symptoms and signs. Where these are discussed in relevant chapters of this book and the *OHCS*, reference is given. See also Chapter 2: *History and physical examination* p20–36.

Abdominal distension

Causes: The famous five Fs—*fat*, *fluid*, *faeces*, *fetus*, or *flatus*. Also *food* (eg in malabsorption). Specific groups:

Air:	*Ascites:*	*Solid masses:*	*Pelvic masses:*
Gastrointestinal	Malignancy*	Malignancy*	Bladder: full or Ca
obstruction	Hypoproteinaemia	Lymph nodes	Fibroids; fetus
(incl. faecal)	(eg nephrotic)	Aorta aneurysm	Ovarian cyst
Aerophagy (air	R heart failure	*Cysts:* renal	Ovarian cancer
swallowing)	Portal hypertension	pancreatic	Uterine cancer

*The cell of origin may be of any intra-abdominal organ, eg large bowel, stomach, pancreas, liver, kidney.

Air is resonant on abdominal percussion.

Ascites is demonstrated by shifting dullness (p31) or by eliciting fluid thrill: place patient's hand firmly on his abdomen in sagittal plane and flick one flank with your finger while your other hand feels on the other flank for a fluid thrill.

The characteristic feature of *pelvic masses* is that you cannot get below them (ie their lower border cannot be defined).

Causes of *right iliac fossa masses*: appendix mass or abscess (p114); kidney mass; Ca caecum; a Crohn's or TB mass; intussusception; amoebic abscess—or any pelvic mass (see above).

Causes of *ascites with portal hypertension*: cirrhosis; portal nodes; inferior vena cava or portal vein thrombosis.

See causes of hepatomegaly (p48) and splenomegaly (p152).

Abdominal pain

of GI origin varies greatly depending on the underlying cause. Examples include: irritation of the mucosa (acute gastritis), smooth muscle spasm (acute enterocolitis), capsular stretching (liver congestion in CCF), peritoneal inflammation (acute appendicitis), direct splanchnic nerve stimulation (retroperitoneal extension of tumour). The *character*, *duration* and *frequency* depend on the mechanism of production. The *location* and *distribution* of referred pain depend on the anatomical site. *Time of occurrence* and *aggravating or relieving factors* such as meals, defecation and sleep also have special significance related to the underlying disease process. The site of the pain may provide a clue as to the cause. Evaluation of the '*acute abdomen*' is considered on p112.

Amaurosis fugax

See p684 & p436.

Anaemia

may be assessed from the skin creases and conjunctivae (pale if Hb<9g/dl). Koilonychia and stomatitis suggest Fe-deficiency. Anaemia with jaundice suggests malignancy or haemolysis.

39

Apex beat This is the point furthest from the manubrium where the heart can be felt beating. Normally this is in the 5th intercostal space in the mid-clavicular line (5th ICS MCL). Displacement laterally may be due to cardiomegaly or mediastinal shift. Assess its character: a *pressure loaded* (hyperdynamic) apex is a forceful sustained impulse that is not displaced. A *volume overloaded* (hyperkinetic) apex is forceful but not sustained which is displaced downwards and laterally. *Tapping* in mitral stenosis (palpable 1st heart sound); *dyskinetic* after anterior MI or with left ventricular aneurysm; *double* or *triple impulse* in HOCM.

40

Apraxia See Dyspraxia.

Back pain See p676.

Blackouts See p418.

Breath sounds and **added sounds** See p324.

Breathlessness See p44.

Cachexia Severe generalized muscle wasting which usually implies serious disease, principally carcinoma and cardiac failure. CNS causes: Alzheimer's disease and any other cause of prolonged inanition. In the Third World, the cause may be malnutrition (*OHCS* p186) or infection—eg TB, enteropathic AIDS ('slim disease') associated with *Cryptosporidium*).

Carotid bruit Bruit due to stenosis (of at least 30%) often near the origin of internal carotid artery. Heard best behind the angle of jaw, the usual cause of the stenosis is atherosclerosis. The significance of symptomless carotid bruits is that there is a small increased risk of stroke, myocardial infarct and death. Consider aspirin prophylaxis 75mg/24h PO and Doppler studies (p432–6).

Chest deformities *Barrel chest:* AP diameter↑, tracheal descent and poor expansion, seen in chronic hyperinflation (eg severe asthma/COAD). *Pigeon chest (pectus carinatum):* Prominent sternum and flat chest seen in chronic childhood asthma and rickets. *Funnel chest (pectus excavatum):* Local sternum depression (lower end). *Kyphosis* is increased forward spinal convexity *Scoliosis* is lateral curvature (*OHCS* p620)—the last two may cause a restrictive ventilatory defect. *Harrison's sulcus* is a groove deformity of lower ribs at the diaphragm attachment site, suggesting chronic childhood asthma or rickets.

Chest pains See p256.

Cheyne–Stokes respiration Breathing becomes progressively deeper and then shallower (there may be a period of apnoea) in cycles. Causes: brainstem lesions or compression (stroke, ICP↑). If the cycle is very long (eg 3mins) the cause may be a prolonged lung-to-brain circulation time (eg in chronic pulmonary oedema, poor cardiac output). It is enhanced by narcotic use. CNS causes often signify a bad prognosis (eg imminent respiratory arrest).

Chorea Continuous flow of jerky movements, flitting randomly from one limb or part to another. Each movement looks like a fragment of a normal movement. Caused by abnormalities of basal ganglia: Huntington's (p702); Sydenham's (p304); SLE (p672); Wilson's (p712); kernicterus; polycythaemia (p612); neuroacanthocytosis (a familial association of acanthocytes in peripheral blood with chorea, oro-facial dyskinesia and axonal neuropathy);[1] thyrotoxicosis (p542); drugs (L-dopa, contraceptive steroids). Treat with dopamine antagonists, eg haloperidol 0.5–1.5mg/8h PO, or tetrabenazine 25–50mg/8h PO.

Chvostek's sign Tapping over the facial nerve beneath the zygoma causes spasms of the facial musculature in hypocalcaemia.

---

1 R Hardie 1991 *Brain* **114** 13–49

Clubbing The finger nail has an exaggerated longitudinal curvature, there is a loss of the angle between the nail and the nail bed, and the nail-fold feels boggy.

Thoracic causes:
- Bronchial carcinoma (usually *not* small cell)
- Chronic lung suppuration
  —empyema, abscess
  —bronchiectasis
  —cystic fibrosis
- Fibrosing alveolitis
- Mesothelioma

GI causes:
- Inflammatory bowel (esp. Crohn's disease)
- Cirrhosis
- GI lymphoma
- Malabsorption (coeliac dis)

Cardiac causes:
- Cyanotic congenital heart disease
- Infective endocarditis

Rare causes
Familial (usually before puberty)
Unilateral in—brachial arterio-venous malformation
—axillary artery aneurysm
Thyrotoxicosis (thyroid acropachy)

Constipation See p488.

Cough This symptom is a non-specific reaction to irritation anywhere from the pharynx to the lungs. The patient can sometimes help with the anatomical localization if it is of proximal origin. Some characteristics of the cough may help diagnostically: even if the patient is not expectorating, listening to the cough may tell you that it is *productive*. Other features may help: a *brassy cough*, described as hard and metallic, may be associated with pressure on the trachea; a cough in which the patient complains of a retrosternal pain like a hot poker occurs in *tracheitis*; a *bovine cough* is prolonged and of deep pitch, and occurs in abductor paralysis of the vocal cords; croup is a harsh and hoarse cough of laryngitis; a hacking, irritating, frequent cough occurs in pharyngitis, tracheobronchitis and early in pneumonia. Some coughs are psychogenic. See also **Haemoptysis** (p48).

Chronic cough: Think of pertussis, TB, foreign body, asthma (eg nocturnal).

Cramp This is painful muscle spasm. Cramp in the legs is common, especially at night. It may also occur after exercise. It is only occasionally a symptom of disease, in particular: salt depletion, muscle ischaemia, myopathy. Forearm cramps suggest motor neurone disease. Night cramps in the elderly may respond to quinine bisulphate 300mg at night PO twice weekly. Writer's cramp is the specific complaint of an inability to perform the motor act of writing. The pen is typically gripped firmly. This is a form of dystonia. There is normally no neurological deficit. The condition rarely responds either to drugs or to psychological management. Similar specific dystonias may apply to other motor tasks.

Cyanosis This may be central or peripheral.

Central cyanosis represents a concentration of reduced Hb of at least 2.5g/dl; hence it occurs more readily in polycythaemia than anaemia. Causes can be divided into:

1 Lung disease resulting in inadequate oxygen transfer (eg COAD, severe pneumonia)—usually correctable by increasing the inspired O_2.
2 Shunting from pulmonary to systemic circulation (eg R–L shunting VSD, patent ductus arteriosus, transposition of the great arteries)—cyanosis *not* reversed by increasing inspired oxygen.
3 Conditions that do not allow adequate oxygen uptake—(eg met-, or sulph-haemoglobinaemia).

Peripheral cyanosis will occur in causes of central cyanosis, but may also be induced by changes in the peripheral and cutaneous vascular systems in patients with normal oxygen saturations. It occurs in the cold, in hypovolaemia, and in arterial disease, and is therefore not a specific sign.

Dictionary of symptoms and signs

Deafness See p424.

Dehydration Dry skin and mucous membranes (<5%, 2.5 litre deficit); cool peripheries, decreased skin turgor, tachycardia (5–8%, 4 litre deficit); sunken eyes, postural hypotension, oliguria (<400ml/24h) (9–12%, 6 litre deficit); moribund with signs of shock (>12%, >6 litre deficit). Distinguish water from saline loss (p628).

Diarrhoea See p486.

Dizziness usually means one of three things: vertigo, imbalance or faintness. *Vertigo* is the subjective sensation of the environment spinning around the patient (p420). *Imbalance* implies difficulty walking in a straight line and may result from disease of peripheral nerves, posterior columns, cerebellum or other central pathways. *Faintness* is the feeling of being about to pass out. It occurs with some seizure disorders and a variety of non-neurological conditions (p418).

Dysarthria See p430.

Dysdiadochokinesis See p456.

Dyspepsia and **indigestion** These are broad terms, used more by patients than their doctors to signify epigastric or retrosternal pain (or discomfort) which is usually related to meals. ▶ Find out exactly what your patient means. 30% have no clear abnormality on endoscopy (p498). Of the positive findings:

Oesophagitis alone 24%	Gastritis 9%	≥2 'lesions' 23%
Duodenal ulcer (DU) 17%	Duodenitis 6%	Bile reflux 0.7%
Hiatus hernia 15%	Gastric ulcer 5%	Gastric cancer 0.2%

Dyspnoea is the subjective sensation of shortness of breath, often exacerbated by exertion. Try to quantify exercise tolerance (eg dressing, distance walked, climbing stairs). May be due to:
- *Cardiac*—eg mitral stenosis or left ventricular failure of any cause; LVF is associated with *orthopnoea* (dyspnoea worse on lying flat; 'How many pillows?') and *paroxysmal nocturnal dyspnoea* (PND; dyspnoea waking the patient from sleep). There may be ankle oedema also.
  ▶ Any patient who is shocked may also be dyspnoeic—and this may be shock's presenting feature.
- *Pulmonary*—both airway disease and interstitial disease. May be difficult to separate from cardiac causes; asthma may also wake the patient from sleep as well as cause early morning dyspnoea and wheeze. Focus on the circumstances in which dyspnoea occurs (is it due to exposure to an occupational allergen?).
- *Anatomical*—ie diseases of the chest wall, muscles or pleura.
- *Miscellaneous*—thyrotoxicosis, ketoacidosis, anaemia, psychogenic. Look for other clues—dyspnoea at rest not associated with exertion may be psychogenic; look for features of respiratory alkalaemia (peripheral and perioral paraesthesiae and carpopedal spasm).

Speed of onset may help:

Acute	Subacute	Chronic
Foreign body	Asthma	COAD and chronic
Pneumothorax	Parenchymal disease	parenchymal diseases
Acute asthma	eg alveolitis	Non-respiratory cause
Pulmonary embolus	effusions	eg cardiac failure
Acute pulmonary oedema	pneumonia	anaemia

Dysuria is painful micturition and reflects urethral or bladder inflammation (infection most commonly; also urethral syndrome, p374). *Strangury* is the distressing desire to pass something that will not pass (eg bladder calculi).

Epigastric pain *Acute, severe:* Peritonitis; pancreatitis; GI obstruction; biliary colic; peptic ulcer; ruptured aortic aneurysm; myocardial infarction and other medical conditions (p280).

Chronic causes: Peptic ulcer; gastric cancer; chronic pancreatitis; aortic aneurysm; root pain (referred from the spine).

Exophthalmos (proptosis) This term refers to significant, undue prominence of one or both eyes. Examination is best carried out by looking either at the patient's profile or looking down on the face from above. The commonest cause is in association with Graves' disease (p542–3) when it is usually bilateral; but Graves' disease is also the commonest cause of unilateral exophthalmos. In unilateral exophthalmos also consider a space-occupying lesion in or behind the orbit (eg a tumour). A pulsating eye indicates an arterio-venous (caroticocavernous) fistula behind the eye.

Faecal incontinence Faecal incontinence is common in the elderly and may account for a high proportion of patients investigated for 'diarrhoea'. By far the most common cause is 'spurious diarrhoea': a failure of peristalsis leads to a physical block (liquid stool trickles uncontrollably past 'rectal rocks')—or the rectum balloons with soft faeces which also leak out. *Diagnosis* is made on rectal examination. *Treatment* usually requires enemas (eg sodium phosphate or larger volume soap-and-water enemas on alternate days to empty the rectum). If the cause is neurogenic, give codeine phosphate (30mg/12h PO) to constipate, with twice weekly enemas giving planned evacuation. *Prophylaxis:* exercise; high fibre diet; bulk-formers (p488).

CNS causes: Parkinson's disease; autonomic neuropathy (eg DM); over-stretching of nerves (from roots S3–4) to puborectalis by childbirth or prolonged straining; cord damage (examine sacral dermatomes, p410).
Other causes: rectal prolapse; sphincter laxity; severe piles.

Faints See Blackouts p418.

Fatigue This feeling is so common that it is a variant of normality. Do not miss depression which commonly presents in this way. Even if the patient is depressed, screening history and examination is important to rule out chronic disease. *Investigation* should include FBC, ESR, U&E, plasma glucose, TFTs and CXR. Arrange for follow-up to see what develops.

Fever and night sweats While moderate night sweating is common in anxiety states, drenching sweats requiring several changes of night-clothes is a more ominous symptom associated with infection or lymphoproliferative disease. The pattern of fever may be relevant (see p186). Rigors are uncontrolled, sometimes violent episodes of shivering which occur with some causes of fever (eg pyelonephritis, pneumonia).

Flatulence 400-1300ml of gas are expelled PR per day, and if this, coupled with belching (eructation) and abdominal distension, seems excessive to the patient, he may complain of flatulence. Excessive eructation may occur in those with a hiatus hernia—but most patients complaining of flatulence have no demonstrable GI disease. The most likely cause is air-swallowing (aerophagy).

Frequency (urinary) means an increased frequency of micturition and it is important to differentiate an increase in urine production (as in diabetes insipidus p566, DM, polydipsia, diuretics, renal tubular disease, adrenal insufficiency and alcohol) from frequent passage of small amounts of urine (eg cystitis, urethritis, neurogenic bladder, extrinsic bladder compression (eg pregnancy), bladder tumour, prostatic enlargement).

'Funny turns' See Blackouts p418.

Guarding This is reflex contraction of the abdominal musculature detected as an involuntary response to the examiner's gentle pressure on the abdomen. It signifies local or general peritoneal inflammation (p112).

Gynaecomastia p556. Haematemesis p494. Haematuria p370.

Haemoptysis Distinguish between this and haematemesis. The former is blood coughed up, usually frothy, alkaline and bright red, in a patient who may have a history of chest disease. It often contains haemosiderin-laden macrophages. Note: melaena may be present if enough blood is swallowed. Haematemesis is acidic and usually dark (coffee grounds) and mixed with food.

Causes of haemoptysis:

1 Respiratory

Traumatic	(contusion, bronchial rupture, post-intubation, foreign body)
Infective	(acute bronchitis*, pneumonia*, lung abscess, bronchiectasis, TB, fungal or parasitic conditions)
Neoplastic	(primary* or secondary)
Vascular	(pulmonary infarction, vasculitis eg Wegener's, RA, SLE, AV fistula, anomalous vessels)
Parenchymal	(diffuse interstitial fibrosis, sarcoidosis, haemosiderosis, Goodpasture's syn, cystic fibrosis)

2 Cardiovascular (Pulmonary oedema*, mitral stenosis, aortic aneurysm)

3 Bleeding diatheses

*Denotes a leading cause in UK (TB and parasitic conditions common abroad).

Halitosis (foetor oris, oral malodour) results from gingivitis (Vincent's angina, p710), metabolic activity of bacteria in dental plaque, or sulphide-yielding food putrefaction. As well as poor hygiene, also consider causes such as smoking, alcohol, drugs (disulfiram, isosorbide). Delusional halitosis is also common. **Treatment** includes trying to eliminate anaerobes: ● Stand nearer the toothbrush ● Dental floss ● 0.2% aqueous chlorhexidine gluconate (*BMJ* 1994 i 217)

Heartburn This is an intermittent, gripping pain usually (retrosternal) worsened by: stooping, lying down, large meals, pregnancy. It indicates oesophageal disease—usually reflux oesophagitis.

Hepatomegaly Abnormally large liver. *Causes: Right heart failure* (may be pulsatile in tricuspid incompetence). *Infections:* Eg infectious mononucleosis; malaria. *Malignancy:* Metastatic or primary, myeloma, leukaemia, lymphoma. *Others:* Sickle-cell disease, other haemolytic anaemias, porphyria.

Hyperpigmentation See *Skin discoloration* (p56).

Hyperventilation is over-breathing that may be either fast (tachypnoea—ie >20 breaths/min) or deep (hyperpnoea—ie tidal volume ↑). Hyperpnoea may not be perceived by the patient (unlike dyspnoea), and is usually 'excessive' in that it produces a metabolic alkalosis. This may be appropriate (Kussmaul respiration) or inappropriate—the latter results in palpitations, dizziness, faintness, tinnitus, chest pains, perioral and peripheral tingling (plasma Ca^{2+} ↓). The commonest cause for hyperventilation is anxiety; others include fever and brainstem lesions.
● *Kussmaul respiration* is deep, sighing breathing that is principally seen in metabolic acidoses—diabetic ketoacidosis and uraemia.
● *Neurogenic hyperventilation* is the sustained, rapid breathing produced by pontine lesions.

Itching (Pruritus) is a common and very unpleasant symptom. Causes may be divided into primary skin conditions or systemic disease. (See table p50.) Secondary skin changes include excoriations, lichenification and shiny finger nails. Investigations if the cause is not obvious: FBC, U&E, LFTs, T_4, ESR and serum iron. Treat the underlying cause. Symptomatic measures include avoidance of hot and sweaty states, aqueous creams, calamine lotion, steroid cream for inflammation. Most drugs are unhelpful but terfenadine 60mg/12h PO or other antihistamines may help.

Primary skin causes:	*Systemic diseases causing itch:*
Eczema	Liver disease (bile salt deposition)
Scabies	Chronic renal failure
Lichen planus	Lymphomas
Drug reactions	Polycythaemia
Atopic dermatitis	Pregnancy
Dermatitis herpetiformis	Iron deficiency
Thyroid disease	

50

Jugular venous pulse and pressure See p258.

Lassitude See Fatigue (p46).

Left iliac fossa pain *Acute:* Gastroenteritis; ureteric colic; UTI; diverticulitis; torted ovarian cyst; salpingitis; volvulus; pelvic abscess; pain from or cancer in an undescended testis.
Chronic or subacute: Constipation; irritable bowel syndrome; large bowel carcinoma; inflammatory bowel disease; hip pathology.

Left upper quadrant pain *Causes:* Large kidney or spleen; gastric or colonic (splenic flexure) cancer; pneumonia; subphrenic or perinephric abscess; renal colic; pyelonephritis.

Lid lag is the lagging behind of the lid as the eye looks down. It is seen in thyrotoxicosis and anxiety states.

Lid retraction is the static state of the upper eyelid traversing the eye *above* the iris, rather than transecting it. It is seen in thyrotoxicosis and anxiety states.

Loin pain *Causes:* Pyelonephritis; hydronephrosis; renal calculus; renal tumour; perinephric abscess; pain referred from the vertebral column.

Lymphadenopathy Causes may be divided into:
1 *Reactive; Infective:* bacterial (pyogenic, TB, Brucella), fungal (coccidiomycosis), viral (EBV, CMV, HIV), parasitic (syphilis, toxoplasma). *Non-infective:* sarcoid, connective tissue disease (rheumatoid), dermatopathic (eczema, psoriasis), drugs (phenytoin), berylliosis.
2 *Infiltrative: Benign:* Sinus histiocytosis, lipidoses. *Malignant:* lymphoma (p606 & p608), metastases.

Musculoskeletal symptoms The 3 main complaints are: *pain, deformity* and *limitation of function.*

Pain: *Degenerative arthritis* generally produces an aching pain worse with exercise and relieved by rest. Discomfort may be more in certain positions or motions. Cervical or lumbar spine degeneration may also produce subjective changes in sensation not following dermatome distribution. Both inflammatory and degenerative joint disease produce *morning stiffness* in the affected joints but in the former this generally improves during the day, while in the latter the pain is worse at the end of the day. The pain of *bone erosion* due to tumour or aneurysm tends to be deep, boring and constant. The pain of *fracture* or *infection* of the bone is severe and throbbing and is increased by any motion of the part. *Acute nerve compression* causes a sharp, severe pain radiating along the distribution of the nerve. Joint pain may be referred eg hip disorder to anterior and lateral aspect of the thigh or to knee; shoulder to the lateral aspect of the humerus; cervical spine to the interscapular area, medial border of scapulae or tips of shoulders and lateral aspect of arms. (*Back pain*—p676.) *Limitation of function:* Causes: pain, bone or joint instability, or restriction of joint movement (eg due to muscle weakness, contractures, bony fusion or mechanical block by intracapsular bony fragments or cartilage).

Nail disorders See p25.

Nausea See Vomiting (p60).

Nodules (subcutaneous) Rheumatoid nodules; polyarteritis nodosa; xanthomata; tuberose sclerosis; neurofibromata; sarcoid; granuloma annulare; rheumatic fever.

Oedema (p92). Via Starling's Law, this may be from *increased venous pressure* (eg venous thrombosis or obstruction, right-heart failure) or *lowered intravascular oncotic pressure* (plasma protein↓, eg cirrhosis, nephrosis, malnutrition or protein-losing enteropathy—here water moves down the osmotic gradient into the interstitium to dilute the solutes there, increasing entropy, according to the laws of thermodynamics). On standing, venous pressure at the ankle rises according to the height of blood from the heart (~100mmHg). This increase is short-lived if leg movement pumps blood through valved veins; but if venous pressure rises, or valves fail, capillary pressure rises, fluid is forced out of them (oedema), the PCV rises locally, and microvascular stasis occurs.

Non-pitting oedema (ie non-indentible with a finger) implies poor lymphatic drainage (lymphoedema), eg primary (Milroy's syndrome p704) or secondary (radiotherapy, malignant infiltration, infection, filariasis).

Oliguria is defined as a urine output of <400ml/24h. This occurs in extreme dehydration, severe cardiac failure, urethral or bilateral ureteric obstruction, acute and chronic renal failure.

Orthopnoea See **Dyspnoea** above.

Pallor is a non-specific sign and may be racial, familial, or cosmetic. Pathology suggested by pallor includes anaemia, shock, Stokes–Adams attack, vasovagal faint, myxoedema, hypopituitarism, and albinism.

Palmar erythema Associations: cirrhosis, pregnancy and polycythaemia.

Palpitations represent to the patient the sensation of feeling his heart beat; to the doctor, the sensation of feeling his heart sink. Have the patient tap out the rate and regularity of the palpitations. Enquire into the circumstances of the symptom: people often feel their (normal) heart beat in the anxious nocturnal silence of the bedroom. Most importantly ask about any associated symptoms such as chest pain, dyspnoea, or faintness which may imply haemodynamic compromise. Often a clinical diagnosis of awareness of a normal heart beat may be made. If not, proceed to ambulatory ECG monitoring (p270).

Paraphimosis occurs when a tight foreskin is retracted and then then becomes irreplaceable as the glans swells. Treat by squeezing the glans for about 30 mins.

Paroxysmal noctural dyspnoea (PND) See **Dyspnoea** p44.

Pelvic pain *Causes:* UTI; retention of urine; bladder stones; menstruation; labour; pregnancy (eg retroverted uterus); endometriosis (OHCS p52); salpingitis; endometritis (OHCS p36); ovarian cyst (eg torted). Cancer of: rectum, colon, ovary, cervix, bladder.

Percussion note See p29 and p324.

Phimosis The foreskin occludes the urinary meatus, and obstructs the passage of urine. Treat by circumcision.

Polyuria is the passing of excessive volumes of urine. Urine output depends on fluid intake and body losses. Up to 3.5 litres per day is typical. Causes of polyuria include diabetes mellitus, cranial diabetes insipidus, nephrogenic diabetes insipidus (inherited or secondary to renal disease), hypercalcaemia, polydipsia and chronic renal failure.

Prostatism The symptomatic manifestations of prostatic enlargement are collectively termed 'prostatism'. These are of two types: **1** *irritative bladder symptoms* (frequency, urgency, dysuria) **2** *obstructive symptoms* (eg decrease in size and force of urinary stream, hesitancy and interruption of stream during voiding). Obstructive symptoms may also be produced by a urethral stricture, tumours, urethral valve or bladder neck contracture—which is why prostatism is a misleading term.

Pruritus See Itching (p48).

Ptosis is drooping of the upper eyelid associated with the inability to elevate the lid completely. It is best observed with the patient sitting up, his head held by the examiner. The third cranial nerve innervates the main muscle concerned (levator palpebrae). However, nerves from the cervical sympathetic chain innervate the superior tarsal muscle, and a lesion of these nerves will cause a mild ptosis which can be overcome if the patient is asked to look up. *Causes:* 1 Third nerve lesions usually causing unilateral complete ptosis. Look for other evidence of third nerve lesion (ophthalmoplegia with outward deviation of the eye, pupil dilated and unreactive to light and accommodation). 2 Sympathetic paralysis usually causes unilateral partial ptosis. Look for other evidence of sympathetic lesion (constricted pupil, lack of sweating on same side of the face—Horner's syndrome). 3 Myopathy (dystrophia myotonica, myasthenia gravis). These usually cause bilateral partial ptosis. 4 Congenital (present since birth). May be unilateral or bilateral, is usually partial and is not associated with other neurological signs. 5 Syphilis.

Pulses 1 *Rate and rhythm* may be assessed by the radial pulse. An irregularly irregular pulse will be AF or multiple ectopics. Regularly irregular may be 2nd degree heart block (p264). A bounding pulse occurs in CO_2 retention, liver failure and sepsis. Shock gives a small volume thready pulse. 2 *Character and volume* are best assessed at the brachials or carotids. Feel the radial and femoral pulse simultaneously if the patient is hypertensive (?radio-femoral delay of coarctation).

Character	*Cause(s)*
Small volume, slow-rising	Aortic stenosis
Bisferiens (collapsing and slow-rising)	Mixed aortic valve disease
Collapsing	Aortic incompetence
	Hyperdynamic circulation
	Patent ductus arteriosus
Pulsus alternans	Left ventricular failure
Jerky	HOCM
Pulsus paradoxus (weakens in inspiration by more than 10mmHg)	Asthma Tamponade/constrictive pericarditis

Pupillary abnormalities Are the pupils equal, central, circular, dilated or constricted, and do they react to light, directly and consensually, and on convergence/accommodation?

Irregular pupils are caused by iritis, syphilis, or globe rupture.

Dilated pupils are caused by third cranial nerve lesions and mydriatic drugs. But always ask: is this pupil dilated, or is it the other which is constricted?

Constricted pupils are associated with old age, sympathetic nerve damage (Horner's syndrome—p700, and see **Ptosis** above), opiates, miotics (eg pilocarpine eye-drops for glaucoma), and pontine damage.

Unequal pupils (anisocoria) may be due to a unilateral lesion or drug application, syphilis, or be a Holmes–Adie pupil (see below). Some inequality is normal.

Reaction to light: Test by covering one eye and shining light into the other obliquely. Both pupils should constrict (one by the direct, the other by the consensual or indirect light reflex). The site of the lesion may be deduced by knowledge of the pathway: from the retina the message passes up the optic nerve to the superior colliculus (in the midbrain) and thence to the third nerve nuclei bilaterally. The third nerve causes pupillary constriction.

Dictionary of symptoms and signs

Reaction to accommodation/convergence: If the patient first looks at a distant object and then at the examiner's finger held a few inches away, the eyes will converge and the pupils constrict. The neural pathway involves a projection from the cortex to the nucleus of the third nerve.

Holmes–Adie (myotonic) pupil: This is a benign condition, which occurs usually in women and is unilateral in about 80% of cases. The affected pupil is normally moderately dilated and is poorly reactive to light, if at all. It is slowly reactive to accommodation; wait and watch carefully: it may eventually constrict more than a normal pupil. It is often associated with diminished or absent ankle and knee reflexes, in which case the Holmes–Adie syndrome is present.

Argyll Robertson pupil: This occurs in neurosyphilis, but a similar phenomenon may occur in diabetes mellitus. The pupil is constricted. It is unreactive to light, but reacts to accommodation. The iris is usually patchily atrophied and depigmented.

Hutchinson pupil: This refers to the sequence of events resulting from rapidly rising unilateral intracranial pressure (eg in intracerebral haemorrhage). The pupil on the side of the lesion first constricts then widely dilates. The other pupil then goes through the same sequence.

Rebound abdominal pain If, on the sudden removal of pressure from the examiner's hand, the patient feels a *momentary increase* in pain, then rebound tenderness is said to be present. It signifies local peritoneal inflammation, manifest as pain as the peritoneum rebounds after being gently displaced.

Rectal bleeding *Causes:* diverticulitis; colonic cancer; polyps; haemorrhoids; radiation proctitis; trauma; fissure-in-ano; angiodysplasia (this common cause of bleeding in the elderly is due to arteriovenous malformation).

Regurgitation Gastric and oesophageal contents are regurgitated effortlessly into the mouth—without contraction of the abdominal muscles and diaphragm (the feature distinguishing it from true vomiting). Regurgitation is rarely preceded by nausea, and when due to gastro-oesophageal reflux, it is often associated with heartburn.

Right iliac fossa pain *Causes:* All causes of left iliac fossa pain (see p48) plus appendicitis but usually excluding diverticulitis.

Right upper quadrant (hypochondrial) pain *Causes:* Gallstones; hepatitis; appendicitis (a common site in pregnancy); colonic cancer at the hepatic flexure; right kidney pathology (eg renal colic; pyelonephritis); intrathoracic conditions (eg pneumonia); subphrenic or perinephric abscess.

Rigors are the uncontrolled, sometimes very violent episodes of shivering which occur as a patient's temperature rises from normal to pyrexia.

Skin discoloration Generalized hyperpigmentation due at least in part to melanin may be genetic, or due to radiation, Addison's disease (p552), chronic renal failure (p388), pregnancy, oral contraceptive pill, any chronic wasting disease (eg TB, Ca), malabsorption, biliary cirrhosis, haemochromatosis, chlorpromazine, or busulphan. Hyperpigmentation due to other causes occurs in jaundice, haemochromatosis, carotenaemia, and gold therapy.

Splenomegaly Abnormally large spleen. *Causes:* See p152.

Sputum Smoking is the commonest cause of increased sputum production and it may have black specks from inhaled carbon. Yellow-green sputum is due to cell debris (bronchial epithelium, neutrophils, eosinophils) and is not always infected. Bronchiectasis results in large quantities of greenish sputum. Blood-stained sputum (*haemoptysis*) always requires further investigation.

57

Dictionary of symptoms and signs (continued)

Stridor is an inspiratory sound due to partial obstruction of the upper respiratory tract. That obstruction may be due to something within the lumen (eg foreign body, tumour), within the wall (eg oedema, laryngospasm, tumour, croup) or extrinsic (eg goitre, lymphadenopathy). It is a medical (or surgical) emergency if the airway is compromised.

Tactile vocal fremitus See p28.

Tenesmus This is a sensation felt in the rectum of incomplete emptying following defecation—as if there was something else left behind, which cannot be passed. It is very common in the irritable bowel syndrome (p518), but can be caused by a tumour.

Tinnitus See p424.

Tiredness See **Fatigue** (p46).

Tremor is rhythmic movement of parts of the body. There are 3 types:
1 *Resting tremor*—worst at rest; feature of parkinsonism, but the tremor is more resistant to treatment than bradykinesia or rigidity.
2 *Postural tremor*—worst when eg arms outstretched. May be an exaggerated physiological tremor (eg anxiety, thyrotoxicosis, alcohol, many drugs), metabolic (eg hepatic encephalopathy, CO_2 retention), due to brain damage (eg Wilson's disease, syphilis) or *benign essential tremor* (BET). This is usually a familial (autosomal dominant) tremor of arms and head which may present at any age and is suppressed by moderately large amounts of alcohol. Rarely progressive. Propranolol (40–80mg/12h PO) helps $\frac{1}{3}$ of patients as does primidone (50–750mg/24h PO).
3 *Intention tremor*—is worst during movement and occurs in cerebellar disease (eg in MS). No effective drug has been found.

Trousseau's sign The hands and feet go into spasm (carpopedal spasm) in hypocalcaemia. The metacarpophalangeal joints become flexed and the interphalangeal joints are extended.

Urinary changes *Cloudy urine* may suggest pus in the urine (infection) but is often normal phosphate precipitation in an alkaline urine. *Pneumaturia* (bubbles in the urine as it is passed) occurs with UTI due to gas-forming organisms or may signal an enterovesical fistula caused by inflammatory or neoplastic disease of the GI tract. *Nocturia* is seen in prostatism, DM, UTI and reversed diurnal rhythm as occurs with renal and cardiac insufficiency. *Haematuria* (blood in the urine) must be considered to be due to neoplasm until proven otherwise (p370).

Urinary frequency See **Frequency** (p46).

Vertigo See p420.

Visual disturbances may be due to eye disease or drugs but organic lesions of the CNS may present with a history of visual field loss (stroke), double vision (MS, trauma, tumour, basilar artery insufficiency, chronic basilar meningitis), flashing lights (migraine, seizure disorder), visual hallucinations (seizure disorder, drugs) or transient blindness (vascular lesions, migraine). **Changes in smell and taste** suggest focal lesions of the olfactory nerves or chorda tympani. **Disturbance of speech** may be noted by the patient or the doctor. Try to assess if difficulty is with articulation (*dysarthria*) or of word command (*dysphasia*) (p430).

Vocal resonance See p29 and p324.

Dictionary of symptoms and signs

Vomiting Ask about amount, colour, sour, old or recent food, blood, 'coffee grounds', relationship to eating (see table).

This can cause a metabolic alkalaemia, hyponatraemia, and hypokalaemia—as well as haematemesis from a Mallory–Weiss tear of the oesophagus.

GI causes:	CNS causes:	Metabolic:
Food poisoning (p190)	Intracranial pressure ↑	Uraemia
Gastroenteritis	eg tumours,	Pregnancy
GI obstruction (p130)	meningoencephalitis	Drugs:
Appendicitis	Ménière's disease	Alcohol
'Acute abdomen' (p112)	Labyrinthitis	
	Head injury	
	Cerebellar and brainstem disease	
	Migraine	
	Psychiatric disorder	

Waterbrash Saliva suddenly fills the mouth. It typically occurs after meals, and may denote oesophageal disease. It is suggested that this is an exaggeration of the oesophago-salivary reflex.

Weight loss is a non-specific feature of chronic disease and depression. It is associated with malnutrition, chronic infections and infestations (eg TB, enteropathic AIDS), malignancy, DM and hyperthyroidism (typically in the presence of increased appetite). Severe generalized muscle wasting is also seen as part of a number of degenerative neurological and muscle diseases and in cardiac failure (cardiac cachexia) although in the latter right heart failure may not make weight loss a major complaint.

61

4 Geriatric medicine

Relevant pages in other chapters:
Stroke (p 432); prevention and investigation of stroke (p 434).

The elderly patient in hospital

▶ *Any deterioration is due to treatable disease until proved otherwise.*

Beware of ageism

1 Contrary to stereotype, most old people are fit.[1] 95% of those over 65yrs, and 80% of those over 85yrs do not live in institutions; about 70% of the latter can manage stairs and can bath without help.

2 If there is a problem, find the cause. Do not accept problems as the inevitable result of ageing. Look for: disease; loss of fitness; and social factors, which may be amenable to therapy.

3 Do not restrict treatment simply because a person is old. Old people vary. Age alone is a poor predictor of outcome for an individual and should not be used as a substitute for careful clinical assessment of each patient's potential for benefit and risk.[2]

Characteristics of disease in old age There are differences of emphasis in the approach to old people compared with young people.[3]

1 *Multiple pathology:* Several disease processes may coincide: some may be irrelevant to the presenting illness.

2 *Multiple aetiology:* One problem may have several causes. Treating each alone may do little good; treating all may be of much benefit.[4]

3 *Non-specific presentation of disease:* Some presentations are common in old people. In particular the 'geriatric giants':[5] incontinence (p74); immobility; instability (falls, p68); and confusion (p70 & p446). Virtually any disease may present with these. Furthermore, typical signs and symptoms may be absent (eg myocardial infarction without chest pain, chest infection without cough or sputum).

4 Rapid deterioration can occur if disease untreated.

5 Complications are common.

6 More time required for recovery. Points 4–6 reflect impairment in homeostatic mechanisms and loss of 'physiological reserve'.

7 Impaired metabolism and excretion of drugs. Doses may need lowering.

8 Social factors are central in aiding recovery and return to home.

Special points *In the history:*
- Assess disability (p66).
- Obtain home details (eg stairs, access to toilet? can alarm be raised?).
- Medication (What? When? Assess understanding and 'compliance').
- Social network (regular visitors; family and friends).
- Care details (what services are involved? Home help; meals delivered; nurse. Who else is involved in the care?).
- Speak to others (relatives; neighbours; carers; GP).

On examination:
- Check BP lying and standing (postural hypotension).
- Rectal examination (constipation causes incontinence).
- Detailed CNS examination is often needed, eg if presentation is non-specific. This may tire the patient, so consider doing in small batches.

1 C Victor 1991 *Health and Health Care in Later Life* Open University Press   2 Working Group of the Royal College of Physicians 1991 *J Roy Col Phys* 25 197–205   3 After J Grimley Evans 1992 *Oxford Textbook of Geriatric Medicine*, OUP 703–6   4 D Fairweather 1991 *J Roy Col Phys* 25 105   5 B Isaacs 1992 *The Challenge of Geriatric Medicine* OUP

Geriatric medicine

The elderly patient in hospital

Impairment, disability, and handicap

▶ Focus on patients' problems not doctors' diseases.

Doctors are at home with diagnosing disease and identifying impairment. But we are slow to see the patient's perspective. Three key concepts can help: impairment; disability; handicap.

Impairment refers to systems or parts of the body that do not work. 'Any loss or abnormality of psychological, physiological or anatomical structure or function.'[1] For example, following stroke, paralysis of the right arm or dysphasia would be impairment.

Disability refers to things people cannot do. 'Any restriction or lack (resulting from an impairment) of ability to perform an activity in the manner, or within the range, considered normal for a human being.'[1] For example, following stroke, difficulties in 'activities of daily living' (eg with dressing, walking).

Handicap refers to the inability to carry out social functions. 'A disadvantage for a given individual, resulting from an impairment or disability, that limits or prevents the fulfilment of a rôle ... for that individual'.[1] For example, following stroke, being unable to go out to work or to go to tea with a friend.

Making use of these distinctions Two people with the same impairment (eg paralysed arm) may have different disabilities (eg one able to dress, the other unable). The disabilities are likely to determine the quality of the person's future. Treatment may usefully be directed at reducing disabilities. For example, Velcro® fasteners in place of buttons may enable a person to dress.

Three stages of management

1 Assessment of disability and handicap:[2] Traditional 'medical' clerking focuses on disease and impairment. Full assessment requires, in addition, a thorough understanding of disability and handicap. For structured assessment of disability we recommend the Barthel index (p67). Discuss further detailed assessments with expert therapists, and the patient, and relatives will help define their problems.

2 Who can help? As a hospital doctor you are part of a large team, but you may need to inform others that their help is needed. Other relevant professionals are: nurses (including community liaison nurses), occupational therapists, physiotherapists, speech therapists, social workers, the general practitioner, geriatrician and psychogeriatrician. There are self-help organizations for most chronic diseases aimed both for patients and their carers (p804).

3 Generate solutions to problems: A list of disabilities is the key. To plan rehabilitation look at each disability (eg unable to dress). Work out origin of disability (in terms of disease and impairment). Goals need agreeing with patient and relatives in the light of their wishes and professional predictions. Record agreed goals and ensure all work together with mutual understanding. Review, renew and adapt goals.

1 WHO 1989 Technical Report Series No 779, *Health of the Elderly*, Geneva
2 E Dickinson 1990 *Lancet* **337** 778

Barthel's index of activities of daily living (BAI)[1,2]

Bowels	0	Incontinent (or needs to be given enema)
	1	Occasional accident (once a week)
	2	Continent
Bladder	0	Incontinent, or catheterized and unable to manage
	1	Occasional accident (max once per 24 hours)
	2	Continent (for more than seven days)
Grooming	0	Needs help with personal care: face, hair, teeth, shaving
	1	Independent (implements provided)
Toilet use	0	Dependent
	1	Needs some help but can do something alone
	2	Independent (on and off, wiping, dressing)
Feeding	0	Unable
	1	Needs help in cutting, spreading butter, etc.
	2	Independent (food provided within reach)
Transfer	0	Unable—as no sitting balance
	1	Major help (physical, one or two people), can sit
	2	Minor help (verbal or physical)
	3	Independent
Mobility	0	Immobile
	1	Wheelchair-independent, including corners, etc.
	2	Walks with help of one person (verbal or physical)
	3	Independent
Dressing	0	Dependent
	1	Needs help but can do about half unaided
	2	Independent (including buttons, zips, laces, etc.)
Stairs	0	Unable
	1	Needs help (verbal, physical, carrying aid)
	2	Independent up and down
Bathing	0	Dependent
	1	Independent (bath: must get in and out unsupervised and wash self. Shower: unsupervised/unaided)

NB: the aim is to establish the degree of independence from any help.

1 Joint report of the Royal College of Physicians (RCP) and the British Geriatrics Society (BGS) 1992, London, RCP and BGS
2 D Wade 1992 *Measurement of Neurological Rehabilitation*, OUP

Stroke, falls, postural hypotension

Rehabilitation after stroke See p432–4 for stroke's acute phase.
▶ Good care consists in attention to detail. The principles of rehabilitation are those of any chronic disease (p66). However, a number of points need emphasizing—*Special points in early management:*
● Watch patient drinking a glass of water. If he chokes, he may require *nil by mouth* for some days (and IV fluids), then semi-solids (eg jelly; avoid soups and crumbly food). Avoid early NG tube (needed only in those few patients with established chronic swallowing problems).
● Avoid damaging shoulders of patient through careless lifting.
● Ensure good bladder and bowel care through frequent toiletting. Avoid early catheterization which may prevent return to continence.
● Position the patient to minimize spasticity.
● Often 'pseudo-emotionalism' is a problem after stroke, eg sobbing, unprovoked by sorrow (failure of limbic system inhibition by cortex). It may respond to tricyclics,[1] or fluoxetine (*OHCS* p341).
Special tests: ● Asking to point to a named part of the body tests for *perceptual dysfunction* ● Copying patterns of matchsticks tests *spacial ability.* ● Dressing or drawing a clock face tests for *apraxia.* ● Picking out and naming common objects from a pile tests for *agnosia.*

Falls are common and have a plethora of causes. 50% follow tripping up, or an accident; about 10% are related to loss of consciousness or dizziness. For most of the rest there is no clear cause.[2]
History: Exact circumstances of fall. Find out about any CNS, cardiac or musculoskeletal abnormalities. Drugs (eg causing: postural hypotension (below); sedation; arrhythmia, eg tricyclics; parkinsonism, eg neuroleptics including metoclopramide and prochlorperazine). Alcohol.
Examination: For causes: Postural hypotension; arrhythmias; carotid bruits; detailed neurological examination including cerebellar signs, parkinsonism, proximal myopathy, gait, watch patient get up from chair, test for Romberg's sign (p33); instability in knees.
For consequences: Cuts; bruises; fractured ribs (± pneumonia) or limbs, especially neck of femur; head injury (± subdural haematoma).
Management: NB: Confidence may be shattered even if no serious injury. Look for causes and treat. Consider physiotherapy—to include learning techniques to get up from floor. Occupational therapists may advise on reducing household hazards, and adding aids or a personal alarm radio-linked by phone to 24h cover—*so* reassuring to those living alone.

Postural hypotension is important because it is common and a cause of falls and poor mobility. Typical times are after meals, on exercise, and on getting up at night. May be transient with illness (eg flu).
Causes: Leg vein insufficiency; autonomic neuropathy (p463); drugs (diuretics, nitrates, antihypertensives, antidepressants, sedatives). The red cell mass may be low.[3] *Management:* Reduce or stop drugs if possible. Counsel patient to stand up slowly in stages. Try compression stockings (*OHCS* p598). Reserve fluid-retaining drugs (eg indomethacin ≤75mg/24h PO or fludrocortisone 0.1mg/24h PO increasing as needed) in those severely affected when other measures fail.

---

1 J van Gijn 1993 *Lancet* ii 816   **2** D Prudham 1981 *Age & Ageing* **10** 141   **3** R Hoeldtke 1993 *NEJM* **329** 611

Nicer interventions

The foregoing pages of this chapter paint a bright and optimistic picture of geriatric medicine. The message is: list all disabilities and handicaps, and work down the list curing or ameliorating everything that you can, and avoiding dismissing items on the list as the inevitable consequence of ageing. The trouble is, neither over-pressed doctors nor our bewildered patients may survive this approach, unless something else is added. A true story illustrates this: an old man with chronic but not too severe obstructive airways disease went to hospital for assessment. The doctor was very thorough, and the whole process (including various investigations and waits for these in rather poor surroundings) took more than 3 hours. The exhausted patient returned home rather more breathless than usual. Later he began to cough, and within 48 hours of his appointment he died. The assessment and the doctors' care was in many ways ideal—but not for the patient. It is very easy to over-investigate patients, to raise their expectations—and then to let them down.

A different approach to the exhaustive problem list (if it shows signs of exhausting either the doctor or the patient) is to develop the skill of recognizing opportunities for minimal, highly selected interventions which act as nudges to move the patient in a healthier direction, without interrupting the delicate matrix which allows the patient to function in the first place. For example, if a patient with Parkinson's disease is troubled by insomnia you could attempt to combine objective assessment of your changes to his antiparkinsonian treatment (how long does it take you to do X?) with treatment of the insomnia (fresh air) and with attempts to keep him fit (repetitive exercise)—by simply suggesting that he notes down how long it takes him to walk to the letter-box. This intervention will, incidentally, also give you a written record of any tendency to micrographia. So the message is now 'How many nails can I hit on the head with one blow of the hammer?' or 'Of all the interventions which are possible, which is the nicest one?' (using 'nicest' in both its senses).

Only experience can teach when to opt for complete problem lists and multiple interventions, and when to opt for minimalism. The advantages of developing a flair for this is that time will be released for other activities, such as listening to the patient, and giving him or her time to share fears. You may also give yourself time for a little rest and recuperation—an exhausted, uncommunicative doctor is usually a bad one. It is by having the time and the inclination to explore patients' fears that you may be of most use to them. Almost any elderly patient coming into hospital will have fears (for example of dying—and in geriatric medicine this fear will often not be groundless). You may vastly reassure the patient if you explain that one of your main jobs is to ensure that when death comes, it is as good a death as possible. The fears will often not centre on the patient himself. His worry may be that his handicapped daughter will be forced out of her lodgings when he is gone, or that his son's marriage will fail again if he is not there. Sympathetic listening may be all you can offer, but do not undervalue how much patients appreciate this.

Dementia

▶ Ensure no treatable cause has been missed.
▶ Assume confusion is due to acute illness until proved otherwise (p446).

Dementia is a syndrome, with many causes, of global impairment of cognition in clear consciousness (cf acute confusional state).

Clinical features The key to diagnosis is a good history (usually requiring an informant) of progressive impairment of memory and other cognitive functioning, together with objective evidence of such impairment. The history should go back at least several months, and usually several years.

Typically the patient has become increasingly forgetful, and has carried out the normal tasks of daily living (eg cooking, shopping) with increasing incompetence. Sometimes the patient appears to have changed personality eg by becoming uncharacteristically rude and aggressive. For objective evidence, carry out a test of cognitive functioning (p72).

Epidemiology Increasing incidence with age. Very rare below 55yrs. 5–10% prevalence above 65yrs. 20% prevalence above 80yrs.

Commonest causes *Alzheimer's disease* now includes onset at any age (previously used for presenile onset ie <65yrs only). The commonest cause at any age. About 65% of all cases. Defined pathologically (*senile plaques* and *neurofibrillary tangles* especially in cortex). Cortical and subcortical reduction in many neurotransmitters including acetylcholine, noradrenaline and 5-hydroxytryptamine. Cause unknown. ~70% female. Increased familial incidence notably if early age of onset. Familial cases sometimes associated with mutations on gene(s) or chromosome 21. Usually insidious onset, slow deterioration and mean survival 5–7yrs (OHCS, p742).

Multi-infarct dementia: About 25% of all cases. Essentially multiple small strokes. Usually evidence of vascular pathology (high BP, previous strokes); sometimes of focal neurological damage; onset sometimes sudden and deterioration often stepwise.

Less common causes Chronic alcohol abuse; Huntington's (p702); Jakob–Creutzfeldt disease (p702); Parkinson's disease (p452); Pick's disease; HIV infection (p212); subacute sclerosing panencephalitis (p440); progressive multiple leucencephalopathy. **Ameliorable causes** Hypothyroidism; B₁₂/folate↓; syphilis; thiamine deficiency (usually from alcohol abuse); operable cerebral tumour (eg parasagittal meningioma); subdural haematoma; CNS; cysticercosis (p248); arsenic poisoning; normal pressure hydrocephalus (dilation of ventricles without signs of raised pressure, due possibly to partial obstruction of CSF flow from subarachnoid space. A CSF shunt may reverse intellectual deterioration. Suggested by incontinence early in dementia, and gait dyspraxia).

Investigations Examine for focal neurological signs, FBC and film, ESR; U&E (including Ca²⁺); LFT; syphilis serology; B₁₂; folate; thyroid function; CXR; urine ward test. Consider: CT scan (atypical history, young onset, head injury); EEG. After counselling carers, test for HIV if at risk. (Consider zidovudine (p212 & p214) if +ve.)

Management Treat any treatable cause. Treat concurrent illnesses (these may contribute significantly to confusion). In most people the dementia remains and will progress. The approach to management is that of any chronic illness (p66). Involve relatives and relevant agencies. Alzheimer's Disease Society (p804).

Dementia
[*OTM* 21.42; 25.14]

The mini-mental state examination

If dementia is suspected test memory and other intellectual abilities formally. The mini-mental state examination[1] is one of many similar tests; it has been more fully studied than most.

- What day of the week is it? [One point]
- What is the date to-day? Day; Month; Year [One point each]
- What is the season? (Allow flexibility when testing during times of seasonal change). [One point]
- Can you tell me where we are now? What country are we in? [One point]
- What is the name of this town? [One point]
- What are two main streets nearby? [One point]
- What floor of the building are we on? [One point]
- What is the name of this place? (or what is this address?) [One point]
- Read the following, then proffer the paper: 'I am going to give you a piece of paper. When I do, take the paper in your right hand. Fold the paper in half with both hands and put the paper down on your lap.' [Give one point for each of the three actions.]
- Show a pencil and ask what it is called. [One point]
- Show a wristwatch and ask what it is called. [One point]
- Say (once only): 'I am going to say something and I would like you to repeat it after me: No ifs, ands or buts.' [One point]
- Say: 'Please read what is written here and do what it says.' [Show card with: CLOSE YOUR EYES written on it]. Score only if action is carried out correctly. If respondent reads instruction but fails to carry out action, say: 'Now do what it says'. [One point]
- Say: 'Write a complete sentence on this sheet of paper.' Spelling and grammar are not important. The sentence must have a verb, real or implied, and must make sense. 'Help!', 'Go away' are acceptable. [One point]
- Say: 'Here is a drawing. Please copy the drawing.' [See drawing opposite.] Mark as correct if the two figures intersect to form a four-sided figure and if all angles are preserved. [One point]
- Say: 'I am going to name three objects. After I have finished saying all three I want you to repeat them. Remember what they are because I am going to ask you to name them again in a few minutes.' [Name three objects taking 1 second to say each, eg APPLE; TABLE; PENNY.] Score first try [one point each object] and repeat until all are learned.
- Say: 'Now I would like you to take 7 away from 100. Now take 7 away from the number you get. Now keep subtracting until I tell you to stop.' Score 1 point each time the difference is 7 even if a previous answer was incorrect. Continue until 5 subtractions (eg 93, 86, 79, 72, 65). [Five points]
- Say: 'What were the three objects I asked you to repeat a little while ago?' [One point each object]

Interpreting the score
The maximum is 30. Score 28–30 does not support the diagnosis of dementia. Score 25–27 is borderline. Score <25 suggests dementia but consider also *acute confusional state* and *depression*. ~13% of over-75s in the general population have scores <25.

1 M Folstein 1975 *J Psychiatric Research* **12** 189–198

The mini-mental state examination

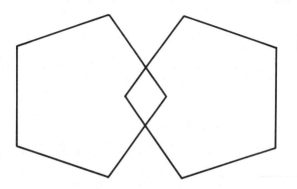

Close your eyes

Urinary incontinence

▶ Think twice before inserting a urinary catheter. ▶ Carry out rectal examination to exclude faecal impaction. ▶ Is bladder palpable after voiding (retention with overflow)?

Anyone might 'wet themselves' on a long coach journey. Do not think of people as either dry or incontinent but as incontinent in certain circumstances. Attending to these circumstances is as important as focusing on the physiology.

Incontinence in men Enlargement of the prostate is the major physiological cause of incontinence: urge incontinence (see below) or dribbling may result from the partial retention of urine. Transurethral resection of the prostate may weaken the bladder sphincter and cause incontinence. Troublesome incontinence therefore needs specialist surgical assessment.

Incontinence in women

1 *Functional incontinence* ie when physiological factors are relatively unimportant. The patient is 'caught short' and too slow in finding the toilet because of eg immobility or unfamiliar surroundings.

2 *Stress incontinence:* Leakage of urine due to incompetent sphincter. Leakage typically occurs when intra-abdominal pressure rises (eg coughing, laughing). The key to diagnosis is the loss of small (but often frequent) amounts of urine when coughing etc. Examine for pelvic floor prolapse. Look for cough leak, when patient standing and with full bladder. Stress incontinence is common during pregnancy and following childbirth. It occurs to some degree in about 50% of postmenopausal women. In elderly women, pelvic floor prolapse is the most common cause.

3 *Urge incontinence* is the most common type seen in hospital practice. The urge to pass urine is quickly followed by uncontrollable complete emptying of the bladder as the detrusor muscle contracts ('detrusor instability'). Large amounts of urine flow down the patient's legs. In the elderly it is usually related to organic brain damage. Look for evidence of: stroke; Parkinson's; dementia. Other causes: urinary infection; diabetes; diuretics; 'senile' vaginitis and urethritis.

In both sexes incontinence may result from diminished awareness due to confusion or sedation. Occasionally incontinence may be purposeful (eg preventing admission to an old people's home); or due to anger.

Management *Check for:* UTI; DM; excessive diuretics; faecal impaction. Investigate renal function (plasma urea and creatinine). Stress incontinence may be reduced by pelvic floor excercises. A ring pessary is usually helpful if uterine prolapse. Surgery to reduce uterine prolapse is sometimes required but must be preceded by cystometry and urine flow rate measurement to ensure there is no detrusor instability or sphincter dyssynergia. If urge incontinence, examine for spinal cord and CNS pathology (including cognitive test, p72); and for vaginitis (treat with dienoestrol 0.01% cream). The patient (or carer) should complete an 'incontinence' chart for 3 days to obtain pattern of incontinence. Maximize access to toilet; advise on toiletting régime (eg every four hours). The aim is to keep bladder volume below that which triggers emptying. Drugs may help reduce night-time incontinence (see table opposite) but are generally disappointing. Consider aids (absorbent pad, Paul's tubing).

Urinary incontinence
[*OTM* 11.39]

Agents for detrusor instability	Symptoms which they may improve
Oxybutynin 2.5mg/12h-5mg/12h PO	Frequency, urgency, urge incontinence
Imipramine 50mg PO at night	Nocturia, enuresis, coital incontinence
Propantheline 15mg/8h PO 1h ac	Frequency
Desmopressin nasal spray 20µg as a night-time dose	Nocturia/enuresis (↓urine production)
Oestrogens	Postmenopause urgency/frequency/nocturia
	?by raising bladder sensory threshold
Surgery, eg clam ileocystoplasty	The bladder is bisected, opened like a clam and 25cm of ileum is sewn in
Hypnosis/psychotherapy	(This requires good motivation)

75

Principal sources: G Wilcock 1991 *Geriatric Problems in General Practice*, OUP; L Cardoso 1991 *BMJ* ii 1453

Hypothermia

▶ Have a high index of suspicion and a low-reading thermometer.[1]

Most patients are elderly and do not complain of feeling cold. It is because they do not feel the cold that they have not tried to warm themselves up. In younger adults hypothermia is due, usually, either to exposure to extreme cold or it is secondary to impaired conscious level (eg following excess alcohol or drug overdose).

Definition Core (rectal) temperature <35°C.

In the elderly hypothermia is often caused by a combination of causes:

1 Impaired homeostatic mechanisms: usually age related.
2 Low room-temperature: poverty, poor housing.
3 Disease: Impaired thermoregulation (pneumonia, MI, heart failure). Reduced metabolism (immobility, hypothyroid, DM). Autonomic neuropathy (DM, Parkinson's). Increased heat loss (psoriasis). Reduced awareness of cold (dementia, acute confusional state). Increased exposure to cold (falls especially at night when cold). Drugs (major tranquillizers, antidepressants, diuretics). Alcohol.

Clinical features If temperature above 32°C, these may be mild: pallor; apathy; tachycardia. If temperature below 32°C they may include: bradycardia with irregularities (including VF); hypotension; impaired consciousness and coma. The patient's abdomen can feel 'colder than clay'.

Diagnosis Check oral or axillary temperature. If ordinary thermometer shows <36.5°C, use a low-reading thermometer to see whether the rectal temperature is <35°C.

Investigations Urgent U&E, blood glucose and serum amylase. Thyroid function tests; FBC; blood cultures. Consider blood gases. The ECG may show J-waves.

Management
● Take urgent bloods (as above).
● IVI (for access or to correct electrolyte disturbance).
● Cardiac monitor (both VF and AF can occur during warming).
● Consider antibiotics to prevent pneumonia (p336). Give these routinely in patients >65yrs with temperature <32°C.
● Consider urinary catheter (to assess renal function).
● *Slowly re-warm.* The danger is to reheat too quickly leading to peripheral vasodilation, shock and death. Aim at $\frac{1}{2}$°C increase per hour. Old conscious patients should sit in a warm room taking hot drinks. Thermal blankets may cause too rapid warming in old patients. The first sign of too rapid warming is falling BP. Treat by allowing patient to cool down slightly.
● Rectal temperature, BP, pulse and respiration every $\frac{1}{2}$ hour.

Complications Cardiac arrhythmias (if there is a cardiac arrest continue with resuscitation for 40mins as cold brains are less damaged by hypoxia); pneumonia; pancreatitis; acute renal failure; intravascular coagulation.

Prognosis Depends on age and degree of hypothermia. If age >70yrs and T° <32°C then mortality >50%.

Before discharge from hospital Anticipate problems. Will it happen again? What is her network of support? Review medication (could you stop tranquillizers)? How is progress to be monitored? Liaise with GP, and possibly with social worker.

1 M Hall 1986 *Medical Care of the Elderly* 2 ed, John Wiley

Hypothermia

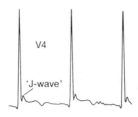

Kindly supplied by Drs Richard Luke and EM McLachlan.

Geriatric medicine

Notes

5 Surgery

(*These are the 3 conditions where the promptest surgery is essential. ▶ Notify your registrar *at once*.)

Relevant pages in other chapters:
▶ See the *Gastroenterology* contents page (p480).

Signs and symptoms: Abdominal distension (p38); epigastric pain (p46); faecal incontinence (p46); flatulence (p46); guarding (p46); heartburn (p48); hepatomegaly (p48); LIF and LUQ pain (p50); rebound tenderness (p56); RIF and RUQ pain (p56); splenomegaly (p56); tenesmus (p58); vomiting (p60); waterbrash (p60); weight loss (p60).

Urology: Urine (p370–1); haematuria (p370); urological malignancy (p392); gynaecological (*OHCS* p70); urinary tract obstruction (p378).

The language of surgery

Incisions have their names

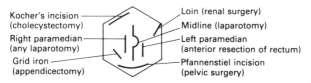

Kocher's incision (cholecystectomy)
Right paramedian (any laparotomy)
Grid iron (appendicectomy)
Loin (renal surgery)
Midline (laparotomy)
Left paramedian (anterior resection of rectum)
Pfannenstiel incision (pelvic surgery)

Abdominal areas

1,2: Right & left upper quadrants.
3,4: Right & left flanks.
5,6: Right & left iliac fossae.
  7: Epigastrium.
  8: Central abdominal area.
  9: Suprapubic area.

-ostomy An artificial opening usually made to create a new connection either between two conduits or between a conduit and the outside world—eg colostomy: the colon is made to open onto the skin (p96). Stoma means a mouth.

-plasty This is the refashioning of something to make it work, eg pyloroplasty relieves pyloric obstruction.

-ectomy Cutting something out—eg appendicectomy.

-otomy Cutting something open—eg laparotomy (the opening of the abdomen).

-oscope An instrument for looking into the body—eg cystoscope (a device for looking into the bladder).

-lith- Stone eg nephrolithotomy (cutting the kidney open to get to a stone).

-chole- To do with gall or bile.

-cyst- A fluid-filled sac.

-gram A radiological image, often using a radio-opaque contrast medium.

-docho- To do with ducts.

-angio- To do with tubes or blood vessels.

Per- Going through a structure (invasive).

Trans- Going across a structure.

Use the above to understand *choledochojejunostomy* and *percutaneous transhepatic cholangiogram*.

A fistula is an abnormal communication between two epithelial surfaces (or endothelial in arterio-venous fistulae), eg gastrocolic fistula (stomach/colon). Fistulae commonly close spontaneously but will not do so in the presence of chronic inflammation, distal obstruction, epithelialization of the track, foreign bodies or malignant tissue. External intestinal fistulae are managed by protection of the skin, replacement of fluid and electrolytes, parenteral nutrition and, if this fails, operation.

A sinus is a blind ending track which opens on to a surface.

Pre-operative care

1 To ensure that the right patient gets the right surgery. Have the symptoms and signs changed? If so, inform the surgeon.
2 To assess and balance the risks of anaesthesia.
3 To ensure that the patient is as fit as possible.
4 To decide on the type of anaesthesia and analgesia.
5 To allay all anxiety and pain.

The pre-operative visit Assess cardiovascular and respiratory systems, exercise tolerance, existing illnesses, drug therapy and allergies. Assess past history—of myocardial infarction, diabetes mellitus, asthma, hypertension, rheumatic fever, epilepsy, jaundice. Assess any specific risks—eg is the patient pregnant?

Have there been any anaesthetic problems (nausea, DVT)? Has he had a recent GA? (Do not repeat halothane within 6 months.)

Family history Ask questions about malignant hyperpyrexia (p86); dystrophia myotonica (p470); porphyria; cholinesterase problems; sickle-cell disease (test if needed). Does the patient have any specific fears or worries?

Drugs (see also p86) Ask about allergy to any drug, antiseptic, plaster.

Antibiotics: neomycin, streptomycin, kanamycin, polymixin and tetracyclines may prolong neuromuscular blockade.

Anticoagulants: Beware regional anaesthesia. Check clotting.

Beta-blockers: Continue up to and including the day of surgery as this precludes a labile cardiovascular response.

Digoxin: Continue up to and including morning of surgery. Check for toxicity and do plasma K^+. Suxamethonium increases serum K^+ and can lead to ventricular arrhythmias in the fully digitalized. Potentiation of the vagotonic effects of digitalis by halothane and neostigmine may be precipitated by an increase in calcium and low potassium.

Diuretics: Beware hypokalaemia. Do U&E (and bicarbonate).

Insulin: If on long acting preparations stabilize pre-operatively on short acting. IV 5% dextrose with KCl and insulin the morning of theatre allows accurate control. See p108.

Levodopa: Possible arrhythmias when the patient is under GA.

Lithium: Stop 3 weeks before surgery. It may potentiate neuromuscular blockade and cause arrhythmias. Beware post-operative toxicity and U&E imbalance. See p797.

MAOI: Stop 3 weeks before surgery. Interaction with narcotics & anaesthetics may lead to hypotensive/hypertensive crisis.

Ophthalmic drugs: Anticholinesterases used to treat glaucoma (eg ecothiopate iodine) may cause sensitivity to, and prolong duration of, drugs metabolized by cholinesterases, eg suxamethonium.

Oral hypoglycaemics: No chlorpropamide 24h pre-op (long $t_{1/2}$).

Steroids: If the patient is on or has recently taken steroids give extra cover for the peri-operative period (p110).

Tricyclics: These enhance adrenaline, exerting anticholinergic effects causing tachycardia, arrhythmias and low BP.

Pre-operative examination and tests: a guide for house-surgeons

Reassure, inform and and obtain written consent from the patient (eg remember to get consent for orchidectomy in orchidopexy procedures; also inform thyroidectomy patients of the risk of nerve damage).

Be alert to systemic problems, chronic lung disease (cough), hypertension, arrhythmias, and murmurs (SBE prophylaxis needed?).

Be alert to mechanical problems: neck mobility (consider lateral neck x-ray) to warn of difficult intubations.

Tests Be guided by the history and examination.

● U&E, FBC and ward test for blood glucose in most patients. If Hb <10g/dl tell anaesthetist. Investigate/treat as appropriate. U&E are particularly important if the patient is on diuretics, a diabetic, a burns patient, or has hepatic or renal diseases, or has starved, has an ileus, or is parenterally fed.

● Crossmatching: group and save for mastectomy, cholecystectomy. Crossmatch 2 units for Caesarean section; 4 units for a gastrectomy, and >6 units for abdominal aortic aneurysm surgery.

● Specific blood tests: LFT in jaundice, malignancy or alcohol abuse. Amylase in acute abdominal pain.
Blood glucose in diabetic patients (p108). Drug levels as appropriate (eg digoxin).
Clotting studies in liver disease, DIC, massive blood loss, already on warfarin or heparin. Contact lab as special bottles are needed.
HIV, HBsAg in high risk patients—after appropriate counselling.
Sickle test in those from Africa, West India or Mediterranean—and others whose origins are in malarial areas.
Thyroid function tests in those with thyroid disease.

● CXR: if known cardiorespiratory disease, pathology or symptoms.

● ECG: those with poor exercise tolerance, or history of myocardial ischaemia, hypertension or rheumatic fever.

Preparation ► Fast the patient.

● Check on any bowel or skin preparation needed.

● Start any DVT prophylaxis as indicated. One régime is heparin 5000U SC pre-op, then every 12h SC until ambulant.

● Catheterize and insert nasogastric tube (before induction and its attendant risk of vomiting) as indicated.

● Book any pre-, intra-, or post-operative x-rays as needed. Make sure that the pathologist knows if there is to be a frozen section.

● Book post-operative physiotherapy.

ASA classification (American Society of Anesthesiologists)

1 Normally healthy.
2 Mild systemic disease.
3 Severe systemic disease that limits activity; not incapacitating.
4 Incapacitating systemic disease which poses a threat to life.
5 Moribund. Not expected to survive 24h even with operation.

You will see a space for an ASA number on most standard anaesthetic charts. It is an index of health at the time of surgery. The prefix E is used in emergencies.

Prophylactic antibiotics in gut surgery

▶ *The major error of antibiotic therapy is unnecessary usage.*

Wound infections These occur in 20–60% of those undergoing GI surgery. Sepsis may lead to delayed mobilization, haemorrhage, wound dehiscence, and so start a fatal chain of events. Prophylaxis substantially reduces infection rates.

Rules for success:
● Give the antibiotic just before (eg 1h) surgery.
● Give it IV or IM (or as a suppository for metronidazole).
● Usually the antibiotic can be stopped after 24h.
● Use antibiotics which will kill anaerobes and coliforms (eg metronidazole and gentamicin, p181).

Bowel preparation is important before colorectal surgery. Typically the patient is admitted 3 days prior to surgery and put on a 'low residue' diet. 24h prior to surgery only free fluids are allowed. Mechanical preparation is needed and involves use of laxatives and washouts. One sachet of Picolax® (10mg sodium picosulphate with magnesium citrate) may be given on the morning before surgery and a second sachet during the afternoon before surgery; but these should be used with care if there is any risk of perforation. Washouts should be continued until evacuation is clear.

Antibiotic régimes There is usually a local preference but suitable examples are:

Biliary surgery: Broad spectrum penicillin (eg ampicillin 500mg IV/8h for 3 doses) or cephalosporin (eg cefuroxime 750mg/8h, 3 doses IV/IM).

Appendicectomy: A 3-dose régime of metronidazole suppositories 1g/8h, + cefuroxime 750mg/8h IV or gentamicin IV.

Colorectal surgery: Cefuroxime 750mg/8h, 3 doses IV/IM plus metronidazole 500mg/8h, 3 doses IV.

Sutures

Sutures (stitches) are obviously central to the art of surgery. The range of types available may appear confusing. Generally they are absorbable or non-absorbable and their structure may be divided into monofilament, twisted or braided. Examples of absorbable sutures are catgut, polyglactin (Vicryl®) and polyglycolic acid (Dexon®). Non-absorbable sutures include silk, nylon and prolene. Monofilament sutures are quite slippery but minimize infection and produce less reaction. Braided sutures have plaited strands and provide secure knots, but they may allow infection to occur between their strands. Twisted sutures have 2 twisted strands and similar qualities to braided sutures.

    The time of suture removal depends on the site and the general state of the patient. Face and neck sutures tend to be removed after 2–3 days, scalp and back of neck after 5 days and abdominal incisions after 5–8 days. In patients with poor wound healing, eg on steroids, with malignancy, infection or cachexia, the sutures may need 14 days or longer.

Surgery

Anaesthesia

Before anaesthesia explain to the patient what will happen, and where he will wake up, otherwise the recovery room or ITU will be frightening. Explain that he may feel ill on waking. The pre-medication aims to allay anxiety and to make the anaesthesia itself easier to conduct. Examples:

- Omnopon 20® (morphine 13.44mg/ml, papaverine & codeine); adults: 0.5–1ml SC/IM. Children over 1yr: 0.01–0.015ml/kg SC/IM. If aged between 1 month and 1yr, use half-strength paediatric ampoules (Omnopon 10®, 0.015–0.02ml/kg).

Give 400µg hyoscine (scopolamine) for amnesia and secretion drying (child 15µg/kg) or 600µg atropine (child 20µg/kg) IM.

- Temazepam 10–20mg PO.
- For children (2–7yrs), trimeprazine tartrate 2mg/kg as syrup (30mg/5ml).

Give pre-medication drugs 1h before surgery.

The side-effects of anaesthetic agents
Hyoscine, atropine: Tachycardia, urinary retention, glaucoma.
Opiates: Respiratory depression, absent cough reflex, constipation.
Thiopentone: (For rapid induction of anaesthesia) Laryngospasm.
Halothane: Vasodilatation, arrhythmias, hepatitis.

The complications of anaesthesia are due to loss of:
Pain sensation: Urinary retention, diathermy burns.
Consciousness: Cannot communicate 'wrong leg'.
Muscle power: No respiration, no cough (leading to pneumonia and atelectasis), corneal abrasion. Cannot phonate—eg 'I feel aware of . . .' when paralysed and in pain, and unable to communicate.

Local anaesthesia If unfit for general anaesthesia, local nerve blocks or spinal blocks (contraindication: anticoagulation) using long-acting local anaesthetics (eg bupivacaine) may be indicated. Intercostal blocks aid post-operative pain control.

Drugs complicating anaesthesia ▶ Inform anaesthetist.

Anticoagulants: Epidural and spinal blocks are contraindicated.

Anticonvulsants: Give the usual dose up to 1h before surgery.
Give drugs IV (or by NGT) post-op, until the patient can take oral drugs. Sodium valproate: an IV form is available now (give the patient's usual dose). Phenytoin: give IV slowly (<50mg/min). IM phenytoin absorption is unreliable.

Contraceptive steroids: These may increase the risk of DVT (particularly high-dose 'pill'). However, if stopped well before surgery there is risk of pregnancy. We recommend stopping about 1 month before surgery, if major, or orthopaedic surgery; otherwise do not stop. However, check local practice. Encourage early mobilization and consider anticoagulation post-op.

For other drugs, see p82.

Malignant hyperpyrexia This is a rare complication. There is a rapid rise in temperature (eg 1°C/5mins, up to 43°C) and acidosis due to rigidity. It may respond to prompt treatment with dantrolene. Give 1mg/kg/5mins IV—up to 10mg/kg in total. See *OHCS* p722.

Post-operative complications

During the first post-operative day
- Primary haemorrhage: ie continuous bleeding, starting during surgery. Replace blood loss. If severe, return to theatre for adequate haemostasis. Treat shock vigorously (p720).
- Reactive haemorrhage: haemostasis appears secure until BP rises, and bleeding starts. Replace blood and re-explore wound.
- Basal atelectasis: minor lung collapse leads to cough, spiking pyrexia and basal changes on CXR. Treat by physiotherapy and control of pain to allow coughing.
- Shock suggests blood loss, myocardial infarction, pulmonary embolism or septicaemia. *Treatment:* see p720.
- Pain (p90). Unexplained hypertension is often a sign of pain.
- Low urine output (p100).

During the next few days
- Paralytic ileus: check U&E, use IV fluids and NGT suction. (NG-tubes are uncomfortable and increase reflux and risk of aspiration. Use only if necessary—vomiting, gastric distension).
- Secondary haemorrhage: occurs as a result of infection.
- Pneumonia: treat by physiotherapy and amoxycillin 250mg/8h PO (or ampicillin 500mg/6h IV). See p332.
- Wound dehiscence (the stitches pull out of the tissues, and the wound falls apart): risk factors may be divided into pre-op (poor nutritional state, malignancy, obesity, re-operation), intra-op (poor technique, long procedure time) and post-op (infection, haematoma). It may be preceded by a pink serous discharge. Replace guts, cover with a sterile dressing, allay anxiety, give parenteral opiates and return to theatre.
- Deep venous thrombosis: see p92.
- Urinary retention: this may require catheterization if simple measures such as adequate privacy or a warm bath fail.
- Urinary tract infection: send MSU. Give trimethoprim 200mg/12h PO if treatment is needed before sensitivities are known.
- Pyrexia: consider pneumonia, wound infection, UTI, DVT, infected drip site, septicaemia and abscesses (eg subphrenic or pelvic).
- Wound infections: antibiotics are often of marginal significance and only used if there is cellulitis. Begin when a swab culture is known or use flucloxacillin empirically, 250mg/6h PO/IV.
- Confusion: think of hypoxia, abnormal U&E, infection, drugs, alcohol withdrawal, urinary retention, pain.

Late post-operative complications Adhesions between portions of the bowel which have been handled; incisional hernias (through a defect in the abdominal wall) in which the guts come to lie subcutaneously.

Surgical drains

The decision when to insert and remove drains may seem to be one of the great surgical enigmas—but there are basically two types. Most are inserted to drain the area of surgery and are often put on gentle suction. These are removed when they stop draining. The other type of drain is used to protect sites where leakage may occur in the post-operative period such as bowel anastomoses. These form a tract and are removed after about one week.

'Shortening a drain' means withdrawing it (eg by 2cm/day). This allows the tract to seal up, bit by bit.

Post-operative complications
Surgical drains

The control of pain

▶ *Don't forget how painful pain is.* Don't let him put up with it bravely.

▶ Before giving analgesia try to elucidate the cause of pain and how it may be removed. Then consider which analgesic will be appropriate and use it accordingly. Reassurance and comfort will contribute to any analgesic effect. This process begins before procedures are embarked upon.

Rules for success ● Chart each pain. ● Regular doses and assessment.

1 Non-narcotic analgesia Aspirin: 300–600mg/4h PO pc (if IM NSAID needed, try diclofenac, eg 75mg/12–24h IM (up to 4 doses). Good for musculoskeletal pain. CI: children, peptic ulcer, haemophilia, on anticoagulants. Cautions: asthma, renal/hepatic impairment, pregnancy.

Paracetamol: 0.5–1.0g/4h PO (up to 4g daily). Avoid if liver impairment.

2 Narcotics for moderate pain Codeine phosphate: 30–60mg/4h PO is good for headache (eg if ICP↑). Dihydrocodeine 30mg/4h PO is stronger. SE: see below.

3 Narcotic ('controlled') drugs for severe pain *Diamorphine* (5–10mg/4h—but you may use much more) is usually the drug of choice. It may be given PO, SC or IV. It is highly soluble so that an injection need never be more than 1ml. *Morphine* is an alternative (eg 10–15mg/4h PO/IM/IV). *Pethidine* (50–100mg/4h IM) is said to be good for colic (intestinal, renal, biliary). It excites smooth muscle less than other narcotics. *Buprenorphine* 200–400µg/8h PO (useful as it is given sublingually) or 300µg/8h IM. Avoid use with other opiates—their effect is reduced as buprenorphine is a partial antagonist. NB: it is only partially reversed by naloxone.

Problems with narcotics: An anti-emetic is usually needed with each dose (eg prochlorperazine 12.5mg/6h IM). Other side-effects include respiratory depression, constipation, cough suppression, urinary retention and hypotension. They cause sedation which is a disadvantage in hepatic failure, and head injury. Dependency is rarely a problem.

Epidural analgesia Opioids and anaesthetic are given into the epidural space by infusion or as boluses—and can be very effective in acute pain. Ask the advice of the Acute Pain Service (if available).[1] SE: Opiate-mediated respiratory depression (less of a problem with lipid-soluble drugs, eg fentanyl); local anaesthetic-induced autonomic blockade (BP↓).[1]

Other techniques of pain control Continuous patient-regulated subcutaneous infusion of narcotics. If brief pain relief is needed (eg for changing a dressing or exploring a wound, try inhaled nitrous oxide (with 50% O_2—as Entonox®) with an 'on demand' valve. Radiotherapy and cytotoxic drugs often relieve pain in malignancy. Treat conditions which exacerbate pain (eg constipation, depression, anxiety). Local heat, local or regional anaesthesia and neurosurgical procedures may be useful.

Guidelines for success
● Chart and review all pains.
● Try oral medication before parenteral (if possible).
● Treat underlying pathology wherever possible.

For pain relief in terminal care, see *OHCS* p442–4.

1 *Drug Ther Bul* 1993 **31** 9

The control of pain

Deep vein thrombosis (DVT)

Seen in ~30% of surgical patients, DVTs also often present in non-surgical patients. 65% of below-knee DVTs are asymptomatic (these rarely embolize to the lung—the major complication).

Risk factors Age, pregnancy, oestrogen drugs, surgery (especially pelvic), previous DVT, malignancy, obesity, immobility.

Signs are unreliable. There may be calf tenderness and mild fever. Look for pitting oedema, increased warmth and distended veins. Swelling >2cm compared to the other side is significant. Homans' sign (↑resistance to forced foot dorsiflexion, which may be painful) is of dubious value, and may dislodge thrombus.

Differential diagnosis Cellulitis may be impossible to exclude, and may coexist with DVT. Ruptured Baker's cyst (p694).

Tests Venography is definitive; Doppler studies are less good.

Prevention Mobilize early; heparin 5000U/12h SC. Use support hosiery, OHCS p598 (in major surgery, intermittent pneumatic pressure, continued for 16h post-op; stop contraceptive pills for major procedures and HRT (OHCS p64, OHCS p18) 4wks pre-op).

Treatment Calf thrombi <10cm long may be treated with compression stockings (OHCS p598) and heparin 5000U/12h SC. Larger thrombi need full heparinization IV via pump eg 15,000U/12h adjusted according to APTT (p596). Start warfarin at the same time and stop heparin when INR is 2.0–3.0. Treat for 3 months.

Swollen legs

Bilateral oedema usually implies systemic disease; oedema reflects either increased venous pressure (right heart failure) or decreased intra-vascular oncotic pressure (any cause of hypoalbuminaemia—so test the urine for protein in bilateral oedema). It is 'dependent' (affecting lowest part of body, because of gravity), which is why legs are usually affected. However, in severe cases, oedema may extend above the legs. In the bed-bound, or after a night's sleep, the fluid redistributes to the new dependent area, causing a sacral pad. The exception to this is the local increase in venous pressure which occurs in IVC or bilateral iliac vein obstruction: the swelling neither extends above the legs nor redistributes. In all of these situations both legs need not necessarily be affected to the same extent.

Causes: Right heart failure—also raised JVP and hepatomegaly.
Hypoalbuminaemia—liver failure, nephrotic syndrome, malnutrition, malabsorption, and protein losing enteropathy.
Ovarian or uterine mass (including pregnancy and puerperium).

Unilateral oedema usually implies either localized venous obstruction, or a soft tissue cause of a swelling.

Causes: DVT; unilateral iliac vein obstruction; cellulitis; ruptured Baker's cyst (p694).

Management Try to remove the cause. Simply giving a diuretic indiscriminately is *no* answer to oedema. First ameliorate dependent oedema by elevating the legs (put ankles higher than hips—do not just use foot stools), and elevate the foot of the bed (~30cm). Correctly fitting graduated support stockings (OHCS p598) are also helpful. Complications: venous stasis, leg ulcers.

Non-pitting oedema results from inadequate lymphatic drainage (lymphoedema) eg Milroy's syndrome (p704), filariasis.

Deep vein thrombosis (DVT)
Swollen legs

93

Specific post-operative complications

Biliary surgery After exploration of the common bile duct (CBD) a T-tube is usually left in the bile duct draining freely to the exterior. A T-tube cholangiogram is performed at 8–10 days and if there are no retained stones the tube may be pulled out.

Retained stones may be removed by ERCP (p498), further surgery or instillation of stone-dissolving agents via the T-tube to dissolve the stone. If there is distal obstruction in the CBD, fistula formation may occur with a chronic leak of bile.

Other complications of biliary surgery are stricture of the CBD; cholangitis; bleeding into the biliary tree (haemobilia) which may lead to biliary colic, jaundice and haematemesis; pancreatitis and chronic leak of bile causing biliary peritonitis. In the jaundiced patient it is important to maintain a good urine output as the danger is the hepato-renal syndrome (p110).

Thyroid surgery Laryngeal nerve palsy (hoarseness), hypoparathyroidism (p546), hypothyroidism, thyroid storm (p740). Tracheal obstruction due to haematoma in the wound may occur. ►► Relieve by immediate removal of stitches or clips.

Prostatectomy Retrograde ejaculation, epididymo-orchitis, impaction of blood clot in the urethra, causing urinary retention ('clot retention'), urethral stricture and incontinence. Urokinase release may promote devastating reactive haemorrhage. Hypervolaemia may be a problem during TURP as >1 litre of irrigating fluid may be absorbed.[1]

Haemorrhoidectomy Infection, stricture of anus, haemorrhage.

Mastectomy Arm oedema; necrosis of skin edge; 'My husband won't cuddle me any more'.

Aortic surgery Bleeding, thrombosis, embolism, trauma to ureters and anterior spinal artery (leading to paraplegia), gut ischaemia, renal failure, respiratory distress, aorto-enteric fistulae.

Colonic surgery Sepsis, ileus, fistulae, anastomotic leaks, obstruction, haemorrhage, trauma to ureters or spleen.

Tracheostomy Stenosis, mediastinitis, surgical emphysema.

Splenectomy Thrombocytosis, pneumococcal septicaemia (prevent by giving pneumococcal vaccine post-operatively); other infections.

Genitourinary surgery Septicaemia (from instrumentation in the presence of infected urine), urinoma—rupture of a ureter or renal pelvis leading to a mass of extravasated urine.

Gastrectomy See p164.

Specific post-operative complications

Nasogastric (Ryle's) tubes

These are tubes which are passed into the stomach via either the nose or the mouth. Size 16 is large, 12 medium, and 10 is small.

When to use them ● To empty the stomach before anaesthesia.
● Bowel obstruction, acute pancreatitis, and paralytic ileus.
● For irreversible dysphagia (eg motor neurone disease).
● For feeding ill patients (use a special fine-bore tube).

Problems Discomfort, ↑gastric secretion, reflux, aspiration.

How to pass a nasogastric tube Nurses are the experts and will ask you (who may have never passed one before) to do so only when they fail. Proceed as follows: 1 Don a plastic apron and observe the nurse having one more attempt. This will give you a rough idea of how it is done—she may even succeed. 2 Tell the patient what you want to do and take a new, cool (hence less flexible) tube. 3 Lubricate it with jelly (lignocaine) and place it in the nostril, advancing gently towards the occiput (not upwards). 4 Advance the tube into the oesophagus during a swallow. If this fails try the other nostril, then oral insertion. 5 Use litmus paper to confirm the contents are gastric. 6 Drain into a dependent catheter bag. If you fail: ▶ call your registrar—and do not allow plans to pass the tube under anaesthesia: the great danger is fatal inhalation of vomit on induction.

Stoma care

A stoma is an artificial union constructed either between two conduits (eg a choledochojejunostomy), or, more commonly between a conduit and the outside world—eg a colostomy, in which faeces are made to pass through a hole in the anterior abdominal wall into an adherent plastic bag, hopefully as 1–2 formed motions per day. (Mouths and anuses are natural stomas.) The physical and psychological aspects of stoma care may be undervalued. Many hospitals have a stoma nurse who is expert in fitting secure, odourless devices. Ask her to talk to the patient *before* surgery—or else he may reject his colostomy and never attend to it. He may even take his own life. Explain what a colostomy is, why it is needed, where it will be and what it will look like.

1 *Loop colostomies:* A loop of colon is exteriorized, opened and sewn to the skin. A rod under the loop prevents retraction and may be removed after 7 days. This is often called a defunctioning colostomy but this is not strictly true as faeces may pass beyond the loop. A loop colostomy is used to protect a distal anastomosis or to relieve distal obstruction.

2 *End colostomy:* The bowel is divided; the proximal end brought out as a stoma. The distal end may be: ● Resected, eg abdominoperineal excision of the rectum. ● Closed and left in the abdomen (Hartman's procedure). ● Exteriorized, forming a 'mucous fistula'.

3 *Double-barrelled (Paul–Mikulicz) colostomy:* The colon is brought out as a double-barrel. It may be closed using an enterotome.

Ileostomies protrude from the skin and emit fluid motions.
End ileostomy usually follows proctocolectomy, typically for UC.
Loop ileostomy is sometimes used to protect distal anastomoses.

Complications: Prolapse, stricture, ischaemia, hernia, plasma $K^+\downarrow$, refraction, bleeding, ileostomy diarrhoea.

Nasogastric (Ryle's) tubes
Stoma care

Putting in an IV cannula (a 'drip')

▶ Even the most skilled will sometimes be defeated by a 'drip'.

1 **Set up a tray** Swab to clean skin. Find: 3 cannulas, a 2ml syringe with 1% lignocaine (1ml) and 3 fine needles (orange), cotton wool to stop bleeding from unsuccessful attempts, tape to fix the cannula.

2 **Set up a 'drip'-stand** with first bag of fluid (carefully checked with a nurse); 'run through' a giving set (a nurse will show how).

3 **Ask a nurse to help** until you are experienced. Nurses prefer helping to changing the bed because of spilt blood.

4 **Explain** procedure to patient. Place the tourniquet around top of arm.

5 **Search hard for the best vein** (palpable rather than merely visible). Don't be too hasty. Rest the arm below the level of the heart to aid filling.

6 **Sit comfortably** with the apparatus ready to hand.

7 **Tapping on the vein** often brings it to prominence. Avoid sites spanning joints: these are uncomfortable and easily 'tissue'.

8 **Place a paper towel** under arm to soak up any blood.

9 **Clean the skin** around the site you have chosen. Use local anaesthetic: it is much kinder. Use a fine needle to raise a bleb of lignocaine—to look like a nettle sting—just to the vein's side. Wait 15sec. Insertion technique is best taught at the bedside by an expert.

After the 'drip' is in: 1 Connect fluid tube and check it is running well. **2** Fix cannula firmly with tape. **3** Bandage a loop of the tube to the arm. If the 'drip' is across a joint use a splint. **4** Check the flow speed. Write a fluid chart (p100). Does the nurse understand it? **5** Explain to the patient that there is no needle left in the arm. Explain the care that needs to be taken. **6** When the 'drip' comes down remove the IV cannula. (A patient once asked at the first follow-up appointment if he still needed 'this green plastic thing on the back of my hand'.)

If you fail after 3 attempts ▶ Shocked patients need fluid quickly: if you are having trouble putting in a 'drip' call your senior. The advice below assumes that the 'drip' is not immediately life-saving.

Experienced doctors, like drivers, forget they had to learn about drips. Ask to be taught and ask for help when help is needed. Is this the right needle for the right job? What is the 'drip' for? If the patient may need blood quickly use a large needle (eg brown or grey colour-coded). A green needle is suitable for clear fluids. For those with fragile veins who will not need fluids quickly, use a small needle (pink). Proceed as follows:

1 Explain to the patient that veins are difficult.

2 Fetch a bowl of warm water. This gives you time to calm down.

3 Immerse the patient's arm in the warm water for 2mins.

4 Use a blood pressure cuff at 80mmHg as a tourniquet—and try again.

If you still can't get the 'drip' in You are now getting downcast, so call your senior—it may pique your pride but there is nothing that makes your registrar happier than to succeed with a 'drip' where you have failed. Calling him could make both of you and your patient happy. Not many things do that! If you are too frightened of your senior, ask another house officer—they are much more likely to succeed than you at this juncture. If you can't find another person to help, have a cup of coffee and return an hour later. Veins are capricious: they come and go.

'The drip has tissued' Ask yourself: • Is the 'drip' still needed? • Is there fluid in bag and giving-set? • Are the control taps open? • Inspect the cannula: take bandage off. • Are there kinks in the tube?

Putting in an IV cannula (a 'drip')

If the 'drip' site is inflamed the 'drip' needs resiting. If the site is healthy, gently infuse a 2ml syringe of sterile 0.9% saline through the cannula. If resistance prevents this, the 'drip' needs resiting.

(Needle-stick injury: see p213.)

IV fluids on the surgical ward

Pre-operative fluids A shocked or dehydrated patient should rarely be rushed to theatre before adequate resuscitation. Anaesthesia may compound shock by causing vasodilatation and depressing cardiac contractility. The exceptions are exsanguination from a ruptured ectopic pregnancy or aortic aneurysm, where blood is lost faster than it can be replaced.

Post-operative fluids A normal requirement is 2–3 litres/24h which allows for urinary, faecal and insensible loss.

A standard régime: (One of many) 2 litres 5% dextrose with 1 litre 0.9% saline/24h. Add K^+ post-op (20mmol/l). See p628 for another example.

When to increase the above régime:
● Dehydration: this may be ≥ 5 litres if severe. Replace this slowly.
● Shock.
● Losses from gut: replace NGT aspirate volume with 0.9% saline.
● Transpiration losses: feverish patients and burns.
● Pancreatitis: there are large pools of sequestered fluid which should be allowed for.
● Low urine output (the night after surgery) is almost always due to inadequate infusion of fluid. Treat by increasing IVI rate unless patient is in heart or renal failure, or profusely bleeding (when blood should be transfused). The rate may be increased to 1 litre/h for 2–3h. Only if output does not increase should a diuretic be used. Consider a CVP line if estimation of fluid balance is difficult. A normal value is 0–5cm of water relative to the sternal angle.

NB: if not catheterized exclude retention, but otherwise do not catheterize until absolutely necessary.

When to decrease the above régime: Acute renal failure—give 500ml plus the previous day's output (► with no K^+).
In heart failure—halve the volume (1–1.5 litres/24h).

Guidelines for success
● Be simple. Chart losses and replace them. Know the urine output. Aim for 60ml/h; 30ml/h is the minimum ($\frac{1}{2}$ml/kg/h).
● Measure plasma U&E if the patient is ill. Regular U&ES are not needed on young fit people with good kidneys.
● Start oral fluids as soon as possible. They are safer.

What fluids to use ► *Haemorrhagic/hypovolaemic shock* (see p720): Insert 2 large IV cannulae, for fast fluid infusion. Start with crystalloid (eg 0.9% saline) or colloid (eg Haemaccel®) until fresh blood available. The advantage of crystalloids is that they are very cheap—but they do not stay as long in the intravascular compartment as colloids, because they equilibrate with the total extracellular volume (dextrose is useless for resuscitation as it rapidly equilibrates with the enormous intracellular volume). In practice, the best results in resuscitation are achieved by combining crystalloids and colloids. Aim to keep the haematocrit at ~0.3, and urine flowing at >30ml/h. Monitor pulse & BP often.

Septicaemic shock: Use a plasma-like substance (eg Haemaccel®).

Heart or liver failure: Avoid sodium loads: use 5% dextrose.

Excessive vomiting: Use 0.9% saline: replace losses.

IV fluids on the surgical ward

Blood transfusion and blood products

General points
- Find out the local procedures used to ensure that the right blood gets to the right patient. Use them.
- Take blood for crossmatching from only one patient at a time. Label immediately. This minimizes risk of wrong labelling of samples.
- When giving blood monitor TPR and BP every half-hour.
- Do not use giving sets which have contained dextrose or Haemaccel®.

Whole blood *Indication:* Exsanguination. Use crossmatched blood if possible, but if not, use 'universal donor' group ORh−ve blood; change to crossmatched blood unless >3U given—in which case go on with group ORh−ve blood as isoagglutinins may cause haemolysis in the new blood. ► Blood >2 days old has no effective platelets. Complications of massive transfusion: platelets↓, Ca²⁺↓, clotting factors↓, K⁺↑, hypothermia.

Packed cells (Haematocrit ~70%) Use for correcting anaemia. Each unit will raise Hb by 1–1.5g/dl. If the patient is prone to heart failure give each unit over 4h and give frusemide (eg 40mg slow IV/PO) with alternate units. Keep a careful eye on the JVP and listen for pulmonary oedema.

Platelet transfusions (p610) Not usually needed in thrombocytopenia if: there is no bleeding; the count is >20 × 10⁹/l; no surgery is planned.

Fresh frozen plasma (FFP) Use for correcting clotting defects: eg DIC, warfarin overdosage where vitamin K would be too slow, liver disease, thrombotic thrombocytopenic purpura. It is expensive and carries all the risks of blood transfusion. It is probably not needed routinely after large transfusions—only when a suspected clotting defect is confirmed in the lab. FFP should not be used simply as a volume expander.

Human albumin solution (Plasma protein fraction) is produced as 4.5% or 20% protein solution and is basically albumin to use for plasma expansion or protein replacement. There are no compatibility requirements. Both of the solutions have much the same Na⁺ content and 20% albumin can be used temporarily in the hypoproteinaemic (eg liver disease; nephrotic) who is fluid overloaded, without giving an excessive salt load.

Others Cryoprecipitate (eg factor VIII); coagulation concentrates (for self-injection in haemophilia), immunoglobulin (eg anti-D, OHCS p109).[1]

Complications of transfusing red cells
- ABO incompatibility leading to anaphylactic shock (p720).
- Mild allergic reactions: itching, urticaria, bronchospasm and fever. *Management:* give hydrocortisone 100mg IV and chlorpheniramine 10mg IV. Consider slowing or stopping the transfusion.
- Heart failure, if the transfusion is too rapid.
- Transfer of viruses (eg hepatitis B and C; HIV), bacteria or protozoa.

NB: a rapid spike of temperature (>40°C) at the start of a bottle indicates that the transfusion should be stopped (suggests intravascular haemolysis—see opposite). For a slowly rising temperature (<40°C) slow the IVI—this is most frequently due to antibodies against white cells (if recurrent, use leucocyte depleted blood or white cell filter).

1 *Drug Ther Bul* 1993 **31** 89   2 *Handbook of Transfusion Medicine* 1989 HMSO, London

Blood transfusion and blood products

▶ Avoid transfusions unless Hb is dangerously low, or there is risk of immediate further haemorrhage (eg active peptic ulcer). Transfuse up to 8g/dl. If Hb <~5g/dl with heart failure, transfusion is vital to restore Hb to safe level, eg 6–8g/dl, but must be done with great care. Give packed cells slowly with 40mg frusemide IV/PO with alternate units (do not mix with blood). Check for ↑JVP and basal crepitations; consider CVP line. If CCF gets worse, and immediate transfusion is essential, try a 2–3U exchange transfusion, removing blood at same rate as transfused.

Action on suspected transfusion reaction[2]

- STOP the transfusion
- Keep line open with 0.9% saline
- Check patient identity against unit
- Send blood unit + giving set to blood bank
- Send 20ml clotted and 5ml EDTA (FBC bottle) samples to blood bank
- Ask advice of blood bank if urgent transfusion still needed

Nutritional support in hospital[1]

Almost 50% of hospital in-patients may be malnourished.

Hospitals can become so focused on curing disease that they ignore the foundations of good health. Malnourished patients will recover more slowly, and experience more complications, than those well fed.

Why are so many hospital patients malnourished?

1 Increased nutritional requirements (eg sepsis, burns, surgery).
2 Increased nutritional losses (eg malabsorption, output from stoma).
3 Decreased intake (eg dysphagia, sedation, coma).
4 Effect of treatment (eg nausea, diarrhoea).
5 Enforced starvation (eg prolonged periods *nil by mouth*).
6 Missing meals through being whisked off, eg for investigations.
7 Difficulty with feeding and no one available to give enough help.
8 Unappetizing food: 'They feed me stuff I wouldn't give my cat'.

Identifying the malnourished patient

History: Recent weight change; recent reduced intake; diet change (eg recent change in consistency of food); nausea, vomiting, pain, diarrhoea which might have led to reduced intake.

Examination: Examine for state of hydration (p636): dehydration can go hand in hand with malnutrition, and overhydration can mask the appearance of malnutrition. Evidence of malnutrition: skin hanging off muscles (eg over biceps); no fat between fold of skin; hair rough and wiry; pressure sores; sores at corner of mouth.

Investigations: Generally unhelpful. Low albumin suggestive (but is affected by many things other than nutrition).

Prevention of malnutrition Assess nutritional status, and weight, on admission and regularly (eg weekly) thereafter. Identify those at risk (see above). Ensure that investigations and treatment do not interfere substantially with nourishment. Provide food appetizing to patient.

Nutritional requirements Most patients can be adequately nourished with 2000–2500kCal and 7–14g nitrogen/24h. Even catabolic patients rarely need more than 2500kCal. Very high calorie diets (eg 4000kCal/24h) can lead to fatty liver. If patient requires nutritional support seek help from dietitian.

Approximate energy contents Glucose 4kCal/g; fat 10kCal/g. To convert kCal to kJ multiply by 4.2.

Enteral nutrition (ie nutrition given into gastrointestinal tract)
If at all possible give nutrition by mouth. An all-fluid diet can meet requirements (but get advice from dietitian). If danger of choking (eg after stroke) consider semi-solid diet before abandoning food by mouth.

Tube feeding: This is giving liquid nutrition through enterally placed tube (eg nasogastric, nasoduodenal, directly into stomach via gastrostomy). Use nutritionally complete commercially prepared feeds. Standard feeds (eg Nutrison standard®, Osmolite®) normally contain 1kCal/ml and 4–6g protein per 100ml. Most people's requirements are met in 2 litres/24h. Specialist advice from dietitian is essential. Nausea and vomiting is less of a problem if feed given continuously with pump, but may have disadvantages compared with intermittent nutrition.

---

1 J Lennard-Jones 1992 *A positive approach to nutrition as treatment*. Kings Fund Report

Nutritional support in hospital

Guidelines for success
- Use fine bore nasogastric feeding tube when possible.
- Check position of nasogastric tube before starting feeding.
- Build up feeds gradually to avoid diarrhoea and distension.
- Weigh weekly, check blood glucose and plasma electrolytes (including zinc and magnesium if previously malnourished).
- Close liaison with dietitian is essential.

105

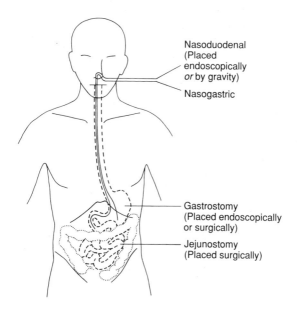

Nasoduodenal
(Placed endoscopically *or* by gravity)

Nasogastric

Gastrostomy
(Placed endoscopically or surgically)

Jejunostomy
(Placed surgically)

Parenteral (intravenous) feeding

Do not undertake parenteral feeding lightly: it has risks.
Specialist advice is vital.

Only consider parenteral feeding if the patient is likely to become malnourished without it. This normally means that the gastrointestinal tract is not functioning (eg due to bowel obstruction), and is unlikely to function for at least 7 days. Parenteral feeding may supplement other forms of nutrition (eg in active Crohn's disease when insufficient nutrition can be absorbed in the gut) or be used alone (total parenteral nutrition—TPN).

Administration Nutrition is normally given through a central venous line as this usually lasts longer than if given into a peripheral vein. Insert under strictly sterile conditions and check its position on x-ray.

Requirements There are many different régimes for parenteral feeding. Most provide about 2000kCal and 10–14g nitrogen in 2–3 litres; this amount to be given daily. About half the calories are provided by fat and half by carbohydrate. The régime should include vitamins, minerals, trace elements and electrolytes; these will normally be included by the pharmacist.

Complications
Sepsis: (eg *Staph epidermidis* and *aureus*; *Candida*; *Pseudomonas*; infective endocarditis). Look for spiking pyrexia and examine wound at tube insertion point.
Metabolic imbalance: Electrolytes; plasma glucose; deficiency syndromes.
Mechanical: Pneumothorax; embolism of IVI tip.

Guidelines for success
● Liaise closely with nutrition team.
● Meticulous sterility. Do not use central venous lines for uses other than feeding. Remove line if you suspect infection. Culture its tip.
● Review fluid balance at least twice daily, and requirements for energy and electrolytes daily.
● Check weight, fluid balance, and urine glucose daily throughout period of parenteral nutrition. Check plasma glucose, creatinine and electrolytes (including calcium and phosphate), and full blood count daily until stable and then three times a week. Check LFTs and lipid clearance three times a week until stable and then weekly. Check zinc and magnesium weekly throughout.
● Do not rush. Achieve maintenance régime by small steps.

Parenteral (intravenous) feeding

Diabetic patients on surgical wards

Insulin-dependent diabetes mellitus
- The stress of an intercurrent illness increases basal insulin requirement (see p535).
- Always try to put the patient first on the list (surgery, endoscopy, bronchoscopy, etc.). Inform the surgeon and anaesthetist early.
- Stop all long-acting insulin the night before. Ensure IV access is available should it be required urgently. If surgery is in the morning, stop *all* SC morning insulin. If surgery is in the afternoon, have the usual short-acting insulin in the morning with breakfast. No medium or long-acting insulin.
- Check U&E on the morning of surgery, and start an IVI of 1 litre of 5% dextrose with 20mmol (1.5g) KCl/8h to continue till the patient is eating normally. Check saline needs, but some dextrose must be infused constantly to maintain the blood glucose.
- Add 50U short-acting insulin (eg Actrapid®, Humulin S®, Velosulin®) to 50ml 0.9% saline to give via an infusion pump—give according to a sliding scale (p109) adjusted in the light of blood glucose.
- Check blood glucose hourly. Aim for 7–11 mmol/l during surgery.
- Continue IV insulin and dextrose till the patient eats his second normal meal. Then switch to usual SC insulin regimen.

Practical hints:
- If blood glucose persistently <4mmol/l, decrease each item by 0.5–1.0U/h; if always >11mmol, increase by 0.5–1U/h.
- For type I DM, aim for some insulin to be infused all the time as the patients are insulin deficient; eg a patient usually on Mixtard® 40U/35U needs 75U insulin/24h, ie ~3U/h basal requirement. Adjust sliding scales accordingly. ● Some prefer to add soluble insulin to 500ml bags of 5% dextrose with 20mmol KCl/4–6h and adjust the amount of insulin added according to the mean glucose level in the previous 4h. This is useful if close monitoring (hourly glucose ward tests) is not possible, as there is no risk of giving insulin without dextrose.

Non-insulin dependent diabetes mellitus These patients are usually controlled on oral hypoglycaemics (see p530). Perform a fasting blood glucose: if >10mmol/l, treat as for IDDM. Otherwise halve the dose of long-acting sulphonylurea (eg chlorpropamide) the day before surgery and omit the tablet on the day of surgery. Check a fasting glucose on the morning of surgery: if >15mmol use insulin, either IV as above or SC sliding scale (see p109). Use half the normal drug dose (and supplement with SC soluble insulin before meals) until a normal diet is established. Convert to IV insulin for major surgery.

Diet controlled diabetes Usually there is no problem, but the patient may become temporarily insulin-dependent.

Diabetic patients on surgical wards

IV Insulin Sliding Scale
(A guide only)

Fingerprick glucose (mmol/l)	IV soluble insulin (units/h)
<2	None (give 50ml of 50% glucose)
2.0–6.4	0.5U/h
6.5–8.9	1.0U/h
9–10.9	2.0U/h
11–16.9	3.0U/h
17–28	4.0U/h
>28	8.0U/h

NB: Check glucose hourly and adjust insulin accordingly.

Subcutaneous Insulin Sliding Scale
(A guide only)

Fingerprick glucose (mmol/l)	SC soluble insulin (Units)
<4.0	None
4.0–6.9	2 Units
7.0–10.9	4 Units
11.0–16.9	6 Units
17.0–28.0	8 Units

NB: Check fingerprick glucose 4-hourly if the patient is NBM or check glucose pre-meals if SC insulin being used to supplement other hypoglycaemic agents.

Jaundiced patients undergoing surgery

Patients with obstructive jaundice are particularly prone to develop renal failure after surgery (the hepato-renal syndrome).

This may be because of the toxic effect of bilirubin on the kidney. In practice this means that a good urine output must be maintained in such patients around the time of surgery. The following régime is one way of ensuring this.[1]

Pre-operative preparation
● Avoid morphine in the pre-medication.
● Insert IV line and run in 1 litre of 0.9% saline over 30–60mins following pre-med (unless the patient has heart failure).
● Insert a urinary catheter.
● Give 500ml of 10% mannitol IV over 20mins 1h before surgery.
● Commence renal dose dopamine (2–5µg/kg/min) IV.

During operation ● Measure urine output hourly.
● Give 100ml of 10% mannitol IV if the urine output <60ml/h.
● Give 0.9% saline IV to match the urine output.

For 48h after operation ● Measure urine output every 2h.
● Give 100ml 10% mannitol IV over 15mins if the urine output is <100ml in 2h.
● Give 0.9% saline at rate to match urine output and fluid lost through NGT; and give 2 litres of dextrose-saline every 24h.
● Measure urea and electrolytes daily.
● Give 20mmol of potassium per litre of fluid after 24h post-op if urine output good.

Surgery in patients taking steroids

Patients taking steroid drugs or those with Addison's disease need extra steroid cover to cope with the stress of surgery.

Major surgery Give hydrocortisone 100mg IM with the pre-medication and then every 6h IV/IM for 3 days. Then return to previous medication.

Minor surgery Prepare as for major surgery except that hydrocortisone is given for 24h, not 3 days.

The major risk with adrenal insufficiency is hypotension, so if this is encountered without an obvious cause, it may be worthwhile giving a dose of 100mg hydrocortisone IV.

Thyroid disease and surgery

Thyroid surgery for hyperthyroidism If severe give carbimazole until euthyroid (p542). Arrange operation date and 10–14 days before this start aqueous iodine oral solution (Lugol's solution), 0.1–0.3ml/8h PO well diluted with milk or water. Continue up until surgery.

Mild hyperthyroidism Give propranolol 80mg/8h PO at the first consultation. 10 days before surgery give Lugol's solution as above. Stop on the day of surgery but continue propranolol for 5 days post-op.

1 Based on the régime used by the anaesthetic department at the Royal United Hospital, Bath.

Surgery

The acute abdomen

Essence A patient who becomes acutely ill and in whom symptoms and signs are chiefly related to the abdomen has an 'acute abdomen'. Very rarely is laporotomy needed immediately. Clinical signs and repeated examination is the key to making the decision.

There are 2 clinical syndromes which may well require laparotomy:

1 Rupture of an organ (Spleen, aorta, ectopic pregnancy) Shock is the leading sign. Abdominal swelling may be seen. Note any history of trauma. Peritonism may be surprisingly mild. NB: *delayed* rupture of the spleen may occur some weeks after trauma.

2 Peritonitis (Perforation of a peptic ulcer, diverticulum, appendix, bowel or gall bladder) Signs: prostration, shock, lying still. There is generalized peritonitis (tenderness, guarding and rebound tenderness), board-like abdominal rigidity and absent bowel sounds. An erect CXR may show air under the diaphragm. NB: *acute pancreatitis* (p116) produces this syndrome but does *not* require a laparotomy—so do not omit doing a serum amylase.

Syndromes which may not require a laparotomy

1 *Local peritonitis:* Seen in diverticulitis, cholecystitis, salpingitis and appendicitis. (The latter *will* need surgery.) If abscess formation is suspected (swelling, swinging fever and WCC↑) a laparotomy for drainage is needed. Look for 'meteorism' on the plain AXR (p776).

2 *Colic:* This is a pain which waxes and wanes, regularly and frequently. It is caused by muscular spasm in a hollow viscus (gall bladder, gut, ureter, uterus). The patient is restless with the pain (unlike in peritonitis).

Obstruction of the bowel See p130.

Pre-operative care ▶ Do not rush the patient to theatre. Resuscitate adequately first, as anaesthesia compounds shock. The only exceptions to this rule are exsanguination from a ruptured ectopic pregnancy (OHCS p24) or a leaking abdominal aneurysm (p118)—where blood is being lost faster than it can be replaced.

Put to bed. Nil by mouth. IVI (0.9% saline). Treat shock. Relieve pain (p90). Take blood for crossmatch, U&E, FBC, amylase, blood cultures. Do CXR and ECG in patients over 50yrs. A plain abdominal film may be helpful diagnostically: consider erect or decubitus films although these are rarely helpful. Give antibiotics for peritonitis, eg cefuroxime 750mg/8h IV with metronidazole 500mg/8h IV/PR. Consent for surgery.

Medical causes of acute abdominal symptoms

Myocardial infarction	Pneumonia	Sickle-cell crises
Gastroenteritis or UTI	Tabes	Malaria
Diabetes mellitus	Zoster	Typhoid fever
Epidemic myalgia	PAN, TB	Cholera
Pneumococcal peritonitis	Porphyria	*Yersinia enterocolitica*
Henoch–Schönlein	Thyroid	Phaeochromocytoma
purpura (p700)	storm	Heroin withdrawal
Lead colic	Pethidine	
	addiction	

The acute abdomen

Causes of air under the diaphragm

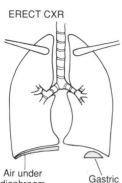

ERECT CXR

Air under diaphragm

Gastric air bubble

- Perforation of bowel
- Gas-forming infection
- Pleuroperitoneal fistula
  (carcinoma, TB, trauma)
- Crohn's disease
- Iatrogenic (surgery, laparoscopy)
- *Per vaginam*
  (post partum, waterskiiers)
- Interposition of bowel between liver and diaphragm

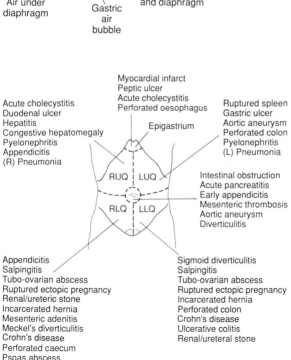

Myocardial infarct
Peptic ulcer
Acute cholecystitis
Perforated oesophagus

Epigastrium

Acute cholecystitis
Duodenal ulcer
Hepatitis
Congestive hepatomegaly
Pyelonephritis
Appendicitis
(R) Pneumonia

Ruptured spleen
Gastric ulcer
Aortic aneurysm
Perforated colon
Pyelonephritis
(L) Pneumonia

RUQ | LUQ

RLQ | LLQ

Intestinal obstruction
Acute pancreatitis
Early appendicitis
Mesenteric thrombosis
Aortic aneurysm
Diverticulitis

Appendicitis
Salpingitis
Tubo-ovarian abscess
Ruptured ectopic pregnancy
Renal/ureteric stone
Incarcerated hernia
Mesenteric adenitis
Meckel's diverticulitis
Crohn's disease
Perforated caecum
Psoas abscess

Sigmoid diverticulitis
Salpingitis
Tubo-ovarian abscess
Ruptured ectopic pregnancy
Incarcerated hernia
Perforated colon
Crohn's disease
Ulcerative colitis
Renal/ureteral stone

Acute appendicitis

The most common surgical emergency (lifetime incidence: 6%).

Pathogenesis Gut organisms invade the appendix wall, eg after lumen obstruction from lymphoid hyperplasia, or impaired ability to prevent invasion, brought about by improved hygiene (so less experience in dealing with gut pathogens). This 'hygiene hypothesis' explains the rise in appendicitis rates in the early 1900s and its later decline (as pathogen exposure dwindles further).[1]

Symptoms As the inflammatory process begins, there is colicky, central abdominal pain. Once the peritoneum becomes involved with the inflammatory process the pain shifts to the right iliac fossa. Anorexia is almost invariable but vomiting rarely prominent. Constipation is usual but diarrhoea may occur.

Signs	Lying still	RIF signs:
Tachycardia	Flushed	Tenderness, guarding
Fever 37.5–38.5°C	Coughing hurts	Rebound tenderness
Furred tongue	Shallow breaths	PR painful on right

Special tests: Rovsing's sign (pain more in the RIF than the LIF when the LIF is pressed). In women, do a vaginal examination: does she have salpingitis (+ve cervical excitation; *OHCS* p50)?

Variations in the clinical picture ● The schoolboy with vague abdominal pain who will not eat his favourite food.
● The infant with diarrhoea and vomiting.
● The shocked, confused octogenarian, who is not in pain.

Hints and pitfalls ● Don't rely on tests (eg WCC; urinoscopy).
● If the child is anxious use his hand to press his belly.

▶ Expect your diagnosis (both of 'appendicitis' and 'not appendicitis') to be wrong half the time. This means that those who seem not to have appendicitis should be re-examined often. Laparoscopy may be helpful.

Differential diagnosis	Diverticulitis	Perforated ulcer
Mesenteric adenitis	Salpingitis	Cystitis
Food poisoning	Cholecystitis	Crohn's disease

Treatment Prompt appendicectomy. Metronidazole 1g/8h + cefuroxime 750mg/8h, 3 doses IV starting 1h pre-op, reduces wound infections.

Complications Perforation with peritonitis with later infertility in girls (so have a lower threshold for surgery in girls); appendix mass (inflamed appendix surrounded by omentum); appendix abscess.

Treatment of an appendix mass There are 2 schools of thought: *conservative* and *early surgery*. Try the former initially—NBM and antibiotics (eg cefuroxime 750mg/8h IV and metronidazole 500mg/8h IV). Mark out the size of the mass and proceed to surgery if it enlarges or the patient gets more toxic (pain↑; temperature↑; pulse↑; WCC↑). If the mass resolves it is usual to do an interval (ie delayed) appendicectomy. Exclude a colonic tumour in the elderly.

Treatment of an appendix abscess Surgical drainage.

Appendicitis in pregnancy (1/2000 pregnancies) Pain and tenderness are higher due to displacement of the appendix by the uterus. Appendicectomy is well tolerated[2] but fetal mortality approaches 30% after perforation—so prompt assessment is vital.

1 D Barker 1988 *BMJ* i 953   **2** 1985 *Arch Surg* **120** 1362

Acute appendicitis

Acute pancreatitis

▶ This unpredictable disease (mortality 5–10%) is often managed on surgical wards, but because surgery is often not involved it is easy to think that there is no acute problem. There is.

Essence There is self-perpetuating acute inflammation of pancreas and other retroperitoneal tissues. Litres of extracellular fluid are trapped in the gut, peritoneum and retroperitoneum. There may be rapid progress from a phase of mild oedema to one of necrotizing pancreatitis. In fulminating cases the pancreas is replaced by black pus. Death may be from shock, renal failure, sepsis or respiratory failure, with contributory factors being protease-induced activation of complement, kinin, and the fibrinolytic and coagulation cascades.

Causes 'GET SMASHED': Gallstones, Ethanol, Trauma, Steroids, Mumps, Autoimmune (PAN), Scorpions[1], Hyperlipidaemia (↑Ca^{2+}, hypothermia), ERCP, Drugs (azathioprine, diuretics, didanosine]. Often none found.

Symptoms Gradual or sudden severe *epigastric* or *central abdominal pain* (radiates to back); vomiting is prominent. Sitting forward may help.

Signs (may be mild in serious disease!) Pulse, fever, jaundice, shock, ileus, rigid abdomen ± local/generalized tenderness and periumbilical discoloration (Cullen's sign, or at the flanks, Grey Turner's sign).

Diagnosis Serum amylase >1000U/ml (lesser rises in cholecystitis, mesenteric infarction, perforated peptic ulcer, renal failure; amylase may be normal even in severe pancreatitis.) Abdominal films: absent psoas shadow (due to retroperitoneal fluid), 'sentinel loop' of proximal jejunum (a solitary air-filled dilatation). Do ultrasound (for gallstones) and CRP.

Differential diagnosis: Any acute abdomen (p112), myocardial infarct.

Management Obtain expert help.
1 Set up IVI and give plasma expanders (eg Haemaccel®) and 0.9% saline until vital signs are satisfactory and urine flows at >30ml/h. If shocked or elderly consider CVP. Check daily weight.

2 Analgesia: pethidine 100mg/4h & prochlorperazine 12.5mg/8h IM.

3 Hourly pulse, BP and urine flow; daily FBC, U&E, Ca^{2+}, glucose, amylase, blood gas. Repeat ultrasound may show peripancreatic fluid.

4 Nil by mouth. Insert a nasogastric tube—eg if persistent vomiting.

5 If deteriorating, admit to ITU, and in suspected abscess formation or pancreas necrosis (eg on contrast-enhanced CT), consider laparotomy and debridement.
NB: antibiotics and peritoneal lavage are unproven.

Prognosis (see table). C-reactive protein can be a helpful marker. Some patients suffer *recurrent oedematous pancreatitis* so often that near-total pancreatectomy is contemplated. Evidence is accumulating 🔲 that 'oxidant' stress is important here, and anti-oxidants apparently do work—eg selenium, methionine and β-carotene: seek expert help.[2]

Complications *—Early:* Circulatory collapse, shock lung, renal failure, DIC, Ca^{2+}↓(10ml of 10% calcium gluconate IV slowly is, rarely, necessary; albumin replacement has also been tried), glucose↑ (transient; 5% need insulin). *Later:* (>1 week) Pancreatic necrosis, pseudocyst (fluid in the lesser sac, eg after ≥6 weeks)—presents as palpable mass, persistent ↑amylase or ↑LFTs, and fever. It may resolve or need drainage into bowel. Abscess: this requires drainage. Bleeding is caused by elastase eroding a major vessel. Skilled angiographists can stop bleeding, and allow planned surgery under more favourable circumstances. Thrombosis may occur in the splenic and gastro-duodenal arteries, or in the colic branches of the superior mesenteric artery, causing bowel necrosis.

Ranson's criteria of severity of acute pancreatitis

At presentation	During the first 48h
Age >55 years	Haematocrit fall >10%
WBC >16 × 10⁹/litre	Urea rise >10mmol/litre
Glucose >11mmol/litre	Serum Ca^{2+} <2mmol/litre
LDH >350 iu/litre	Base excess greater than −4
AST >60 iu/litre	PaO_2 <8kPa
	Estimated fluid sequestration >6 litres

Mortality correlates as follows:
0–2 criteria, mortality 2%; 3–4 = 15%; 5–6 = 40% and 7–8 = 100%.

[1] The venom is the most common cause in Trinidad (Erlewyn Lajeunesse) [2] *Bandolier* 1994 5 6

Aneurysms

These are abnormal dilatations in blood vessels. They may be fusiform or sac-like (eg Berry aneurysms in the circle of Willis). Common sites: aorta, iliac, femoral and popliteal arteries. Atheroma is the usual cause; also penetrating injuries and infections (eg SBE, syphilis).

Complications from aneurysms: rupture; thrombosis; embolism; pressure on other structures, infection.

Dissecting thoracic aortic aneurysms Blood splits the aortic media, causing sudden tearing chest pain radiating to the back. The cause may be cystic medial degeneration of the artery. As dissection advances, branches of the aorta may occlude leading sequentially to hemiplegia (carotid), unequal arm pulses and BP, paraplegia (anterior spinal artery)—and anuria (renal arteries). Aortic incompetence and inferior MI may develop if dissection extends proximally. ▶▶ Action: crossmatch blood (10u); do ECG and CXR (expanded mediastinum?). Ask a cardiologist to decide if surgery is appropriate. Do echo, CT, angiography.

Management: ▶▶ 1 Diamorphine 2.5–5mg IV PRN.
2 ECG, CXR (PA±lateral), X-match (10u), echo, CT or angiography.
3 Transfer to ITU; arterial line and urinary catheter.
4 Hypotensive treatment (aim for systolic BP 100–120mmHg and urine output >30ml/h); labetalol (p302); nitroprusside 10µg/min IV (↑ as required) + propranolol 1mg/min IV, repeated every 2mins up to 10mg—if β-blocker contraindicated, use the ganglion blocker trimetaphan (10mg/ml at eg 3–4mg/min IVI, ↑ as required).

Ruptured abdominal aortic aneurysm Incidence/yr rises with age: 22/100,000 of those aged 60–64; 177/100,000 if 80–84.

Presentation: Intermittent or continuous abdominal pain (radiating to the back, iliac fossae, or groins), collapse, and an *expansile* abdominal mass (ie the mass expands and contracts: swellings that are simply pulsatile merely transmit the pulse—eg nodes overlying the aorta). It may be possible to aspirate blood from the peritoneum. The main differential diagnosis is pancreatitis. If in doubt, proceed as for a ruptured aneurysm.

If the aneurysm ruptures into the GI tract, there is overwhelming haematemesis and a swift death.

Management: ▶▶ Summon the most experienced surgeon and anaesthetist available. Warn theatre. Put up several large IVIs. Treat shock with ORh−ve blood, but keep systolic BP ≤100mmHg (but note: a *raised* BP is common early on). Take blood for amylase, Hb, crossmatch (10–40u may eventually be needed). Take the patient straight to theatre. Do not waste time doing x-rays: fatal delay may result. Catheterize the bladder. Give prophylactic ampicillin + flucloxacillin—both 500mg IV. Surgery involves clamping the aorta above the leak, and inserting a Dacron® graft (eg a trouser graft, with each 'leg' attached to an iliac artery). Mortality with treatment: 21–70%; without treatment: 100%.

Unruptured abdominal aortic aneurysms Prevalence: 3% of those >50yrs. They may be symptomatic, or cause abdominal or back pain. They may be discovered on incidental abdominal examination (this misses ~⅓ of even quite big aneurysms). Ultrasound, plain abdominal films (with a lateral) and CT are good tests, which may be repeated to keep small aneurysms (<6cm across) under review: but small aneurysms *do* rupture. The aim is to operate before the 30% which are destined to rupture actually do so. Elective operative mortality is 2–10%—this makes informed consent (the Rees rules, OHCS p430) the key issue in developing ultrasound-based screening of 'healthy' people.

Diverticular disease

A *diverticulum* is an outpouching of the wall of the gut. The term *diverticulosis* means that diverticula are present, whereas *diverticular disease* implies they are symptomatic. *Diverticulitis* refers to inflammation within a diverticulum. Although diverticula may be congenital or acquired and can occur in any part of the gut, by far the most important type are acquired colonic diverticula, to which this page refers.

Pathology Most occur in the sigmoid colon with 95% of complications at this site but right-sided diverticula do occur. Lack of dietary fibre is thought to lead to high intraluminal pressures which force the mucosa to herniate through the muscle layers of the gut. One-third of the population in the Western world has diverticulosis by the age of 60 years.

Diagnosis Rectal examination, sigmoidoscopy, barium enema, colonoscopy.

Complications of diverticulosis

1 *Painful diverticular disease:* There may be altered bowel habit; pain—usually colicky, left-sided and relieved by defecation; nausea and flatulence. A high-fibre diet (wholemeal bread, fruit and vegetables) with added bran is usually prescribed. 5–30g/day of bran may be needed, the correct dose found by trial and error. Antispasmodics such as mebeverine 135mg/8h PO may be helpful.

2 *Diverticulitis:* This is recognized by the features above plus signs of inflammation: pyrexia and raised WCC and ESR. The colon is tender and there may be localized or generalized peritonitis. Treatment: bed rest, NBM, IV fluids and antibiotics—cefuroxime 750mg/8h IV with metronidazole 500mg/8h IV/PR are suitable until the results of cultures are available. Most settle on this régime but there may be abscess formation (necessitating drainage) or perforation. The latter requires peritoneal lavage and formation of a proximal colostomy. The perforation may be resected at the first operation or later. Mortality following perforation approaches 40%.

3 *Haemorrhage* is usually sudden and painless. This is a common cause of large rectal bleeds. Bleeding usually stops with bed rest but transfusion may be required. If bleeding is uncontrollable embolization or colonic resection may be needed after angiography or colonoscopy has located the bleeding point.

4 *Fistulae* may form between the colon and the bladder (so causing pneumaturia and intractible UTIs), vagina or small bowel. Treatment is surgical, including colonic resection.

5 Post-infective strictures may form in the sigmoid colon.

Barium enema examination

This should be preceded by rectal and sigmoidoscopic examination. It is useful in the diagnosis of polyps, carcinoma, diverticulosis and inflammatory bowel disease. It should not be done if there is suspected perforation of the colon or toxic megacolon (p514).

*Preparation:*Avoid high fibre food (fruit, vegetables, cereals) for 2 days. Give clear fluids only for 1 day; laxatives for 1–2 days; then an enema and/or colonic washout before the examination.

Diverticular disease

Gallstones

Bile contains cholesterol, bile pigments (from broken-down haemo-globin) and phospholipids. If the concentrations of these vary, different kinds of stones may be formed.[1] *Pigment stones:* small, friable, irregular and radiolucent. Risk factors: haemolysis. *Cholesterol stones:* large, often solitary, radiolucent. Risk factors: female sex, age, obesity, clo-fibrate. *Mixed stones:* faceted (calcium salts, pigment and cholesterol): 10% radio-opaque. *Gallstone prevalence:* 9% of those >60yrs.

Stones may cause acute or chronic cholecystitis, biliary colic, pancrea-titis or obstructive jaundice (p484).

Acute cholecystitis follows impaction of a stone in the cystic duct, which may cause continuous epigastric or RUQ pain, vomiting, fever, local peritonism, or a gall bladder (GB) mass. The main difference from biliary colic is the inflammatory component (local peritonism, fever, WCC↑). If the stone moves to the common bile duct (CBD) jaundice may occur. *Murphy's sign:* lay 2 fingers over the RUQ. Ask the patient to breathe in. This causes pain and arrest of inspiration as the inflamed GB impinges on your fingers. The sign is only +ve if a similar manoeuvre in the LUQ does not cause pain.

Tests: WCC, ultrasound (thickened GB wall, pericholecystic fluid and stones), HIDA cholescintigraphy (to reveal a blocked cystic duct).

Treatment: NBM, pain relief, IVI and cefuroxime 750mg/8h IV. In suitable candidates cholecystectomy is performed within 48h. If it is delayed relapses occur in 18%.[2] Mortality 0–1%.[3] Otherwise perform cholecystectomy after 3 months.

Chronic cholecystitis Stones cause intermittent colic and chronic inflam-mation. Vague abdominal discomfort, distension, nausea, flatulence and intolerance of fats may also be caused by reflux, ulcers, irritable bowel syndrome, chronic relapsing pancreatitis and tumours (eg stomach, pancreas, colon, GB).

Ultrasound is the best way to demonstrate stones. Oral cholecysto-grams (contrast given orally is concentrated in a healthy GB) and IV cholangiograms (IV contrast outlines the CBD) are less reliable.

Treatment: Cholecystectomy (laporoscopically is now a possibility). No comparative trials favour lithotripsy.

Biliary colic RUQ pain (radiates to back) ± jaundice. *Treatment:* pain control (pethidine 50–100mg IM/4h + prochlorperazine); cholecystec-tomy. *Differential diagnosis* can be hard as the above may overlap. Urinoscopy, CXR and ECG help exclude other diseases.

Other presentations include ● *Cholangitis* (bile duct infection) causing RUQ pain, jaundice and high fevers with rigors. Treat with eg cefuroxime 750mg/8h IV and metronidazole 500mg/8h PR. ● *Pancreatitis.* ● *Empyema:* the obstructed gall bladder fills with pus. ● *Gallstone ileus:* a stone perforates the GB, ulcerating into the duodenum. It may pass on to obstruct the terminal ileum. x-ray: air in CBD, small bowel fluid levels and a stone.

Silent stones: many advise elective surgery. An alternative for small cholesterol-rich radiolucent stones in a functioning GB is medical dissolu-tion, eg ursodeoxycholic acid 8–15mg/kg/24h PO for up to 2yrs (causes less diarrhoea than chenodeoxycholic acid), but high recurrence on lithotripsy.

1 L Bennion 1978 *NEJM* **299** 1221   **2** A Mitchell 1982 *BMJ* i 27   **3** P Jones 1982 *BMJ* ii 1376

Complications of gallstones
1. In the gall bladder
    a. biliary colic
    b. acute and chronic cholecystitis
    c. empyema
    d. mucocele
    e. carcinoma
2. In the bile ducts
    a. obstructive jaundice
    b. pancreatitis
    c. cholangitis
3. In the gut
    a. gallstone ileus.

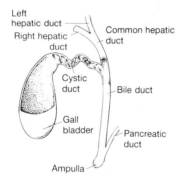

Reproduced from *Cunningham's Manual of Practical Anatomy* 14 ed, OUP 1977, Vol. 2, p138, with kind permission.

Gangrene

Gangrene is the death of tissue occurring as a result of damage to vascular supply. The tissues appear black and may slough off. If tissue death and infection occur together wet gangrene is said to exist. Dry gangrene implies death without infection. Note the line of demarcation between living and dead tissue. Amputation of the part is required and should always be covered by IV or IM benzylpenicillin 600mg/6h starting 1h pre-op and continuing for 5 days to prevent gas gangrene.

Gas gangrene is a myositis produced by *Clostridium perfringens*. Risk factors: diabetes mellitus, trauma, malignancy. Toxaemia, delirium and haemolytic jaundice occur early. There is surface oedema, crepitus (usually) and bubbly brown pus. *Treatment:* remove all devitalized tissue (may involve amputation) and give benzylpenicillin as above (metronidazole if hypersensitive).

Abscesses

▶ If there is pus about, let it out.

Essence Abscesses are cavities containing pus. They present as a mass with swinging pyrexia, malaise and neutrophilia.

Cutaneous abscesses Staphylococci are the most common organisms. Haemolytic streptococci are only common in hand infections. Proteus is a common cause of non-staphylococcal axillary abscesses. Below the waist faecal organisms are common (aerobes and anaerobes).
Treatment: Incision and drainage alone usually cures.

Perianal/ischiorectal abscesses These are usually caused by bowel organisms (occasionally staphylococci, rarely TB). Incision and drainage should be performed. Associations are diabetes, Crohn's disease, and malignancy. Occasionally a fistula results, especially in Crohn's disease and further surgery may be required. Antibiotics should not be used routinely.

Pelvic abscesses These may often be felt rectally. A swinging fever and the passage of mucus suggests the diagnosis. They may be drained per rectum.

Subphrenic abscesses may complicate any intra-abdominal sepsis. They rarely give rise to localizing features so giving rise to the aphorism: 'Pus somewhere, pus nowhere = pus under the diaphragm'. Ultrasound or x-ray screening on respiration is useful. *Treatment:* try antibiotics if early but be prepared to attempt ultrasound-guided drainage (if unilocular) or operation.

Cerebral abscess This may follow ear, sinus, dental or periodontal infection; skull fracture; congenital heart disease; SBE; bronchiectasis.
Signs: Seizures, fever, localizing signs or signs of ICP↑.
Tests: CT/MRI (eg nonenhancing area); stereotactic biopsy. ESR↑, WCC↑.
Treatment: Urgent neurosurgical referral; prompt attention to rising intracranial pressure (p734). If frontal sinuses or teeth are the source of the abscess, the likely organism will be *Str milleri* (micro-aerophilic), followed by oropharyngeal anaerobes. In abscesses from the ear *B fragilis* or other anaerobes are most common. Bacterial abscesses are often peripheral, whereas toxoplasma are deeper (eg basal ganglia).

Ischaemia of the gut

Limb embolism and ischaemia

Chronic ischaemia This is almost always due to atherosclerosis.

Symptoms: Claudication, rest pain and gangrene. Claudication (from the Latin, meaning *to limp*) is pain felt in the leg or buttock on exercise. The calves are most commonly affected. After walking for a fairly consistent distance (the claudication distance), cramping pain forces the patient to stop. He is able to resume exercise after rest. Pain at rest signifies critical ischaemia (eg burning pain at night relieved by hanging legs over side of bed).

Signs: Absent pulse(s); temperature and colour differences between limbs; atrophic skin; ulcers, and gangrene. Assessment: Doppler pressures, angiography, digital subtraction angiography.

Investigations: Hb; wcc; esr; u&e; serum proteins; syphilis serology; glucose; ecg. In particular to look for evidence of anaemia, collagen disease, renal disease, cardiac ischaemia. *Ankle-brachial pressure index:* Normal = 1. Claudication = 0.5. Rest pain = 0.3. Impending gangrene ≤0.2.[1]

Management: Most claudicants improve with weight reduction, exercise and stopping smoking. Treat diabetes, hypertension and hyperlipidaemia. Vasodilators rarely help but naftidrofuryl and oxpentifylline are often tried. Although quite a few trials have given encouraging results, few have been placebo-controlled, and, in those that have been, there is evidence of reporting bias (ie journals only report positive findings: if you do enough trials some will be positive, and this may be the reason for the observation that it is only in the smaller trials that the larger effects are seen). Note that ischaemic vascular beds are usually fully dilated, and added generalized dilatation may steal flow away from them.[2]

Arteriography: Use if limbs are in jeopardy (rest pain, early gangrene) or if quality of life is badly affected. If distal vessels are adequate (good 'run off'), reconstruction with grafts may be tried.

Percutaneous transluminal angioplasty is a recent development: a balloon is inflated at the site of stenosis. It is particularly good for iliac and renal arteries. Sympathectomy may help. It may be surgical or chemical (phenol injected under radiographic control). Amputation is the last resort (done under penicillin cover, p124). ► Start early rehabilitation.

Acute ischaemia This may be due to an embolus, thrombosis or trauma. There is little difference in presentation and so when the cause is in doubt it is vital to perform angiography.

Signs and symptoms: Six Ps—the part is **p**ale, **p**ulseless, **p**ainful, **p**aralysed, **p**araesthetic and '**p**erishing with cold'. The onset of fixed mottling of the skin implies irreversible changes. Emboli commonly arise from the heart (infarcts, af) or an aneurysm.

Management: ► If embolism is suspected, embolectomy should be done as early as possible—eg under local anaesthetic. Post-operative care: heparin (15,000U/12h ivi). Ischaemia following trauma demands urgent angiography and surgery. Acute thrombosis may be shown on angiography and urgent reconstructive treatment may be helpful.

1 S Yao 1970 *Br J Surgery* **57** 761   **2** *Drug Ther Bul* 1990 **28** 1

Examples of arterial reconstruction: (% = legs saved in survivors)

Femoropopliteal bypass grafts	80%
Femorotibial bypass grafts	60%
Grafts for rest pain in the elderly	80%
Grafts for gangrene (toes only)	65%
Grafts for gangrene of the foot	30%

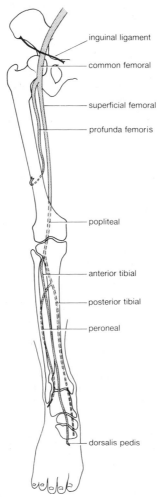

inguinal ligament

common femoral

superficial femoral

profunda femoris

popliteal

anterior tibial

posterior tibial

peroneal

dorsalis pedis

Obstruction of the bowel

Features of obstruction Anorexia, nausea, vomiting with relief. Colicky abdominal pain with distension. Constipation (neither faeces nor flatus passed)—need not be absolute if obstruction is high. Active, 'tinkling' bowel sounds.

On AXR (plain abdominal x-ray) look for abnormal gas patterns (gas in the fundus of stomach and throughout the large bowel is normal). On erect AXR look for horizontal fluid levels.

The key decisions 1 *Is the obstruction of the small or large bowel?* In small bowel obstruction vomiting occurs earlier, distension is less, pain is higher in the abdomen. ▶ *The supine* AXR *plays a key role in diagnosis.* In small bowel obstruction the AXR shows central gas shadows and no gas in the large bowel. Small bowel is identified by valvulae conniventes which completely cross the lumen (large bowel haustra do *not* cross all the lumen's width).

In large bowel obstruction pain is more constant and often over a distended caecum, AXR shows gas proximal to the block (eg in caecum) but not in the rectum.

2 *Is there ileus?* (functional obstruction due to reduced bowel motility) or *mechanical obstruction?* In ileus there is no pain and bowel sounds are absent.

3 *Is the bowel strangulated?* Signs: the patient is more ill than you would expect. There is a sharper and more constant pain than the central colicky pain of the obstruction and it tends to be localized. Peritonism is the cardinal sign. There may be fever and wcc↑.

Causes *Small bowel:* Adhesions, herniae external and internal, intussusception, Crohn's, gallstone ileus, tumour, swallowed foreign bodies (eg smuggled packs of cocaine). *Large bowel:* Tumour, volvulus, faeces.

Management The site, speed of onset and completeness of obstruction determine therapy. Strangulation and large bowel obstruction require surgery soon. Paralytic ileus and incomplete small bowel obstruction can be managed conservatively, at least initially.

Conservative: Pass NGT and give IV fluids to rehydrate and correct electrolyte imbalance (p100)—ie 'drip and suck'.

Surgery: Strangulation requires urgent surgery (within 1 hour), as does large bowel obstruction with gross dilatation (>8cm) and tenderness over the caecum—as perforation is close. Usually large volumes of IV fluid must be given. For less urgent large bowel obstruction there is time to give one enema in an attempt to clear the obstruction and to correct fluid imbalance.

Sigmoid volvulus occurs when the bowel twists on the mesentery and can produce severe, rapid strangulated obstruction. There is a characteristic AXR with an 'inverted U' loop of bowel. It tends to occur in the elderly, constipated patient and is often managed by sigmoidoscopy and insertion of a flatus tube.

Pseudo-obstruction resembles mechanical obstruction of the large bowel but no cause for obstruction is found. The condition should be treated conservatively. Clinical features resemble mechanical obstruction and there is a case for investigating the cause by colonoscopy or water soluble contrast enema in most cases of suspected mechanical obstruction.[1]

1 H Dudley 1986 *BMJ* **i** 1157

Obstruction of the bowel

Paediatric surgical emergencies

Congenital hypertrophic pyloric stenosis This usually presents not at birth but in the first 3–8 weeks as projectile vomiting. The baby is malnourished and always hungry and the diagnosis is made by palpating the pyloric mass in the RUQ during a feed. The baby can be severely depleted of water and electrolytes and this imbalance needs correction before surgery. Pass a nasogastric tube. Treatment is by Ramstedt's pyloromyotomy which involves incision of the muscle down to the mucosa.

Intussusception The small bowel telescopes, as if swallowing itself. The patient may be any age (usually 5–12 months) and presents with inter-mittent crying associated with vomiting and drawing the legs up. Blood may be passed PR—classically 'red-currant jelly'. This is less common in older patients who are more likely to have a long history (>3 weeks) and contributing pathology (lymphoma, Henoch–Schönlein purpura, Peutz–Jeghers' syndrome; cystic fibrosis, ascariasis or nephrotic syndrome).[1]

A sausage-shaped abdominal mass may be felt and the patient may quickly become shocked and moribund. ~60% of early intussusceptions (<24h) may be reducible by hydrostatic pressure during a barium enema.[2] If this fails surgery is needed. Absolute CI to barium enema: peritonitis; perforation; severe hypovolaemia.

Torsion of the testis

The aim is to recognize this condition before the cardinal signs and symptoms are fully manifest, as prompt surgery saves testes. ▶ If in any doubt, surgery is required.

Cardinal symptoms Sudden onset of pain in one testis—which makes walking most uncomfortable. (Pain in the abdomen, nausea and vomiting are common.)

Cardinal signs Inflammation of one testis—it is tender, hot and swollen. (The testis tends to lie high and transversely.) Doppler USS may demon-strate blood flow (or lack of) to testis.

Torsion may occur at any age but is most common at 15–30yrs.

Treatment: ▶ Ask the patient's consent for possible orchidectomy and for *bilateral* fixation (orchidopexy). At surgery expose and untwist the testis. If its colour looks good, return it to the scrotum and fix *both* testes to the scrotum.

Differential diagnosis: The main one is epididymitis but here the patient tends to be older, there may be symptoms of urinary infection, infected urine, and more gradual onset of pain. Also consider tumour, trauma and an acute hydrocele.

1 S Miller 1988 *BMJ* **i** 518   **2** *Lancet* 1985 **ii** 250

132

Paediatric surgical emergencies
Torsion of the testis

Urinary retention and catheters

Retention is the inability to empty the bladder completely and follows failure of bladder pressure to overcome urethral resistance. Thus it may be seen in bladder weakness or urethral obstruction. It is subdivided into acute and chronic forms.

Acute retention There is pain and complete failure of micturition. The bladder usually contains around 600ml of urine. The cause in men is nearly always prostatic obstruction. In the *history*, ask about *obstructive* (hesitancy, poor stream, dribbling) and *irritative* (frequency, urgency, dysuria) prostatic symptoms. Retention may be precipitated by anticholinergic drugs, 'holding on', constipation, alcohol ingestion, or infection (p374).

Examination: Abdomen, PR, perineal sensation (to exclude cauda equina compression, p676).

Investigations: MSU, U&E, FBC, consider acid phosphatase & IVU.

Treatment: Catheterization ± an elective prostate procedure (below).

Chronic retention is more insidious. The bladder capacity may be >1.5 litres. Presentations: overflow incontinence, acute on chronic retention, a lower abdominal mass, UTI or renal failure. Prostatic enlargement is the common cause. Others: pelvic malignancy, CNS disease.

Only catheterize the patient if there is pain, urinary infection or renal impairment (eg urea >12mmol/l). Institute definitive treatment promptly.

Catheters *Size* (in French gauge): 12 is small; 16 is large. Usually a 12 or 14 is right. Use the smallest you can. Latex is soft; *simplastic* rather firmer. A *silastic* (silicone) catheter is suitable for long-term use, but costs more. Foley is the typical shape. Coudé catheters have an angled tip to ease around the prostate but are potentially more dangerous. Teeman catheters have tapered ends for a similar reason.

Problems with catheters: 1 Infection (do not use antibiotics unless systemically unwell). Consider bladder irrigation—eg chlorhexidine 1/10,000. 2 *Bladder spasm* may be painful. Try reducing the water in the balloon or using anticholinergic drugs, eg propantheline 15mg/8h PO.

Treating benign prostatic hypertrophy (BPH)
1 Transurethral resection of the prostate (TURP) is a common operation with few problems. ≤14% will become impotent. Crossmatch 2 units. Consider antibiotics eg cefuroxime 750mg/8h IV, 3 doses. Beware of excessive bleeding post-op and clot retention. ~20% of TURs will need redoing within 10yrs. Cardiovascular mortality appears to be raised many months after TUR (compared with open surgery).[2]
2 Retropubic prostatectomy needs less skill, but is an open operation.
3 Prostatic balloon dilatation.[3]
4 Transurethral microwave therapy (TUMT, a 90min out-patient procedure).[4]
5 Transurethral laser induced prostatectomy (TULIP).
6 α-blockers: indoramin 20mg/12h PO, increased by 20mg/wk to 50mg/12h.
Interactions: potentiation of alcohol, sedatives, β-blockers, diuretics.
Side-effects: depression; dizziness; dry mouth; ejaculatory failure; extrapyramidal effects; nasal congestion; weight gain.
Cautions: heart failure; depression; Parkinson's; hepatic/renal failure.

1 *Lancet* Ed 1988 i 1083   2 *Lancet* Ed 1991 ii 606   3 J Fitzpatrick 1992 *Non-surgical Treatment of BPH*, Churchill Livingstone, Edinburgh   4 AS Bdesha 1993 *BMJ*, i 1293

Advice to patients following a TURP
- Avoid driving for 2 weeks.
- Avoid sex for 2 weeks. Then get back to normal. The amount you ejaculate may be reduced. This is because it flows backwards into the bladder. This is not harmful, but may make the urine cloudy. It means that you may be infertile.
- Expect to pass blood in the urine for the first 2 weeks. A small amount of blood colours the urine bright red. Do not be alarmed.
- At first you may need to urinate *more* frequently than before. Do not be despondent. In 6 weeks things should be much better.
- If you get hot or cold (fevers), or if urination hurts, take a sample of urine to your doctor.

Bladder catheterization

1. Per-urethral catheterization

This may be used to relieve urinary retention, to monitor urine output in critically ill patients or to collect an uncontaminated urine specimen for diagnosis. It is *contraindicated* in urethral disruption (eg pelvic fracture) and acute prostatitis. Catheterization introduces bacteria into the bladder, so aseptic technique is essential. Women are often catheterized by nurses but you should be able to catheterize patients of either sex.
- Have the patient lie supine in a well-lit area: women with knees flexed wide and heels together. In women, use one gloved hand to prep the urethral meatus in a pubis to anus direction, holding the labia apart with the other hand. With uncircumcised men, retract the foreskin to prep the glans; use a gloved hand to hold the penis still and off the scrotum. The hand used to hold the penis or labia should not touch the catheter to insert it. Use a set of forceps if necessary.
- Apply sterile lignocaine to the tip of the catheter and insert some into the urethra. Stretch the penis perpendicular to the body to eliminate any internal folds in the urethra that may lead to a false passage.
- Steady *gentle* pressure should be used to advance the catheter and any significant obstruction encountered should prompt withdrawal and re-insertion. In men with prostatic hypertrophy, a Coudé tip catheter may manoeuvre past the prostate.
- Insert the catheter to the hilt and wait till urine emerges before inflating the balloon and remember to check the capacity of the balloon before inflation. Pull the catheter back so that the balloon comes to rest at the bladder neck.
- Remember to reposition the foreskin in uncircumcised men to prevent massive oedema of the glans after the catheter is inserted.

Suprapubic catheterization is sometimes necessary and may be preferred. Ensure the bladder is distended; then clean the skin. Infiltrate with local anaesthetic down to the bladder, nick the skin and then insert the catheter down vertically above the symphysis pubis. After urine is draining advance the catheter over the trocar and tape it down securely.

Bladder tumours

What appear as benign papillomata rarely behave in a purely benign way, and they are almost certainly indolent transitional cell (urothelial) malignancies. Adenocarcinomas and squamous cell carcinomas are rare in the West, but the latter may follow chronic infestation with *Schistosoma haematobium*.

Incidence 1:5000 in the UK.

Histology is important for prognosis: grade 1—differentiated; grade 2—intermediate; grade 3—poorly differentiated.

Presentation Painless (or painful) haematuria; recurrent UTIs.

Associations Smoking; aromatic amines (rubber industry).

Investigations Urine: Culture, microscopy and cytology (bladder papillomas are one cause of 'sterile pyuria'); FBC.
An IVU may show filling defects and ureteric involvement.
EUA is required to assess tumour spread.
Cystoscopy with biopsy is diagnostic.
CT or lymphangiography may show involved pelvic nodes.

TNM staging (EUA = examination under anaesthesia)

T1 Tumour in mucosa or submucosa	Not felt at EUA
T2 Superficial muscle involved	Rubbery thickening at EUA
T3 Deep muscle involved	EUA: mobile mass
T4 Invasion beyond bladder	EUA: fixed mass

Treatment of transitional cell carcinoma

T1: (80% of all cases). Diathermy via cystoscope. Consider intravesical chemotherapeutic agent for multiple small tumours.

T2: Diathermy via cystoscope. Consider radiotherapy for poorly differentiated tumours.

T3: This is controversial; possibilities are:
• Radical radiotherapy
• Cystectomy
• Combination of radiotherapy and cystectomy.
Complications of cystectomy include disruption of sexual and urinary function.

T4: Usually palliative radiotherapy. Chronic catheterization and urinary diversions may help to relieve pain.

Chemotherapy is playing an increasingly important role and may allow more conservative surgery. One toxic but effective régime uses methotrexate, vinblastine, doxorubicin and cisplatin.[1]

Follow up History, examination and cystoscopy every 6 months.

Tumour spread Local—to pelvic structures; lymphatic—to iliac and para-aortic nodes; haematogenous—to liver and lungs.

Survival This depends on age at surgery. For example, the 3-yr survival after cystectomy for T2 and T3 tumours is 60% if 65–75yrs old, falling to 40% if 75–82yrs old (in whom the operative mortality is 4%).[2] With unilateral pelvic node involvement only 6% of patients survive 5yrs. The 3-yr survival with bilateral or para-aortic node involvement is nil.

Massive bladder haemorrhage: This may complicate treatment. Treat by alum solution bladder irrigation (safer than formalin).[3]

1 F Skinner 1984 *J Urol* **131** 1065   2 C Sternberg 1984 *Proc Am Soc Clin Oncology* **3** 160
3 N Bullock 1985 *BMJ* ii 1522

Bladder tumours

Breast lumps and breast carcinoma

▶ All solid lumps need histological or cytological assessment.

History Previous lumps, family history, pain, nipple discharge, change in size related to menstrual cycle, parous state, drugs.

Examination Inspect (arms up and down). Note position, size, consistency, mobility, fixity and local lymphadenopathy. Is nipple discharge present? (from a single duct?); is the nipple inverted? If so, is this recent? Is the skin involved (*peau d'orange*)?

Management If cystic, aspirate to dryness. Send the fluid for cytology. If solid and discrete, consider one of: fine needle aspiration cytology, Tru-cut® or open biopsy. Consider ultrasound or mammography depending on age and clinical suspicion.

Causes of lumps Fibroadenoma, cyst, cancer, fibroadenosis (diffuse lumpiness, often in upper outer quadrant), periductal mastitis (often secondary to duct ectasia), fat necrosis, galactocele, abscess, 'non-breast' lumps—eg lipomas, sebaceous cysts.

Causes of discharge Duct ectasia (green/brown/red, often multiple ducts and bilateral), intraduct papilloma/adenoma/carcinoma (red discharge, often single duct), lactation. Management involves diagnosing the underlying cause, for example by mammography, ductogram, microdochectomy or total duct excision, and then treating it as needed.

Breast carcinoma *Stage by TNM classification:* T1 <2cm. T2 2–5cm. T3 >5cm. T4 Fixity to chest wall or *peau d'orange*. N1 Mobile ipsilateral nodes. N2 Fixed ipsilateral nodes. M1 Distant metastases.

Risk-factors: Nulliparity; 1st pregnancy at aged >30yrs; early menarche, late menopause, +ve family history, not breast feeding, past breast cancer. ?Contraceptive steroids, ?hormone replacement therapy.

Treating early cancer: Surgery: wide local excision + DXT, simple mastectomy, and modified radical mastectomy give equivalent survival. Local recurrence may be commoner with smaller procedures. ▶ *Fully discuss all options with the patient and her partner.* Consider size of lump and breast, and site of lump. Chemotherapy+tamoxifen (oestrogen antagonist) much improves survival (not just if elderly) in both node +ve and node −ve women (it saves ~12 lives/100 women treated).[1] NB: there is evidence that tamoxifen may occasionally cause endometrial carcinoma.[2] Local radiotherapy reduces local recurrence rates but has no effect on survival.

Distant disease: Assess ESR, LFTs (Alk phos↑), Ca^{2+}, CXR, skeletal survey, bone scan, liver USS/CT. DXT to local, painful bony lesions. For widespread disease tamoxifen (eg 20mg/24h PO) is the 1st choice. If there is initially success but subsequently relapse, consider chemotherapy.

Preventing breast cancer deaths: Education and self-examination. Well woman clinics. Mammography (using negligible radiation)—cancer pickup rate: 5 per 1000 'healthy' women over 50yrs. Yearly 2–view mammograms could reduce mortality by 40%—but the price is the serious but needless alarm caused: there are ~10 false +ve mammograms for each true +ve result.[3] In premenopausal women, breasts are denser and mammograms hard to interpret, and screening appears not to save lives.[4]

1 Early Breast Cancer Trialists' Collaborative Group 1992 *Lancet* **i** 1 & 71   **2** F van Leeuwen 1994 *Lancet* **343** 448   **3** J Reidy 1988 *BMJ* **ii** 932   **4** *JAMA* 1994 **271** 152

Breast lumps and breast carcinoma

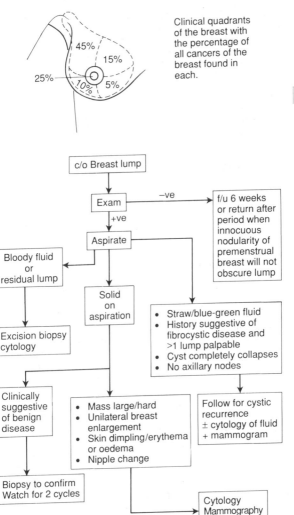

Clinical quadrants of the breast with the percentage of all cancers of the breast found in each.

45% 15%
25%
10% 5%

139

c/o Breast lump

Exam —ve→ f/u 6 weeks or return after period when innocuous nodularity of premenstrual breast will not obscure lump

+ve

Aspirate

Bloody fluid or residual lump

Solid on aspiration

• Straw/blue-green fluid
• History suggestive of fibrocystic disease and >1 lump palpable
• Cyst completely collapses
• No axillary nodes

Excision biopsy cytology

Clinically suggestive of benign disease

• Mass large/hard
• Unilateral breast enlargement
• Skin dimpling/erythema or oedema
• Nipple change

Follow for cystic recurrence ± cytology of fluid + mammogram

Biopsy to confirm Watch for 2 cycles

Cytology Mammography Lumpectomy

Colorectal carcinoma

These adenocarcinomas are the 2nd most common cause of cancer deaths in the UK (20,000 deaths/yr). 56% of presentations are in those >70yrs old.

Predisposing factors Neoplastic polyps, ulcerative colitis (and, to a lesser extent, Crohn's disease), family history, familial polyposis, previous cancer and, possibly, low fibre diet. NSAIDs may be *protective*.

Presentation depends on site: *Left-sided:* Bleeding PR; altered bowel habit; tenesmus; mass PR (60%). *Right:* Weight↓; Hb↓; abdominal pain. *Either:* Abdominal mass; obstruction; perforation; haemorrhage; fistula.

Investigations FBC, proctoscopy, sigmoidoscopy, barium enema or colonoscopy. Consider LFTs, CT scan, liver ultrasound.

Staging is by Dukes' classification:

	Treated survival:
A: Confined to bowel wall	90% 5-yr survival
B: Extension through bowel wall	65% 5-yr survival
C: Involvement of regional lymph nodes	30% 5-yr survival

If there are distant metastases only 5% survive 5yrs.

Spread This is local, lymphatic, haematogenous (to liver 75%, lung and bone) or transcoelomic.

Treatment Surgery may be performed either to attempt cure (removing the draining lymphatic field) or to relieve symptoms: *Right hemicolectomy* for tumours in the caecum, ascending and proximal transverse colon. *Left hemicolectomy* if in distal transverse colon or descending colon.
Sigmoid colectomy for tumours of sigmoid colon.
Anterior resection if in low sigmoid or high rectum. Anastamosis is achieved at the first operation. Stapling devices are helpful.
Abdomino-perineal (A-P) resection for tumours low in the rectum (<~8cm from anal canal): permanent colostomy and removal of rectum and anus.
Radiotherapy may be of some value in Ca rectum and decreases local recurrence. *Chemotherapy:* Post-op hepatic vein IVI of 5-FU may be tried (¼ of those in whom a curative resection seems possible will die from the effects of micro-metastases, usually to the liver)—but metastases >0.5mm are dependent on hepatic artery flow, and may survive this onslaught.[1]

Prognosis 60% are amenable to radical surgery, and 75% of these will be alive at 7yrs (or will have died from non-tumour-related causes).[2]
Screening with home tests for faecal occult blood:[3] This is known to improve prognosis, but false +ve rates are high (10% of those screened), prompting so many needless and unpleasant colonoscopies that it is doubtful if the criteria for valid screening are met (OHCS p430).[4]

Polyps are lumps that appear above the mucosa. There are several types:
1 *Inflammatory:* Ulcerative colitis, Crohn's, lymphoid hyperplasia.
2 *Hamartomatous:* Juvenile polyps, Peutz-Jeghers' syndrome.
3 *Neoplastic:* Tubular and villous adenomata. These have malignant potential, especially if villous, >2cm and dysplastic.

Symptoms of polyps: The passage of blood and mucus PR.

They should be biopsied and removed if they show malignant change. Most can be reached by the flexible colonoscope and diathermy can avoid the morbidity of partial colectomy. Check resection margins are clear of tumour.

1 SG Archer 1989 *Br J Surg* **76** 545 2 N Gordon 1993 *BMJ* ii 707 3 JS Mandel 1993 *NEJM* **329** 1365 4 DA Ahlquist 1993 *NEJM* **329** 1351

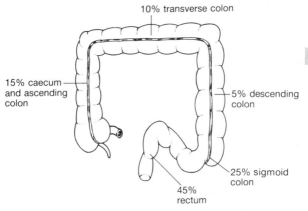

Anatomical location of cancer of the large bowel.

Carcinoma of the stomach

The incidence of gastric carcinoma has shown a sharp decline in Western countries in the last few years but it remains a tumour notable for its gloomy prognosis and non-specific presentation.

Incidence 33/100,000 in the UK but there are unexplained wide geographical variations, being especially common in Japan.

Associations ● Blood group A ● Atrophic gastritis ● Social class IV–V ● *H pylori* (p490) ● Adenomatous polyps ● Pernicious anaemia

Pathology The adenocarcinoma may be polypoid, ulcerating or leather bottle type (linitis plastica). Some are confined to mucosa and submucosa—so-called early gastric carcinoma. 50% involve the pylorus, 25% the lesser curve and 10% the cardia.

Symptoms (Often non-specific) Dyspepsia (p490) lasting longer than a month in patients aged ⩾40yrs demands gastrointestinal investigation. Others: weight loss, vomiting, dysphagia, anaemia.

Signs suggest incurable disease—eg epigastric mass, hepatomegaly, jaundice, ascites, an enlarged left supraclavicular (Virchow's) node (Troissier's sign), acanthosis nigricans (p686).

Spread is local, lymphatic, blood-borne and transcoelomic eg to ovaries (Krukenberg tumour).

Investigations Gastroscopy with multiple biopsies to the ulcer edge. ▶ *All gastric ulcers must be biopsied as even malignant ulcers may appear to heal on drug treatment.*

Treatment Metastasis is obvious in up to 60% of patients and is a contraindication to curative surgery—which involves wide excision of tumour (5cm margins) and lymph nodes. For tumours in the distal two-thirds, a partial gastrectomy may suffice, whereas more proximal tumours usually need total gastrectomy.

Palliation is often needed for obstruction, pain or haemorrhage and involves judicious use of drugs, surgery and radiotherapy.

Five year survival <10% overall, but much better for 'early' gastric carcinoma which is confined to the mucosa and submucosa.

Carcinoma of the oesophagus

This causes ~4000 deaths/year in the UK. There is striking geographical variation in incidence. Premalignant associations: Barrett's oesophagus, achalasia, Plummer–Vinson syndrome (p492).

20% occur in the upper, 50% in the middle, and 30% in the lower oesophagus. Most are squamous cell carcinomas but adenocarcinoma can arise from areas of columnar lining (Barrett's oesophagus).

Spread is *direct*, by extensive submucosal infiltration, and local spread, to *lymph nodes* and, late, via the *blood* stream.

Presentation is usually by poorly localized dysphagia, weight loss, retrosternal chest pain, hoarseness or lymphadenopathy. Symptoms progress rapidly.

Differential diagnosis Any cause of dysphagia (p492).

Tests Barium swallow, CXR, oesophagoscopy with biopsy/brushings.

Treatment We do not know how to treat this disease. Radiotherapy and surgery compete for the most dismal statistics (6% *vs* 4% 5-yr survival). Radiotherapy is cheaper and safer. The only prospective comparative trial failed from low recruitment. Palliation: p498.

Carcinoma of the stomach
Carcinoma of the oesophagus

Lumps

▶ Examine the regional lymph nodes as well as the lump. If the lump is a node, examine its area of drainage.

History How long has it been there? Does it hurt? Any other lumps? Is it getting bigger? Ever been abroad? Otherwise well?

Physical examination Remember the six Ss: site, size, shape, smoothness, surface and surroundings. Other questions: does it transilluminate (see below)? Is it fixed to skin or underlying structures? Is it fluctuant? Lumps in certain sites call to mind particular pathologies (see lumps in groin and scrotum p146). Remember to feel if a lump is pulsatile; this may seem to be a minor detail until you are faced with a surprise on a minor operations list.

Transilluminable lumps Find a dark room and a bright, thin 'pencil' torch, and place it on the lump, from behind, so the light is shining through the lump towards your eye. If the lump glows red it is said to transilluminate. Fluid-filled lumps such as hydrocoeles are good examples of transilluminable swellings.

Lipomas These benign fatty lumps, occurring wherever fat can expand (not scalp or palms), have smooth, imprecise margins and a hint of fluctuance. They only cause symptoms via pressure. Malignant change is very rare (suspect if rapid growth, hardening or vascularization occurs).

Sebaceous cysts These are intradermal, so it is impossible to draw the skin over the lump. There may be a characteristic punctum which marks the blocked sebaceous outflow. Infection is quite common, and foul pus exits through the punctum.

Causes of lymph node enlargement Infections: glandular fever; brucellosis; TB; pre-AIDS; toxoplasmosis; actinomycosis; syphilis. Others: malignancy (carcinoma, lymphoma); sarcoidosis.

Boils (furuncles) These are abscesses which involve a hair follicle and its associated glands.

A carbuncle is an area of subcutaneous necrosis which discharges itself onto the surface through multiple sinuses.

Rheumatoid nodules These are collagenous granulomas which appear on the extensor aspects of joints—especially the elbows. They occur in established cases of rheumatoid arthritis.

Ganglia These are degenerative cysts from an adjacent joint or synovial sheath commonly seen on the dorsum of the wrist or hand and dorsum of the foot. They may transilluminate. 50% will disappear spontaneously. For the rest, the treatment of choice is excision rather than the traditional blow from your bible (the Oxford Textbook of Medicine!).

Fibromas These may occur anywhere in the body, but most commonly they are under the skin. These whitish benign tumours contain collagen, fibroblasts and fibrocytes.

Dermoid cysts These contain dermal structures. They are often found in the midline.

Malignant tumours of connective tissue These include the fibrosarcoma, liposarcoma, leiomyosarcoma (smooth muscle) and rhabdomyosarcoma (tumour of striated muscle).

Lumps

Lumps in the groin and scrotum

Diagnosis of groin lumps Think of structures in the area—femoral and inguinal herniae, saphena varix (p162—both have cough impulse), lymph nodes, femoral aneurysm, psoas abscess, ectopic testis, skin lumps.

Diagnosis of lumps in the scrotum 1 *Can you get above it?* If not, it is an inguinoscrotal hernia (inguinal hernia extending into scrotum).
2 *Is it separate from the testis?*
3 *Is it cystic or solid? (Does it transilluminate?, p144)*

Separate and cystic—epididymal cyst.
Separate and solid—epididymitis (may also be orchitis).
Testicular and cystic—hydrocoele.
Testicular and solid—tumour, orchitis, granuloma, gumma.

Ultrasound may help in sorting out testis tumours from other pathologies. Do not assume that an injured testis was normal before the injury: this is not a rare mode of tumour presentation, and ultrasound may help here.

Epididymal cysts These usually develop in adult life and contain either clear or milky (spermatocoele) fluid. They lie above and behind the testis. Remove if they are symptomatic.

Hydrocoeles (Fluid within the tunica vaginalis) may be primary (idiopathic) or secondary. Primary hydrocoeles are more common, larger and usually develop in younger men. Secondary hydrocoeles may follow testicular tumours, trauma or infection. Hydrocoeles may be treated by aspiration (may need repeating) or surgery.

Epididymo-orchitis This may be due to *E coli*, mumps, gonococcal infection or TB. The area is usually tender. Take a urine sample and look for urethral discharge. Note: a 'first catch' will be more likely to show abnormalities than a mid-stream sample.

Testicular tumours are now the commonest malignancies of young men. *Varieties:* seminoma (30–40yrs); teratoma (20–30yrs); tumours of Sertoli or Leydig cells. ~10% of malignancies occur in undescended testes, even after orchidopexy. A contralateral tumour is found in 1 in 20 instances.[1] *Presentation:* usually as a painless lump in the testis but some are noticed after complaints of pain, trauma or infection. *Spread* is initially lymphatic to the para-aortic nodes and then to the mediastinum. Haematogenous spread is usually to the lungs. *Staging* is essential: 1 No evidence of metastasis. 2 Infradiaphragmatic node involvement. 3 Supradiaphragmatic node involvement. 4 Lung involvement.

Tests: Directed towards staging: CXR, IVU, CT or lymphangiogram, excision biopsy. α-fetoprotein and β-human chorionic gonadotrophin (β-HCG) are useful tumour markers and of value in diagnosis and monitoring of treatment. Do levels before and during treatment.

Treatment: Do an orchidectomy via an inguinal incision, with the spermatic cord occluded before mobilization (less risk of intraoperative spread). Options are constantly being refined (surgery, radiotherapy, chemotherapy). Seminomas are more radiosensitive than teratomas. Stage 1 tumours may be treated simply by orchidectomy as above, with close follow up to detect relapse. Chemotherapy is very effective in achieving cure—eg cisplatinum, vinblastine and bleomycin.

5-year survival: This is good (>~70%—more for early disease).
Prevention of late presentation: Self-examination.[2]

1 N Bullock 1987 *BMJ* ii 456   **2** J Thornhill 1986 *BMJ* i 480

Diagnosis of scrotal masses

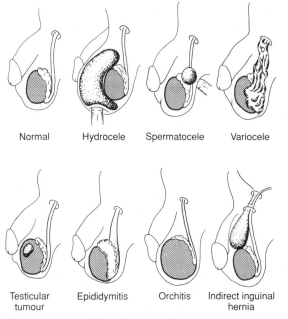

After RD Judge, GD Zeidema, FT Fitzgerald 1989 *Clinical Diagnosis* 5 ed, Little Brown, Boston.

Lumps in the neck

Do not biopsy these lumps until tumours within the head and neck have been excluded by an ENT surgeon. Culture all biopsied lymph nodes for TB.

Diagnosis Complete head and neck examination; ultrasound shows lump consistency. CT defines masses in relation to their anatomical neighbours. Do virology and Mantoux test. Consider fine needle aspiration (FNA).

Common causes Sebaceous cysts; lipomas; nodes; thyroid lumps (p150).

Others Muscle tumour; TB; aneurysms; laryngocoele; pharyngeal pouch.

Thyroglossal cyst: Fluctuant midline swellings moving on tongue protrusion, developing in cell rests in thyroid's migration path. Treatment: surgery.

Causes of lymphadenopathy: TB, tonsillitis, other infections (detailed on p144); sarcoidosis; malignancy (80% are metastatic squamous cell carcinomas, and 90% of these have their primaries in the head and neck.)

Branchial cysts: These feel like a half-filled hot water bottle bulging from beneath the anterior border of sternomastoid into the carotid triangle. They are due to non-disappearance of the cervical sinus (where the 2nd branchial arch grows down over 3rd and 4th). Lined by squamous epithelium, their fluid contains cholesterol crystals. Treat by excision.

Cystic hygroma: These arise from the jugular lymph sac. They transilluminate brightly. Treat by surgery or hypertonic saline sclerosant.

Carotid body tumour (Chemodectoma): This slow-growing tumour splays out the carotid bifurcation. It is usually firm, occasionally is soft and pulsatile. It does not usually cause bruits. It may be bilateral, familial and malignant (5%). This tumour should be suspected in tumours just anterior to the upper third of sternomastoid. Diagnose by digital computerized angiography. *Treatment:* extirpation by a vascular surgeon.

The salivary glands There are 3 pairs of major salivary glands (parotid, submandibular and sublingual) as well as numerous minor glands.
History: lumps; swelling related to food; pain; taste; dry eyes.
Examination: note external swelling; look for secretions; bimanual palpation for stones. Examine VIIth nerve and regional nodes. Cytology: FNA.

Recurrent unilateral pain and swelling is likely to be due to a stone. 80% are submandibular. The classical story is of pain and swelling on eating—with a red, tender, swollen but uninfected gland. The stone may be seen on plain x-ray or by sialography. Distal stones are removed via the mouth but deeper stones may require excision of gland.
Chronic bilateral symptoms may coexist with dry eyes and mouth and autoimmune disease—eg Sjögren's or Mikulicz's syndrome (p708, p704).
Fixed swellings may be from tumours or sarcoid—or are idiopathic.

Salivary gland tumours: '80% are in the parotid, 80% of these are pleomorphic adenomas, 80% of these are in the superficial lobe.'
▶ Any salivary gland swelling must be removed for assessment if present for >1 month. VIIth nerve palsy signifies malignancy.

Benign or malignant:	*Malignant:*	*Malignant:*
Cystadenolymphoma	Mucoepidermoid	Squamous carcinoma
Pleomorphic adenoma	Acinic cell	Adenocarcinoma
		Adenoid cystic carcinoma

Pleomorphic adenomas often present in middle age; grow slowly; and are removed by parotidectomy. Adenolymphomas: usually elderly men; soft; treat by enucleation. Carcinomas: rapid growth; hard fixed mass; pain; facial palsy. Treatment: surgery+radiotherapy. Surgery complications: 1 Facial palsy—often briefly. Arrange a facial nerve stimulator to be in theatre to aid identification. 2 Salivary fistula—this will often close spontaneously. 3 *Frey's syndrome* (gustatory sweating)—here tympanic neurectomy may help.

Thyroid lumps

Examination Give the patient a glass of water and watch the neck during swallowing. Stand behind and feel the thyroid for size, shape (smooth or nodular), tenderness and mobility. Feel for nodes. Percuss for retrosternal extension. Listen for a bruit.

If the thyroid is enlarged (goitre), ask these 3 questions:
1 Is the thyroid smooth or nodular?
2 Is the patient euthyroid, thyrotoxic or hypothyroid?
  Smooth, non-toxic goitre: Endemic (iodine deficiency), congenital, goitrogens, thyroiditis, physiological, autoimmune goitre (including Hashimoto's thyroiditis).
  Smooth, toxic goitre: Graves' disease
3 If thyroid is nodular are there many nodules or a single lump?
  Multinodular goitre: Usually euthyroid but hyperthyroidism may develop. Hypothyroidism and malignancy are rare.

The *single thyroid lump* is a common surgical problem, and about 10% will prove to be malignant. *Causes:* cyst, adenoma, discrete nodule in multinodular goitre, malignancy. First ask: is the patient thyrotoxic?
● Do T_3 & T_4. If thyrotoxic: ● Do an iodine or technetium thyroid scan. This will either show one 'hot' (uptake↑) nodule (a toxic adenoma—usually treated by radioiodine) or multiple 'hot' nodules (which should be treated by thyroidectomy).

If the patient is *not* thyrotoxic, the next test to do is: ● *Ultrasound*, to see if the lump is solid or cystic. Then: ● *Scan* the lump: is it 'hot' or 'cold'? If 'hot', malignancy is unlikely. An alternative is to do: ● *Needle aspiration* and send the fluid for cytology. Unless these tests indicate that the lump is definitely benign, arrange surgery. If it *is* benign: leave it alone.

Thyroid carcinoma There are 4 types:
1 *Papillary:* 70%. Often young, spread by lymph nodes. Treatment: if no lymph node spread, do a lobectomy and give thyroxine to suppress TSH. If lymphatic spread has occurred, do a total thyroidectomy and regional node clearance.
2 *Follicular:* 25%. Middle-aged, spreads early via blood (eg to bone, lungs). Well differentiated. Treat by total thyroidectomy and T_4 suppression or radioiodine (^{131}I) ablation.
3 *Anaplastic:* Rare. Elderly, poor response to any treatment.
4 *Medullary:* 5%. Either sporadic or part of MEN syndrome (p546). May produce calcitonin. Treat by thyroidectomy and node clearance (after first screening for a phaeochromocytoma).

Thyroid surgery *Indications:* Pressure symptoms, hyperthyroidism (p542), carcinoma, cosmetic reasons. The patient must be rendered euthyroid pre-op, by antithyroid drugs and/or propranolol. Check vocal cords by indirect laryngoscopy. Complications:
  Early: Recurrent laryngeal nerve palsy, haemorrhage (► may cause airway compression needing instant removal of clips for evacuation of clot), hypoparathyroidism (check plasma Ca^{2+} daily, usually transient), thyroid storm (symptoms of severe hyperthyroidism—treat by propranolol PO or IV, antithyroid drugs and iodine, p740). *Late:* hypothyroidism.

Abdominal masses

As with any mass ask about size, site, shape and surface. Find out if it is pulsatile and if it is mobile. Examine supraclavicular and inguinal nodes. Is the lump ballottable?

Right iliac fossa masses

Appendix mass	Crohn's disease	Actinomycosis (p223)
Appendix abscess	Intussusception	Kidney malformation
Caecum carcinoma	TB mass	Tumour in an
Pelvic mass (see below)	Amoebic abscess	undescended testis

Abdominal distension Flatus, fat, fluid, faeces and fetus. Fluid may be outside the gut (ascites) or sequestered in loops of bowel (obstruction or ileus). To demonstrate ascites elicit the signs of fluid thrill and shifting dullness.

Causes of ascites:
Malignancy
Low albumin (eg nephrosis)
CCF, pericarditis
Infection—especially TB

Ascites with portal hypertension:
Cirrhosis
Portal nodes
Budd–Chiari syndrome (p696)
IVC or portal vein thrombosis

Tests: aspirate ascitic fluid (paracentesis) for cytology, culture and protein estimation using an 18G needle in the RIF; ultrasound.

Left upper quadrant mass Think of spleen, stomach, kidney, colon, pancreas and then rare causes (eg neurofibroma). Cysts of the pancreas may either be true cysts (congenital; cystadenomas; retention cysts of chronic pancreatitis or cystic fibrosis) or pseudocysts: fluid in the lesser sac seen with acute pancreatitis.

Splenomegaly Think of infections eg glandular fever, malaria and infective endocarditis, portal hypertension (p496), leukaemia, lymphoma, SLE, rheumatoid arthritis and haemolytic anaemia.

Massive splenomegaly: Malaria, myelofibrosis, kala-azar, CML.

Smooth hepatomegaly Hepatitis, CCF, sarcoid. In tricuspid incompetence the liver is enlarged and pulsatile.

Craggy hepatomegaly Secondaries or a primary hepatoma. NB: nodular cirrhosis more often leads to a small, shrunken liver, rather than an enlarged craggy one.

Hepatosplenomegaly Leukaemia, thalassaemia, sickle-cell anaemia, glandular fever, portal hypertension and amyloidosis.

Pelvic masses Determine if truly pelvic by trying to palpate below it. It is pelvic if you cannot 'get below it'. Causes: fibroids, fetus, bladder, ovarian cysts, ovarian malignancies.

Investigation of the lump There is much to be said for performing an early CT body scan (if available). This will save time and money compared with leaving the test to be the last in a long chain of investigations.[1] If CT is not available, ultrasound should be the first test. Other tests: IVU, liver and spleen isotope scans, Mantoux test (p200). Routine tests: FBC (with film), ESR, U&E, LFTs, proteins, Ca²⁺, CXR, AXR, biopsy tests; a tissue diagnosis may be made using a fine needle guided by ultrasound or CT control.

1 J Husband 1985 *BMJ* i 527

Around the anus

Pruritus ani itch occurs if anus is moist or soiled, eg fissures, incontinence, poor hygiene, tight pants, threadworm, fistula, dermatoses. Other causes: lichen sclerosis, anxiety, contact dermatitis.

Treatment:
- Scrupulous hygiene
- Cotton gloves at night
- No steroid/antibiotic creams
- Moist wipe post-defecation
- No spicy foods
- Try anaesthetic cream

Fissure-in-ano This is a midline longitudinal split in the squamous lining of the lower anus—often, if chronic, with a mucosal tag at the external aspect—the 'sentinel pile'. Most are posterior.
- Most are due to hard faeces. They make defecation very painful.
- Rare causes: syphilis, herpes, trauma, Crohn's, anal Ca, psoriasis.
- Examine with a bright light. Do a PR ± sigmoidoscopy. Groin nodes suggest a complicating factor (eg immunosuppression from HIV).
- *Treatment:*[1] try 5% lignocaine ointment, extra dietary roughage and careful anal toilet. Chronic fissures (ie >6wks old) are unlikely to heal, so day-case *lateral partial internal sphincterotomy* (or, less successfully, *manual anal dilatation* under GA) may be needed.

The perianal haematoma (also called a thrombosed external pile). Both names are wrong because it is a clotted venous saccule.[2] It appears as 2–4mm 'dark blue berry' under the skin. It may be evacuated via a small incision under local anaesthesia or left alone if >1 day old.

Pilonidal sinus This is not strictly an anal condition but is a pit in the natal cleft caused by a penetrating hair. (Barbers get this between their fingers.) Treatment is excision of the sinus tract ± primary closure but is often unsatisfactory.

Rectal prolapse The mucosa, or rectum in all its layers, may descend through the anus. This leads to incontinence in 75%. It is due to a lax sphincter and prolonged straining. Treatment: fix the rectum to the sacrum (± rectosigmoidectomy, with no abdominal wound, as exposure is via amputating the prolapse[3]) or encircle the anus with a Thiersch wire.

Anal ulcers are rare: consider Crohn's disease, TB, syphilis.

Skin tags seldom cause trouble but are easily excised.

Ischiorectal/perianal abscess See p124. **Piles** See p156.

Anal carcinoma[3] *Incidence:* 300/yr in the UK. Risk factors: syphilis, anal warts, anoreceptive homosexual men (often young).

Types of neoplasia: Mostly squamous cell cancers; also basaloid tumours (arising just above the dentate line), small cell cancers (highly malignant), mucoepidermoid cancers, and adenocarcinomas.

Clinical features: Bleeding, pain, bowel habit change, pruritus ani, masses (ulcerated or submucosal), stricture formation.

Differential diagnosis: Condyloma acuminata, leucoplakia, lichen sclerosus, Bowen's disease, Crohn's disease.

Treatment: If local excision is not possible, radiotherapy (for squamous lesions) is as effective as anorectal excision with colostomy—and 75% of patients retain normal anal function.[4]

---

1 *Drug Ther Bul* 1990 **28** 3 **2** W Thomson 1982 *Lancet* ii 467 **3** *Lancet* Ed 1991 ii 605
4 R Talbot 1988 *BMJ* ii 239

Examination of the anus and rectum Explain what you are about to do. Have the patient on his left side, his knees brought up towards the chest. Use gloves and lubricant. Part the buttocks and inspect the anus. Ask the patient to bear down. Press your index finger against the side of the anus. Ask the patient to breathe deeply and insert your finger with a twisting motion. Feel for masses—such as the cervix or prostate. Are there any other masses? Note stool or blood on the glove and test for occult blood. Wipe anus. Consider proctoscopy (for the anus) or sigmoidoscopy (which mainly inspects the rectum).

Haemorrhoids (piles)

Piles (haemorrhoids) are not simply dilated veins: they are cushions of vascular tissue (like the lips) occurring in the rectal mucosa.[1] They are drained by tributaries of the superior haemorrhoidal veins which drain into the inferior mesenteric (thence portal) vein. With the patient in lithotomy position (on his back with thighs fully abducted and hips flexed) piles take up 3 characteristic positions—at 3, 7 and 11 o'clock.

Classification[2] *First-degree piles* remain in the rectum.

Second-degree piles prolapse through the anus on defecation but spontaneously reduce.

Third-degree piles As for second-degree but require digital reduction.

Fourth-degree piles remain persistently prolapsed.

Piles are not painful unless they thrombose when they protrude and are gripped by the anal sphincter, blocking venous return.

Differential diagnosis Other causes of anal pain include fissure, perianal haematoma, abscess, tumour, proctalgia fugax (idiopathic, intense, stabbing rectal pain). Never ascribe bleeding to piles without adequate examination or investigation.

Causes Constipation with prolonged straining is the key factor. Congestion from a pelvic tumour, pregnancy, CCF or portal hypertension are important in only a minority of cases.

Pathogenesis There is a vicious circle: vascular cushions protrude through a tight anus, become further congested, so hypertrophying to protrude again more readily. These protruding piles may then strangulate.

Clinical picture The patient notices bright red rectal bleeding—often coating motions, or dripping into the pan after defecation. There may be mucus discharge and *pruritus ani*. Anaemia may occur.

In all cases of rectal bleeding perform:
- An abdominal examination (to rule out other diseases).
- A rectal examination: prolapsing piles are obvious. Internal haemorrhoids are not palpable.
- Proctoscopy, to see the internal haemorrhoids.
- Sigmoidoscopy to identify pathology higher up.

Treatment A high-fibre diet may help. Soft paraffin soothes. Correct anal spasm (if present) by manual dilatation under general anaesthesia. Prevent mucosal prolapse by fixing it to the bowel wall using an injected sclerosant (2ml of 5% phenol in oil) or rubber band ligation—which produces an ulcer to anchor the mucosa (complications: pain, bleeding, infection). Photocoagulation and freezing are also used. These may obviate the need for haemorrhoidectomy (excision of the pile with ligation of the vascular pedicle) which requires admission and may be complicated by haemorrhage or stenosis.

Treatment of thrombosed piles is with analgesia, and bed rest. The pain usually resolves in 2–3 weeks. Some surgeons advocate early operation.

1 W Thomson 1975 *Br J Surg* **62** 542   **2** A Dennison 1989 *Am J Gastroenterology* **84** 475

Haemorrhoids (piles)

Hernias

Definitions Any structure which passes through another structure and ends up in the wrong place is a *hernia*. Hernias involving bowel are said to be *irreducible* if they cannot be pushed back into the right place. This does not mean that they are either obstructed or strangulated: incarcerated ≡ irreducible. They are *obstructed* if bowel contents cannot pass through them—the classical features of intestinal obstruction soon appear. They are *strangulated* if ischaemia occurs—the patient becomes toxic and requires urgent surgery.

Inguinal hernias These are the commonest and are fully described on p160.

Femoral hernias The bowel enters the femoral canal and presents as a mass in the upper medial thigh or above the inguinal ligament where it points down the leg, unlike an inguinal hernia which points into the groin. They occur more commonly in females than males and are likely to be irreducible and to strangulate. The *neck* of the hernia is felt below and lateral to the pubic tubercle (inguinal hernias are above and medial to this point).

The boundaries of the femoral canal are *anteriorly* and *medially* the inguinal ligament; *laterally* the femoral vein and *posteriorly* the pectineal ligament. The canal contains fat and Cloquet's node.

Surgical repair is recommended.

Paraumbilical hernias These occur just above or below the umbilicus. Risk factors are obesity and ascites. Omentum or bowel herniates through the defect. Surgery involves repair of the rectus sheath.

Epigastric hernias These pass through linea alba above the umbilicus.

Incisional hernias These follow breakdown of muscle closure after previous surgery. The patient is often obese, so repair is not easy, and is often not recommended.

Spieghelian hernias These occur at the lateral edge of the rectus sheath, below and lateral to the umbilicus.

Lumbar hernias These occur through one of the lumbar triangles.

Richter's hernia This involves bowel wall only—not lumen.

Obturator hernias These occur through the obturator canal. Typically there is pain along the medial side of the thigh in a thin lady.

Other examples of hernias
- Of the nucleus pulposus into the spinal canal (slipped disc).
- Of the uncus and hippocampal gyrus through the tentorium (tentorial hernia) in space-occupying lesions.
- Of the brainstem and cerebellum through the foramen magnum (Arnold–Chiari malformation).
- Of the stomach through the diaphragm (hiatus hernia, p490–1).

Hernias

Inguinal hernias

Indirect inguinal hernias pass through the internal inguinal ring and, if large enough, out through the external ring. Direct hernias pass through a defect in the abdominal wall into the inguinal canal. Predisposing conditions: chronic cough, constipation, urinary obstruction, heavy lifting, ascites, previous abdominal surgery.

Landmarks *The internal ring* is at the *mid-point* of the inguinal ligament, $1\frac{1}{2}$cm above the femoral pulse (which crosses the mid-inguinal point).

The external ring is just above and medial to the pubic tubercle which is a bony prominence forming the medial attachment of the inguinal ligament.

The relations of the inguinal canal
Floor: The inguinal ligament.
Roof: Fibres of transversalis and internal oblique.
Front: The aponeurosis of external oblique, with internal oblique for the lateral third.
Back: Laterally, transversalis fascia; medially, the conjoint tendon. The canal contains the spermatic cord (the round ligament in the female) and the ilioinguinal nerve.

Examining the patient
1 Is the lump visible? If so, ask the patient to reduce it—if he cannot, make sure that it is not a scrotal lump. Ask him to cough. Inguinal hernias appear inferomedial to the external ring.

2 If no lump is visible, feel for a cough impulse.

3 If there is no lump, ask the patient to stand and repeat the cough.

Always look for previous scars, feel the other side and examine the external genitalia.

Differentiating direct from indirect hernias This exercise is loved by examiners but is of little clinical use. The best method is to reduce the hernia and occlude the internal ring with two fingers. Ask the patient to cough—if the hernia is restrained it is indirect, if it pops out it is a direct hernia.

Indirect hernias:	Direct hernias:	Femoral hernias:
Common (80%)	Less common (20%)	Often in females
	Reduce easily	Often irreducible
Can strangulate	Rarely strangulate	Often strangulate

Repair of inguinal hernias Advise the patient to diet and stop smoking pre-operatively. Various methods of repair are used, eg the Bassini repair, the 'nylon darn', and the Shouldice method.[1] The latter includes a multi-layered suture involving both anterior and posterior walls of the inguinal canal. As its recurrence rate is less than that of other methods (eg <2% vs 10%), it is the suture repair advocated by the hernia working party of the Royal College of Surgeons.

Local anaesthetic techniques and day-case 'ambulatory' surgery may halve the price of surgery. This is important because this is one of the most common operations (70,000/yr in UK).

Returning to work Eg at 4 weeks, with full activity in 8–12 weeks.

---

1 J Mansberger 1992 *American Surgeon* **58** 211–2

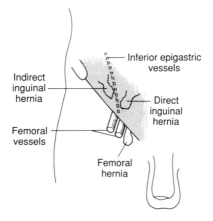

Varicose veins

Essence Blood from superficial veins of the leg passes into the deep veins by means of perforator veins (they perforate the deep fascia) and at the sapheno-femoral junction. Normally, valves prevent blood from passing from deep to superficial veins. If they become incompetent (venous hypertension from prolonged standing, occlusion by fetus, fibroids, ovarian tumour, previous DVT) dilatation (*varicosities*) of the superficial veins occurs. See *Oedema*, p52.

Varicose veins are common, leading to 50,000 admissions/yr in the NHS.

Symptoms Veins unsightly, aching, oedema, eczema, ulceration, haemorrhage, thrombophlebitis.

Method of examination 1 Ask the patient to stand.

2 Note signs of poor skin nutrition: ulcers usually above the medial malleolus (varicose ulcers, OHCS p596) with deposition of haemosiderin causing brown edges, eczema and thin skin.

3 Feel for a cough impulse at sapheno-femoral junction (signifying incompetence). The percussion test: tap the top of a vein and feel how far down its length you can feel the repercussions, which are normally interrupted by competent valves.

4 The Trendelenburg test determines if the valve at the sapheno-femoral junction is competent: lie the patient down and raise the leg to empty the vein. Place two fingers on the sapheno-femoral junction (5cm below and medial to the femoral pulse). Ask the patient to stand, keeping fingers in place. Do the varicosities fill? If so, there are deep or communicating valve leaks. Then release fingers—if the veins fill quickly the sapheno-femoral valve must be incompetent. In the latter case the Trendelenburg operation (disconnection of the saphenous and femoral veins) should be of benefit.

5 Doppler ultrasound probes, to listen to flow through incompetent valves (eg the sapheno-femoral junction, or the short saphenous vein behind the knee). If such incompetence is not identified and treated at the time of surgery, varicosities will return.[1]

6 Before surgery, ensure that all varicosities are indelibly marked.

Treatment 1 Patient education. Avoid prolonged standing; support stockings; lose weight; regular walking.

2 Injection therapy—especially for varicosities below the knee if there is no gross sapheno-femoral incompetence. Ethanolamine is injected and the vein compressed for a few weeks to avoid clotting (intravascular granulation tissue obliterates the lumen). Unsuitable for perforation sites. Max 10–12ml can be used and care should be taken to prevent extravasation.

3 Operative. There are several choices, depending on the anatomy of the veins and surgical preference, eg sapheno-femoral disconnection; multiple avulsions; stripping from groin to upper calf (stripping to the ankle is not needed, and may damage the saphenous nerve).

Post-operatively the legs should be tightly bandaged and the legs elevated for 24h. Thereafter encourage walking 3 miles a day—as many short walks, rather than as one long one.

Saphena varix This is a dilatation (varicosity) in the saphenous vein at its confluence with the femoral vein. It is one of the many causes of a lump in the groin (p146). Because it transmits a cough impulse, it may be mistaken for an inguinal hernia—but on closer inspection, it may have a bluish tinge to it, and it disappears on lying down.

1 W Campbell 1990 *BMJ* i 763; see also D Tibbs 1992 *Varicose Veins and Related Disorders*, Butterworth-Heinemann, Oxford

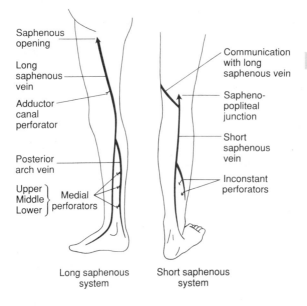

Long saphenous
system

Short saphenous
system

Gastric surgery and its aftermath

Peptic ulcers and gastric carcinomas are the commonest indications for gastric surgery.

Partial gastrectomy (the Billroth operations)

Billroth I: Partial gastrectomy with simple reanastomosis
Billroth II (Polya gastrectomy): Partial gastrectomy. The duodenal stump is oversewn (leaving a blind loop), and anastomosis is achieved by a longitudinal incision further down the duodenum.

STOMACH

Billroth I   DUODENUM   Billroth II (or Polya)

Operations for duodenal ulceration See p166.

Operations for benign gastric ulceration Those near the pylorus may be considered similarly to duodenal ulceration (p166). Away from the pylorus elective operation is rarely needed as ulcers respond well to medical treatment, stopping smoking and avoidance of NSAIDs—but a partial gastrectomy may be needed.

Haemorrhage is usually treated with partial gastrectomy or simple excision of the ulcer. Perforation is treated with excision of the hole for histology, then closure.

Malignant gastric ulceration See p142.

Complications of peptic ulcer surgery
Recurrent ulceration: Symptoms are similar to pre-operatively but complications are more common and response to medical treatment is poor. Further surgery is difficult.

Abdominal fullness: A feeling of early satiety, sometimes with discomfort and distension. It improves with time and advice is to take small, frequent meals.

Dumping syndrome: Fainting and sweating after eating, possibly due to food of high osmotic potential being dumped in the jejunum, and causing oligaemia because of rapid fluid shifts. 'Late dumping' is due to hypoglycaemia and occurs 1–3h after taking food. It is relieved by eating glucose. Both problems tend to improve with time but may be helped by taking more protein and less glucose in meals.

Bilious vomiting: This is difficult to treat—but often improves with time.

Diarrhoea: This can be particularly disabling after a vagotomy. Codeine phosphate may help.

Metabolic complications
Weight loss: This is usually due to poor calorie intake.

Malabsorption and bacterial overgrowth (the blind loop syndrome) may occur.

Anaemia: This is usually due to iron deficiency following hypochlorhydria and stomach resection. B_{12} levels are frequently low but megaloblastic anaemia is rare.

Gastric surgery and its aftermath

Complications of peptic ulcer surgery (These are approximate and depend on the skill of the surgeon.)

	Partial gastrectomy	Vagotomy & pyloroplasty	Highly selective vagotomy
Recurrence	2%	7%	>7%
Dumping	20%	14%	6%
Diarrhoea	1%	4%	<1%
Metabolic	++++	++	0

Operations for duodenal ulcer

Peptic ulcers usually present as epigastric pain and dyspepsia (p490). There is no reliable method of distinguishing clinically between gastric and duodenal ulceration. Although management of both is usually medical in the first instance, surgery still has an important role.

Surgery is considered for:
1 When medical treatment has failed

2 Complications:
   a. *haemorrhage*
   b. *perforation*
   c. *pyloric stenosis*

Several types of operation have been tried but, as whenever considering an operation, one must consider efficacy, side-effects and mortality.

1 **Elective surgery** may be undertaken for symptoms which fail to respond to medical treatment. Operations to consider:
   a. *Gastrectomy* (p164).
   b. *Vagotomy and drainage procedure:* A vagotomy reduces acid production from the stomach body and fundus, and reduces gastrin production from the antrum. However, it interferes with emptying of the pyloric sphincter and so a drainage procedure must be added. The commonest is a pyloroplasty in which the pylorus is cut longitudinally and closed transversely. A gastroenterostomy is an alternative.
   c. *Highly selective vagotomy (HSV).* Here the vagus supply is denervated only where it supplies the lower oesophagus and stomach. The nerve of Laterget to the pylorus is left intact thus gastric emptying is unaffected. The results of this operation are greatly dependent on the skill of the surgeon.

2 a. **Haemorrhage** Bleeding may be controlled endoscopically by injection, diathermy, laser coagulation or heat probe. Operation should be considered for haemorrhage >5 units or rebleeding, especially in the elderly. The ulcer is oversewn to stop the bleeding. The pylorus is opened at operation so it is common to perform a definitive vagotomy and pyloroplasty.

   b. **Perforation** The peritoneal cavity is washed out and the hole oversewn, usually with an omental plug. The decision whether or not to perform a definitive operation usually depends on previous symptoms—if there has been a long history of pain or dyspepsia a vagotomy and pyloroplasty is usually added.

   c. **Pyloric stenosis** This is a late complication, presenting with vomiting of large amounts of food some hours after meals. (Adult pyloric stenosis is a complication of duodenal ulcers, and has nothing to do with congenital hypertrophic pyloric stenosis.) Surgery is always required—usually a vagotomy and drainage procedure. Some may attempt highly selective vagotomy with dilatation. The operation should be done on the next available list, after correction of the metabolic defect—hypochloraemic, hypokalaemic metabolic alkalosis.

Operations for duodenal ulcer

Minimally invasive surgery

The term 'keyhole surgery' or minimal access surgery may be preferred, because these procedures can be as invasive as any laparotomy, having just the same set of side effects—plus some new ones.[1] It is the size of the incision and the use of laparoscopes which marks out this branch of surgery. As a rule of thumb, whatever can be done by laparotomy can also be done with the laparoscope. This does not mean that it *should* be done, but if the patient feels better sooner, has less postoperative pain, and can return to work earlier, then these specific techniques will gain ascendency—provided hospitals can afford the equipment.[2] Established uses for this type of surgery are for oesophageal, gastric, duodenal and colonic procedures—including the removal of cancers—as well as hernia repairs, appendectomies and cholecystectomies.

It is worth noting that advantages do not include time. In upper GI procedures, laparoscopic procedures take longer than open procedures. Also, the patient needs to spend a night in hospital, usually. This has economic implications when comparing laparoscopic hernia repair with its open alternative done under local anaesthesia (after which the patient can go home the same day). The other side of the economic equation for hernia repair is that postoperative discomfort is much less after laparoscopic procedures, and the patient can return to full employment after a week.[1]

Problems with minimal access surgery: for the surgeon

Inspection: Anatomy looks different—eg in hernia repairs which are approached differently in open procedures. (In cholecystectomy, visualization may be facilitated by the laparoscope which can gain access to areas which require difficult traction to reveal in open surgery.)

Palpation is impossible during laparoscopic procedures. This may make it hard to locate colon lesions prior to cutting them out. This means that pre-operative tests may need to be more extensive (eg colonoscopy *and* barium enema). Alternatively colonoscopy can take place during the procedure, or methylene blue can be injected at the first colonoscopy.

Skill: Here the problem is not just that a new skill has to be learned and taught. Old skills may become attenuated if most cholecystectomies are performed via the laparoscope, and new practitioners may not achieve quite the level of skill in either spheres if they try to do both.

Problems for the patient and his general practitioner

Postoperative complications: What may be quite easily managed on a well-run surgical ward (eg haemorrhage) may prove a challenge for a GP and be terrifying for the patient, who may be all alone after early discharge.

Loss of tell-tale scars: After laparoscopy there may only be a few stab wounds in the abdomen, so that the GP or the patient's future carers have to guess at what has been done. The answer here is not to tattoo the abdomen with arrows and instructions, but to communicate carefully and thoroughly with the patient, so that he or she knows what has been done.

Problems for the hospital

Just because minimal access surgery is often cost-effective, it does not follow that hospitals can afford the procedures. Instruments are continuously being refined, and quickly become obsolete—so that many are now produced in disposable single-use form. Because of budgeting boundaries, hospitals cannot use the cash saved by early return to work or by freeing-up bed use to pay for capital equipment and extra theatre time which may be required.

1 JR Monson 1993 *BMJ* ii 1346   2 K Lawrence 1994 *Lancet* **343** 308

6 Infectious diseases

Rarities for reference:

Junior doctors working in the developed world are unlikely to have much experience with many of the following diseases (which may be common abroad). The importance of consulting early with experts cannot be over-emphasized. Specific treatments are given here principally to enable the reader to have an intelligent discussion with the relevant expert.

Relevant pages in other chapters: Prophylactic antibiotics in gut surgery (p84); infective endocarditis and its prevention (p312 & p314); pneumonia (p332); unusual pneumonia (p334); lung abscess (p334); management of pneumonia (p336); bronchiectasis (p338); fungi and the lung (p340); UTI (p374); meningitis (p442 & p444); septic arthritis (p662).

Endometritis (*OHCS* p36); pelvic infection (*OHCS* p50); prenatal infection (*OHCS* p98); perinatal infection (*OHCS* p100); measles, mumps and rubella (*OHCS* p214); parvoviruses (*OHCS* p215); overwhelming infection in the neonate (*OHCS* p238); the ill and feverish child (*OHCS* p248); TB meningitis (*OHCS* p260); orbital cellulitis (*OHCS* p484); ophthalmic shingles (*OHCS* p484); mastoiditis (*OHCS* p534); sinusitis (*OHCS* p554); tonsillitis (*OHCS* p556); skin infections (*OHCS* p590); skin infestations (*OHCS* p600); osteomyelitis (*OHCS* p644).

Notifiable diseases (marked ND on relevant pages). Inform the Consultant in Communicable Disease Control (CCDC)

Anthrax	Plague
Cholera	Poliomyelitis
Diphtheria	Rabies
Dysentery (amoebic or bacillary)	Relapsing fever
Ebola virus disease	Rubella
Encephalitis (acute)	Scarlet fever
Food poisoning	Smallpox
Leprosy	Tetanus
Leptospirosis	Tuberculosis
Malaria	Typhoid fever
Measles	Typhus
Meningitis (acute)	Viral haemorrhagic
Meningococcal septicaemia	fevers (inc. Lassa fever)
Mumps	Viral hepatitis
Ophthalmia neonatorum	Whooping cough
Paratyphoid fever	Yellow fever

Note: reporting AIDS is voluntary—and in strict confidence—to the Director, Communicable Disease Surveillance Centre (PHLS), 61 Colindale Avenue, London NW9 5EQ. Telephone advice: 081 200 6868.

The essence of infectious diseases

First of all define a clinical syndrome. Look for a possible causative organism. Then entertain Koch's (1843–1910) postulates—ie show that 1 you can find the organism in all instances of the disease; 2 Culture the organism; 3 Show that the culture can reproduce the disease when innoculated into an animal; 4 Recover the organism from the experimental disease.

The next step is to clone the organism's genome; then design a molecule to kill it, or, better still, a vaccine to prevent its effects, and so to raise the possibility of extinguishing the organism's life on earth for ever. But before this happy state of affairs exists, there is the hardest hurdle of all for a great discovey: you must make its application available and desired by all members of our species.

There is probably no known example where all elements of this sequence has been gone through in this orderly way, but all the great discoveries in infectious diseases have engaged this paradigm somewhere along its path.

The classification of pathogens

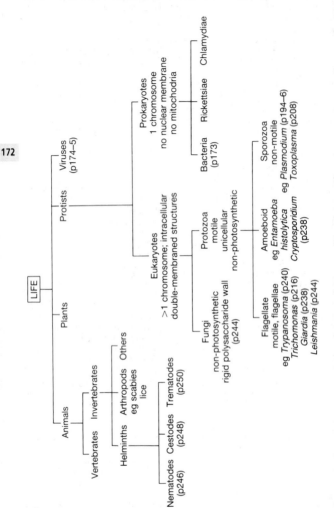

Classification of principal pathogenic bacteria

GRAM +VE COCCI

Staphylococci (p222)
  Staph aureus (coagulase +ve)
  Staph epidermidis (coagulase −ve)

Streptococci* (p222)
  β-haemolytic Streptococci
    Strep pyogenes (group A)
  α-haemolytic Streptococci
    (viridans)
    Strep mitior
    Strep pneumoniae
    Strep sanguis
  Non-haemolytic Streptococci
    Strep bovis
    E faecalis (enterococcus)
    E mutans
  Anaerobic Streptococci

GRAM +VE BACILLI (rods)

Aerobic
  Bacillus anthracis
    (anthrax: p222)
  Corynebacterium diphtheriae
    (diphtheria: p222)
  Listeria monocytogenes (p223)
  Nocardia (nocardiasis: p223)

Anaerobic
  Clostridia
    Clostridium botulinum
      (botulism: p223)
    Clostridium perfringens
      (gas gangrene: p223)
    Clostridium tetani
      (tetanus: p226)
    Clostridium difficile
      (pseudomembranous colitis:
      p486)
  Actinomyces israelii
    (actinomycosis: p223)

ORGANISMS ON THE BORDERLINE BETWEEN VIRUSES AND BACTERIA

  Mycoplasma (atypical pneumonia:
    p334)
  Rickettsia (typhus: p236;
    Coxiella burnetii: p236)
  Chlamydia (p216)

GRAM −VE COCCI

Neisseria meningitidis
  (=meningococcus)—meningitis
  and septicaemia: p444
Neisseria gonorrhoeae
  (=gonococcus)—gonorrhoea:
  p220

GRAM −VE BACILLI (rods)

Enterobacteriaceae (p190, p224)
  Escherichia coli
  Proteus mirabilis
  Serratia marcescens
  Klebsiella aerogenes
  Salmonella
  Shigella
  Enterobacter species
  Yersinia enterocolotica
  Yersinia pestis (plague, p225)
Haemophilus influenzae (p224)
Brucella species
  (brucellosis: p224)
Bordetella pertussis,
  whooping cough: p224
Pasteurella multocida (p225)
Pseudomonas aeruginosa (p224)
Vibrio cholerae
  (Cholera: p229)
Legionella pneumophila
  (atypical pneumonia: p334)
Campylobacter jejuni (food
  poisoning: p190)
Anaerobes
  Bacteroides (wound infections, p84)
Helicobacter pylori (p490)

Mycobacteria
  Mycobacterium tuberculosis
    (TB, p198–200)
  Mycobacterium leprae
    (leprosy: p230)
  'Atypical' mycobacteria
    eg *M bovis*

Spirochaetes (p218, p232)
  Treponema
  Borrelia
  Leptospira

This table is not intended to be comprehensive nor to follow closely bacteriological taxonomy. It is simply a guide for the forthcoming pages.

*The classification of streptococci may be as β-, α-, or non-haemolytic, by Lancefield antigen (mainly β-haemolytic streptococci express Lancefield antigens), or by clinical species (eg *Strep pyogenes*—see also p222). There is much crossover amongst these groups; the above is a generalization for the most important organisms.

Viruses (various)

A. Double-stranded DNA
  (a) Papova virus: papilloma virus—human warts
      progressive multifocal leucoencephalopathy
  (b) Adenovirus: 10% viral respiratory disease
                  7% viral meningitis
                  >30 serotypes
  (c) Herpes viruses
      1. Herpes simplex virus 1 (p202)
      2. Herpes simplex virus 2
      3. Herpes (varicella) zoster virus (p202)
      4. Epstein–Barr virus: infectious mononucleosis (p204)
                  Burkitt's lymphoma
                  Some nasopharyngeal carcinoma
      5. Cytomegalovirus (p208)
      6. Herpes virus 6 (HHV-6): roseola infantum (Exanthem subitum) (OHCS
                  p214).
  (d) Pox virus
      1. Variola—smallpox (now officially eradicated)
      2. Vaccinia—cowpox
      3. Orf—cutaneous pustules in man, contracted from sheep
      4. Molluscum contagiosum: groups of pearly umbilicated papules
  (e) Hepatitis B virus (see p210)
B. Single-stranded DNA
     Parvovirus (only group with single-stranded DNA) (OHCS p214)

A. Double-stranded RNA: Reovirus
  B. Positive single-stranded RNA   (a) Picornavirus   (b) Togavirus
  C. Negative double-stranded RNA   (a) Orthomyxovirus
                                     (b) Paramyxovirus   (d) Rhabdovirus
                                     (c) Arenavirus   (e) Bunyavirus
  D. Positive single-stranded RNA with reverse trancriptase—Retrovirus
A. Reovirus: eg Rotavirus—infantile gastroenteritis
B. (a) Picornavirus:
      1. Rhinovirus—common cold with >90 serotypes infectious from
             2 days before symptoms for up to 3 weeks.
      2. Enterovirus:
         (i) Coxsackie: Group A: 30% viral meningitis
                                 gastroenteritis
                        Group B: pericarditis
                                 Bornholm's dis
         (ii) Poliovirus (p234)   (iii) Echovirus: 30% viral meningitis
         (iv) Hepatitis A (atypical)
  (b) Togavirus: (i) Rubella   (ii) Alphvirus   (iii) Flavivirus: yellow fever
                                                                    dengue
C. (a) Orthomyxovirus: Influenza A, B, C
  (b) Paramyxovirus: Parainfluenza Mumps Measles
                     Respiratory syncytial virus: 20% serious
                     respiratory disease in children
  (c) Arenavirus: Lassa fever
             some viral haemorrhagic fevers
             lymphocytic-choriomeningitis (LCM) virus
  (d) Rhabdovirus: Rabies
  (e) Bunyavirus: certain viral haemorrhagic fevers
D. Retrovirus—HIV I & II (p212 & p214)

Travel advice[1]

▶ Most problems are due to ignorance, indiscretions, and lack of immunity, which are all at least partly amenable to forward planning. Accidents (± alcohol abuse) are the most frequent cause of problems in travellers. HIV, malaria and cholera are the other great threats; their prevention is dealt with elsewhere (p196, p212, p214 & p228).

[1 = live vaccine]	Doses needed	Gap between doses: 1st & 2nd	2nd & 3rd	Booster interval
Yellow fever	1			10 years
Typhoid injection	1			3 years
Typhoid oral[L]	4	2 days	2 days	3–5 days
Tetanus	3	4 weeks	4 weeks	5–10 years
Polio[L]	3	>6 weeks	>6 weeks	5–10 years
Rabies pre-exposure UK	3	7–28 days	6–12 months	10 years
USA	3	7 days	21 days	2 years
Meningococci	1			3 years
Japanese encephalitis	3	1–2 weeks	2–4 weeks	1–4 years
Tick encephalitis	3	1–3 months	9–12 months	5–10 years
Plague	3	1–3 months	3–6 months	6 months
Hepatitis A (Havrix®)	3	2–4 months	6–12 months	5–10 years
(immunoglobulin)	1			2–6 months
Hepatitis B	3	1 month	5 months	2–5 years
if travelling soon	3	1 month	1 month	1 year

If only 1 attendance is possible, all is not lost (?make up *en route*):
Africa: Meningitis, typhoid, tetanus, polio, γ-globulin, yellow fever.
Asia: Meningitis (some areas), typhoid, tetanus, polio, γ-globulin.
S America: Typhoid, tetanus, polio, γ-globulin, yellow fever.[1]

Water ▶ If in doubt, boil all water. Chlorination is an alternative but does not kill amoebae. Tablets are available from pharmacies. In emergency, one drop of laundry bleach (4–6%), or 4 drops of tincture of iodine may be added to 1 litre of water. Filter water before purifying. It is important to distinguish between simple gravity filters and water purifiers (which also attempt to chemically sterilize). Choose a unit which is verified by bodies such as the London School of Hygiene and Tropical Medicine (eg the Travel Well Personal Water Purifier supplied by MASTA[2]). Make sure that all containers are disinfected. Always try to avoid surface water and intermittent tap supplies. Avoid ice (and ice cream). In Africa assume that all unbottled water is unsafe (except in South Africa, and capitals in Botswana, Gabon, Morocco, Nigeria, Senegal). When using bottled water, ensure the rim is clean and dry, and drink from its rim, or carry your own cup.

Food ▶ Hot, well-cooked food is the safest (prawns need ≥8mins boiling). In China, for example, a piping hot rat kebab might be a safer bet than a salad. Peel your own fruit. If you cannot wash your hands, discard the part of the food which you are holding (with bananas, careful unzipping obviates this precaution).

---

1 R Dawood 1992 *Travellers' Health*, OUP   2 Medical Advisory Service for Travellers Abroad (tel 071 631 4408) provides detailed written travel advice for ~£5. Also try Healthline (tel 0898 345 081), the Liverpool School of Tropical Medicine (051 708 9393) and the Royal Geographical Society Expedition Advisory Centre, 1 Kensington Gore, London SW7 2AR

Susceptibility of selected bacteria to certain antibacterial drugs

	Penicillin V/G	Flucloxacillin	Amp./Amoxycillin	Carbenicillin/Ticarcillin/Piperacillin/Azlocillin	Cephradine/Cephalexin/Cefaclor	Cefuroxime/Cephalothin/Cefazolin	Cefotaxime/Ceftazidime	Imipenem	Erythromycin	Lincomycin/Clindamycin	Tetracyclines	Chloramphenicol	Trimethoprim	Aminoglycosides	Vancomycin	Metronidazole	Ciprofloxacin
Staph aureus (penicillin sensitive)	1	0	0	0	0	2	0	0	2R	R	2R	R	2	2	2	R	2
Staph aureus (penicillin resistant)	R	1	R	R	R	2	R	R	2R	R	2R	R	2R	2	2	R	2
Strep (group A)	1	0	0	0	0	2	0	0	R	R	R	R	R	0	R	R	R
Strep pneumoniae	1	0	0	0	2	2	2	0	R	R	R	R	R	0	R	R	R
Enterococcus faecalis	R	1	R	2	0	0	0	0	R	2R	R	R	2R	R	0	R	R
N meningitidis	1	0	2	0	0	2	2	0	0	0	R	R	0	0	R	R	R
Listeria monocytogenes	2	0	1	0	0	0	0	0	2	2	2	R	0	2	0	R	2
H influenzae	R	0	1R	0	R	2	2	0	R	2R	2R	R	R	0	R	R	2
E coli	R	1R	1R	R	2R	2R	2	2	R	R	1R	1R	2R	R	R	R	2
Klebsiella species	R	R	R	R	2R	2R	2	R	R	R	2R	2R	2R	R	R	R	2
Serratia/Enterobacter species	R	R	R	R	2R	2R	2R	R	R	R	2R	2R	2R	R	R	R	2
Proteus species	R	1R	R	R	2R	2R	2R	R	R	R	1R	1R	R	R	R	R	2
Pseudomonas aeruginosa	R	R	R	R	R	R	2R	R	R	R	R	R	R	R	R	R	2
Bacteroides fragilis	R	2R	R	2R	R	2R	R	2	2	2R	2R	2R	R	R	1	0	R
Other Bacteroides species	R	R	R	R	R	R	R	R	R	2R	R	R	R	R	1	1	R

Key: 1 = susceptible, first choice 2 = susceptible, second choice R = resistance likely to be a problem 0 = usually inappropriate

Antibiotics: penicillins

[CC=creatinine clearance, ml/min]

Antibiotic (and its uses):	Usual adult dose:	In renal failure:
Amoxycillin (As ampicillin but better absorbed PO. For IV therapy, use ampicillin.)	250–500mg/8h PO 3g/12h in recurrent or severe pneumonia	Reduce dose if CC <10
Ampicillin Broader spectrum than penicillin G; more active against Gram −ve rods; β-lactamase sensitive. Amoxycillin better absorbed PO. Well excreted in bile.	500mg/4–6h IM/IV	Reduce dose if CC <10
Azlocillin (Antipseudomonal; activity>ticar- or carbenicillin)	2–5g/8h IV	If CC <30 give every 12h
Benzylpenicillin = penicillin G (Most Streps, syphilis, *Neisseria*, tetanus, actinomycosis, gas gangrene, anthrax, many anaerobes.)	300–600mg/6h IV, more in meningitis 1 megaunit = 600mg If dose >1.2g, inject at rate of <300mg/min	If CC <10 max dose is 6g/24h
Carbenicillin (For *Proteus*, *Pseudomonas*; activity < ticarcillin.)	5g/4–6h IV over 4 mins UTI:2g/6h IM	Reduce dose if CC <20
Cloxacillin (for G+ve β-lactamase producers (*Staph aureus*) only. Less is absorbed than fluclox.)	250mg/4–6h IM 500mg/4–6h IV	Dose unaltered.
Co-amoxiclav eg Augmentin® is clavulanic acid+amoxycillin (As for ampicillin, but β-lactamase resistance confers much broader spectrum.)	1–2 tabs/8h PO 1.2g/6–8h IV NB: 1 tab=250mg amoxycillin + 125mg clavulanic acid	CC: Dose/12h: 10–30 1–2 tab 600mg IV <10 1 tab; 300mg IV
Flucloxacillin (As for cloxacillin, and is best PO.)	250mg/6h PO, ½h before meals 250mg/6h IM 250–1000mg/6h IV	Dose unaltered if CC >10
Phenoxymethyl-penicillin = penicillin V (As for Pen G but less active. Used as prophylaxis or to complete an IV course.)	250–500mg/6h PO ½h before food	In severe RF give doses every 12h
Piperacillin (Very broad spectrum—with anaerobes and *Pseudomonas*. Keep for severe infections. Combine with aminoglycoside. Antagonized by cefoxitin.)	4g/6h slowly IV	CC: Max dose: 40–80 16g/24h 20–40 12g/24h <20 8g/24h dialysis:6g/24h
Procaine penicillin (Depot; good in syphilis (S) and gonorrhoea (G).)	S: 0.6–1.2g/24h IM for 10 days G: ♀=3.6g; ♂=2.4g stat	Dose unaltered
Ticarcillin (Wide spectrum; eg *Pseudomonas*, *Proteus*. Use with aminoglycoside. Activity <azlocillin or piperacillin.)	5g/6–8h IV	Reduce dose if CC <20 to 2g/8–12h

Side-effects ● Hypersensitivity: rash (ampicillin rashes do not necessarily indicate penicillin allergy but 'penicillin allergic' implies allergy to *all* penicillins); serum sickness (2%); anaphylaxis (<1:100,000). ● In huge overdose or intrathecal injection: fits and coma. ● Diarrhoea (pseudomembranous colitis is rare). ● U&E imbalance given IV.

Antibiotics: cephalosporins

Spectrum Most cephalosporins are active against *Staphs* (including those producing β-lactamase), streptococci (except group D, *Enterococcus faecalis* and *faecium*), pneumococci, *E coli*, some *Proteus*, *Klebsiella*, *Haemophilus*, *Salmonella*, and *Shigella*. The second generation drugs (cefuroxime, cephamandole) have greater activity against *Neisseria* and *Haemophilus* than cephalothin. The third generation (*) drugs (cefotaxime, ceftazidime, ceftizoxime) have better activity against Gram −ve organisms, at the expense of less Gram +ve activity, especially against *Staph aureus*. *Pseudomonas* is usually susceptible to ceftazidime (reserve for this).

Uses When to use a cephalosporin is controversial and varies according to local practice. The oral cephalosporins (cefaclor, cefadroxil, cephradine, cephalexin and cefuroxime axetil) may be used in lower respiratory tract infections, otitis media, skin and soft tissue lesions, and UTIs, but they are not first line agents. (The effect on *Haemophilus influenzae* varies.) They may be used as second line agents or to complete a course that was started with an IV cephalosporin. The major use of cephalosporins is parenteral, eg as prophylaxis in surgery (p84) and in post-op infections. Suspected life-threatening infections due to Gram −ve bacteria may be treated blindly with a third generation drug. In neutropenic patients, if the organism has been identified, the third generation drugs are very useful, but their place in 'blind' treatment has not yet been clarified. Cephalosporins may also become the drugs of first choice in other special circumstances, such as penicillin sensitivity or where aminoglycosides are better avoided. *Klebsiella* pneumonia is best treated with a third generation drug.

The principal **adverse effect** of the cephalosporins is hypersensitivity. This is seen in <10% of penicillin-sensitive patients. There may be GI upsets, reversible rises in transaminases or alk phos, eosinophilia, rarely neutropenia, nephrotoxicity and colitis. There are some reports of clotting abnormalities, and there may be false +ve reactions for glucose in urinalysis, false +ve Coombs' test, or a disulfiram reaction to alcohol. There is an increased risk of nephrotoxicity when first generation cephalosporins are co-administered with frusemide, gentamicin, and vancomycin, but second and third generation drugs are probably safe. Cephamandole may potentiate warfarin.

[CC = creatinine clearance in ml/min. RF = renal failure. CC^M = CC/1.73 m²]

Antibiotic	Usual adult dose:	Notes:
Cefaclor	250mg/8h PO Max: 4g/24h	Oral route. No dose change in RF
Cefadroxil	½–1g/12h PO	RF load with 1g, then: CC^M 26–50: 1g/12h;11–25: 1g/24h; 0–10: 1g/36h;
Cefixime	Syrup=100mg/5ml ½–1yr:3.75ml/day 1–4yrs:5ml/day 5–10yrs: 10ml/day 200mg/12–24h PO	Active against streps, coliforms, *Proteus*, *haemophilus* (eg β-lactam +ve), Anaerobes, staphs, *E faecalis* and *pseudomonas* are resistant. Renal failure: normal dose if CC>20ml/min

Antibiotic	Usual adult dose:	Notes:
Cefotaxime*	1–2g/8–12h IV/IM Max: 12g/24h eg in *Pseudomonas*	If CC <5, halve dose. Wide spectrum. Variable activity against *Pseudomonas*. Not for *E faecalis*, *Bacteroides* or *Listeria*. For serious infections only including meningitis.
Cefoxitin	1–2g/6–8h IV/IM Max dose: 12g/24h	Active against *Bacteroides*, so useful in bowel surgery and pre-op. RF—CC: CC 30–50: 1–2g/8–12h; CC 5–9: ½–1g/12–24h; CC 10–29: 1–2g/12–24h; CC <5: ½–1g/24–48h.
Ceftazidime*	UTI: ½–1g/12h Other: 1–2g/8h Max: 2g/8h Route: IV/IM	Wide spectrum, incl. most *Pseudomonas*, but not *E faecalis* or *Bacteroides* For serious infections only. In RF: CC 31–50: 1g/12h; CC 6–15: 0.5g/24h; CC 16–30: 1g/24h; CC ≤5: 0.5g/48h.
Ceftizoxime*	UTI: ½–1g/12h Other: 1–2g/8h Max: 2.7g/8h Route: IV/IM	Wide spectrum, poor against *Pseudomonas*, *E faecalis*, *Bacteroides*. For bad infections only. In RF: CC 50–79: ½–1½g/8h; 5–49: ¼–1g/12h; ≤5: ½–1g/48h.
Ceftriaxone*	Long t½ 1–4g once daily IM/IVI. (IM route: give ≤1g at one IM site; dilute with lignocaine 1%, 3.5ml per g of ceftriaxone) Gonorrhoea: ¼g stat IM Pre-colorectal surgery: 2g IM + an agent against anaerobes Children: *OHCS* p248	Many Gram −ve & +ve infections. OK in renal failure unless CC <10ml/min (limit dose to 2g/day or less here; do levels, also do levels if on dialysis)
Cefuroxime	¼–½g/12h PO ¾g/8h IM/IV Severe: 1½g/8h Max IV: 3g/8h Max IM: ¾g/8h	Wide spectrum & good Gram −ve activity. Very useful in surgery for prophylaxis and post-op infections. In RF: CC 10–20: 750mg/12h; <10: 750mg/24h.
Cephalexin	Minor infection: ½g/8h PO Max: 4g/24h PO	An oral cephalosporin. Reduce dose proportionately in RF if CC<60. If CC<10: ½g/24h
Cephamandole	½–2g/4–8h IM or IV	Similar to cefuroxime. In RF load with 1–2g; then CC 50–80: ¾–2g/6h CC 25–50: ¾–2g/8h; 10–25: ½–1g/8h 2–10: ½–1g/12h; <2: ¼–¾g/12h
Cephazolin	Minor infection ½g/8–12h IM/IV	Less stable to β-lactamase than cefuroxime In RF, load with ½–1g; then: CC 40–70: 250–1250mg/12h CC 20–40: 125–600mg/12h CC 5–20: 75–400mg/24h CC <5: 37.5–200mg/24h
Cephradine	¼–½g/6h PO or ½–1g/12h PO or ½–2g/6h IM/IV	Less active than cefuroxime. In RF load with 750mg; then 500mg at frequency dictated by CC[M]: >20: 6–12h; 15–19: 12–24h; 10–14: 24–40h; 5–9: 40–50h; <5: 50–70h.

Note: * signifies a third-generation drug.

Antibiotics: others

Antibiotic (and uses):	Adult dose:	Notes
Amikacin* (See gentamicin.)	7.5mg/kg/12h IV Dose↓ in RF	Resistance is less common than with gentamicin
Chloramphenicol (Used in typhoid fever and *Haemophilus* infection. Also in 'blind' treatment of meningitis if *Haemophilus* could be cause. Avoid late in pregnancy and when breast feeding. Toxicity is overstated.)	500mg/6h PO 12.5mg/kg/6h IV (PO is the best route)	SE:marrow depression, optic & peripheral neuritis; GI upset(all rare).Reduce dose in hepatic failure.In RF, avoid if CC <10ml/min. Avoid long or repeated courses. Do FBC often. Interacts with warfarin, phenobarbitone, phenytoin, rifampicin, sulphonylureas.
Chlortetracycline	250–500mg/6h PO	See tetracycline.
Ciprofloxacin (Useful in adults with cystic fibrosis, typhoid, other enteric infections, *Campylobacter*, prostatitis and serious multiple or resistant infections. Avoid overuse.)	200–400mg/12h IVI over $\frac{1}{2}$h or 250–750mg/12h PO UTI: 250–500mg/ 12h PO	A newer drug. The only good oral antipseudomonal agent. Reduce dose by 50% if CC <20. β-lactamase resistant. It potentiates theophylline. SE: rashes, D&V, LFTs↑.
Clindamycin (Active against Gram +ve cocci incl. penicillin resistant Staph, and anaerobes.)	150–300mg/6h PO or 0.2– 0.9g/8h IV/IM	May cause pseudomembranous colitis (p486). Used in Staph bone and joint infections/peritonitis.
Co-trimoxazole (Sulphamethoxazole 400mg+trimethoprim 80mg ∴=480mg) Used in COAD; prostatitis; pneumocystis (p334) *Haemophilus* bone & joint infections. Consider trimethoprim alone esp. in UTI. (Also use in traveller's diarrhoea.)	960mg/12h PO 960mg/12h IM/IV increase to 1.44g /12h in severe infections. Higher still in *Pneumocystis*. (60mg/kg/12h) Reduce to 480mg/ 12h if treating for > 14 days.	SE: jaundice; Stevens–Johnson; marrow depression; folate↓; Reduce dose in renal failure CC 15–25:halve dose after 3 days. Avoid if CC <15, unless haemodialysis is available. CI: G6DP↓. Interactions: warfarin,phenytoin,sulphonylureas,methotrexate. Most SE due to sulphonamide.
Demeclocycline	300mg/12h PO	See tetracycline.
Doxycycline (Useful in traveller's diarrhoea, leptospirosis, and (with rifampicin) brucellosis).	200mg PO 1st day then 100mg/24h	As for tetracycline,but may be used in RF if necessary.
Erythromycin (Similar to penicillin but spectrum wider, so used in penicillin allergy. Good in pneumonia (esp. *Legionella*); *Campylobacter* enteritis; mycoplasma; chlamydia—so a good drug in pelvic inflammatory disease). (OHCS—p50)	250–500mg/6h PO (≤4g/day in legionella). IVI dose: 6.25– 12.5 mg/kg/6h (adult and child)	SE: GI disturbance. Painful phlebitis (IV route). Potentiates warfarin, theophylline, ergotamine, carbamazepine, cyclosporin. Note: the estolate form is obsolete; use lactobionate form IV and the stearate or 'phEur' enteric-coated pellets (Erymax® 250mg capsules).
Fusidic acid (Good against Staph incl. penicillin-resistant; very helpful in osteomyelitis.)	500mg/8h PO 500mg/8h IVI over 6h	Combine with another antiStaph drug. SE: GI upset, reversible changes in LFTs.

180

Antibiotic (and uses):	Adult dose:	Notes
Gentamicin* (Spectrum wide but poor against streptococci and anaerobes, so use with penicillin and/or metronidazole.) For serious infections only or prophylaxis in SBE.	0.7–1.7mg/kg/8h IV Load with 2mg/kg CC Dose example: 70 80mg/8h IV 50 80mg/12h IV 20 80mg/24h IV 10 80mg/48h IV <5 80mg/4 days after dialysis	▶ See nomogram—p756. In renal failure, give normal loading dose, then reduce frequency. Avoid prolonged treatment or giving frusemide simultaneously. Avoid in pregnancy, myasthenia gravis. SE: oto- and nephrotoxicity.
Imipenem (+cilastatin) (Very broad spectrum:Gram +ve &–ve,anaerobes (including B fragilis)+aerobes; β-lactamase stable.)	250mg/6h-1g/8h IVI CC Dose example: 31–70 500mg/6–8h 21–30 500mg/8–12h 5–20 250–500mg/12h	Avoid in pregnancy & lactation. SE: fits;D&V; myoclonus, eosinophilia, WCC↓, Coombs' +ve; LFTs abnormal.
Metronidazole (Drug of choice against anaerobes, Gardnerella, Entamoeba histolytica, and Giardia. Also in pseudomembranous colitis—by PO route p486.)	400mg/8h PO. PR dose: 1g/8h for 3 days then 1g/12h. IVI dose: 500mg/8h for ≤7 days	Disulfiram-reaction with alcohol, interacts warfarin, phenytoin, cimetidine. CI: liver failure. Pregnancy & breast feeding: avoid high dose regimes.
Minocycline (Spectrum > tetracycline. 2nd line meningococcal prophylaxis.)	100mg/12h PO	As for tetracycline but can cause dizziness and vertigo.
Nalidixic acid (Used only for UTIs.)	1g/6h PO for 7 days, reducing to 500mg/6h PO If CC <20, avoid	SE: allergy; D&V; muscle pain; weakness; phototoxicity; jaundice; fits; disturbed vision. Avoid if G6PD↓ and lactation; care in liver disease. May potentiate warfarin.
Netilmicin* (Spectrum broader than that of gentamicin but may be less active against Pseudomonas. Route: IM or IV.)	2–3mg/kg/12h; max 7.5mg/kg/day (<48h) CC Max dose (IV): 80 2.0mg/kg/8h 50 1.37mg/kg/8h	SE: as for gentamicin, but less toxic. Complete deafness has not been reported.
Nitrofurantoin (Used only for UTIs.)	50–100mg/6h PO with food	CI: CC <50; G6PD↓. SE: D&V, neuropathy, fibrosis
Oxytetracycline	250–500mg/6h PO ac	See tetracycline.
Rifampicin (Myobacteria, some Staphs, prophylaxis in some meningitis contacts.)	450–600mg/24h PO ac	↓Dose in liver disease. Interferes with contraceptive pill. SE: p200.
Tetracycline (Used in chronic bronchitis; 1st line in chlamydia, Lyme disease, mycoplasma, brucellosis (in combination), rickettsia.)	250–500mg/6h PO ac 100mg/4–12h IM 500–1000mg/12h IV**	Avoid below age 12, in pregnancy or RF if CC <50. Absorption↓ by Fe, milk and antacids. SE: photosensitivity, D&V.
Tobramycin* (As gentamicin, better against Pseudomonas.)	1–1.6mg/kg/8h IV Dose↓ in RF.	As for gentamicin, but possibly less toxic.
Trimethoprim (Use in UTI and in some COAD. Dose in prophylaxis: 100mg/24h PO.)	200mg/12h PO 150–250mg/12h IV	SE: folate↓, marrow, depression, D&V. Lower dose if CC <20.
Vancomycin* (PO: pseudomembranous colitis; IV: resistant Staph and other Gram +ve organisms.)	125mg/6h PO 500mg/6h IVI over 60mins or 1g/12h IVI over 100mins	CC dose/d CC dose/d 100 21mg/kg 10 2mg/kg 50 11mg/kg 5 1mg/kg SE: renal & oto-toxicity

*Monitor blood levels (p797). **Not in liver disease. RF = renal failure; d = day; CC = creatinine clearance (ml/min).
NB: if dialysis is being used, special considerations apply.

'Blind' treatment of infection

For the doses of antibiotics, see p176–81.

Infection	Likely cause	Blind treatment*
PUO *(See p187) Only after very exhaustive investigation, it may be necessary to resort to a therapeutic trial, guided by a travel history	Tuberculosis Malaria Enteric fever Brucellosis Kala-azar Endocarditis ??	see p200 quinine or mefloquine ciprofloxacin see p224 see p242 see below steroids + anti-TB drugs
Septicaemia (no clues)	??	cefuroxime + gentamicin
Septicaemia (abdo or pelvis)	Gram −ve or anaerobes	benzylpenicillin + metronidazole + gentamicin
Septicaemia if immunocompromised	Consider *Pseudomonas*	ceftazidime + metronidazole + gentamicin OR piperacillin + gentamicin
Septicaemia from a UTI	Coliforms	cefuroxime + gentamicin
Septicaemia with a purpuric rash	*Meningococcus* rarely *Pneumococcus* or other bacteria	benzylpenicillin (high dose)
Septicaemia from skin or bone	Gram positive organisms	benzylpenicillin + flucloxacillin
Meningitis	*Pneumococcus* *Meningococcus* *Haemophilus*	benzylpenicillin (add chloramphenicol for *Haemophilus* if <12yrs) or cefotaxime
Pneumonia	See p332–7	
Osteomyelitis/ Septic arthritis	*Staph aureus*	benzylpenicillin + flucloxacillin or cefotaxime
Endocarditis	Streptococci	benzylpenicillin or ampicillin (high dose) + gentamicin (low)
Endocarditis (on a prosthetic valve)	*Staph epidermis* Streptococci	Vancomycin + gentamicin
Cellulitis	Streptococci	benzylpenicillin + fluctoxacillin
UTI	Coliforms	trimethoprim

Tests ▶ If possible, culture before treating. 3 sets of blood cultures, swabs, MSU, faeces for pathogens, sputum culture+immediate Gram stain, CXR, serum for virology. FBC, ESR, U&E, blood gases, coagulation studies.

Prognosis Poor if: very old or young, BP↓, WCC↓, PaO_2↓, DIC, hypothermia.

Sources (for p176–82): *British National Formulary*; *Data Sheet Compendium* 1993–4; OTM 2 ed 1987 OUP

Using a side-room laboratory

Any side room with a sink and a power point is suitable for conversion into a small laboratory. By building up your repertoire of techniques, as your patients present you with material, quite a degree of skill is possible. This enables you to have intelligent conversations with laboratory staff, and, with experience, may obviate the need to call them at night. It gives you a feel for the limitations of a given technique, and keeps you in touch.

Blood Adhere to local policies regarding specimens that could be HBsAg or HIV positive. Watch a technician make a good blood film, then try yourself. First allow the thin film to dry, then stain as follows. Cover with 10 drops of Leishman's stain. After 30sec add 20 drops of water. Leave for 15mins. Pick up the slide with forceps (to avoid purple fingers) and rinse in fast-flowing tap water—for 1sec only. Allow to dry. Now examine under oil immersion. Note red cell morphology. Do a differential white cell count. Polymorphs have lobed nuclei. Lymphocytes are small (just larger than red cells) and round, and have little cytoplasm. Monocytes are larger than lymphocytes, but similar, with kidney-shaped nuclei. Eosinophils are like polymorphs, but have prominent pink-red granules in the cytoplasm. Basophils are rare, and have blue granules. Learn to use a white cell counting chamber for the WCC—but do not expect this to be as accurate as electronic methods.

Malaria: Clinicians working in the UK are unlikely to gain enough experience, but for those using this book in the tropics, Field's stain is easy to use and gives good results. It also will allow detection of trypanosomiasis and filaria. Dip a thick film in solution A for 5sec and B for 3sec. Dip in tap water for 5sec after each staining. Stand to dry. Examine the film for at least 5mins before saying that it is negative.

Pus and the Gram stain Make a smear and perform a Gram stain. Fix by gentle heat. Flood the slide with cresyl violet for 30sec. Wash in running water. Flood with Lugol's iodine for 30sec. Wash again with running water. Decolorize with acetone for 1–3sec until no blue colour runs out. Over-zealous decolorization may result in misleading results. Counterstain for 30sec with neutral red or safranin. Wash again, and dry. Gram positive organisms appear blue-black, and Gram negative ones look red, but are easier to miss.

Aspirates (eg ascites) and CSF require too much skill for the amateur.

Urine should be examined under the microscope as described on p370. Dipstick analysis should also be performed.

Drug abuse and infectious diseases

▶ Always consider this, especially in younger patients. Ask direct questions: 'Do you use any drugs? Have you ever injected any drug? Does your partner use any drugs? Do you share needles? Have you ever had a HIV test? How do you finance your drugs (eg male or female prostitution)? Ever in prison?' List drugs used, and prescribed drugs, with names of prescriber. Behavioural clues:

● 'Just passing through your area . . .'; no GP; evasive answers.
● Erratic behaviour on the ward; unexplained absences; mood swings.
● Unrousable in the mornings; agitation from day 2; petty theft.
● Demands analgesia/anti-emetics (cyclizine). Knows pharmacopeia well.
● Heavy smoking; strange smoke smells (cannabis, cocaine, heroin).

Physical clues:

● Acetone or glue smell on breath (solvent abuse).
● Small pupils (opiates) enlarging on day 2 if no narcotics supplied.
● Needle tracks on arms, groin, legs, between toes; IV access hard.
● Abscesses and lymphadenopathy in nodes draining injection sites.
● Signs of drug associated illnesses (eg SBE, p312; AIDS, p214).
● Odd tattoos: eg over veins; dots around the neck; self-harm signs (eg cigarette burns, wrist lacerations); scars from accidents or violence.

Common presentations

Unconscious (see p716)	Narcotics (naloxone, p748); barbiturates; solvents; benzodiazepines (if in ITU consider flumazenil 0.2mg IV over 15sec, then 0.1mg/min as needed, up to 2mg).
Psychosis or agitation	'Ecstasy' = MDMA (feels omnipotent), LSD, amphetamines, benzodiazepines. Give haloperidol (OHCS p360).
Asthma/dyspnoea	Is there opiate-induced pulmonary oedema? NB: asthma may follow the smoking of heroin.
Lung abscess	Rt-sided endocarditis (Staph) until proved other.
Fever/PUO	Is it SBE, eg with no cardinal signs (p312)?
Shivering & headache	After a 'bad hit' (chemical/organism contamination). Do blood cultures; start eg gentamicin (p180).
Hyperpyrexia	MDMA. Beware myoglobinuria, DIC, renal failure.
Abscesses	If over injection site, then often of mixed organisms.
DVT (p92)	Especially on injecting suspended tablets into groin. Is there compression damage (compartment syndrome)? Do CK.
Pneumonia	Pneumococcus, haemophilus, TB, pneumocystis.
Tachyarrhythmia	(if young); cocaine, amphetamines, endocarditis.
Jaundice	Hepatitis A, B, C, or D.
Osteomyelitis	Including spinal. *Staph aureus*/Gram−ve organisms.
Constipation	If severe, opiate abuse may be the cause.
Blindness	Consider fungal ophthalmitis ± endocarditis.
Runny nose	Opiate withdrawal (+colic/diarrhoea); cocaine use.
Neuropathies	(And any odd CNS signs.) Consider solvent abuse.

General management A non-judgemental approach will produce better cooperation and may avoid self-discharge. Establish firm rules of acceptable ward behaviour. NSAIDs are useful for pain relief. Do not prescribe benzodiazepines or chlormethiazole.

Prostitutes need STD screen, speculum examination (OHCS p2), and cervical cytology as carcinoma-*in-situ* is common (OHCS p35). Screen for hepatitis B (vaccination, p512); safe sex and safe injection advice. HIV testing (p212). ▶ Liaise with community teams. See OHCS p362.

The vocabulary of drug-abusers[1]

The first step in helping a drug abuser is to communicate. To understand what he or she is telling you, the following may be helpful.

Amphetamines	Speed; whiz; Billy; pink champagne
Amyl-nitrate	Goldrush; poppers; snappers
Barbiturates	Barbs; idiot pills
Cocaine	Coke; Charlie; uncle; the white; the nice; snow
(free base)	Rock; crack; nuggets; wash; gravel
Dihydrocodeine	Dfs
Drug-induced sleep	Gauching; nodding; going on the nod
Drug intoxication	Stoned; off it; bladdered; ripped; wiped out
Heroin (± cocaine)	Smack; the nasty; gear; brown; scag
Ecstasy (MDMA)	E; x; echo; disco biscuit; XTC
Febrile reaction	Bad hit
Filter	A bud (usually a cigarette tip, through which drugs are drawn before being injected).
Injecting	Hitting up; jacking up; cranking; having a dig
subcutaneous	Skin popping
IM	Muscle popping
subclavian	Pocket shot
failed	Miss
LSD	Acid; trips; cardboard
Marijuana	Weed; pot; draw; ganja; grass; resin; Mary
Methadone	Mud
Needles	Spikes; nails
Obtaining drugs	Score (selling drugs = deal)
PCP	Angel dust; KJ; ozone; missile
Physeptone ampoules	Amps
Prostitution	Working the block/square; doing business; on the game; on the batter; flogging one's golly.
Prostitute's client	Mush; punter
Shop-lifting	Grafting
Smoking cocaine	Bonging
Smoking heroin	Chasing the dragon
Syringes	Works; tools (barrel of a syringe = gun)
Temazepam	Temazies; eggs; jellies
Tourniquet	Key
Wanted by police	'On me toes'; keeping head down
White heroin	China white
Withdrawing from opiates	Turkeying; clucking

185

1 D Stockley 1992 *Drug Warning*, Optima Books, London

Pyrexia of unknown origin (PUO)

Most pyrexias result from self-limiting viral infections. Prolonged fever which has resisted diagnosis requires thorough investigation in order to resolve the enigma of the fever chart.

Causes[1] Infection (23%); multisystem diseases, eg connective tissue diseases (22%); tumours (7%); drug fever (3%); miscellaneous diseases (14%); it is often impossible to reach a diagnosis (25% in this series).

- Infections Abscesses (subphrenic, perinephric, pelvic); TB (CXR may not show miliary or extrapulmonary TB); other granulomata (eg actinomycosis, toxoplasmosis); parasites (eg amoebic liver abscess, malaria—including 'baggage' malaria, p188); bacteria (brucella, salmonella); abscesses (subphrenic, perinephric, pelvic); infective endocarditis (some organisms are not readily cultured, eg Q fever, *Chlamydia psittaci*, fungi); viruses (eg EBV, CMV, hepatitis); HIV-related infection.

- Neoplasms Especially *lymphomas* (any temperature pattern: the Pel–Ebstein fever, p606, is rare). Occasionally *solid tumours* (especially GI and renal cell carcinoma). Frequently, the patient does not feel hot during fevers. The fever associated with *leukaemia* is usually infective.

- Connective tissue diseases RA, SLE, PAN, giant cell arteritis, acute rheumatic fever, Still's disease.

- Other causes—Drug reaction (may begin months after starting drug, but fever drops within days of stopping it); sarcoid; pulmonary emboli; inflammatory bowel disease; intracranial pathology; familial Mediterranean fever (genetic recurrent polyserositis with fever, abdominal pain, pleurisy and arthritis; treat with colchicine); factitious.

History Note especially foreign travel, contacts with animals and infected people, bites, cuts, surgery, rashes, mild diarrhoea, drugs (including non-prescription), immunization, sweats, weight loss, lumps, itching.

Examination should be repeated several times. NB: teeth, rectal and vaginal exams, skin lesions, lymphadenopathy, hepatosplenomegaly.

Investigations *Stage 1 (the first days):* FBC, ESR, U&E, LFTs, CRP, WBC differential, blood cultures (several, from different veins, at different times of day) baseline serum for virology, sputum microscopy and culture (specifying also for TB), urine dipstick, microscopy and culture stool for microscopy (ova, cysts, and parasites), CXR. If patient is very ill, consider blind treatment as in septicaemia (p182).

Stage 2: Repeat history and examination. Rheumatoid factor, ANA, antistreptolysin titre, Mantoux, ECG, bone marrow, lumbar puncture. Consider withholding drugs, one at a time for 48h each.

Stage 3: Investigate the abdomen: ultrasound, Gallium scan, indium labelled white cell scan, abdominal CT, barium studies, IVU, liver biopsy, exploratory laparotomy.

Stage 4: Consider treating for TB, endocarditis, vasculitis.

See also *Post-operative complications* (p88) and *Diagnosing the tropical traveller* (p188).

---

1 DC Knockaert 1992 *Arch Int Med* 152 51–55

Pyrexia of unknown origin (PUO)
[*OTM* 5.605]

Diagnosing the tropical traveller

Useful telephone numbers Schools of tropical medicine: Glasgow 041 946 7120, Liverpool 051 708 9393, London 071 636 8636.

In every ill traveller consider: 1 *Malaria* (p194–6): The best way to exclude this is for an expert to examine thick films (p183). Very prolonged microscopy may be needed with thin films. NB: the tropical traveller can sometimes be mosquitoes, not the patient: they may stowaway in luggage causing malaria in non-tropical destinations.[1].
2 *Typhoid* (p228): Fever, relative bradycardia, abdominal pains, cough, and constipation are suggestive; an enlarged spleen and rose spots more so. Diagnose by blood or bone marrow culture.
3 *Amoebic liver abscess* (p238). 4 *Dengue fever* (p235).

Jaundice Hepatitis, malaria, leptospirosis, yellow fever, typhoid, liver abscess, thalassaemia, sickle-cell disease, alcohol, G6PD deficiency. Look for anaemia.

Hepatosplenomegaly Malaria, kala-azar, schistosomiasis, typhoid, brucellosis.

Gross splenomegaly Malaria, kala-azar.

Diarrhoea and vomiting (p190 & p486) Examine fresh stool. *E coli* commonest (toxin-producing). Consider *Salmonella*, *Shigella*, *Campylobacter*, *Cholera*, *Giardia* and *E histolytica*. Also *Vibrio parahaemolyticus*, *Staph aureus*, *Clostridium perfringens*, *Plesiomonas*.

Erythema nodosum TB, leprosy, fungi, sulphonamides, sarcoid. Also UC, Crohn's, post-streptococcal, contraceptive pill, pregnancy.

Anaemia Malaria, hookworm, kala-azar; malabsorption.

Skin lesions Onchocerciasis (itchy nodules), scabies (burrows), leprosy (ulcers, depigmented areas, insensitive areas), tropical ulcers, typhus ('eschar' = scab), leishmaniasis (ulcers or nodules).

Acute abdomen Sickle-cell crisis, ruptured spleen, perforation of a typhoid ulcer, toxic megacolon amoebic or bacillary dysentery.

Rarities to consider ▶ Use local emergency isolation policy.
- *Yellow fever* (p235) Antibodies appear in the second week.
- *Lassa fever:* suspect in rural travellers from Nigeria, Sierra Leone, or Liberia, presenting with fever, sore throat (exudative), facial oedema, and prostration. Diagnosis: EM, serology. Isolate and refer.
- *Marburg and Ebola virus:* fever, myalgia, D&V, hepatitis, and bleeding tendency. (Sudan, Zaïre, Kenya). Because of known incubation periods it is possible to exclude Ebola, Marburg and Lassa fever if the patient fell ill >3 weeks after leaving Africa.
- *Viruses causing haemorrhage*, eg haemorrhagic fever with renal syndrome (HFRS, eg with oliguria), Crimea-Congo haemorrhagic fever.
- *Rabies* (p234). • *Viral encephalitis.*

The world is big: ultimately only experts have the geographical knowledge necessary to construct a differential diagnosis, and to know local drug resistances. Even within one region risks vary. The urban visitor is at less risk than the intrepid hiker. The details of the travel history are very important, even if you cannot interpret them yourself.

---

1 F Castelli 1993 *Tr Roy Soc Trop Med Hyg* **87** 394

Diagnosing the tropical traveller

Gastroenteritis

The ingestion of certain bacteria, viruses and toxins (bacterial and chemical) is a common cause of diarrhoea and vomiting (p60, p486). Contaminated food and water are common sources, but often the specific origin is never identified. The history should include details of food and water taken, how it was cooked, time until onset of symptoms, and whether fellow-diners were affected. The food eaten, incubation period, and clinical picture give clues as to the causative organism. Note: food poisoning is a notifiable disease (p171).

Organism/Source	Incubation	Symptoms	Food
Staph aureus	1–6h	D&V, P, hypotension	Meat
B cereus (emetic)	1–5h	D&V	Rice
Red beans	1–3h	D&V	
Heavy metals, eg zinc	5 mins–2h	V, P	
Scombrotoxin	10–60 mins	D, flushing, sweating erythema, hot mouth	Fish
Mushrooms	few mins–24h	D&V, P, fits, coma, renal and liver failure	
Salmonella	12–48h	D&V, P, T°↑, septicaemia and local infections.	Meat Eggs, poultry
Clostridium perfringens	8–24h	D, P, no fever	Meat
Clostridium botulinum	12–36h	V, paralysis	Processed food
V parahaemolyticus	12–24h	Profuse D; P, V	Seafood
B cereus (diarrhoeal)	12–24h	D, P, no fever	Rice
Campylobacter	2–5 days	bloody D, P, T°↑	Milk, water°
Small round structured viruses	36–72h	D&V, T°↑, malaise	Any food
E coli	12–72h	like cholera or dysentery°	
Y enterocolitica	24–36h	D, P, T°↑	Milk°
Cryptosporidium	4–12 days°		
Giardia lamblia	1–4 weeks	See p238	Water°
Entamoeba histolytica	1–4 weeks	See p238	°
Rotavirus	1–7 days	D&V, T°↑, malaise	°
Shigella	2–3 days	bloody D, P, T°↑	Any food

NB: V = vomiting (or nausea); D = diarrhoea; P = abdominal pain; T°↑ = fever; bold type indicates a characteristic symptom. °Rarely food- or water-borne.

Tests *Stool microscopy & culture* (p487) *if:* patient has been abroad, is from an institution or at day care, or an outbreak is suspected. In these circumstances, culture of the food source is also useful, hence the value of early notification of the local Public Health doctor.

Prevention Basic hygiene; longer cooking and rewarming times (to a core T° ≥70°C for ≥2mins) with prompt consumption.

Management Most cases are self-limiting. Give fluids PO (level teaspoon of salt to 2 pints of water with glucose, or 4.5% glucose in half normal saline) or IV. Give *anti-emetic* (eg prochlorperazine 12.5mg/6h IM) and antidiarrhoeal (eg codeine 30mg PO/IM) if the symptoms are severe—but not in dysentery (p228). Give antibiotics in:

- Cholera: tetracycline reduces transmission.
- Severe campylobacter: erythromycin 250–500mg/6h PO or ciprofloxacin PO.
- Salmonella bacteraemia: ciprofloxacin 200mg/12h IV.
- Severe shigella infection: ciprofloxacin 500mg/12h PO.
- Traveller's diarrhoea: use trimethoprim 200mg/12h or doxycycline 100mg/12h (or ciprofloxacin).

Often other antibiotics may be more appropriate; check sensitivities in individual patients.

Gastroenteritis

Immunization

Active immunization usually stimulates antibody production. BCG stimulates native cell-mediated immunity. *Passive immunization* provides pre-formed antibody (non-specific or antigen-specific).

Immunizations The new suggested DoH schedule[1] (L=live vaccine)

3 days:	BCG[L] (if TB in family in last 6 months). See below.
	Hepatitis B vaccine if mother HBsAg+ve. See *OHCS* p208.
2 mths:	'Triple' (pertussis, tetanus, diphtheria) 0.5ml SC;
	HiB = *Haemopilus influenzae type b*, *OHCS* p209; oral polio[L].
	Note: If premature still give at 2 postnatal months.
3 mths:	Repeat 'triple', HiB and polio[L].
4 mths:	Repeat 'triple', HiB and polio[L].
12–18 mths:	Measles/Mumps/Rubella[L] (MMR$_{II}$® vaccine); 0.5ml SC.
4–5yrs:	Tetanus, diphtheria & polio[L] booster. MMR$_{II}$® dose 2.
10–14yrs:	BCG[L]. (Also Rubella[L] for girls who have missed MMR$_{II}$®.)
15–18yrs:	(school leaving): Tetanus-with-low-dose Diphtheria 'Td[L]' + polio[L] 'boosters'.
Later:	Boosters: Tetanus (p226), polio, rubella (*OHCS*, p214).

▶ Acute febrile illness is a contraindication to any vaccine. Give live vaccines either together, or separated by ≥3 weeks. Do not give live vaccines if there is a *primary* immunodeficiency disorder, or if taking steroids (>60mg/day prednisolone, >2mg/kg/day if a child). *HIV infection and AIDS* (*OHCS* p217). Give these children all usual immunizations (including live vaccines) except for BCG in areas where TB prevalence is low.

▶ Contraindications to vaccines are given in full in *OHCS* p208.

BCG (bacille Calmette–Guérin) live attenuated anti-TB vaccine (effective in up to 80% subjects for ~10yrs). If TB is rife, or if past TB in family in the last 6 months, give BCG (*intradermally*—0.05ml for neonates; 0.1ml if older). Make a 7mm blanched wheal (for volumes of ~0.1ml) between the top and middle thirds of the arm (deltoid's insertion), or, for cosmetic reasons, the upper, outer thigh. Use a short needle with a short bevel. Expect to feel marked resistance as the injection is given. If, during the injection, propagation of the wheal stops, the vaccine is going too deep: re-insert. A swelling appears after 2–6 weeks, developing into a papule or small ulcer. Avoid air-occluding dressings (air access helps). SE: pain, local abscess. CI: pyrexia, oral steroids (not inhaled), sepsis or eczema at vaccination site, immune pareses (eg HIV, malignancy).

Mantoux test p200. Offered in UK to those at risk of TB, eg TB contacts, health workers and those aged 10–13yrs, to find those needing BCG.

Travel immunization See p175. Take expert advice (Liverpool School of Tropical Medicine, tel. 0151 708 9393—UK). Agents available: *Yellow fever* (single dose) gives immunity after 10 days for 10yrs. *Monovalent (whole cell) typhoid* (2 doses, 4 weeks apart; the first is 0.5ml SC/IM, and the second may be 0.1ml intradermally—ie producing a raised bleb in the skin). If the Vi polysaccharide typhoid vaccine is used only one dose SC/IM is needed. Repeat after 3yrs. If live oral form[2] used, give 3 doses (1 capsule on alternate days, 1h before food with a cool drink). *Cholera* vaccine is no longer recommended (p229). *Polio* booster or course if not previously vaccinated. Consider also *tetanus, diphtheria, rabies, human immunoglobulin* (hepatitis A). Advice to travellers is more important than vaccination: eg malaria prophylaxis, simple hygiene and protective techniques (eg avoiding insect bites).

Other vaccines Hepatitis B (p512), anthrax, botulism, influenza, pneumococcal (p618), typhus, meningitis, rabies, Japanese encephalitis.

193

1 *Immunization Against Infectious Disease*, DoH, 1996   2 *Drug Ther Bul 1993* **31** 11

Malaria[ND]: clinical features and diagnosis

Plasmodium protozoa injected by anopheline mosquitoes invade RBCs, causing haemolysis and cytokine release.

P falciparum malaria Incubation: ~7–14 days (up to 1yr if semi-immune or after prophylaxis). Most travellers present within 2 months. After a prodrome of headache, malaise, myalgia and anorexia, there may be paroxysms lasting 8–12h: sudden coldness, a severe rigor for up to 1h, then high fever, flushing, vomiting and drenching sweats. ▶ Classical tertian and subtertian periodicity (paroxysms separated by 48 and 36h) are quite rare; daily (quotidian) or irregular paroxysms are commoner. Signs: anaemia, jaundice, and hepatosplenomegaly without lymphadenopathy or rash. There are no relapses after successful cure.

Complications: *Cerebral malaria* accounts for 80% of malaria deaths (mortality 20%). Patients are unrousable and may have fits. Focal signs are uncommon, but there is hypertonia and hyperreflexia. ▶ Exclude hypoglycaemia and meningitis.
Hypoglycaemia occurs in severe malaria especially during quinine therapy and in pregnancy. There may be extensor posturing and oculogyric crises if it complicates cerebral malaria.
Acute renal failure may complicate hypotension and haemoglobinuria (which may be precipitated by antimalarial drugs in G6PD deficiency). Haemoglobinuria may also occur in severe malaria in patients with normal G6PD: this is 'blackwater fever'.
Pulmonary oedema may be due to over-zealous rehydration, uraemia or ARDS (see p350).
'Algid' malaria is the complex of shock complicating a serious manifestation of malarial infection, eg pulmonary oedema, acidosis, or bacterial infection. ▶ Also consider splenic rupture.

Maternal and fetal death both have high incidences in malaria. Use chemoprophylaxis in pregnant women in areas of transmission.

Benign malaria has a very low mortality although the acute illness is very similar to that of *falciparum* malaria. Incubation periods are longer: *P ovale*, 15–18 days; *P vivax* 12–17 days; *P malariae*, 18–40 days; 5–10% of *malariae* malaria presents over one year after infection. Benign tertian malaria is caused by *P vivax* or *P ovale*: fever occurs every 3rd day (ie days 1, 3, 5, etc.). Benign quartan malaria is caused by *P malariae*: fever recurs every 4th day (ie days 1, 4, 7, etc.), but fever regularity is often not seen early in the infection. Relapse occurs as parasites lie dormant in the liver (*P vivax* and *ovale*) or blood (*P malariae*).

Complications: Nephrotic syndrome due to glomerulonephritis occurs in chronic *P malariae* infection in children.

Diagnosis Repeated thin & thick film microscopy. In partially treated patients, examine bone marrow smears if blood smears are −ve. Serology is only helpful epidemiologically. *Other tests:* FBC (anaemia, WCC often normal), platelets (usually low), glucose, U&E (hyponatraemia, uraemia), urinalysis, blood culture. Always ask yourself: Have I excluded *falciparum*? If it is falciparum, what is the parasite rate?

Malaria: clinical features and diagnosis

Poor prognostic signs (*falciparum*) Age <3yrs or pregnant; deep coma, fits, no corneal reflex, decerebrate rigidity, retinal haemorrhage, pulmonary oedema, blood glucose <2.2mmol/l, hyperparasitaemia (>5% of RBCs or 250,000/μl), WCC >12 × 10^9/l, Hb <7g/dl, DIC, creatinine >265μmol/l, high lactic acid (>6mmol/l CSF or plasma), plasma 5′-NT↑, and anti-thrombin III low.[1] If ≥20% (or >10^4/μl) of parasites are mature trophozoites or schizonts, the prognosis is bad, whatever the degree of parasitaemia (reflects critical mass of sequestered RBCs).[2]

1 H Gilles 1991 *Management of Severe and Complicated Malaria*, WHO **2** N White 1993 *Trans Roy Soc Trop Med Hyg* **87** 436

Malaria: treatment, prophylaxis and pitfalls

▶ Pitfalls ● Believing that preventives always work. They may not.
● Failing to take a full travel history, including stop-overs in transit.
● Failure to examine enough thick and thin films.
● Delay while seeking lab confirmation. ▶▶ Prompt action saves lives.
● Relying on regular fever or splenomegaly to make diagnosis.
● Not observing falciparum patients closely for the first few days.

Treatment ▶ If species unknown or mixed infection treat as *P falciparum* ▶ Nearly all *P falciparum* are now resistant to chloroquine—treat with quinine unless sure is chloroquine sensitive ● Chloroquine is the drug of choice for benign malarias (p194) in most parts of the world—but chloroquine-resistant *P vivax* is emerging in Papua New Guinea and Indonesian Irian Jaya.[1] ● Never treat with chloroquine if used as prophylaxis. ▶ For advice in UK on treatment phone 071 387 4411 (London) or 051 708 9393 (Liverpool).

Falciparum malaria: If can swallow and no complications (p194) give 600mg quinine salt*/8h PO for 7 days; then Fansidar® (pyrimethamine + sulfadoxine) 3 tabs stat, or if Fansidar®-resistant, give tetracycline 250mg/6h for 7 days with quinine. If patient seriously ill give quinine salt 20mg/kg (max 1.4g) IV over 4h, then after 8–12h give 10mg/kg (max 700mg) over 4h every 8–12h. Give IV until can swallow and complete oral course. Use Fansidar®/tetracycline as above. Note*: correct for hydrochloride/dihydrochloride/sulphate, but bisulphate has less quinine.

Benign malarias: Give chloroquine PO as 600mg base, 300mg 8h later then 300mg/24h for 2 days. Primaquine 15mg/24h for 14 days (21 days if Chesson strain from SE Asia/W Pacific) after chloroquine if vivax/ovale to destroy liver phase and prevent relapse (screen for G6PD deficiency first). Note: 150mg chloroquine base = 250mg chloroquine phosphate (PO) = 200mg chloroquine sulphate (IV).

Other treatment: Minimize pyrexia with tepid sponging and paracetamol. Transfuse if severe anaemia. Consider exchange transfusion if parasitaemia >10% and patient deteriorating. If cerebral malaria give prophylactic phenobarbitone 3.5mg/kg IM: avoid steroids and osmotic diuretics. Treat 'algid' malaria as malaria + bacteraemic shock (p720).

Monitor TPR, BP, urine output, blood glucose frequently. Daily parasite count, platelets, U&E, plasma creatinine and bilirubin.

Prophylaxis[2] ● For advice in UK phone 071 388 9600 (London) or 051 708 9393 (Liverpool). Prophylaxis does not afford full protection. Measures must still be taken to reduce insect bites (nets; repellent; long-sleeved clothes in evening). ● Take drugs from 1 week before travel to 4 weeks after return. ● None required if just visiting cities of Hong Kong, Thailand and Singapore. ● No adequate protection for parts of SE Asia.

Guidelines: Low risk areas (eg N Africa, Middle East, Central and parts of S America): chloroquine base 300mg/week PO + proguanil 200mg/24h PO. Folate supplements with proguanil if pregnant (also with Maloprim®).

High risk areas: Sub-Saharan Africa, Indonesia (outside Bali), Laos, Vietnam, rural China, Myanmar (Burma), rural Philippines, Cambodia: use mefloquine (1 tab/week for non-pregnant adults). If poor access to medical care anticipated, also carry treatment course: eg Fansidar® 3 tabs stat or halofantrine 500mg/6h PO, 3 doses, repeated at 1 week (CI: pregnancy). Alternative: Maloprim® (pyrimethamine+dapsone) 1 tablet/wk.

1 G Murphy 1993 *Lancet* i 96 See also H Gilles 1991 *Management of Severe and Complicated Malaria*, WHO, and M Molyneux 1993 *BMJ* i 1175   2 D Bradley 1993 *BMJ* i 1247

Side-effects: Chloroquine: headache, D&V, pruritus, psychosis, retinopathy in chronic use. Fansidar®: Stevens–Johnson sy, haemopoeisis↓. Primaquine: haemolytic anaemia in G6PD deficiency, methaemoglobinaemia. Mefloquine: GI upset, dizziness, fits, neuropsychiatric disturbance, long $t_{1/2}$—so do not give if history or family history of epilepsy, psychosis, or if delicate tasks are to be performed (eg pilots), or if pregnant or risk of pregnancy within 3 months of last dose. Interaction: quinidine. Maloprim®: haemolysis, agranulocytosis. Halofantrine: GI upset.

Children's doses: Chloroquine: to 2 weeks old, $\frac{1}{8}$ adult dose; to 4 months, $\frac{1}{5}$; to 1yr, $\frac{1}{4}$; to 3yrs, $\frac{1}{3}$; to 7yrs, $\frac{1}{2}$; to 12yrs, $\frac{3}{4}$ then full dose. Fansidar®: avoid below 6 weeks; to 4yrs, $\frac{1}{4}$ adult dose; to 8yrs, $\frac{1}{2}$; to 14yrs, $\frac{3}{4}$; then full dose.

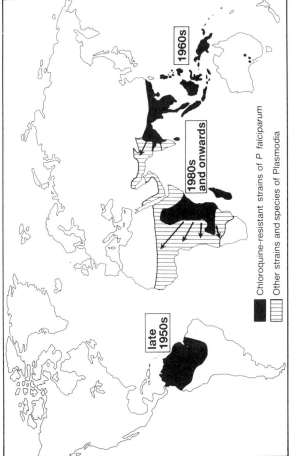

After GC Cook 1988 *Lancet* i 32

Tuberculosis: clinical syndromes

Primary TB The initial infection with *Mycobacterium tuberculosis*, often in childhood, is usually pulmonary (by droplet spread). A peripheral lesion forms, and its draining nodes are infected. There is an early spread of the bacilli throughout the body, but immunity rapidly develops, and the infection becomes quiescent at all sites. The commonest non-pulmonary *primary* infection is GI, most commonly affecting the ileocaecal junction and associated lymph nodes.

Post-primary TB Any form of immunocompromise may allow reactivation, eg malignancy; diabetes mellitus; steroids; debilitation, especially HIV or old age. The lung lesions progress and fibrose (usually upper lobe). Any other site may become the main clinical problem. In the elderly or Third World children, dissemination of multiple tiny foci throughout the body (including back to the lungs) results in miliary TB. Dissemination also occurs in the immunocompromised.

Clinical features

Primary TB is usually asymptomatic, but there may be fever, lassitude, erythema nodosum, cough, sputum, or phlyctenular conjunctivitis. AFBs (acid-fast bacilli) may be identified in sputum. CXR may help.

Post-primary TB is usually symptomatic, but often symptoms are non-specific. It may cause weight loss, anorexia, night sweats, and malaise. But there may be more organ-specific symptoms and signs.

- *Pulmonary TB* may cause cough, sputum, haemoptysis (may be massive), pneumonia, pleural effusion and chest pain. Again CXR and sputum are important. When a pleural effusion is tapped parietal pleural biopsy is taken at the same time as biopsy has a higher yield. A mycetoma (p340) may form in the cavities.
- *Miliary TB* is associated with a characteristic CXR. Clinical features may be non-specific. Look for retinal TB (*OHCS* p500). Biopsy of liver, bone marrow, lymph nodes, or lung parenchyma may yield AFBs or granulomata.
- *TB meningitis* See p442. Treatment of TB meningitis: *OHCS* p260.
- *Genitourinary TB* causes dysuria, haematuria, and frequency. Classically, there is a sterile pyuria (ie white cells but no growth). Take 3 early morning urine samples (EMUs) for AFBs. Renal ultrasound and IVU are very helpful. Renal TB may spread to bladder, seminal vesicles, epididymis, or the Fallopian tubes. For endometrial TB, see *OHCS* p36.
- *TB bone* usually affects the spine: adjacent vertebrae collapse (Pott's vertebra) in association with a paravertebral abscess. x-ray and biopsy. (See *OHCS* p644.)
- *TB peritonitis* is associated with abdominal pain and GI upset. Look for AFBs in ascites, but laparotomy may be required.
- *TB pericarditis* may present with effusion, tamponade, dry or constrictive pericarditis, or pericardial calcification.

Additional points in all those with TB ▶ Consider HIV.
▶ Ensure regular monitoring to check for compliance and drug toxicity.
▶ Notify the CCDC (consultant in communicable disease control), to arrange contact tracing, and screening (preferably by a chest physician).

Tuberculosis: clinical syndromes

TB with AIDS

This is a common, serious, but treatable complication of HIV infection (p214). It is estimated that 30–50% of those with AIDS in the developing world also have TB.[1] Interactions of HIV and TB are as follows:[2]

- Mantoux tests may be falsely negative.
- There is increased reactivation of latent TB.
- Presentation may be atypical.
- Previous BCG vaccination may not stop those with HIV from getting TB.[3]
- Smears may be negative for acid-fast bacilli. This makes culture all the more important—and is vital to characterize drug resistance.
- Those smears which are positive tend to contain less AFBs.
- Atypical CXR: lobar or bibasilar pneumonia, hilar lymphadenopathy.
- More pleural, pericardial, miliary, and meningeal involvement.
- There is good response to drugs, but they tend to be more toxic.
- Lifelong chemoprophylaxis with isoniazid is controversial,[3] but everyone agrees that regular clinical monitoring is important.

▶ Use full respiratory isolation procedures if TB patients are near HIV+ve patients. Serious nosocomial (ie hospital-acquired) outbreaks of multi-drug resistant TB are now problems in USA and France, and affect both HIV+ve and HIV−ve populations. Mortality is about 50%. Test TB cultures against first and second line chemotherapeutic agents.[4]

First line agents:		*Second line agents:*
Isoniazid	Streptomycin	Ofloxacin, Ciprofloxacin
Rifampicin	Amikacin	Cycloserine
Pyrazinamide	Kanamycin	Ethionamide
Ethambutol	Capreomycin	Aminosalicylic acid

1 M Merson 1992 *BMJ* i 209   2 British Thoracic Society Guidelines 1992 *BMJ* i 1231   3 E Ong 1992 *BMJ* i 1567   4 M D Iseman 1993 *NEJM* **329** 784.   See also: British Thoracic Society Guidelines for TB Treatment: L Oimerod 1990 *Thorax* **45** 403–8

Tuberculosis: diagnosis and treatment

Diagnosis *Microbiology:* Auramine-stained *Mycobacterium tuberculosis* resists acid-alcohol decolorization ('acid [–alcohol] fast bacilli', A[A]FBs). The clinical picture dictates the most useful specimens.

Histological or radiological support for the diagnosis may come from the characteristic pathology—the caseating granuloma (caseating means like soft cheese), calcification, and cavitation (forming cavities).

Immunological evidence of TB infection: The tuberculin test.
- This TB antigen is injected intradermally and the cell-mediated response at 48–72h is recorded.
- In the Mantoux test, 0.1ml of 1 in 10,000, 1 in 1000, and 1 in 100 dilutions provide 1, 10, and 100 tuberculin units (TU) respectively. The test is +ve if it produces ≥10mm induration, and −ve if it <5. Always state tuberculin concentration and mm of induration.
- The Heaf and Tine tests are for screening and consist of a circle of primed needles which inject the tuberculin.[1]
- A +ve tuberculin test proves only that the patient has immunity not active infection. It may indicate previous exposure or BCG. A strong positive more likely means active infection.
- If active TB is strongly suspected, use 1 TU. If it is +ve, infection is likely. Otherwise, interpret in the clinical context.
- False-negative tuberculin tests occur in immunosuppression, including miliary TB, sarcoid, AIDS, lymphoma.

Treatment of pulmonary TB *Before treatment*, stress the importance of compliance (checked by looking at a urine sample: rifampicin turns it orange-red). Assess liver and kidney function. Test colour vision before and during treatment as ethambutol may cause ocular toxicity which is reversible if stopped quickly.
- *Initial phase* (8 weeks on 3–4 drugs) **1** Rifampicin 600-900mg (child 15mg/kg) PO ac 3 times/wk. **2** Isoniazid 15mg/kg PO 3 times/wk. **3** Pyrazinamide—child: 50mg/kg; adult: 2.5g PO (2g if <50kg) 3 times/wk. **4** If resistance possible, add ethambutol 25mg/kg/24h PO (not children) or streptomycin 0.75–1g/24h IM (child 15mg/kg/24h). Do LFTs weekly.
- *Continuation phase* (4 months on 2 drugs) rifampicin and isoniazid at same doses. (2 Rifinah 300® tablets contain rifampicin 600mg and isoniazid 300mg.) As poor communities cannot buy much rifampicin (it is the best anti-TB drug), some régimes use twice weekly rifampicin (600mg) + isoniazid (900mg) to good effect.[2] If resistance is a problem, ethambutol 15mg/kg/24h PO may be used.
- Give pyridoxine 10mg/24h throughout treatment.

Main side-effects ▶ Seek help in renal or hepatic failure, or pregnancy.
Rifampicin: Hepatitis (stop if bilirubin rises; a small rise of AST is not significant), orange urine, contact lens staining, contraceptive steroid inactivation, flu-like syndrome with intermittent use.
Isoniazid: Neuropathy, hepatitis, pyridoxine deficit, agranulocytosis.
Ethambutol: Optic neuritis (colour vision is the first to deteriorate).
Pyrazinamide: Arthralgia (gout is a contraindication), hepatitis.

Preventing TB in HIV +ve people Consider isoniazid 300mg/day PO (children 5mg/kg, max 300mg) for 1yr—if the patient has not had BCG and the Mantoux test is >5mm (Heaf 1–4). If BCG vaccinated, only use prophylaxis if Mantoux >10mm (Heaf 3–4). If prophylaxis is not used, monitor clinical state and CXR.[3]

1 See DoH video (No. UK 6257, 5-day loans are free)   **2** A Castelo 1989 *Lancet* ii 1173
3 British Thoracic Society recommendation 1992 *BMJ* i 1231

Tuberculosis: diagnosis and treatment
[*OTM* 5.278]

Herpes infections

Varicella zoster Chickenpox (see *OHCS* p216) is the primary infection with this virus. After the infection, it remains dormant in dorsal root ganglia. Reactivation causes shingles. Shingles affects 20% adults at some time. High risk groups: the elderly or immunosuppressed.

Clinical features of shingles: Pain in a dermatomal distribution precedes malaise and fever by a few days. Some days *later* macules, then papules and vesicles, develop in the same dermatome. Lower thoracic dermatomes and the ophthalmic division of trigeminal nerve (ophthalmic shingles, *OHCS* p484) are most vulnerable. If the sacral nerves are affected, urinary retention may occur. Occasionally motor nerves may be affected, causing paralysis. If the conjunctiva is affected, beware iritis. Measure acuity often. Tell to report any visual loss at once (*OHCS* p484). Recurrence is suggestive of HIV.

Tests: Rising antibody titres; viral culture. Neither is usually needed.

Treatment: It is not clear if everyone with shingles needs acyclovir (to reduce viral replication)—but if it is going to be used, use it as early as possible: 800mg/5h PO for 5 days (if immunocompromised: 10mg/kg/8h by slow IVI for 7 days and monitor renal function). Control pain with oral analgesic or low-dose amitriptyline. There is evidence to support a 4-week course of prednisolone to reduce post-herpetic neuralgia. If the conjunctiva is affected, apply 3% acyclovir ointment 5 times a day. Beware iritis. Measure acuity often. Tell to report any visual loss at once (*OHCS* p484). SE of oral acyclovir: vomiting, urticaria, fatigue and encephalopathy.

Complication: Post-herpetic neuralgia, affecting the same dermatome, can persist for years, dominating and destroying the patient's life. It may be very difficult to treat. Try carbamazepine and then phenytoin. As a final step, ablation of the appropriate ganglion may be tried (often with no success) and it leaves the patient with an anaesthetic dermatome.

Herpes simplex virus (HSV) Manifestations of primary infection:

1 *Systemic infection*, eg with fever, sore throat and lymphadenopathy may pass unnoticed. If immunocompromised it may be life-threatening with fever, lymphadenopathy, pneumonitis, and hepatitis.

2 *Gingivostomatitis:* Ulcers filled with yellow slough appear on tongue and buccal mucosa.

3 *Herpetic whitlow:* The virus gains access to the digit through a breach in the skin and local irritating vesicles form.

4 *Traumatic herpes (herpes gladiatorum):* Vesicles develop at any site where HSV is ground into the skin by brute force.

5 *Eczema herpeticum:* HSV erupts on eczematous skin. It may be extensive.

6 *Herpes simplex meningitis:* This is uncommon and is self-limiting (typically HSV II in women during a primary attack).

Herpes infections
[OTM 5.59]

7 Herpes simplex encephalitis: (Usually HSV 1) There may be fever, fits, headaches, and personality change, or the patient may present in coma. High mortality. Seek expert help: admit to ITU; careful fluid balance to minimize cerebral oedema (consider mannitol, p734); use high dose (10mg/kg/8h IV for 10 days) acyclovir.

8 Genital herpes: ♂: Anus, penis shaft/glans develop small, grouped vesicles and papules (±palm, feet or throat involvement). Also pain, fever, dysuria. ♀: OHCS p30. Give analgesia+acyclovir, eg 200mg 5 times daily for 10 days. Recommend condoms. If frequent (eg ≥6/yr) or severe recurrences, try continuous acyclovir ≤400mg/12h PO.

9 HSV keratitis: These are corneal dendritic ulcers. See OHCS p480.

Investigations: Rising antibody titres in primary infection; viral culture. In HSV encephalitis, MRI (or CT) may show a temporal lobe cerebritis. LP with PCR on CSF for HSV allows rapid diagnosis where available. Brain biopsy now rarely performed—treat empirically.

Recurrent HSV may lie dormant in ganglion cells and be reactivated by eg any intercurrent illness, immunosuppression, menstruation, or even sunlight. Cold sores (perioral vesicles) are one manifestation—and the response to acyclovir cream is often disappointing.

203

Infectious mononucleosis (glandular fever)

This is a common disease of young adults (although patients may be of any age) which may pass unnoticed or cause acute illness—rarely followed by lethargy for months. It is caused by the Epstein–Barr virus (EBV), which preferentially infects B lymphocytes. There follows a proliferation of T-cells (the 'atypical' mononuclear cells—see below) which are cytotoxic to EBV-infected cells. The latter are 'immortalized' by EBV infection, and, very rarely, proliferate to form a picture indistinguishable from immunoblastic lymphoma in immunodeficient individuals whose supressor T-cells fail to check multiplication of these B-cells.

The incubation period is uncertain, but may be about 4–5 weeks. Spread is thought to be by saliva or droplet.

Symptoms and signs Sore throat, fever, anorexia, malaise, lymphadenopathy, petechiae on palate, splenomegaly, hepatitis, and haemolytic anaemia. Occasionally encephalitis, myocarditis, pericarditis, neuropathy. Rashes may occur—particularly if the patient is given ampicillin (this does not indicate lifelong ampicillin allergy).

Investigations Blood film shows a lymphocytosis with many atypical lymphocytes (eg 20% of all WBCs). Such cells may be seen (usually in fewer numbers) in many viral infections (especially CMV), toxoplasmosis, drug hypersensitivity, leukaemias, lymphomas, and lead intoxication.

Heterophil antibody tests (eg Monospot® or Paul–Bunnell): Heterophil antibodies develop in the early stages in most patients and disappear after about 3 months. These antibodies agglutinate sheep red blood cells. They can be absorbed (and thus agglutination is prevented) by ox red cells, but not guinea-pig kidney cells. This pattern distinguishes them from other heterophil antibodies. These antibodies do not react with EBV or its antigens. *False +ve* Monospot® tests occur (rarely) in hepatitis, lymphoma, leukaemia, rubella, malaria, carcinoma of pancreas, SLE.

Virology: EBV-specific IgM implies current infection, IgG shows that the patient has been previously infected.

Differential diagnosis Cytomegalovirus, viral hepatitis, HIV seroconversion illness, toxoplasmosis, leukaemia, streptococcal sore throat (may coexist), diphtheria.

Treatment Bed rest. Avoid alcohol. Consider prednisolone PO for severe symptoms or complications (80mg, 45mg, 30mg, 15mg, and 5mg on successive days, then stop).

Complications Low spirits, depression, and lethargy which may persist for months or even years. Also thrombocytopenia, ruptured spleen, upper airways obstruction (may need observation on ITU), secondary infection, pneumonitis, aseptic meningitis, Guillain–Barré syndrome, lymphoma and autoimmune haemolytic anaemia. All are rare.

Other EBV-associated diseases EBV (augmented by holoendemic malaria and HIV infection) is a cause of Burkitt's lymphoma (p608). EBV+ve large cell (B-cell) lymphomas occasionally appear in the immunocompromised.[1] The EBV genome has been found in the Reed–Sternberg cell of Hodgkin's lymphoma, and EBV is also associated with nasopharyngeal carcinoma and there are some pointers to its presence in SLE.

---

1 *Lancet* Ed 1989 i 1171

Infectious mononucleosis (glandular fever)

[*OTM* 5.72]

Influenza

This disease may be so mild as to pass unnoticed in some people, while in others—the elderly and the debilitated—it causes much mortality. In pandemics, millions have died. The virus (RNA orthomyxovirus) has 3 types (A, B, and C). Subtyping (for type A) is by haemagglutinin (H) and neuraminidase (N) characteristics.

Mutations occur readily, especially in A viruses, and new strains have differing antigenic properties. The WHO classification denotes type, host origin, geographical origin, strain number, year of isolation and subtype, eg A/Swine/Taiwan/2/87/(H3,N2).

Spread is by droplets. Incubation period 1–4 days. Infectivity 1 day before to 7 days after symptoms start. Immunity Those attacked by one strain are immune to that strain only.

Pathology An acute necrotizing (and even haemorrhagic) inflammatory process involves the upper and lower respiratory tract.

Clinical features After incubating for 48h there may be abrupt fever, malaise, headache and myalgia—also prostration, vomiting and depression. Convalescence may be slow.

Diagnosis This is easy in epidemics (but a fatal pitfall is not asking about travel, so missing malaria). Serology may help (rising titres over 2–3 weeks) if a firm diagnosis must be made. Culture is possible.

Complications Viral or secondary bacterial (esp. *Staph aureus*) bronchopneumonia, sinusitis, otitis media, encephalitis, pericarditis. There is some evidence from the 1957 epidemic suggesting that *in utero* infection is associated with later development of schizophrenia.

Treatment Bed rest and aspirin are the most common remedies. Amantadine 100mg/12h PO for 5 days markedly reduces symptoms in some outbreaks (H2,N2; H3,N2; H1,N1).[1] In those few who develop pneumonia (with or without secondary infection) most authorities recommend rapid transfer to ITU as dyspnoea and anoxia frequently progress very fast to circulatory collapse and death.[2]

Prevention Use whole *trivalent vaccine* (from inactivated viruses), reserving split (fragmented virus) for those <13yrs old. It is prepared from current serotypes. Indications: diabetes; chronic lung, heart or renal disease; immunosuppression; medical staff (in epidemics). Dose: 0.5ml SC once; children repeat after 4 weeks (½ dose if <3yrs old). SE: mild pain or swelling (17%). Guillain–Barré syndrome and pericarditis are rare. Fever, headaches and malaise are often reported (10%), but no more so than in those given a placebo injection.[3] 6 weeks of *amantadine* 100mg/24h PO can give good cover (eg 91% for H2,N3, and H1,N2 subtypes of A virus). SE: insomnia, nervousness. Uraemia and immunosuppression are reasons for vaccine failure—so consider oral prophylaxis.

Upper respiratory infections

The common cold (coryza) The rhinovirus is the commonest cause, with >80 distinct strains. It produces initially a watery nasal discharge which becomes mucopurulent over a few days. Incubation: 1–4 days. Complications: (eg 6% in children) otitis media, pneumonia, febrile convulsions. Treatments (eg cough suppressants, antibiotics) do not influence its course. If nasal obstruction in infants hampers feeding, 0.9% saline drops may help.

1 A Connolly 1993 *BMJ* ii 1452   2 L Van Voris 1981 *JAMA* 245 1128–31   3 T M Goveart 1993 *BMJ* ii 988

Toxoplasmosis

The protozoan *Toxoplasma gondii* infects man via broken skin, gut, or lungs. The definitive host, the cat, excretes oocysts, but eating poorly cooked infected meat is probably as important as contact with cat faeces. In the human, the oocysts release trophozoites, which migrate throughout the body, with a predilection for eye, brain and muscle. Toxoplasmosis is distributed world-wide, but is common in the tropics. Infection is life-long. HIV may reactivate it.

Clinical features The vast majority of infections are asymptomatic: in the UK about 50% have been infected by age 70. Symptomatic acquired toxoplasmosis resembles infectious mononucleosis, and is usually self-limiting lasting a few months. Ocular infection, usually congenital, presents with posterior uveitis causing blurring and ache, often in the second decade of life. Encephalitis and myocarditis occur only rarely, except in the immunocompromised host (eg AIDS), where it may be fatal.

Diagnosis 4-fold rise in antibody titre over 4 weeks or presence of specific IgM implies acute infection (unreliable if HIV +ve). The 'dye test' was the first serological test used. Parasite isolation is difficult, but lymph node or brain biopsy may be diagnostic.

Treatment Often self-limiting without treatment. If the eye is involved, or in the immunocompromised use pyrimethamine 25–50mg/8h PO for 5 days then 25–50mg/24h PO for 4 weeks, plus sulphadiazine 4–6g/24h PO (alone if pregnant).

Congenital toxoplasmosis (OHCS p98) This is acquired transplacentally and may induce abortion. Alternatively, any combination of seizures, chorioretinitis, hydrocephalus, microcephaly, and cerebral calcification may occur. Early infection has a worse prognosis.

Cytomegalovirus (CMV)

CMV may be acquired eg by kissing, sexual intercourse, blood transfusion, or organ transplants. Like other herpes viruses, it lies latent after acute infection for the rest of the patient's life, and may reactivate at time of stress or immunocompromise.

Clinical features The vast majority of CMV infections pass unnoticed. 50% women of child-bearing age have been infected. Symptomatic CMV infection resembles infectious mononucleosis. It is a more serious infection in the immunocompromised (eg transplant recipients or AIDS), when it may cause pneumonia, CNS infection, chorioretinitis, colitis, hepatitis, leucopenia, or thrombocytopenia.

Diagnosis Isolation and growth of the virus is slow. Also there may be prolonged CMV excretion following distant infection, making diagnosis of acute CMV infection difficult. Serology is much more helpful. Specific IgM implies acute infection (unreliable if HIV +ve).

Treat serious infections with ganciclovir 5mg/kg/12h IVI over 1h. Active and passive immunization is being explored. Alternative: foscarnet, p214.

Toxoplasmosis

[*OTM* 5.506]

Cytomegalovirus (CMV)

[*OTM* 5.76]

Prevention post-transplantation If seropositive pre-op, ganciclovir (5mg/kg/12h IV for the 1st 2 post-op weeks, then 6mg/24h for 5 days a week for 2 more weeks) may reduce infection rates by 4–5-fold.[1]

NB: be sure to use CMV-free blood or filtered blood when transfusing transplant recipients, leukaemia or HIV patients.

Congenital CMV (*OHCS* p98) The neonate may suffer acute jaundice, hepatosplenomegaly, and purpura. Chronic defects include mental retardation, cerebral palsy, epilepsy, deafness, and eye problems.

209

1 T Merigan 1992 *NEJM* **326** 1182

The A to E of viral hepatitis

Hepatitis A is spread by the faecal–oral route, often in institutions or travellers. Most infections pass unnoticed in childhood.

Clinical features: 2–6 weeks' incubation and a prodrome of anorexia, nausea, joint pain, and fever, precede the jaundice. Tender hepatomegaly, splenomegaly, and lymphadenopathy may occur.

Investigations in hepatitis A infection: Serum transaminases rise 22–40 days after exposure. Specific antibodies rise from 25 days: IgM signifies recent infection, and IgG remains detectable for life.

Treatment: Supportive. Avoid alcohol until LFTs normal.

Prevention: After infection immunity is probably lifelong. Normal human immunoglobulin (0.02ml/kg IM) gives <3 months' passive immunity to those at risk (eg travellers; *all* household contacts) and during incubation.

Hepatitis A vaccination with Havrix® (an inactivated protein vaccine from hepatitis A virus, grown in human diploid cells) Dose: 2 IM doses (into deltoid), 2–4 weeks apart, if >16yrs old. This lasts 1 year (up to 10yrs if a booster is given at 6 months). Use Havrix Junior® if 1–15 yrs old.

Prognosis: Usually self-limiting. (Rarely fulminant.) There is no carrier state, and chronic liver disease does not occur.

Hepatitis B *Spread:* Blood products, secretions, and sexual intercourse. Risk groups: homosexuals, haemophiliacs, health workers, heroin (or other IV drug) abusers, haemodialysis, and those in 'homes'/institutions— eg prison (the 6 Hs). Endemic in tropics and Mediterranean area.

Clinical features: As above, with 1–6 months' incubation. Urticaria and arthralgia are more common with hepatitis B.

Tests: HBsAg (surface antigen) is present from 1–6 months after exposure. The presence of HBeAg (e-antigen) implies high infectivity; it is usually present for 1½-3 months after the acute illness. The persistence of HBsAg for >6 months defines carrier status. This follows 5–10% of infections. Antibodies to HBcAg (core antigen, ie anti-HBc) imply past infection; to HBsAg (ie anti-HBs) alone implies vaccination.

Vaccination may be universal or just for high-risk groups (see p512). Passive immunization with specific antihepatitis B immunoglobulin may be given to non-immune contacts after high-risk exposure.

Treatment: Supportive. Use local infection-control rules. No alcohol. Chronic hepatitis B may respond to IFN-α. Immunize contacts.

Prognosis: Usually good; 5–10% develop chronic hepatitis which may lead to cirrhosis±liver carcinoma. Rare complication: glomerulonephritis.

Hepatitis C is spread similarly to HBV. ~50% develop chronic hepatitis; 5% develop cirrhosis and 15% of these risk hepatocellular carcinoma.[1] Antibody tests become +ve after 3–6 months, and do not indicate if the patient is infective or infected. Newer tests (eg c200 and c22-3 antigens, ± 5-1-1, c-100-3, c33c,) are +ve earlier.[2] Transfusion blood is screened for HCV—offer HCV +ve donors expert advice about the need for liver biopsy. Interferon (IFN-α, eg 3,000,000U, 3 doses/week SC for 6 months) suppresses chronic infection. Monitor with ALT ± PCR. Retreatment is possible for the 50–80% whose remission is temporary[1] (£1300/course). Gammaglobulin provides some protection before unscreened transfusions. As HCV infection is usually subclinical, generalized screening has been suggested.[3]

Markers of hepatitis B virus infection

Markers of HBV infection	Acute infection [marker peak, mths from infection]	Carriers: Infectivity high	Carriers: Infectivity low	Past infection
BeAg	Present early then absent [4½]	Present*	Present*	Absent
BsAg	Present [5]	Present	Present	Absent
anti-HBc IgM	High titre [7]	Absent or ↓	Absent	Absent
anti-HBc IgG	Moderate titres	High titres	Moderate	Moderate
anti HBe	Absent early [9]	Absent	Present	Present
HBs	Absent [>10]	Absent	Absent	Absent
DNA polymerase	Present	Present	Absent	Absent
LFTs	↑↑↑	↑	↔	↔

* If e Ag (a crude marker of viral replication) and anti-HBe are both undetectable, this counts as high infectivity.

Hepatitis D (delta agent) is an incomplete RNA virus which only exists in the presence of hepatitis B, and is spread by the same routes. It increases the incidence of both fulminant hepatic failure and cirrhosis. A serological test is available. Interferon (IFN-α) has limited success in treatment of delta infection. Prevention: as for hepatitis B.

Hepatitis E is similar to hepatitis A in transmission and clinical features with no apparent risk of chronic liver disease. In some areas it is the commonest viral cause of hepatitis in adults and older children. A high mortality (20–39%) has been noticed in pregnancy. Diagnostic tests are being developed. The virus is detectable in stools by electron microscopy. Prevention: none (gammaglobulin is ineffective; no vaccine).

Differential diagnosis *Acute hepatitis:* Drugs, toxins, alcohol, infectious mononucleosis, Q fever, leptospirosis, syphilis, malaria, yellow fever, obstructive jaundice. *Chronic hepatitis:* Alcohol, drugs, chronic active and persistent hepatitis, Wilson's disease.

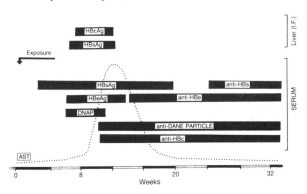

Virological events in acute hepatitis B in relation to serum amino-transferase (AST) peak. I.F.=immunofluorescence; Ag=antigen; HBs=hepatitis B surface; HBe=hepatitis B e; DNAP=DNA polymerase.

1 *Drug Ther Bul* 1993 **31** 61   **2** J Lau 1993 *BMJ* i 469   **3** C Seymour 1994 *BMJ* i 670

Human immunodeficiency virus (HIV)

AIDS (defined on p214) represents the end of a spectrum associated with HIV infection and immunodeficiency. The US Centre for Disease Control (CDC) asserts that HIV positivity with low CD4 lymphocyte count (<200 ×10⁶/l) is sufficient for diagnosis of AIDS. HIV-1 (a retrovirus) is responsible for most cases world-wide, with the related HIV-2 producing a similar illness, perhaps with a longer latent period. ~11 million people are HIV+ve, and ~3 million have AIDS; most are women and children in sub-Saharan Africa. In the UK, incidence is now *not* rising (~30/10⁶/yr).

Transmission is by sexual intercourse (vaginal and anal: 75% of cases worldwide), infected blood or blood products and from mother to fetus. (Screening of blood donors and sterilization of blood products reduce risks from these to minimal levels in developed countries.) In the developing world, heterosexual spread is commonest with equal numbers of men and women affected, and increasing numbers of infected children born to seropositive mothers (vertical transmission, OHCS p98).

Pathology HIV binds to CD4 receptors on helper T-lymphocytes (CD4 cells), monocytes and neural cells. After a long latent period (8–10 years) CD4 cell counts begin to decline, and the immunosuppression increases risk of many opportunistic infections and tumours. HIV also causes direct CNS and renal effects (eg dementia, glomerulosclerosis).

  Billions of virions are produced and destroyed each day in HIV infection—and the daily turnover of infected CD4 cells is also in the billions.

Diagnosis is based on the detection of anti-HIV antibodies in serum. The majority of infected individuals develop antibodies within 3 months of exposure. Detection of viral antigen may provide diagnosis during sero-conversion illness, as may newer techniques including detection by PCR.

Preventing HIV infections ● Promote *life-long* safer sex, barrier contraception and reduction in the number of partners. Videos, followed by discussions, is one way to double the use of condoms.
● Warn heterosexuals about the dangers of sexual tourism/promiscuity.
● Tell drug users not to share needles. Use needle exchange schemes.
● Vigorous control of other STDs can reduce HIV incidence by 40%.
● Strengthen awareness of clinics for sexually transmitted diseases.
● Reduce unnecessary blood transfusions.
● Encourage pregnant women to take an HIV test. (Caesarean sections and zidovudine during birth prevents vertical transmission, OHCS p98).

Counselling *Pre-HIV testing*: Take time to interview the patient to:
● Discuss implications if +ve (job, insurance, mortgage, and for family).
● Determine the level of risk.
● Do the test only if consent is informed.
● Fix post-test counselling (eg to re-emphasize ways to ↓ risk exposure).
Counselling throughout HIV illness: Psychologists may be best placed to do initial counselling; get in expert help as needed. A key issue when a couple are dying from HIV is guardianship of their children; a lawyer's help may be needed here, and also with housing and employment. Drawing up advance directives also needs special skill. Later in the illness, hospices and domiciliary teams from departments of sexual health (genito-urinary medicine) help with terminal care. The GP also has a central rôle.

The human immunodeficiency virus (HIV)

(Seroconversion rate: ~0.4% for HIV; 30% for Hepatitiis B—if HBeAg+ve)

- Wash well; encourage bleeding; do not suck or immerse in bleach.
- Note name, address and clinical details of the 'donor'.
- Report incident to Occupational Health. Fill in an accident form.
- Store blood from both parties. If possible, ascertain HIV and HBsAg of both. Immunize (active and passive) against hepatitis B at once, if needed. Counsel (HIV risk <0.5% if 'donor' is HIV +ve) and test recipient at 3, 6 and 8 months (seroconversion may take this long). **213**
- Antiviral drugs (eg AZT) are controversial; question local guidelines.[1]

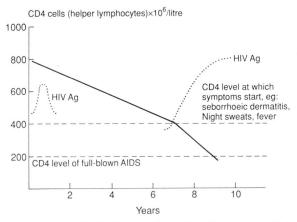

Rough time-course for the detection of HIV antigen (Ag) and the depletion in absolute numbers of CD4 cells.

1 J-P Aboulker 1993 *Lancet* **341** 889

Clinical syndromes of HIV infection

Infection with HIV can be asymptomatic or associated with a variety of clinical syndromes.

- **Group I: Acute seroconversion illness** A self-limiting, glandular fever-like illness: ~6wks post infection. Oral ulcers, encephalopathy, myelopathy and meningitis may occur. Patient is highly infectious.
- **Group II: Asymptomatic infection** Most (~60%) patients will remain fit and well for up to 10yrs after initial infection. Towards the end of this latent period some develop minor skin problems eg seborrhoeic dermatitis, psoriasis, folliculitis and oral hairy leucoplakia (CD4 count ↔ or starting to fall below $400 \times 10^6/l$).
- **Group III: Persistent generalized lymphadenopathy** Defined as nodes >1cm diameter at $\geqslant 2$ extra-inguinal sites and persisting for 3 months or longer. Biopsy if there are symptoms or obvious asymmetry. Most otherwise asymptomatic (CD4 count ↔ or just $<400 \times 10^6/l$).
- **Class IV:** Includes AIDS [Case definition: development of one or more conditions from a defined list of opportunistic infections and tumours for which there is *no other explanation*, in the presence of evidence of HIV infection]. Subgroups (A to E) for constitutional disease, CNS disease, opportunistic infections and tumours including lymphomas and Kaposi's sarcoma (CD4 count usually $<200 \times 10^6/l$).

Opportunistic infections

- **Oropharynx**—*Candida albicans* affects 80% of patients at some time during their illness. Sore mouth, angular cheilitis, dysphagia (if oesophagus involved). Initially use nystatin suspension or amphotericin lozenges; easier treated systemically (eg fluconazole 50mg/24h PO for 14 days). It is likely to relapse so give standby courses.
- **Pulmonary**—*Pneumocystis carinii*: ▶ p334. If CD4 $<200 \times 10^9/l$ give prophylaxis with co-trimoxazole 980mg/24h PO, dapsone or pentamidine. *M tuberculosis:* Multi-drug resistance is a problem: ▶ See p199–200. Late problems: fungi (*Aspergillus* or *Cryptococcus*); nocardia. CMV pneumonitis—use ganciclovir 5mg/kg/12h IV. In documented ganciclovir resistance use foscarnet 20mg/kg IV over 30mins, then 21– 200mg/kg/24h IVI according to renal function for 2–3 weeks (SE: rash vomiting, rash, headache, renal failure, fits, $Ca^{2+}\downarrow$, hypoglycaemia).
- **CNS** *Toxoplasma gondii* is the commonest cause of focal CNS disease in AIDS, producing multiple brain abscesses visible as ring-enhancing lesions on CT or MRI. Treat with pyrimethamine 25–50mg/24h + sulphadiazine 100mg/kg/24h PO for 6 months. (Clindamycin is an alternative agent.) Follow by lifelong secondary prophylaxis. *Cryptococcus neoformans* (p244) causes an insidious meningitis in up to 8% of patients with advanced HIV disease. Fluconazole IV or PO may be an alternative to amphotericin/flucytosine and lifelong secondary prevention with fluconazole 100–200mg/day PO is needed. CMV causes a necrotizing retinitis, presenting insidiously with visual deterioration. Early diagnosis and treatment with IV ganciclovir or foscarnet, continued indefinitely, are needed to prevent progression.
- **Gut** Chronic diarrhoea due to *Cryptosporidium* can be debilitating—no specific treatment. Other causes include *Isospora belli* (use cotrimoxazole), *Microsporidium*, CMV colitis (abdominal pain suggestive), *Mycobacterium avium* complex and HIV enteropathy.

- PUO Important, treatable PUOs in HIV having no clinical focus may be from mycobacteria, pneumococci, salmonellae, fungi or toxoplasma.

Tumours and HIV
- Lymphomas are usually non-Hodgkin's of B-cell type, occurring at any site, predominantly small bowel and CNS. Usually aggressive and respond poorly to treatment. • Kaposi's sarcoma (p702).

Direct HIV effects include seroconversion illness (above), small bowel enteropathy ± partial villous atrophy (weight ↓; diarrhoea; malabsorption), CNS illness (dementia; myelopathy; neuropathy, often sensory and asymmetrical). May respond temporarily to zidovudine. Renal effects: fibrosis, glomerulosclerosis, mesangial proliferation, tubule dilation.

Anti-HIV agents It appears that the best results are obtained with combinations of agents such as zidovudine (AZT), ddi (didanosine), zalcitabine or lamivudine[1] (resistance and other problems are common). Confer with experts. New agents to consider adding into regimens are selective protease inhibitors such as ritonavir (600mg/12h PO with food).[2] HIV protease mediates the final development of HIV, and inhibiting it slows cell-to-cell spread, and lengthens the time to the first clinical event.

▶ In many ways *all* anti-HIV treatment is experimental—so perhaps the best question to ask is not *What is the best treatment for HIV?*—but *Which is the most appropriate trial to enter my patient into?* Discuss these issues with your patient and consultants in infectious diseases.

Note on using ddi, and cautions: The rôle of ddi (didanosine) in combination with AZT in symptomatic HIV infection is being elucidated. It also has a rôle if AZT is not tolerated. Its use for chemoprophylaxis after needle-stick injury has not been demonstrated.[3] *Cautions:* Be alert to:[4]
- Pancreatitis: (possible great danger here). Stop if plasma amylase rises. It may be safe to restart very cautiously once pancreatitis has been excluded, and the amylase is normal, and beware concurrent treatment with pentamidine isethionate and other pancreatoxic agents. Triglycerides are also pancreatoxic, so check these too.
- Liver failure (monitor LFTs, stop if significant elevation).
- Renal failure. • Peripheral neuropathy. • Hyperuricaemia.

Avoid ddi if: Breast-feeding; also on tetracycline (at same time of dose).

Side-effects of ddi: Pancreatitis; peripheral neuropathy; hyperuricaemia (suspend treatment if measures to reduce uric acid concentration fail); diarrhoea; vomiting, confusion; insomnia; fevers; headache, pain, rash, pruritus, seizures. *Adult dose example:* If body weight <60kg, 125mg/12h PO. If ≥60kg: 200mg/12h. Chew thoroughly—or crush and disperse in water (± clear apple juice for flavour).

215

1 Delta committee 1996 *Lancet* 348 283 2 H Nelson 1996 *Lancet* 347 383
3 *Lancet Reviews* 1995 346 s12 & BA Larder 1995 *Science* 269 696–9 4 BNF 1996 (i)

Sexually transmitted disease (STD)

▶ Refer early to Genito-Urinary Medicine Clinic (GUM clinic), for comprehensive microbiology and partner notification. Some clinics provide an on-call service for out of hours advice and most will see patients immediately during the working day. Avoid giving antibiotics until seen in GU clinic or at least discussed.

Presentation
1 *Genital ulcers and lumps* (p218);
2 *Urethritis/dysuria* and *urethral discharge* (p220);
3 *Vaginal discharge* and *vulvovaginitis:* Candida, trichomoniasis, bacterial vaginosis, cervical polyp, atrophic vaginitis (post/peri-menopausal);
4 *Inguinal lymphadenopathy* and *ulceration: Lymphogranuloma venereum* (LGV), chancroid and granuloma inguinale.

History Sexual contacts, timing of last intercourse; duration of relationship; sexual practices and orientation; contraceptive method; contact history; past STD, menstrual and medical history; antimicrobial therapy.

Examination Detailed examination of genitalia including inguinal nodes and pubic hair for lice/nits. Scrotum, subpreputial space and male urethra. PR and proctoscopy (if indicated); PV and speculum examination.

Investigation Ideally refer to GU clinic. Ulcers—HSV cultures (viral transport medium) and dark ground microscopy of smears for *T pallidum*.
Male patients: urethral smear for Gram stain and culture for *N gonorrhoeae* (send quickly to lab in Stuart's medium). *Female patients:* high vaginal swab in Stuart's for microscopy and culture (*Candida albicans*, *Gardnerella vaginalis*, anaerobes, *Trichomonas vaginalis*—may need specific culture medium depending on local facilities); take endocervical swab for *Chlamydia trachomatis* (special medium required). Urine dipstix analysis and send MSU if symptoms suggest cystitis.
Blood tests: Syphilis serology, hepatitis B in high risk groups (homosexual/bisexual men, injecting drug users (IDU) and those from countries where it is endemic), HIV antibody testing may be indicated—ideally leave until patient is seen in GU clinic for counselling.

Symptoms	Causes
● Dysuria/discharge (male)	*C trachomatis*/*N gonorrhoeae*
● Epididymo-orchitis	<35 years *C trachomatis* commonest >35 *E coli*, other coliforms, *N gonorrhoeae*
● Vaginal discharge	*C albicans*, bacterial vaginosis, *T vaginalis*
● Vulvo-vaginitis	*C albicans*, *T vaginalis*
● Pelvic pain/PID	*C trachomatis*, *N gonorrhoeae*.
● Genital ulcers	*Herpes simplex virus*, syphilis, tropical STD. Non-infective eg aphthosis, drugs, Behçet's, carcinoma

The STDs (Also *Syphilis* p218, *gonorrhoea* p220, *Herpes simplex* p202, HIV p212–14, *hepatitis B* p210)

Chlamydia trachomatis	Commonest cause of non-specific or non-gonococcal urethritis (NSU/NGU). Usually asymptomatic in women but if untreated can cause pelvic inflammatory disease (PID) and infertility. Now also the commonest cause of ophthalmia neonatorum (*N gonorrhoeae* less common). In the tropics causes evanescent genital ulcer and persistent inguinal lymphadenopathy (*Lymphogranuloma venereum*). May be confused with inguinal abscess. Avoid surgery as causes chronic sinus formation. Treat with oxytetracycline 500mg/12h × 14 days in uncomplicated infection, doxycycline 100mg/12h or minocycline 100mg/12h × 14 days for epididymo-orchitis or PID. Erythromycin 500mg/12h × 14 days where there is a pregnancy risk. Trace and treat partners.
Human papilloma (HPV)	Genital warts (condylomata acuminata)—2nd commonest STD in the UK. Associated with cervical smear abnormalities in women—(recommend annual smears for at least 5 years from diagnosis). Treat initially with topical podophyllin once weekly, refer to GU clinic if no response in 3–4 weeks. Full STD screen at diagnosis. See OHCS p30.
Trichomonas vaginalis	Flagellate protozoan, common cause of vulvo-vaginitis and vaginal discharge. Usually asymptomatic in men. Full STD screening with tracing and treatment of partners is indicated. Treat with metronidazole 400mg/12h × 5 days or 2g PO stat. Nimorazole 2g PO stat with food if metronidazole ineffective.
Candida albicans	Common cause of white vaginal discharge and vulval itch. Treat women with single dose pessary eg econazole 150mg or clotrimazole 500mg and topical cream.
Bacterial vaginosis (*Gardnerella*)	Common cause of vaginal discharge with malodour. Diagnose by microscopy. Treat with metronidazole 400mg/12h × 5d PO.
Phthirius pubis (crab louse)	Pubic lice (nits are the eggs). Usually identified by the patient; confirm by microscopy. Treat with Quellada®. Patients need full STD screening with contact tracing.
Sarcoptes scabii	Scabies causes a papular, intensely itchy, rash. Look for burrows in finger clefts. Treat patients and contacts by covering from the neck down with malathion 0.5% liquid, after a bath. Not exclusively sexually acquired. See OHCS p600.
Molluscum contagiosum	(Pox virus.) Small umbilicated white papules. No treatment needed.
Haemophilus ducreyi	Tropical, soft, often multiple painful genital ulcers; inguinal lymphadenopathy (buboes) may ulcerate. Depending on local resistance, treat with co-trimoxazole, erythromycin or co-amoxyclav.
Calymmatobacterium granulomatis (*Donovania*)	(*Granuloma inguinale*) Painless tropical condition; red papules in the genital and inguinal creases form granulomatous ulcers. Use tetracycline 500mg/6h × 21 days PO.

Syphilis

▶ Any genital sore is syphilis until proven otherwise.

The spirochaete *Treponema pallidum* enters via a skin abrasion, during sexual contact. All features are due to endarteritis obliterans. It is commonest in homosexuals—but by trying to avoid AIDS their risk is lessening. Incubation: 9–90d. 4 stages:

Primary syphilis A macule at site of sexual contact becomes a very infectious, painless hard ulcer (*primary chancre*).

Causes of genital sores: Friction burns, zips (OHCS p740), herpes simplex, scabies, Behçet's, cancer, chancroid, *Lymphogranuloma venereum*, granuloma inguinale.

Secondary syphilis Occurs 4–8wks after chancre. Fever, malaise, lymphadenopathy, rash (trunk, palms, soles), alopecia, condylomata lata (anal papules), buccal snail-track ulcers; rarely hepatitis, meningism, uveitis, nephrosis.

Tertiary syphilis follows 2–20yrs latency (when patients are non-infectious): there are *gummas* (granulomas occuring in skin, mucosa, bone, joints, rarely viscera eg lung, testis).

Quaternary syphilis 1 *Cardiovascular:* Ascending aortic aneurysm ± aortic regurgitation. 2 *Neurosyphilis: a Meningovascular:* Cranial nerve palsies, stroke; *b General paresis of insane* (GPI): Dementia, psychoses (fatal untreated; treatment stops decline and may reverse it); *c Tabes dorsalis:* Sensory ataxia, numb legs, chest, and bridge of nose, lightning pains, gastric crises, reflex loss, flexor plantars, Charcot's joints (p476). *Argyll Robertson pupil* (p56).

Serology (2 types): 1 Cardiolipin antibody: detectable in primary disease but wane in late syphilis. They indicate active disease and become negative after treatment. *False +ves* (with negative treponemal antibody): pregnancy, immunization, pneumonia, malaria, SLE, TB, leprosy. *Examples:* VDRL (venereal disease research laboratory slide test), RPR (rapid plasma reagin), WR (Wasserman reaction).
2 Treponeme-specific antibody: positive during primary disease, and remain so despite treatment. *Examples:* TPHA (*T pallidum* haemagglutination test), FTA (fluorescent treponemal antibody test), TPI (*T pallidum* immobilization test). These tests do not distinguish syphilis from the non-venereal yaws, bejel, pinta.

Tests In 1° syphilis, treponemes are demonstrated by dark ground microscopy of chancre fluid; serology at this stage is often −ve. In 2° syphilis, treponemes are seen in the lesions and both types of antibody tests are positive. In late syphilis, organisms may no longer be seen, but both types of antibody test usually remain +ve (cardiolipin antibody tests may wane). In neurosyphilis, CSF antibody tests are +ve. If HIV+ve, serology may be −ve during syphilis reactivation.

Treatment Procaine penicillin 600mg/24h IM for 10 days (tertiary or CVS disease: 14–21 days; CNS disease: 14 days). Beware the Jarisch–Herxheimer reaction (p232), commonest in secondary disease; most dangerous in tertiary. Consider steroid cover. Trace contacts.

Congenital syphilis OHCS p98.

Syphilis
[*OTM* 5.386]

Urethritis

Gonorrhoea *Neisseria gonorrhoea* (gonococcus; GC) can infect any columnar epithelium—urethra, cervix, rectum, pharynx, conjunctiva, but not vagina (which is squamous). Incubation: 2–10d.

Clinical features: similar to *C trachomatis* (p217). (50% have both.) Men: purulent urethral discharge and dysuria; or proctitis, tenesmus and discharge PR in homosexuals. Women: usually asymptomatic, but may have vaginal discharge, dysuria, proctitis. Pharyngeal disease is usually asymptomatic.

Complications—Local: Prostatitis, cystitis, epididymitis, salpingitis, Bartholinitis. *Systemic:* Septicaemia, presenting with petechiae, pustules on hands and feet, arthritis (metastatic or sterile), endocarditis (rare). *Obstetric:* ophthalmia neonatorumND (OHCS p100). *Long term:* Urethral stricture, infertility.

Treatment of gonorrhoea: 1 dose of amoxycillin 3g PO+probenecid 1g PO, or procaine penicillin 2.4g (4.8g for ♀) IM+probenecid 1g PO. Trace contacts. No intercourse or alcohol until cured. For penicillin resistance use ofloxacin 400mg PO stat or ciprofloxacin 250mg stat PO. β-lactamase producing strains: cefuroxime 1.5g IV + probenecid 1g PO. Penicillin allergy: co-trimoxazole (1.92g/12h PO) for 2d or ciprofloxacin.

Non-gonococcal urethritis (NGU) is commoner than GC. The discharge tends to be thinner, and the presentation less acute, but this does not help diagnosis in the individual. Women (again typically asymptomatic) may have cervicitis, urethritis or salpingitis (pain, fever, infertility). Rectum and pharynx are not infected.

Organisms: *Chlamydia trachomatis*, *Ureaplasma urealyticum* a kind of Mycoplasma; *Mycoplasma genitalium*, *Trichomonas vaginalis*, *Gardnerella vaginalis*, Gram negative and anaerobic bacteria, *Candida*.

Complications: The local but not the systemic complications of GC. *Chlamydia* cause Reiter's syndrome and neonatal conjunctivitis.

Treatment of NGU: 1 week of oxytetracycline 250mg/6h PO 1h ac, or doxycyline 100mg/12h PO for 1 week. Trace contacts. Avoid intercourse during treatment and alcohol for 4 weeks. Treatment failures: erythromycin 500mg/12h PO (1 week). *Trichomonas* or *Gardnerella*: metronidazole 400mg/12h PO (1 week).

Differential diagnosis of urethritis Traumatic and chemical urethritis, stricture, carcinoma, foreign body.

Investigation of urethritis Patients should not have passed urine for 2h. *Male:* Microscope urethral discharge and scrapings immediately (for gonococci and *Trichomonas*) and culture both. Culture from rectum and pharynx using charcoal swab. Inoculate immediately or store in Stuart's medium. *Female:* Urethral smears do not help. Take a high vaginal swab and scraping for *Candida*, *Trichomonas* and *Chlamydia*, and an endocervical sample for gonococci (smear and culture). Culture also from urethra, rectum and pharynx. *Both:* Check syphilis serology, HBsAg and HIV (if the patient agrees). Blood cultures in septicaemia and arthritis (and culture joint aspirate).

Most causative organisms are seen or grown. But multiple infection is common and may explain treatment failure in gonorrhoea.

Follow-up Arrange to see patients at 1 week and 3 months, with repeat smears, cultures, and syphilis serology.

Urethritis
[*OTM* 5.409]

Miscellaneous Gram +ve bacterial infections

Staphylococci When causing infections these are usually *Staph aureus*. Most commonly, they cause localized infection of the skin or wounds, but septicaemia also occurs. Deep infections with *Staph aureus*: cavitating pneumonia, osteomyelitis, septic arthritis, endocarditis. These infections are usually acute and severe. The production of β-lactamase which destroys many antibiotics (p176–182) is the main problem in treatment. *Staph aureus* toxins may cause food poisoning (p190) or the toxic shock syndrome (fever, a rash with desquamation, diarrhoea, and shock, which has been associated with the use of infected tampons). *Staph epidermidis* (*albus*) is increasingly recognized as a pathogen in the immunocompromised, particularly in connection with IV lines or any prosthesis. It is often enough to remove the infected line, and it is often multiply resistant to antibiotics. In other circumstances, when isolated from a culture, *Staph epidermidis* can usually be assumed to be a contaminant (unless present in several blood cultures, or in neonates).

Streptococci Group A streptococci (*Strep pyogenes*) are the commonest pathogens, causing wound and skin infections (eg impetigo, erysipelas, OHCS p590), tonsillitis, scarlet fever[ND], and septicaemia. Late complications are rheumatic fever and acute glomerulonephritis. *Strep pneumoniae* (the pneumococcus—in clinical specimens is a diplococcus) causes pneumonia, exacerbations of COAD, otitis media, meningitis, septicaemia, and, rarely, primary peritonitis. *Strep sanguis, mutans* and *mitior* (of the 'viridans' group), *bovis*, and *Enterococcus faecalis* all cause endocarditis. *Enterococcus faecalis* also causes UTI, wound infections, and septicaemia. *Strep mutans* is a very common cause of dental caries. *Strep milleri* forms abscesses in internal organs, especially CNS, lungs, and liver. Most streptococci are sensitive to the penicillins, but *Ent faecalis* and *Ent faecium* may present some difficulties. They usually respond to a combination of ampicillin and an aminoglycoside, eg gentamicin, p180–81.

Anthrax[ND], caused by *Bacillus anthracis*, is very rare. It is usually contracted by contact with contaminated carcasses. Commonest is a localized cutaneous 'malignant pustule'. Generalized oedema is a striking feature of the disease. There is fever ± hepatosplenomegaly. Inhaled and swallowed spores may cause pulmonary and GI anthrax respectively—with massive GI haemorrhage and acute breathlessness. The Gram stain is characteristic (Gram +ve rods). Treat with parenteral penicillin.

Diphtheria[ND] is caused by the toxin of *Corynebacterium diphtheriae*. It usually starts with tonsillitis with a false membrane over the fauces. The toxin may cause polyneuritis, often starting with the cranial nerves. Shock may occur from myocarditis, toxaemia, or disease of the conducting system. Treat with antitoxin (OHCS p276) and erythromycin. Treat contacts with erythromycin. Vaccination has been very effective in the UK. Consider diphtheria in the differential diagnosis of skin lesions in recent visitors to the tropics. There is now much diphtheria in Moscow and St Petersburg. Those born before 1942 visiting Russia or the Ukraine who have *close contact* with local people need a primary *low dose* course (0.1ml IM of the single-antigen paediatric vaccine from Evans Medical Ltd—3 doses, at monthly intervals). Offer a booster (0.1ml IM) to those whose primary immunization was >10yrs ago. Explain that it may not work. NB: a new low-dose formulation is becoming available (Merieux). Material for Schick tests is no longer available.

Miscellaneous Gram +ve bacterial infections

Listeriosis is caused by *Listeria monocytogenes*. It is a Gram +ve bacillus with an unusual ability to multiply at low temperatures. Possible sources are pâtés, milk, raw vegetables and soft cheeses (brie, camembert, and blue vein types). Person-to-person spread has never been reported (except in neonates). It may cause a non-specific flu-like illness, pneumonia, meningoencephalitis, or PUO, especially in the immunocompromised (eg pregnancy, where it may cause miscarriage or stillbirth) and neonates. *Diagnosis:* culture blood, placenta, amniotic fluid, CSF, and any expelled products of conception. ▶ Do blood cultures in any pregnant patient with unexplained fever for ≥48h.[1] Serology, vaginal and rectal swabs do not help (it may be a commensal here). *Treatment:* Ampicillin IV (erythromycin if allergic). *Prevention in pregnancy:* ● Avoid soft cheeses, pâtés and under-cooked meat. ● Observe 'best-before . . .' date. ● Ensure that reheated food is piping hot all the way through (observe standing times when using microwaves), and throw away any left-over.

Nocardia species cause chronic subcutaneous infection (eg Madura foot) in warm climates. In the immunocompromised it may cause chest, cerebral, and other infections. Treat with co-trimoxazole or erythromycin.

Botulism is caused by the neurotoxin of *Clostridium botulinum*, which flourishes in preserved, hence anaerobic food (the Latin *botulus* = sausage). It causes prostration, diplopia, blurred vision, photophobia, ataxia, pseudobulbar palsy, and sudden cardiorespiratory failure. GI symptoms may be absent. Nurse on ITU. Intubate early. Seek expert help. Give botulinum antitoxin IM (eg 50,000U of types A and B with 5000U of type E), as early as possible. Also give it to those who have ingested the toxin but who have not yet developed symptoms. Most people also give penicillin.

Clostridium species cause wound infections and gas gangrene (± shock or renal failure) after surgery or trauma (*C perfringens*). Debridement is the cornerstone of treatment, although benzylpenicillin 1.2–2.4g/6h IV, antitoxin and hyperbaric oxygen may also be given. Amputations may be needed, particularly in theatres of war. *Clostridia* food poisoning: p190. *C difficile*: p486).

Actinomycosis Cause: *Actinomyces israelii*. It is usually a subcutaneous infection which forms chronic sinuses with sulphur granule-containing pus. It most commonly affects the area of the jaw (or IUCDs, OHCS p62). It may cause abdominal masses. Treat with benzyl penicillin (p177) for ≥2 weeks post clinical cure. Remove IUCD.

1 DoH Communications 1992, PL/CMO 92–19 (N Sterling)

Miscellaneous Gram −ve bacterial infections

Enterobacteria Some are normal gut commensals, others environmental organisms. However, they may be pathogenic. They are the commonest cause of UTI and intra-abdominal sepsis, especially post-operatively and in the acute abdomen. They are also a common cause of septicaemia. Unusually, they may cause chest infections (especially *Klebsiella*), meningitis, or endocarditis. These organisms are often sensitive to ampicillin and trimethoprim, but in serious infections, cefuroxime ± an aminoglycoside should be used. *Salmonella* and *Shigella* are discussed on p190 and p228.

Pseudomonas aeruginosa is a serious pathogen in hospital, especially in the immunocompromised and cystic fibrosis. As well as pneumonia and septicaemia, it may cause UTI, wound infection, osteomyelitis, and cutaneous infections. The principal problem is its increasing resistance. Each isolate should have sensitivities assessed, but a good combination is piperacillin or azlocillin plus an aminoglycoside. Ciprofloxacin, ceftazidime and imipenem (p181) are newer agents useful against *Pseudomonas*.

Haemophilus influenzae affects children usually <4yrs old, causing otitis media, pneumonia, meningitis, osteomyelitis, septicaemia, and acute epiglottitis. In adults it may cause exacerbations of chronic obstructive airways disease. It is usually sensitive to ampicillin, but chloramphenicol (p180) is more consistently effective. Cefotaxime is a useful third agent. Capsulated types tend to be much more pathogenic than non-capsulated types. Immunization: see p192.

Brucellosis This zoonosis (domestic animals) is common in the Middle East. Symptoms may last for years, and the main problem is thinking of it. Cause: a Gram −ve coccobacillus—*Brucella melitensis* (the most virulent), *B abortus*, *B suis*, or *B canis*. *Clinical features*:

Fever (PUO); sweats	Anorexia; weight loss	Complications:
Myalgia; backache	Rash; bursitis; orchitis	Osteomyelitis; SBE
Vomiting; malaise	Hepatosplenomegaly	Liver abscess
Depression/lassitude	Arthritis; sacroiliitis	Spleen or lung abscess
Constipation/diarrhoea	Pancytopenia	Meningoencephalitis

Tests: Blood culture (≥6wks, contact lab); serology: if titres equivocal (eg >1:40 in non-endemic zones) do ELIZA±immunoradiometric assays.

Treatment: WHO régime: rifampicin 300–450mg/12h + doxycycline 200mg/24h (co-trimoxazole 30mg/kg/12h PO if a child) both PO for 6 weeks (≥3 months if complications). In uncomplicated brucellosis, replacing the rifampicin with streptomycin 1g/day IM for 2–3 weeks is known to reduce relapse rate (2–10% vs >20%),[1] and this may now be the régime of choice.[2]

Bordetella pertussisND causes whooping cough. This begins with a nonspecific catarrhal phase of fever and cough. Only after a week or so, does the child develop the characteristic paroxysms of coughing and inspiratory whoops. Most children recover without complication—although the illness may last some months, but some, especially the very young, may develop pneumonia and consequent bronchiectasis, or convulsions and brain damage. Erythromycin should be given early, if only to limit spread. See immunization, p192 and OHCS p208.

1 G A Acocella 1989 *J Antimicrob Chemother* **23** 433–9   2 G Luzzi 1993 *Trans Roy Soc Trop Med Hyg* **87** 138

Miscellaneous Gram −ve bacterial infections

Pasteurella multocida is acquired via domestic animals especially cat or dog bites. It can cause cutaneous infections, septicaemia, pneumonia, UTI, or meningitis. Treat with penicillin.

Plague[ND] Cause: *Yersinia pestis*; spread by fleas of rodents or, rarely, by cats. Bubonic plague presents with lymphadenopathy (buboes). Pneumonic plague may also be spread by droplets from other infected humans. *Incubation:* 1–7 days. It may present with lymphadenopathy or a 'flu-like illness leading to dyspnoea, cough, copious, bloody sputum, septicaemia and a haemorrhagic fatal illness. *Diagnosis:* Phage typing of bacterial culture, or a 4-fold rise in antibodies to F antigen of *Y pestis*.[1] *Treatment:* Isolate suspects. 1 week's oxytetracycline (250mg/6h PO 1h ac) or streptomycin (used prophylactically too, but no trials have shown efficacy)—if in 1st trimester, amoxicillin 250–500mg/8h PO; if later in pregnancy, co-trimoxazole 480mg/12h PO; children: co-trimoxazole,[2] *OHCS* (p554). Staying at home, quarantine (inspect daily for 1wk), insect sprays to legs and bedding, and avoiding dead animals helps stop spread. Vaccines give no *immediate* protection (multiple doses may be needed).[2]

225

Yersinia enterocolitica Presentation: In Scandinavia, this is a common cause of reactive, asymmetrical polyarthritis of weight-bearing joints, and, in America and Canada, of enteritis. It is also a cause of uveitis, appendicitis, mesenteric lymphadenitis, pyomyositis, glomerulonephritis, thyroiditis, colonic dilatation, terminal ileitis and perforation, and septicaemia.

Tests: Serology is often more helpful than culture, as there may be quite a time-lag between infection and the clinical manifestations. Agglutination titres >1:160 indicate recent infection.[3]

Treatment: None may be needed. The organism is often sensitive to tetracycline, streptomycin, chloramphenicol, and sulphonamides.

Moraxella catarrhalis is an increasingly recognized cause of pneumonia, exacerbations of COAD, otitis media, sinusitis, septicaemia, endocarditis and meningitis. It is usually sensitive to amoxicillin.

See also **Spirochaetes** p232 and *Legionella* p334.

1 F Charlton 1994 *BMJ* i 1060 **2** D Dennis 1994 *BMJ* ii 893 **3** A Borg 1992 *Q J Med* **84** 575

TetanusND

Essence Tetanospasmin, an exotoxin produced by *Cl tetani*, causes muscle spasms and rigidity—the cardinal features of tetanus (='to stretch').

Incidence 30–40 cases/yr in the UK. Mortality: 40% (80% in neonates).

Pathogenesis Spores of *Clostridium tetani* live in faeces, soil, dust, and on instruments. The tiniest breach in skin or mucosa, eg cuts, burns, ear piercing, banding of piles, is all that is required to admit spores. They may then germinate and produce the exotoxin. This then travels up peripheral nerves and interferes with inhibitory synapses.

Clinical features appear from one day to several months from the (often forgotten) injury. There is a prodrome of fever, malaise and headache, before classical features develop: *trismus* (the patient cannot close his mouth); *risus sardonicus* (a grin-like posture of hypertonic facial muscles); *opisthotonus* (arched body, with hyperextended neck); *spasms*, which at first may be induced by movement, injections, noise, etc., but later are spontaneous; they may cause dysphagia and respiratory arrest; autonomic dysfunction (arrhythmias and wide fluctuations in BP).

Differential diagnosis is dental abscess, rabies, phenothiazine toxicity, and strychnine poisoning. Phenothiazine toxicity usually only affects facial and tongue muscles; if suspected, give benztropine 1–2mg IV.

Bad prognostic signs are short incubation, rapid progression from trismus to spasms (<48h), development post-partum or post infection, and tetanus in neonates and old age.

Management Get expert help; use ITU. Monitor ECG, BP. Careful fluid balance.
- Clean wounds, debride as needed. Use IV penicillin or metronidazole[1] 1g/8h PR for 7 days.
- Give human tetanus immune globulin (HTIG) 500U IM to neutralize free toxin. (If only horse antitetanus serum (ATS) is available, give a small sc test dose before giving 10,000U IV and 750U/d to the cut for 3 days.) Some advocate the intrathecal administration of antitoxin.
- Early ventilation and sedation if symptoms progress.
- Control spasms with diazepam 0.05–0.2mg/kg/h IVI or phenobarbitone 1.0mg/kg/h IM or IV with chlorpromazine 0.5mg/kg/6h IM (IV bolus is dangerous) starting 3h after the phenobarbitone. If this fails to control the spasms, paralyse (tubocurarine 15mg IV) and ventilate.

Prevention Active immunization with tetanus toxoid is given as part of the triple vaccine during the first year of life (p192). Boosters are given on starting school and in early adulthood. Once 5 injections have been given, revaccinate only at the time of significant injury.[2]

Primary immunization of adults: 0.5ml tetanus toxoid IM repeated twice at monthly intervals.

Wounds: Any cut merits an extra dose of 0.5ml toxoid IM, unless already fully immune (a full course of toxoid or a booster in last 10 years). The non-immune will need 2 further injections (0.5ml IM) at monthly intervals. If partially immune (ie has had a toxoid booster or a full course >10 years previously), a single booster is all the toxoid that is needed.[2]

Tetanus

[*OTM* 5.265–5.270]

Human tetanus immunoglobulin: This is required for the partially or non-immune patient (defined above) with dirty, old (>6h), infected, devitalized, or soil-contaminated wounds. Give 250–500 units IM, using a separate syringe and site to the toxoid injection.

▶ If the immune status is unknown, assume that the patient is non-immune. Routine infant immunization started in 1961, so many adults are at risk.

▶ Hygiene education and wound debridement are of vital importance.

227

1 I Ahmadsyah 1985 *BMJ* **i** 648   **2** DoH 1992 *Immunization Against Infectious Disease*, HMSO

Enteric fever[ND]

Typhoid and paratyphoid are caused by *Salmonella typhi* and *S paratyphi* (types A, B, and C) respectively. (Other salmonellae cause diarrhoea and vomiting, but rarely typhoid-like illness; see p190, p486.)

Incubation: 3–21 days. Spread: Faecal–oral. 1% become chronic carriers.

Presentation: Usually with malaise, headache, high fever with relative bradycardia, cough, and constipation (or diarrhoea). CNS signs (coma, delirium, meningism) are serious. Diarrhoea is more common after the 1st week. Splenomegaly may develop. Rose spots occur on the trunk of 40%, but may be very difficult to see. Epistaxis and abdominal pain may occur.

Diagnosis: Blood culture in 1st 10 days; faecal/urine culture in 2nd 10 days. Marrow culture has the highest yield. The Widal test is unreliable.

Treat with fluid replacement and adequate nutrition. Chloramphenicol is still the treatment of choice in most areas of the world: 1g/8h PO until pyrexia diminishes, then 500mg/8h for a week and 250mg/6h to make up 14 days. Ciprofloxacin[1] 500mg/12h PO for 14 days, amoxycillin 1g/6h PO for 14 days and co-trimoxazole 960mg/12h PO for 14 days are effective alternatives. IV therapy is an alternative. In encephalopathy, give a course of dexamethasone,[2] 3mg/kg IV stat, then 1mg/kg/6h for 2 days. Antibiotic resistance is a problem.

Complications: Osteomyelitis (eg in sickle-cell disease); DVT; GI bleed or perforation; cholecystitis; myocarditis; pyelonephritis; meningitis; abscess.

Infection can be said to have cleared when 6 consecutive cultures of urine and faeces are −ve. Treat chronic carriers if they are at risk of spreading the disease (eg food handlers). Amoxycillin 4–6g/day+probenecid 2g/day for 6 weeks may work, but cholecystectomy may be needed.

Prognosis: If untreated, 10% die; if treated, 0.1% die. *Vaccine:* p192.

Bacillary dysentery[ND]

Shigella causes abdominal pain and bloody diarrhoea. There is often associated abrupt onset of fever, headache and occasionally neck stiffness. CSF is sterile. Its incubation period is 1–7 days. School epidemics in Britain are usually mild (often *S sonnei*), but imported dysentery may be severe (often *S flexneri* or *S dysenteriae*). Spread is by the faecal–oral route. The organism should be cultured from the faeces. Treatment is supportive with fluids. Avoid antidiarrhoeal agents. Drugs: ciprofloxacin 500mg/12h PO for 3–5 days. Co-trimoxazole 960mg/12h for 3–5 days is an alternative. Imported shigellosis is often resistant to several antimicrobials: sensitivity testing is important. There may be an associated spondyloarthritis (see p669).

Cholera[ND]

Vibrio cholerae (Gram −ve comma-shaped rod)

Incubation: a few hours–5 days. **Spread:** Faecal–oral. Pandemics may occur. It produces profuse watery ('rice water') stools (eg 1 litre/h), fever, vomiting and rapid dehydration, which is the cause of death.

Diagnosis: Direct stool microscopy; culture. *Epidemics:* Currently in S America and Bangladesh (Bengal *Vibrio cholerae 0139*).[1]

Treatment: Strict barrier nursing. Replace fluid and salt losses meticulously, by IVI if shocked. Oral rehydration using WHO formula (20g glucose/l) is not nearly so effective as cooked-rice powder solution (50–80g/l) in reducing stool volume, perhaps because WHO fluid does not contain enough glucose to enable absorption of all the sodium and water in the rehydration fluid. Its high osmolarity (310mmol/l vs 200mmol/l) is also unfavourable to water absorption.[2] Tetracycline 10mg/kg/6h PO for 2 days reduces fluid losses and diminishes transmission.

Prevention: Only drink treated water. Cook all food thoroughly; eat it hot. Avoid shellfish. Peel all vegetables. Vaccination only protects about 50% of people, and then only for a short time, and it is now not recommended.

1 Cholera Working Group 1993 *Lancet* **342** 387   **2** S Gore 1992 *BMJ* **i** 287

Leprosy[ND]

▶ The diagnosis of leprosy must be considered in all who have visited endemic areas who present with painless disorders of skin and nerves.

Mycobacterium leprae affects some millions of people in the tropics and subtropics. Since the widespread use of dapsone the prevalence has fallen (from 0.5% to 2/1000 in Uganda; from 11% to 4/1000 in parts of India).[1]

The Patient The incubation period is months to years, and the subsequent course depends on the patient's immune response. If the immune response is ineffective, 'lepromatous' disease develops, dominated by foamy histiocytes full of bacilli, but few lymphocytes. If there is a vigorous immune response, the disease is 'tuberculoid', with granulomata containing epithelioid cells and lymphocytes, but few or no demonstrable bacilli. Between these poles lie 'borderline' disease.

Skin lesions: Hypopigmented anaesthetic macules, papules, or annular lesions (with raised erythematous rims). Erythema nodosum occurs in 'lepromatous' disease, especially during the first year of treatment.

Nerve lesions: Major peripheral nerves may be involved, leading to much disability. Sometimes a thickened sensory nerve may be felt running into the skin lesion (eg ulnar nerve above the elbow, median nerve at the wrist, or the great auricular nerve running up below the ear).

Eye lesions: ▶ *Refer promptly to an ophthalmologist.* The lower temperature of the anterior chamber favours corneal invasion (so secondary infection and cataract). Inflammatory signs: chronic iritis, scleritis, episcleritis. There may be reduced corneal sensation (v nerve palsy), and reduced blinking (vii nerve palsy) and lagophthalmos (difficulty in closing the eyes, see *OHCS* p480) as well as absent eyelashes (trichiasis).

Diagnosis is by biopsy of a thickened nerve (*in vitro* culture is not possible—as an incidental curio, armadillo foot-pad culture works, but do not taunt your lab by requesting this test!). Skin or nose-blow smears for AFB are +ve in borderline or lepromatous disease. Classification matters as it reflects the biomass of bacilli, influencing treatment: the more organisms, the greater the chance that some will be drug-resistant. Other tests: neutrophilia, ESR↑, IgG↑, false +ve rheumatoid test.

Treatment[2] Ask a local expert about: ● Resistant patterns, eg to dapsone, when ethionamide may (rarely) be needed ● Using prednisolone for severe complications ● Is surgery ± physiotherapy needed as well as chemotherapy? In the UK, advice from the panel of Leprosy Opinion is essential. In other areas, the administration of some drugs should be supervised (S) whereas others need no supervision (NS). For lepromatous and borderline disease, WHO advises rifampicin 600mg PO monthly (S), dapsone 100mg/24h PO (NS), and clofazimine 300mg monthly (S) + 50mg/24h (NS) for 2yrs. In tuberculoid leprosy, rifampicin 600mg monthly (S) and dapsone 100mg/24h (NS) for 6 months. Dapsone may cause haemolysis. ▶ Beware sudden permanent *paralysis* from nerve inflammation by dying bacilli (± *orchitis, prostration,* or *death*); this 'lepra reaction' may be mollified by thalidomide (*NOT* if pregnant). Liaise urgently with a leprologist.

WHO régimes are problematic as many find it hard to attend for supervised therapy (nomads, jungle-dwellers, those living in remote mountains). Shorter courses of treatment may be practical and effective.[2]

1 K Meeran 1989 *BMJ* i 364   **2** MF Waters 1993 *Trans Roy Soc Trop Med Hyg* **87** 500

Leprosy
[*OTM* 5.305]

Spirochaetes

The endemic treponematoses *Yaws* is caused by *Treponema pertenue*, morphologically and serologically indistinguishable from *T pallidum*. It is a chronic granulomatous disease prevalent in children in the rural tropics. Spread is person-to-person, via skin abrasions, and is promoted by poor hygiene. The primary lesion (an ulcerating papule) appears 3–6 weeks after exposure. Scattered secondary lesions then appear particularly in moist skin, but can be anywhere. These may become exuberant. Tertiary lesions are subcutaneous gummatous ulcerating granulomata, affecting skin and bone. Cardiovascular and CNS complications do not occur.

Pinta, caused by *T carateum*, occurs in Central and South America, affecting only skin.

232 *Endemic non-venereal syphilis (bejel):* This is caused by *Treponema pallidum*, in Third World children, when it resembles yaws. In the developed world, *T pallidum* causes venereal syphilis (see p218).

Diagnosis: No serological or microbiological test can differentiate endemic treponematoses from venereal syphilis or each other. Diagnosis is clinical.

Treatment: Procaine penicillin (p177).

Weil's diseaseND, caused by *Leptospira interrogans* the most virulent serogroup of which is *L icterohaemorrhagiae*, is spread by contact with infected rat urine, eg while swimming. It presents with fever, jaundice, headache, injected conjunctivae, painful calves (myositis), purpura, haemoptysis, haematemesis, or any bleeding. Meningitis, myocarditis, and particularly renal failure may develop. The rise in transaminases may be relatively small. Diagnosis is by complement fixation test, the Schuffner agglutination test, and blood, urine and CSF culture. Treat symptomatically and give benzylpenicillin 600mg/6h for 7d. Doxycycline is useful prophylaxis for high risk groups. *The Jarisch–Herxheimer reaction* is rare in leptospirosis. This is a systemic reaction several hours after the first dose of antibiotic is given. There is fever, tachycardia, vasodilation, and it is occasionally fatal. It is thought to be due to the sudden release of endotoxin.

Canicola fever is an aseptic meningitis caused by *Leptospira canicola* (puppies may be source).

Relapsing feverND This is caused by *Borrelia recurrentis* (louse-borne) or *B duttoni* (tick-borne). It has frequently occurred in pandemics following war or disaster. It has killed 5 million people this century. After 2–10 days' incubation, there is abrupt fever, rigors, and headache. A petechial rash (which may be faint or absent), jaundice, and hepatosplenomegaly may develop. Crises of very high fever and tachycardia occur. When the fever abates, hypotension due to vasodilation may develop. It may be fatal. Relapses occur, but are milder. The organisms are seen on Leishman-stained thin or thick films. Treat with tetracycline 500mg PO or 250mg IV as a single dose (but for 10 days for *B duttoni*). The Jarisch–Herxheimer reaction (above) is fatal in 5%: meptazinol 100mg IV slowly is given as prophylaxis with the tetracycline, repeated 30mins later (with the chill phase) and during the flush phase (if systolic BP <75mmHg). Delouse the patient. Doxycycline (p180) is useful prophylaxis in high risk groups.

Spirochaetes

[*OTM* 5.319]

Lyme disease This is caused by *Borrelia burgdorferi* which is tick-borne. It usually begins with erythema chronicum migrans (see p686). This may be associated with malaise, including arthralgia, neck stiffness, and lymphadenopathy. 15% develop CNS abnormalities, eg meningitis, ataxia, amnesia, cranial nerve palsies, peripheral neuropathy, and lymphocytic meningoradiculitis (Bannwarth's syndrome). 10% develop cardiac conduction disturbances and myopericarditis. Some 60% have recurrent oligoarthritis and up to 10% develop a chronic erosive arthropathy. Diagnosis is serological. Treat the skin condition with tetracycline 250–500mg/6h (amoxycillin or penicillin V in children <12yrs) for 10–20 days, and later complications with high dose IV benzylpenicillin or a 3rd-generation IV cephalosporin, eg cefotaxime or ceftriaxone (unavailable in UK).

233

Polio[ND]

Polio is a picornavirus spread by droplet or the faecal–oral route.

Clinical features: 7 days' incubation period, followed by 2 days' flu-like prodrome. The next ('pre-paralytic') stage, consisting of fever (eg 39°C), tachycardia, headache, vomiting, neck stiffness, and unilateral tremor, proceeds in 65% of those reaching the pre-paralytic stage to the paralytic stage with myalgia and lower motor neurone signs and respiratory failure.

Investigations: CSF: WCC↑, polymorphs then lymphocytes, otherwise normal; paired sera (14 days apart); throat swab and stool culture identify virus.

Natural history: <10% of paralytic cases die, but the anterior horn cell loss is permanent, and the patient will need appropriate support. There may be delayed progression of paralysis. The paralysis is worse if there has been strenuous exercise during the incubation period.

Prophylaxis: Live vaccine by oral route (p192). In the tropics the killed parenteral vaccine is needed to induce adequate immunity. Adults should be revaccinated when their children are vaccinated for the first time as GI passage may increase virulence.

Rabies[ND]

Rabies is a rhabdovirus spread by bites from any infected mammal (usually dogs, cats, foxes, wolves, or bats).

Clinical features: Usually 9–90 days' incubation, therefore take prophylactic measures even several months after a bite. Prodromal symptoms of headache, malaise, abnormal behaviour, fever, and itching at the site of the bite proceeds to 'furious rabies' (hydrophobia) in 80%. In these cases, water provokes muscle spasms often accompanied by profound terror. In 20%, 'dumb rabies' starts with flaccid paralysis in the bitten limb and spreads.

Pre-exposure prophylaxis (eg vets, remote expeditions): Give human diploid cell strain vaccine (1ml IM, deltoid) on days 0, 7, & 28, and again at 2–3yrs if still at risk, then every 1–3yrs if still at risk.

Treatment if bitten (for previously unvaccinated): ▶ Seek expert help (Virus reference lab tel. 081 200 4400). Observe the biting animal if possible, to see if the animal dies (but it is possible that it may not die of rabies before the patient does);[1] asymptomatic carriage occurs but has not (yet) produced rabies in man[2]. Clean the wound. Give the vaccine on days 0, 3, 7, 14, 30, and 90 (1ml IM) and human rabies immunoglobulin (20U/kg half given IM and half locally infiltrated around wound). Rabies is usually fatal once symptoms begin, but survival has occurred, if there is optimum CNS and cardiorespiratory support. Staff in attendance should be offered vaccination.

1 T Hemachudha 1991 *NEJM* **324** 1890   2 H Aghomo 1990 *Vet Microbiol* **22** 17–22

Viral haemorrhagic fevers[ND]

Yellow fever[ND] is an epidemic disease due to an arbovirus transmitted by the *Aedes* mosquito. Immunization is discussed on p192.

Clinical features: The 2–14-day incubation period is followed in 85% cases by a mild syndrome of fever, headache, nausea, albuminuria, myalgia, and a relative bradycardia. In the severe form, 3 days of headache, myalgia, anorexia, and nausea are followed by abrupt onset of fever (~40°C). Then, after a few hours remission, prostration, jaundice, and bleeding (especially haematemesis) supervene. Mortality typically ~10%.

Diagnosis: is serological. *Treatment:* This is symptomatic.

Other viral haemorrhagic fevers include Lassa fever[ND], Ebola virus[ND], Marburg virus[ND], and dengue haemorrhagic fever (dengue fever is the most prevalent arbovirus disease). These diseases may start with sudden onset of headache, backache, myalgia, conjunctivitis, prostration and fever, and after a few days develop haemorrhagic complications. There may be spontaneous resolution, renal failure, encephalitis, coma, or death.
Treatment: This is symptomatic.[1]

In all viral haemorrhagic fevers, use special infection control measures.

235

1 K Johnson 1990 *NEJM* **323** 1139

TyphusND

The typhus fevers are caused by species of rickettsia which are conveyed between hosts by arthropods. The incubation period is 2–23 days. There follows an acute illness of sudden onset which may be associated with severe headache, vomiting, photophobia, deafness, and toxaemia. In some species, an *eschar* may be present. This is a single skin lesion at the site of the initial inoculation, which goes on to ulcerate and form a black scar. Later in the acute illness (5th-6th day) a more generalized skin rash occurs (see below). It may be of a maculopapular nature, resembling measles or rubella, or be haemorrhagic. Asymptomatic infections are common as are mild febrile illnesses.

Pathology Widespread vasculitis and endothelial proliferation may affect any or all organs, and thrombotic occlusion may lead to gangrene. Patients die of shock, renal failure, DIC, or stroke.

Diagnosis is difficult as often the clinical picture is non-specific, the organisms are difficult to grow, and the traditional heterophil antibody Weil–Felix test has low specificity and sensitivity. Immunofluorescence and ELISA tests on paired sera are a great improvement, and skin biopsy can be diagnostic in Rocky Mountain spotted fever.

Treatment (Do not wait for serology.) All types are treated with tetracycline 500mg/6h for 7 days. Chloramphenicol 500mg/6h for 7 days is an alternative.

The major diseases
Louse-borne typhus (epidemic typhus): R prowazeki is carried by the human louse *Pediculus humanus*, the faeces of which are inhaled or pass through skin. It may become latent and recrudesce later (Brill–Zinser disease).

Rocky Mountain spotted fever is tick-borne (R rickettsi) and is the other major killer in this group. A characteristic rash appears by the 4th day of fever (pink macules, starting on palms, soles and wrists). In a few days, this becomes purpuric and slightly papular.

Others include *murine (endemic) typhus*, *tick typhus (fièvre boutonneuse)*, *rickettsial pox*, and *scrub typhus* (which is most common in SE Asia).

Q fever

Coxiella burnetii, another rickettsia, causes about 100 cases of Q fever per year in the UK. It is so named because it was first labelled 'query' fever in workers in an Australian abattoir.

Epidemiology It occurs world-wide, and is usually rural with its reservoir in cattle and sheep. The organism is very resistant to drying and is usually inhaled from infected dust. It can be contracted from unpasteurized milk, directly from carcasses in abattoirs, sometimes by droplet spread, and occasionally from tick bites.

Clinical features Q fever should be suspected in anyone with prolonged PUO or atypical pneumonia. It may present with fever, myalgia, sweats, headache, cough, and hepatitis. If the disease becomes chronic, suspect endocarditis ('culture-negative'). This usually affects the aortic valve, but clinical signs may be absent for many years.

236

Typhus
Q fever
[*OTM* 5.341; 5.353; 5.352]

Investigations CXR may show consolidation, usually lower lobe, or one or more mass lesions. These may persist for months during resolution. Liver biopsy may show granulomata. The diagnosis is made serologically. The presence of phase I antigens indicates chronic infection; phase II antigens indicate acute infection.

Treatment Tetracycline is effective *in vitro* (rickettsiostatic), but has not been shown to speed recovery. It is given in the hope of preventing chronicity. Agents used for chronic disease include rifampicin and tetracycline. 237

Giardiasis

Giardia lamblia is a common GI parasite in the UK. It is a flagellate proto-zoon which lives in the duodenum and jejunum. It is transmitted by faeces. Risk factors include travel, immunosuppression, homosexuality, and achlorhydria. Drinking water systems may become contaminated.

Presentation: Often asymptomatic. The combination of lassitude, bloat-ing, flatulence, loose stools and abdominal discomfort is also common. There may be malabsorption.

Diagnosis: Repeated stool microscopy for cysts or trophozoites may be negative, and duodenal aspiration may be necessary. Serology may be positive. Finally, a blind therapeutic trial may be needed.

Differential diagnosis: Any cause of diarrhoea (p188, 190, 486), tropical sprue (p522), coeliac disease (p522).

Treatment: Scrupulous hygiene is important. Give metronidazole 400mg/8h PO for 5 days or 2g/24h for 3 days, or tinidazole 2g PO once and advise to avoid driving and alcohol; or mepacrine hydrochloride 100mg/8h PO for 5–7 days (cheap, but SEs common). If treatment fails, check com-pliance and consider treating the whole family. If diarrhoea persists, avoid milk as lactose intolerance may persist for 6 weeks.

Other protozoa which cause GI infection are *Cryptosporidium* and *Isospora* (particularly common in AIDS), *Balantidium coli*, and *Sarco-cystis*.

Amoebiasis

Infection with *Entamoeba histolytica* (which is world-wide) is spread by the faecal–oral route. It is necessary to boil infected food to destroy cysts. Trophozoites may remain in the bowel or invade extra-intestinal tissues, leaving 'flask-shaped' GI ulcers. Infection may be asymptomatic, or cause only mild diarrhoea. Dysentery lies at the other extreme of the spectrum.

Amoebic dysentery[ND] may begin years after infection. Diarrhoea may begin gradually, becoming profuse and bloody. High fevers, colic and tenesmus are rare, but an acute febrile prostrating illness does occur, as do remissions and exacerbations. Diagnose by microscopy of fresh stool. Serology indicates whether the patient has at some time been infected.

Differential diagnosis: Amoebic dysentery has a gradual onset with pus, red cells, and trophozoites in the stool. Bacillary dysentery often has a sudden onset and may cause dehydration. Its stools may be more watery and have no trophozoites. Acute ulcerative colitis also has a more gradual onset. Dehydration is rare, and the stools are very bloody with relatively few pus cells. Other causes of bloody diarrhoea include *E coli*, *Campylo-bacter*, *Salmonella*, and pseudomembranous colitis.

Amoebic colonic abscesses may perforate causing peritonitis.

Amoeboma is an inflammatory mass most often found at the caecum, where it must be distinguished from other RIF masses.

Hepatic abscesses are usually single masses in the right lobe, and contain anchovy-sauce like fluid. There is usually a high swinging fever, sweats, and RUQ pain and tenderness. LFTs are often slightly abnormal, although

Giardiasis
Amoebiasis
[*OTM* 5.466; 5.512]

mechanical obstruction may occur. There is usually a leucocytosis. Diagnosis is suggested by ultrasound and CT appearances, and positive serology. Aspiration may be required.

Treatment Give metronidazole 800mg/8h PO for 5 days for acute amoebic dysentery (it is very active against the vegetative amoebae) followed by diloxanide furoate 500mg/8h PO for 10 days to destroy any gut cysts (SEs rare). Diloxanide is the best drug for chronic disease when *Entamoeba* cysts, not vegetative forms, are seen in stools. Amoebic liver abscess: metronidazole 400mg/8h for 10 days repeated after 2 weeks as needed; aspirate abscess if no improvement after 72h and give course of diloxanide after metronidazole.

Trypanosomiasis

African trypanosomiasis (sleeping sickness) *T gambiense* causes a slow wasting illness with a long prepatent period (West African variety). *T rhodesiense* causes a more rapidly progressive illness (rural East Africa). The organism enters the skin following a bite from an infected tsetse fly, and spreads to nodes, blood, spleen, heart, and brain.

Clinical features: A subcutaneous chancre at the site of infection, with fevers, lymphadenopathy (posterior cervical nodes are particularly enlarged with *T gambiense* infection: Winterbottom's sign), headache, rashes, joint pains, weight loss, amenorrhoea/impotence, peripheral and pulmonary oedema, ascites, pericardial effusions, and nephritis. Later in the natural history, some patients develop CNS features (apathy, depression, ataxia, dyskinesias, dementia, hypersomnolence, and coma).

Diagnosis is by demonstration of the organisms in the blood or lymph nodes. Cultures should be taken daily for 12 days. Examine CSF.

Treatment: Seek expert help.
- Treat the anaemia and other infections first.
- Early (haemolymphatic) phase: suramin 20mg/kg IV on 5–10 occasions at 5-day intervals. Give 200mg test dose IV first. Pentamidine 4mg/kg IM days 1–7 is an alternative, but beware severe SEs (BP↓, hypoglycaemia, pancreatitis, arrhythmias).
- CNS disease: a cautious régime involves fractions of a maximum dose of melarsoprol of 3.6mg/kg IV slowly:
  Day 1—suramin 0.2g
  Day 5—suramin 1g
  Days 8 and 10–50% of full dose of melarsoprol
  Days 12, 22, 24, 26, 36, 38, and 40—full dose of melarsoprol.

Suramin predictably causes proteinuria and often increased serum urea and creatinine; therefore check renal function frequently.

Melarsoprol causes lethal encephalopathy in up to 10% of patients, characterized by abnormal behaviour, fits and coma. This is partly preventable with prednisolone 1mg/kg/24h PO (max 40mg) starting the day before the first injection.[1] Seek expert help.

Arseno-resistant trypanosomiasis was uniformly fatal until the introduction of difluoromethylornithine (DFMO) in 1984. Dose example: 100mg/kg/6h IVI over 1h for 14 days, followed by 75mg/kg/6h PO for 3–4 weeks. SE: anaemia; diarrhoea; seizures; leucopenia; hair loss. It is also being assessed in Chagas' disease.

American trypanosomiasis (Chagas' disease) is caused by *Trypanosoma cruzi*. It is spread by Reduviid bugs or blood transfusion. After implantation a nodule (Chagoma) forms which may scar. The patient may present acutely with fever, unilateral bipalpebral oedema and ophthalmia (Romañas' sign) lymphadenopathy, and hepatosplenomegaly; or there may be a long latent period (eg 20 years) followed by signs of multi-organ invasion and damage, affecting especially heart and GI smooth muscle. *GI:* Megaoesophagus, dilated stomach, megacolon.

Heart: Dilated cardiomyopathy, bundle branch blocks, syncope, emboli, LVF, ST elevation, T-wave inversion, left ventricular aneurysm.

Diagnosis: Acute disease: organisms may be seen in or grown from blood, CSF, or node aspirate. Chronic disease: serology (eg Chagas' IgG ELISA).

Treatment: This is unsatisfactory. Nifurtimox or benzidazole are used in acute disease (toxic agents which only eliminate the parasite in 50% of patients[3]). Chronic disease can only be treated symptomatically.

1 C Ellis 1985 *BMJ* **291** 1524   2 J Pepin 1989 *Lancet* i 1246   3 L Kirchhoff 1993 *NEJM* **329** 389

Trypanosomiasis
[*OTM* 5.515]

Leishmaniasis

This intracellular infection, caused by *Leishmania* protozoa, is spread by sandflies, and occurs widely throughout Africa, India, Latin America, the Middle East and Mediterranean. It gives rise to granulomata.

Cutaneous leishmaniasis (Oriental sore) is a major health problem affecting over 300,000 people. Currently there is an epidemic in Sudan. It may be mild and localized, or diffuse and severe. Lesions develop at the site of the bite, beginning as an itchy papule, from which the crust may fall off to leave an ulcer. Most heal spontaneously with scarring, which may be disfiguring if the lesions are extensive. *L mexicana* tends to cause destruction of the pinna. Diagnosis of cutaneous leishmaniasis is by microscopy of material aspirated from the edge of the ulcer. *Treatment:* If no spontaneous healing, use sodium stibogluconate (pentavalent antimony) 20mg/kg/24h IV (max 850mg per day) for 10 days. This is only sometimes effective (local infiltration may be used; seek expert advice).

Mucocutaneous leishmaniasis (espundia) is caused by *L brasiliensis* and occurs in S America. Often the primary skin lesion is complicated by spread to mucosae of the nose, pharynx, palate, larynx and upper lip (in that order). This results in serious scarring and death from pneumonia.

Diagnosis: As the parasites may be scanty, a Leishmanin skin test is often necessary to distinguish the condition from leprosy, TB, syphilis, yaws, and carcinoma. Indirect fluorescent antibody tests are available.

Treatment: This is unsatisfactory once mucosae have become involved, so all cutaneous lesions in areas where *L braziliensis* occurs should be treated with pentavalent antimony.

Kala-azar (visceral leishmaniasis) is caused by *L donovani*. Protozoa spread via the lymphatics from (the often minor) cutaneous lesion. In the reticuloendothelial system, organisms multiply in macrophages (Leishman–Donovan bodies). The clinical picture is usually chronic with sudden fevers (eg 3/day), anaemia, massive splenomegaly, bleeding (often epistaxis), lymphadenopathy, wasting, cough and diarrhoea. Death is usually from secondary infection. Acute presentations occur, often with bleeding.

Investigations: Hypersplenism (anaemia, neutropenia, thrombocytopenia), low albumin, raised IgG. Negative Leishmanin skin test, seropositivity. The diagnosis depends on the identification of the organism in bone marrow (80%), lymph nodes or splenic aspirate (95%). Serology often −ve in those with AIDS.

Treatment: Seek expert help. WHO régime: pentavalent antimony (sodium stibogluconate, Sb^V) 20mg/kg/24h IV or IM, up to 850mg/day, for 30 days. SE malaise (patient curls up under the bed clothes all day, not eating), cough, substernal pain, long Q–T interval, arrhythmias, anaemia, uraemia, hepatitis. Régimes are changing—eg 10mg Sb^V/8h for 10d,[1] without the 850mg limit—as 25% fail to respond or relapse. Here pentamidine may be used: deep IM 3–4mg on alternate days max 10 doses: SEs may be fatal (BP↓, arrhythmias, glucose↓, permanent diabetes in 4%[1]). Other anti-leishmanials: aminosidine (paromomycin), amphotericin B (AmBisome®).[1]

Post kala-azar dermal leishmaniasis, with lesions resembling those of leprosy, may occur months or years following successful treatment.

1 R Davidson 1993 *Trans Roy Soc Trop Med Hyg* **87** 130

Leishmaniasis
[*OTM* 5.524]

Fungi

Fungi may cause disease by acting as airborne allergens, by elaborating toxins, or by direct infection. Fungal infection may be superficial or deep, and both are much commoner in the immunocompromised.

Superficial mycoses Dermatophyte infection (*Trichophyton*, *Microsporum*, *Dermatophyton*), causes tinea (ringworm). Diagnose by microscopy of skin scrapings and treat with topical clotrimazole 1% 12-hourly. Continue for 14 days after healing. In intractable lesions, griseofulvin 0.5–1g/24h PO for up to 18 months was previously required, but itraconazole and terbinafine have emerged as promising alternatives.[1]

Candida albicans causes oral (p500) and vaginal thrush. It is diagnosed clinically by observing pearly white lesions, and confirmed by culture. Treat vaginal thrush with one clotrimazole tablet set high in vagina once only. Oral thrush is treated by chewing nystatin pastilles (100,000U/pastille) every 6h pc, continuing for 48h after lesions have disappeared, or amphotericin lozenges (10mg)/6h for 2 weeks. Consider systemic treatment with fluconazole or itraconazole PO for treatment failures, especially in the immunosuppressed.

Malassezia furfur causes pityriasis versicolor: a macular rash, brown on pale skin and pale on tanned skin. Treat with clotrimazole cream as above.

Some superficial mycoses penetrate the epidermis and cause chronic subcutaneous infections such as Madura foot or sporotrichosis. Treatment is complex and sometimes even requires amputation of the affected limb.

Systemic mycoses *Aspergillus fumigatus* may colonize the lung or cause allergic damage. Invasive systemic aspergillosis occurs in the immunosuppressed. Aspergillosis is discussed fully on p340.

Systemic candidiasis also only occurs in the immunocompromised. *Candida* UTI occurs in diabetes mellitus and *Candida* is a rare cause of infective endocarditis of prosthetic valves. It should be considered in the differential diagnosis of any PUO in an immunocompromised patient. Repeated blood cultures should be taken, and serology may help. If it does not resolve when the predisposing factor (eg IV line) is removed, the treatment is amphotericin B IV (use the régime given on p340) or fluconazole 400mg stat then 200mg/day PO.

Cryptococcus neoformans may cause meningitis, pneumonia, and other infections in the immunocompromised. It is especially common in AIDS, but also occurs in sarcoid, Hodgkin's disease, and those on steroids. The history may be a long one and there may be features suggestive of raised intracranial pressure, eg confusion, papilloedema, and cranial nerve lesions. This may delay CSF examination which is the key investigation. If this diagnosis is suspected, specifically request Indian-ink staining of the CSF. The organism may also be successfully cultured and the antigen detected by the latex test in CSF and blood. Cryptococcus meningitis is treated with 5-flucytosine 25–50mg/kg/6h PO plus amphotericin B IV (p340) for 6 weeks. Response may be monitored clinically and serologically. Cryptococcal pneumonia is the other major manifestation and is commonest in USA. Treatment is similar.

1 *Lancet* 1990 **335** 636

Fungi
[*OTM* 5.447]

A number of other fungi, which are most commonly found in the Americas and Africa, may cause deep infection. *Histoplasma capsulatum*, *Coccidioides immitis*, *Paracoccidioides brasiliensis*, and *Blastomyces dermatitidis* may cause asymptomatic infections, acute or chronic pulmonary disease, or disseminated infection. Acute histoplasma pneumonitis may be associated with arthralgia, erythema nodosum, and erythema multiforme. Chronic disease, which is commoner with the other 3 fungi, may cause upper zone fibrosis or radiographic 'coin lesions'. Diagnosis is by serology, culture, and biopsy. These diseases are treated with amphotericin B (see p340), except *Paracoccidioides*, for which ketoconazole is the treatment of choice (200mg/24h PO with food for 14 days or until symptoms have cleared; SE: hepatitis—may be fatal, rashes; CI: hepatic impairment, pregnancy). Itraconazole is proving to be useful and less toxic than ketoconazole.

Nematodes (roundworms)

Necator americanus and *Ankylostoma duodenale* **(Hookworms)**
Both occur in the Indian subcontinent, SE Asia, Central and N Africa and parts of Europe. Necator is also found in the Americas and sub-Saharan Africa. Numerous small worms attach to upper GI mucosa, causing bleeding (a very important cause of iron-deficiency anaemia). Eggs are passed in faeces and hatch in soil. Larvae penetrate feet, so starting new infections. Oral transmission of *Ankylostoma* may occur. *Diagnose* by stool microscopy. *Treat* with mebendazole 100mg/12h PO for 3 days, and iron.

Strongyloides stercoralis is patchily endemic in the subtropics and tropics. Transmitted percutaneously, it causes a transient, linear, itching erythema. Pneumonitis, enteritis, and malabsorption occur, as do acute exacerbations. *Diagnose* from stool microscopy and culture, serology, or duodenal aspiration. *Treat* with thiabendazole 25mg/kg/12h for 3 days or albendazole 10mg/kg/day for 3 days. Hyperinfestation is a problem in the immunocompromised.

Ascaris lumbricoides occurs world-wide. It looks like (and is named after) the garden worm (*Lumbricus*). An unusual characteristic is that it has 3 finely-toothed lips. Transmission is faecal–oral. It migrates through liver and lungs, and settles in small bowel. It is usually asymptomatic, but death may occur from GI obstruction or perforation. It may grow very long (eg 25cm). Look for: ● Faecal ova (stained orange by bile; 60–45μm across; the rate of egg production being ~200,000/day). ● Worms on barium x-rays. ● An eosinophilia. *Treat* with mebendazole as in hookworm (above); levamisole is better, but not on UK market.

Trichinella spiralis is transmitted by uncooked pork world-wide. It migrates to muscle, and causes myalgia, myocarditis, periorbital oedema, fever and swellings. *Treat* with thiabendazole 25mg/kg/24h for 2–5 days.

Trichuris trichiura **(whipworm)** may cause non-specific abdominal symptoms. Diagnose by faecal microscopy. *Treat* as for mebendazole as for hookworm.

Enterobius vermicularis **(pinworm, threadworm)** is common in temperate climates. It causes anal itch as it leaves bowel to lay eggs on the perineum. Apply sticky tape to the perineum and identify eggs microscopically. Treat with mebendazole 100mg PO (1 dose) repeated at 2 weeks if aged ≥2yrs; if aged 6 months to 2yrs: pyrantel 10mg/kg PO (1 dose); tablets are 125mg. Treat whole family. *Hygiene is more important than drugs*—adult worms die after 6wks and do not multiply in colon. Continued symptoms means *reinfection*.

Toxocara canis **(a roundworm/nematode)** is the commonest cause of *visceral larva migrans* (any other invasive helminth may be implicated). It presents with an ocular granuloma (blindness, squint) or gross visceral involvement (eg hepatomegaly, asthma, cough). *Diagnosis* may require histology. *Treatment:* with thiabendazole or diethylcarbamazine, is often unsatisfactory.

Cutaneous larval migrans: Typically slowly-migrating raised itchy tracks on feet of a tropical traveller. It is caused by contact with soil infected by larva of dog or cat hookworm. Treatment: topical thiabendazole or a 3-day course of oral albendazole.

Nematodes (roundworms)

[*OTM* 5.533]

Dracunculus medinensis (guinea-worm) is a very long worm which slowly extrudes from the skin, causing focal dermatitis and ulceration. *Treat* with metronidazole 400mg/8h for 5 days or niridazole (not on UK market) 12.5–25mg/kg/24h PO for 7 days. It is found in tropical and North Africa, the Middle East and the Indian subcontinent.

Filariasis is very common: Prevalence: 18 million world-wide.

1 *Onchocerca volvulus* causes blindness ('river blindness') in 72% of some communities in tropical Africa and America—causing abandonment of large areas of fertile land (eg with 40% of those >50yrs being blind). A nodule forms at the bite; it sheds microfilariae to distant skin sites which develop altered pigmentation, lichenification, loss of elasticity and poor healing. Eyes may develop keratitis, uveitis, cataract, fixed pupil, fundal degeneration or optic neuritis/atrophy. Lymphadenopathy and elephantiasis also occur. *Diagnosis* is by visualizing microfilaria in eye or skin snips. Remove a fine shaving of clean, unanaesthetized skin with a scalpel. Put on slide with a drop of normal saline and look for swimming larvae after 30mins.

2 Lymphatic filaria (eg *Wuchereria*) cause lymphadenitis, elephantiasis, and tropical pulmonary eosinophilia. It occurs mainly in tropical Asia. *Diagnosis* is by thick film and serology.

3 *Loa loa* causes transient, cellulitis-like 'Calabar' swellings, and may migrate across the conjunctiva. Occurs in West and Central Africa.

Treatment: This needs expert help as the death of the filaria may provoke a severe allergic reaction (Mazzotti reaction) which may even cause blindness. Ivermectin, a semisynthetic macrocyclic lactone, does not precipitate the allergic reaction and is now the drug of choice (1 dose of 150µg/kg PO repeated eg each year for onchocerca, 20µg/kg PO as a single dose for wuchereia[1]). This replaces diethylcarbamazine. Give every 6–12 months until the adult worms die. Mass treatment campaigns may prevent blindness in some communities,[2] and the drug company has made ivermectin freely available to them. Lymphoedema responds to compression garments (hard to use)—or benzopyrone (coumarin) 400mg/24h PO.[3]

1 J Sanford 1992 *Guide to Antimicrobial Therapy*, Antimicrobial Therapy Inc, Dallas, ISBN 0–933775–10–5 **2** D Mabey 1993 *Lancet* i 154 **3** JR Casley-Smith 1993 *BMJ* ii 1037

Cestodes (tapeworms)

These worms attach themselves to small bowel mucosa, but do not suck blood.

Taenia solium infection occurs by eating 'measly' pork, or from contaminated water, and *T saginata* is contracted from uncooked, infected beef. They cause vague abdominal symptoms and malabsorption. Contaminated food and water contain cysticerci which develop into adult worms within the gut. However, if a human swallows eggs of *T solium* (faecal–oral route), they may enter the circulation and disseminate throughout his body becoming cysticerci within the human host (cysticercosis). This tapeworm encysts in muscle; skin; heart; eye; and CNS, causing seizures (a common cause in endemic areas) and dementia. Examine the arms and legs for palpable subcutaneous lesions.

Diagnosis is by faecal microscopy and examination of perianal swabs, but it is important to differentiate the two species: this requires examination of the scolex or a mature proglottid; the eggs are indistinguishable. An indirect haemagglutination test is available. The CSF may show eosinophils in neurocysticercosis, and a CT scan may locate cysts. SXR and x-ray soft tissues of thigh may show calcified cysts.

Treat with niclosamide 2g PO followed 2h later with a purge. It may be helpful to use an anti-emetic on waking. Praziquantel (not on UK market) is the drug of choice in neurocysticercosis: 16.6mg/kg/8h PO for 15 days. Albendazole is an alternative. An allergic response to the dying larvae should be covered by dexamethasone 12mg/24h PO for 21 days.

Diphyllobothrium latum This is acquired from uncooked fish, causes the same symptoms as *Taenia solium* and is treated similarly with niclosamide. It may cause a vitamin B_{12} deficiency as the worm consumes it.

Hymenolepis nana and *H diminuta* (dwarf tapeworms) are rarely symptomatic. Treat with niclosamide 2g first day, then 1g/24h for 6 days. *H. Nana* may be treated with a single dose of praziquantel (25mg/kg PO).

Hydatid disease Cystic hydatid disease is a zoonosis caused by ingesting eggs of the dog parasite *Echinococcus granulosus* (a cestode—ie a tapeworm), by contact with dog faeces—eg in rural sheep-farming regions. Hydatid is an increasing public health problem in parts of China, Russia, Alaska, Wales, and Japan.[1] *Presentation* is usually due to a large cyst forming in the lungs (causing eg dyspnoea, pain, or haemoptysis, or anaphylaxis on its discharge through the large airways) or liver (causing hepatomegaly, obstructive jaundice, cholangitis, or PUO). But it may migrate almost anywhere. It may be an incidental finding (eg it is on the differential diagnosis of solitary, or even multiple shadows, on chest x-rays). It may occasionally occur in other organs (eg CNS). *Diagnosis* is aided by plain x-ray, ultrasound and CT of the cyst. A reliable serological test has replaced the variably sensitive Casoni intradermal test. *Treatment:* Surgical excision for symptomatic cysts. Care should be exercised to avoid spilling cyst contents, as this may provoke an anaphylactic reaction. The drug of choice is albendazole 5mg/kg/12h for 7–60 days (seek expert advice), but its rôle is still unclear—it should be given for several weeks before surgical excision.[2] (Alveolar hydatid disease is caused by *E multilocularis*.)

1 A lto 1993 *Trans Roy Soc Trop Med Hyg* **87** 170   **2** RJ Horton 1989 *Trans Roy Soc Trop Med Hyg* **83** 97

Cestodes (tapeworms)

[OTM 5.561]

Trematodes (flukes)

Schistosomiasis (bilharzia) is the most prevalent disease caused by flukes, affecting 200 million people world-wide. The snail vectors release cercariae which can penetrate the skin, eg during paddling—may cause itchy papular rash ('swimmer's itch'). The cercariae shed their tails to become schistosomules and migrate via lungs to liver where they grow. About 2wks after initial infestation there may be fever, urticaria, diarrhoea, cough, wheeze and hepatosplenomegaly—so-called 'Kata-yama fever' (self-limiting, ?immune complex phenomenon). In 1–3 months mature flukes couple and migrate to resting habitats, ie vesical veins (*haematobium*) or mesenteric veins (*mansoni* and *japonicum*). The eggs released from these sites cause granulomata and scarring. Clinical schistosomiasis is an immunological process on the part of the human host which is known to be due to a type IV hypersensitivity reaction (at least for *S mansoni*) to schistosomal eggs.

Clinical features: *S mansoni* occurs in Africa, the Middle East, and Brazil, and presents with abdominal pain and bowel upset, and, later, hepatic fibrosis, granulomatous inflammation and portal hypertension (transformation into true cirrhosis has not been well documented). *S japonicum*, often the most serious, occurs in South-east Asia, tends to affect the bowel and liver, and may migrate to lung and CNS. Urinary schistosomiasis (*S haematobium*) occurs in Africa, the Middle East, Spain, Portugal, Greece, and the Indian Ocean. It starts with frequency, dysuria, haematuria, and incontinence, and may progress to hydronephrosis and renal failure.

Diagnosis is based on finding eggs in the urine (*haematobium*—collect at midday as diurnal variation in egg output; egg has terminal spine) or faeces (*mansoni*—egg has lateral spine, or *japonicum*) or rectal biopsy (all types). Ultrasound is a good screening test for GU morbidity in *S haematobium* infections as it identifies renal congestion, hydronephrosis, thickened bladder wall (but not calcification).

Treatment is with praziquantel (not on UK market): 40mg/kg PO as 2 doses separated by 4–6h for *S mansoni* and *S haematobium*, and 20mg/kg/8h for 1 day in *S japonicum*. Sudden transitory abdominal pain and bloody diarrhoea may occur shortly after treatment.

Fasciola hepatica is spread by sheep, water, and snails. It causes liver enlargement followed by fibrosis. It causes fever, abdominal pain, diarrhoea, weight loss, and mild jaundice with eosinophilia, and is diagnosed on stool microscopy and serology. Praziquantel (as for schistosomiasis) and bithionol are partially effective; triclabendazole is a promising new agent.

Opisthorchis and *Clonorchis* are liver flukes which are common in the Far East, where they predispose to cholangitis, cholecystitis, and cholangiocarcinoma. They are diagnosed on stool microscopy, and are treated with praziquantel.

Fasciolopsis buski is a large intestinal fluke ~7cm long which may cause ulcers and abscesses at the site of attachment. It is treated with praziquantel.

Trematodes (flukes)
[*OTM* 5.571]

Paragonimus westermani (a fluke) is contracted by eating raw fresh-water crabs or crayfish. The parasite migrates through gut and diaphragm to invade the lungs, where it causes cough, dyspnoea, and haemoptysis. Secondary complications: lung abscess and bronchiectasis. It occurs in the Far East, South America, and the Congo, where it is commonly mistaken for tuberculosis (similar clinical and CXR appearances). The sputum contains ova—will be missed if you do not consider diagnosis and only look for AFB. It is treated with praziquantel (25mg/kg/8h PO for 2 days) or bithionol (20mg/kg/12h PO on alternate days for 14 doses).

Infectious diseases

Notes

Notes

253

7 Cardiovascular medicine

254

Relevant pages in other chapters: Cardiogenic shock (p722); cardio-respiratory arrest (p724); cardiovascular examination (p26); carotid bruit (p40); clubbing (p42); cough (p42); cyanosis (p42); dyspnoea (p44); haemoptysis (p46); nodules (p52); oedema (p52); oliguria (p52); palpitations (p52); pulmonary oedema (p726).

Chest pain

Central pain *Nature:* A constricting quality suggests angina, oesophagitis or anxiety; a *sharp* pain may be from pleura or pericardium (both may be exacerbated by deep inspiration, ie pleuritic). A prolonged, intense pain suggests myocardial infarction (MI) ('I thought I was going to die, doctor'). Features which make cardiac pain unlikely are: ● Stabbing pains; ● Pains lasting <30sec, however intense; ● Well-localized left submammary pain ('In my heart, doctor'); ● Pains of continually varying location.

Radiation: To shoulder, either or both arms, or neck and jaw suggests a lesion of the heart, aorta, or oesophagus. The pain of aortic dissection is often in the back. Epigastric pain may be cardiac.

If the pain comes on with exercise, emotion or palpitations, think of cardiac pain or anxiety; if it is brought on by food, lying flat, hot drinks or alcohol, consider oesophagitis. However, a meal may also precipitate angina.

If it is *made better* by stopping exercise, suspect a cardiac lesion. If it is alleviated by antacids suspect oesophagitis. Glyceryl trinitrate relieves both cardiac and oesophageal pain, but acts much more rapidly in cardiac pain. If the pain improves on leaning forward, think of pericarditis.

Associations: If the pain is associated with dyspnoea, consider cardiac pain, pulmonary embolism, pleurisy or anxiety (ask about the peripheral and perioral paraesthesiae of hyperventilation). Myocardial infarction may be associated with nausea, vomiting and sweating, but so are upper GI lesions.

NB: typical angina does not always mean coronary artery disease. Left ventricular outflow obstructions such as aortic stenosis and hypertrophic obstructive cardiomyopathy (HOCM) also give classical angina, as may anaemia. Acute pericarditis can also cause central chest pain, likewise large pulmonary emboli.

Pain that is not central This still may very well be cardiac, but other conditions enter the differential diagnosis.

Pleuritic pain: This is worse on inspiration causing a restriction of breathing. The patient may 'catch his breath'. It implies inflammation of the pleura and localizes the pathology well. While peripheral pulmonary embolism leads to pulmonary infarction and, some days later, pleuritic pain, massive emboli may cause central chest pain resembling myocardial infarction, often with accompanying cardiovascular collapse.

Varicella zoster (shingles): This pain is classically localized to dermatomes and is unaffected by respiration. An area of hyperaesthesia precedes the rash. See p202.

Fractured rib: Gentle pressure on the sternum makes the pain worse. The pain is felt at the site of the fracture.

Ankylosing spondylitis: Note the limited back movement.

Tietze's syndrome (costochondritis): See p710.

Tabes dorsalis: The pain may be 'like a bolt from the blue'.

Gall bladder and *pancreatic disease* may also mimic cardiac pain.

Chest pain
[*OTM* 15.50]

The jugular venous pressure

The internal jugular vein acts as a manometer of right atrial pressure. Observe two features: the height (jugular venous pressure—JVP) and the wave form of the pulse.

Features helping to distinguish venous from arterial pulses: The venous pulse is not usually palpable. It is obliterated by finger pressure on the vein. The venous pressure rises transiently following pressure on the abdomen below the right costal margin (hepatojugular reflux). The JVP alters with changes in posture. Usually there is a double venous pulse for every arterial pulse.

The height Observe the patient at 45°, with his head turned slightly to the left. Look for the right internal jugular vein, which passes just medial to the clavicular head of the sternocleidomastoid up behind the angle of the jaw to the ear lobes. Do not rely on the external jugular vein.

The JVP is the vertical height of the pulse in cm above the sternal angle. It is raised if >3cm.

The waveform
There is a double impulse:
1 *a* wave (atrial contraction)
2 *v* wave—seen at the end of ventricular systole, as tricuspid valve is about to open. It represents maximum venous filling of the right atrium. Between them lie the *x* and *y* descents. The *c wave* represents ballooning back of the tricuspid valve as it closes.

Abnormalities *Raised JVP with normal waveform:* Right heart failure, fluid overload, bradycardia.

Raised JVP with absent pulsation: SVC obstruction (oedema of the head and neck, collateral veins, and absent venous pulsation).

Large a wave: Tricuspid stenosis, pulmonary hypertension, pulmonary stenosis, some cases of LVH (Bernheim effect).

Cannon wave (large *a* wave with rapid fall due to the atrium contracting against a closed tricuspid valve): atria and ventricles contracting simultaneously, eg in complete heart block, atrial flutter, single chamber ventricular pacing, nodal rhythm, ventricular extrasystole and ventricular tachycardia.

Absent a wave: Atrial fibrillation.

Systolic (cv) waves: Tricuspid regurgitation (ie occurring during ventricular systole, combining *c* and *v* waves)—look for earlobe movement.

Slow y descent: Tricuspid stenosis.

Constrictive pericarditis and pericardial tamponade: High plateau of JVP (which rises on inspiration) with deep *x* and *y* descents.

The jugular venous pressure
[*OTM* 13.93; 13.54; 13.8]

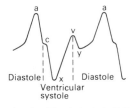

The venous pulse wave. After *Clinical Examination* 4 ed, ed J Macleod 1976, Churchill Livingstone

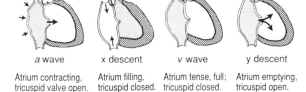

a wave	x descent	*v* wave	y descent
Atrium contracting, tricuspid valve open.	Atrium filling, tricuspid closed.	Atrium tense, full; tricuspid closed.	Atrium emptying, tricuspid open.

Venous pulse. Events occurring in the right atrium are reflected in the jugular waveform.

The heart sounds

▶ Listen systematically: sounds then murmurs.

The heart sounds The first and second sounds are usually clear. Confident pronouncements about other sounds and soft murmurs may be difficult. Even senior colleagues disagree with one another about the more difficult murmurs.

The first heart sound (S_1) represents the closure of mitral and tricuspid valves. It is therefore best heard in mitral and tricuspid areas. It is loud in mitral stenosis (so loud it may be palpable: the 'tapping' apex), if the P–R interval is short and in tachycardia. It is soft in mitral regurgitation and if P–R interval is long. Its intensity is variable in AV block and atrial fibrillation. Splitting in inspiration may be heard and is normal.

The second heart sound (S_2) represents the closure of the aortic and pulmonary valves, producing the A_2 and P_2 components. Splitting in inspiration (ie A_2 and P_2 are heard at very slightly different times) is normal and is due mainly to the variation with respiration of right heart venous return; so it is mainly the pulmonary component which moves, and it is in the pulmonary area that splitting is best heard. Wide splitting occurs in right bundle branch block, pulmonary stenosis, atrial septal defect (wide and *fixed* splitting ie not varying with respiration). Reversed splitting (ie splitting increasing on *expiration*) occurs in systemic hypertension, left bundle branch block, and aortic stenosis (but with aortic stenosis A_2 is often very soft or absent: the single component second sound). A_2 is loud in systemic hypertension. P_2 is loud in pulmonary hypertension, and soft in pulmonary stenosis.

The third heart sound (S_3) occurs shortly after S_2. Normal in young people (<35yrs), it is low pitched and best heard with the bell of the stethoscope. It occurs also in left and right heart failure (heard best in the mitral and tricuspid areas respectively), mitral regurgitation, and constrictive pericarditis (when it is early and more high pitched: the 'pericardial knock').

The fourth heart sound (S_4) occurs just before S_1. Always abnormal, it represents atrial contraction against a ventricle made stiff by any cause, eg aortic stenosis, hypertensive heart disease.

An *ejection systolic click* is heard early in systole in aortic stenosis (but not sclerosis), and systemic hypertension. The right heart equivalent lesions may also cause clicks. Mid-systolic clicks occur in mitral valve prolapse (p306).

An *opening snap* precedes the mid-diastolic murmur of mitral stenosis if the valve is not calcified (listen with diaphragm between apex and lower sternal edge).

Prosthetic sounds are heard from non-biological valves on opening and closing.

Heart sound cadence

To help differentiate between S_3 and S_4, listen to the cadence of the heart sounds.

S_3 sounds like 'Kentucky':
S_1——S_2—S_3
KEN——TU—CKY

S_4 sounds like 'Tennessee':
S_4—S_1——S_2
TE—NNE—–SSEE

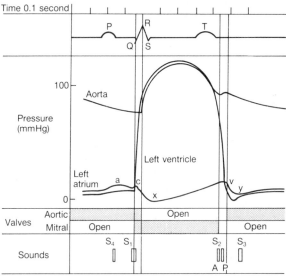

The cardiac cycle. After *Med Intl* **1**(17) 759.

Murmurs

▶ Always consider the other physical signs before listening and answer the question: what do I expect to hear? But do not let your expectations determine what you hear. So while listening, ask yourself: can I hear anything unexpected?

Stenotic murmurs occur when a valve is meant to be fully open, and is not; regurgitant murmurs occur when a valve is meant to be closed, and is not. Think about the cardiac cycle while listening.

Maximize the chance of hearing a murmur by using the stethoscope correctly. The bell should be applied very gently to the skin and is particularly good for hearing low-pitched sounds (eg mitral stenotic murmur). The diaphragm filters out low pitches and makes higher-pitched murmurs easier to detect (eg aortic regurgitant murmur). A bell tightly applied to the skin becomes a diaphragm.

Left heart murmurs are accentuated in expiration and right heart murmurs in inspiration. Other manoeuvres which alter the haemodynamics may be useful. Exercise brings out the murmur of mitral stenosis. The Valsalva manoeuvre makes the murmur softer in mitral regurgitation and aortic stenosis but louder in mitral valve prolapse and hypertrophic obstructive cardiomyopathy; prompt squatting has exactly the opposite effects. Inhalation of the vasodilator amyl nitrate intensifies the murmur of aortic stenosis, but softens that of mitral regurgitation.

Some murmurs will be better heard by bringing the relevant part of the heart closer to the stethoscope eg mitral stenosis in the left lateral position, aortic regurgitation by leaning forward.

Systolic murmurs An *ejection systolic murmur* (ESM; crescendo–decrescendo; diamond-shaped (◆) on phonocardiogram) usually originates from the outflow tract and waxes and wanes with the intraventricular pressures. The louder it is, the more likely it is to be significant. Flow murmurs are common in children and high output states (eg tachycardia, pregnancy) and do not have an associated thrill. Organic causes include aortic stenosis and sclerosis, hypertrophic obstructive cardiomyopathy, and pulmonary stenosis.
A *pansystolic murmur* (PSM) is of uniform intensity and merges with the second sound. It is usually organic and may represent mitral or tricuspid regurgitation (S_1 may also be soft in these), or a ventricular septal defect (p320). Mitral valve prolapse may produce a late systolic murmur.

Diastolic murmurs An *early diastolic murmur* (EDM) is high pitched and easily missed: if you suspect it clinically, then listen in early diastole for 'the absence of silence'. An EDM occurs in aortic (and rarely pulmonary) regurgitation. If the pulmonary regurgitation is secondary to pulmonary hypertension which is itself due to mitral stenosis, then the early diastolic murmur is called a Graham Steel murmur.
Mid diastolic murmurs (MDM) are low pitched and rumbling. They occur in mitral stenosis, rheumatic fever (Carey Coombs' murmur: due to thickening of the mitral valve leaflets) and aortic regurgitation (Austin Flint murmur: due to fluttering of the anterior mitral valve cusp caused by the regurgitant stream).

Murmurs

ECG—a methodical approach

▶ Go through every ECG methodically.

- *Patient's name* and *age*.
- *Date*.
- *Rate:* See p266.
- *Axis:* See p266–7.
- *Rhythm:* If the cycles are not clearly regular, use the 'card method': lay a card along ECG and mark the positions of three successive R-waves. Slide the card to and fro to check all intervals are the same. If there are irregularities, note if different rates are multiples of each other (ie varying block), if changes are abrupt (if so, concentrate on that first complex), or if it is totally irregular (ie AF).
- *The initiation of depolarization* (p291): *Sinus rhythm* is characterized by one normally shaped P-wave preceding each QRS complex. *Atrial fibrillation* (AF) has an irregular baseline and no discernible P-waves; the QRS complexes are irregularly irregular (but can be deceptively regular when the patient is controlled on digoxin). *Atrial flutter* has a baseline of 'sawtooth' atrial depolarization with regular QRS complexes. *Nodal rhythm* has a normal QRS complex but P-waves which are absent or occur just before (ie very short P–R) or within the QRS complex. *Ventricular rhythm* has QRS complexes which are >0.12sec with P-waves following them (p293, paced rhythm p295).
- *P-wave shape:* 'P mitrale' (large LA): bifid P-wave, >0.11sec in II and biphasic in V1. 'P pulmonale' (large RA): tall (>2.5mm) P-wave in II; dominant initial positive vector in P-wave in V1.
- *P–R interval: from start of P-wave to start of QRS complex.* Normal range 0.12–0.21sec. >0.21sec means delay in atrioventricular conduction (heart block, p265). *First degree block* is a constantly prolonged P–R interval in each cycle. In *second degree block* some P-waves are not followed by a QRS; the degree of block may be quantified eg 2:1, 3:1. If the P–R interval increases with each cycle until there is a P-wave which is not followed by a QRS complex, then reverts to a shorter P–R, this is the Wenkebach phenomenon (Mobitz type I). If the P–R interval is constant in 2° block, it is called Mobitz type II. In *third degree block* Ps and QRSs are completely independent of each other. A consistently short P–R interval implies unusually fast AV conduction down an accessory pathway (eg Wolff–Parkinson–White or Lown–Ganong–Levine syndromes, see p288, p291, p704) or nodal rhythm.
- *QRS width:* If >0.12sec suggests conduction defects (p266, p287).
- *QRS height:* Ventricular hypertrophy (p266).
- *Pathological Q-waves:* >25% of succeeding R-waves; >0.04sec wide.
- *Q–T interval:* From start of QRS to end of T-wave. Varies with heart rate. Estimate interval for rate of 60/min using Bazett's formula: Q (corrected) = $(Q-T \text{ interval})/(\sqrt{R-R})$ where R is time between successive R-waves in secs [reference interval for Q–T interval (corrected) is: 0.35–0.43sec].
- *ST segment:* Planar elevation (>1mm) or depression (>0.5mm) implies infarction (p287) and ischaemia (p271) respectively. Widespread saddle-shaped elevation occurs in pericarditis (p319).
- *T-wave* is usually abnormal if negative in I, II, V4–6. It is peaked in hyperkalaemia (p269), and flattened in hypokalaemia.
- *U-waves* may be prominent in hypokalaemia but may be normal.

ECG nomenclature

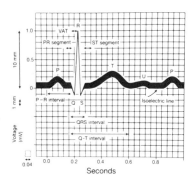

VAT=ventricular
activation time

Heart block

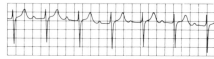

First degree AV block. P-R interval=0.28s

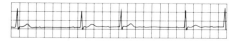

Mobitz type I (Wenkebach) AV block

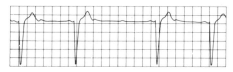

Mobitz type II AV block. Ratio of AV conduction varies from 2:1

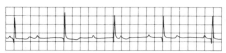

Complete AV block with narrow ventricular complexes.
There is no relation between atrial and the slower
ventricular activity

After Goldman 1976, *Principles of Electrical Cardiography* 9 ed, Lange

ECG—additional points

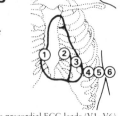

Where to place the chest leads
V1: right sternal edge, 4th intercostal space
V2: left sternal edge, 4th space
V3: half-way between V2 and V4
V4: the patient's apex beat
   All subsequent leads are in the
   same horizontal plane as V4
V5: anterior axillary line
V6: mid- axillary line
(V7: posterior axillary line)

The precordial ECG leads (V1–V6)

Finish a 12-lead ECG with a long rhythm strip in lead II

Rate At usual speed (25mm/sec) each 'big square' is 0.2sec; each 'small square' is 0.04sec. To calculate the rate divide 300 by the R–R interval (in big squares).

The mean frontal axis is the sum of all the ventricular forces during ventricular depolarization. (See p267.)
The 6-limb leads are positioned in the coronal plane (p267).
Normal axis is between −30° and +120° (some say +90°).
Left axis deviation is more negative than −30°.
Right axis deviation is more positive than +120° (some say +90°).

Disorders of the ventricular conducting system
Bundle branch block (ECGs p285 & p287) The evidence of delayed conduction is QRS >0.12sec. Abnormal conduction patterns lasting <0.12sec are incomplete blocks. The area that would have been reached by the blocked bundle depolarizes slowly and late. Take V1 as an example. In V1 right ventricular depolarization is +ve and left ventricular depolarization is −ve. Hence, if the last peak in the QRS complex in V1 is above the isoelectric line (ie positive), it is the right ventricle which is depolarizing late. This is right bundle branch block (RBBB). If the last peak is below the isoelectric line, it is left bundle branch block (LBBB). In V6 the converse is true. ▶ If there is left bundle branch block, no comment can be made on the ST segment or T-wave.

Bifascicular block is the combination of right bundle branch block and left bundle hemiblock (manifest as an axis deviation—eg LAD=left anterior hemiblock).

Trifascicular block is the combination of bifascicular block and first degree heart (atrioventricular) block.

Ventricular hypertrophy can be suggested by unusually large QRS complexes, but there is a great overlap between normals and hypertrophy. The coexistence of a *strain pattern* of ST depression and inverted T-waves in the chest leads (in V1–3 for right ventricular hypertrophy, V4–6 for left) is strong supportive evidence of hypertrophy. The strain may represent ischaemia of the bulky ventricle.
Examples of cited voltage criteria for hypertrophy are:
   Left: the sum of S in V1 and R in V5 or V6 >35mm.
   Right: dominant R in V1 where there is a narrow QRS excluding RBBB.
Differential diagnosis: true posterior infarction (T-wave is upright), or Wolff–Parkinson–White syndrome type A (delta-wave).

How to determine the mean frontal axis

Example 1:

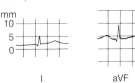

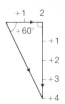

1 In lead I, net sum of forces in QRS complex is +2mm (3mm positive deflection from which subtract 1mm negative deflection);
2 In lead aVF, it is +4mm (5mm positive, 1mm negative);
3 Add the vectors of these two leads together in a nose-to-tail manner both are positive, thus they are oriented *in the direction* of the lead;
4 The mean frontal axis is the resultant, ie approximately 60°.

267

Example 2:

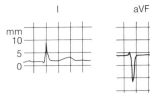

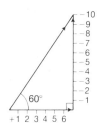

1 In lead I, net sum of forces is +7mm (8mm positive deflection, 1mm negative deflection);
2 In lead aVF, it is −10mm (1mm positive, and 11mm negative);
3 Add these vectors nose-to-tail, taking into account the fact that aVF is negative, and thus goes *away from the direction* of the normal vector of aVF (ie −90° rather than +90°);
4 The mean frontal axis is thus approximately −60°. This is left axis deviation.

Some rules of thumb to determine the ECG axis
● If the complexes in I and II are both predominantly positive, the axis is normal.
● The axis lies at 90° to an isoelectric complex (ie positive and negative deflections are equal in size).

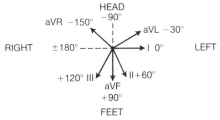

ECG—abnormalities

Myocardial infarction: (For ECG examples, see p281.)
- Within hours, the T-wave becomes abnormally tall, and the ST segment begins to rise.
- Within 24h, the T-wave inverts, as ST elevation begins to resolve.
- Pathological Q-waves also begin to form within a few days.
- Q-waves usually persist, but may resolve completely in 10%.
- T-wave inversion may or may not persist.
- ST elevation rarely persists, except with ventricular aneurysm.

The leads affected reflect the site of the infarct: inferior (II, III, AVF), anteroseptal (V1–4), anterolateral (V4–6, I, AVL), true posterior (mirror image changes in V1–2: tall R and ST↓). Certain infarcts have ST and T changes without Q-waves ('non-Q-wave infarct').

Pulmonary embolism: Sinus tachycardia with an otherwise normal ECG is commonest; may be deep S-waves in I, pathological Q-waves in III, inverted T-waves in III ('SɪQɪɪɪTɪɪɪ'), strain pattern V1–3 (p266), right axis deviation, RBBB, atrial fibrillation.

Digoxin effect: ST depression and inverted T-wave in V5–6 (reversed tick). Changes are more extensive in digoxin *toxicity* (any arrhythmia, but ventricular ectopics and nodal bradycardia are common).

Hyperkalaemia: Tall, tented Ts, wide QRS, absent Ps. (ECG p269.)

Hypokalaemia: Small T-waves, prominent U waves, ventricular bigemini (a premature extrasystole coupled to each sinus beat).

Hypercalcaemia: Shortens Q–T interval.

Hypocalcaemia: Prolongs Q–T interval. Acute pericarditis: p318.

Causes of some ECG abnormalities

Right axis deviation: Right ventricular hypertrophy or strain (eg pulmonary embolism), cor pulmonale, pulmonary stenosis, but not right bundle branch block. Occurring alone, with a normal QRS, it represents left posterior hemiblock.

Left axis deviation: Left ventricular hypertrophy or strain (eg hypertension, aortic stenosis, HOCM), but not left bundle branch block. Occurring alone, with a normal QRS, it represents left anterior hemiblock.

ST elevation: Infarction, variant (Prinzmetal's) angina, pericarditis, ventricular aneurysm.

ST depression: Ischaemia, hypertrophy with strain, digoxin.

Heart block: Ischaemic heart disease, drugs (digoxin, verapamil), myocarditis, cardiomyopathy, fibrosis.

Sinus tachycardia: Exercise, emotion, anxiety, pain, fever, MI, shock, heart failure, drugs (eg atropine, nitrates, adrenaline), constrictive pericarditis, myocarditis, endocarditis, hyperthyroidism, smoking, coffee.

Sinus bradycardia: Athletes, vasovagal attack, drugs (eg β-blockers, verapamil, digoxin), infarction (especially inferior), sick sinus syndrome, hypothyroidism, hypothermia, raised intracranial pressure, jaundice.

ECG—abnormalities

[*OTM* 13.24]

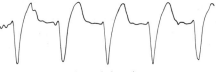

Hyperkalaemia

Exercise ECG testing

This test is used for the diagnosis of ischaemic heart disease, assessment of exercise tolerance, response to treatment, (including surgical revascularization), and prognosis. It is also of value in the assessment of exercise-related arrhythmias.

The patient undergoes a standardized (eg Bruce) protocol of increasing exercise, with continuous 12-lead ECG and blood pressure monitoring. Watch for ST depression, but do not be fooled by J-point depression with steeply upward sloping ST segments: this is associated with a very low incidence of ischaemic heart disease. (The J-point is the junction of QRS complex and ST segment—see p271.)

Stop the test if:
- ST segment depression >2mm.
- The patient is exhausted or in danger of falling.
- There is a ventricular arrhythmia (not just ectopics).
- The blood pressure falls.
- The patient has chest pain (but if it is a diagnostic test, you will have to continue looking for ECG changes—be careful).

Contraindications: Electrolyte disturbance, unstable angina, recent (<7 days) infarction, severe aortic stenosis, pulmonary hypertension, LBBB (p266, you cannot interpret ECG changes—do a thallium scan to assess ischaemia), HOCM, acute myopericarditis, severe hypertension.

Interpreting the test: A positive test only allows one to assess the *probability* that the patient has ischaemic heart disease. 75% with significant coronary artery disease have a positive test, but so do 5% of people with normal arteries. The more positive the result, the higher the predictive accuracy. Down-sloping ST depression is much more significant than up-sloping, eg 1mm J-point depression with down-sloping ST segment is 99% predictive of 2–3 vessel disease.

Mortality: 10 in 100,000; *Morbidity:* 24 in 100,000.

Ambulatory ECG monitoring

Continuous ECG monitoring for 24h (extended to 72h in HOCM) may be used to try and pick up paroxysmal arrhythmias. About one-fifth will have a normal ECG during symptoms: investigate other systems, if appropriate. About 10% will have an arrhythmia coinciding with symptoms.

However, 70% of patients will not have symptoms during the period of monitoring, so little is gained. Give these patients a recorder they can activate themselves during an episode (eg 'Cardiac memo'—least cumbersome, greater monitoring time, and may be able to download information to cardiology department via a telephone).

Recorders may be programmed to detect ST segment depression, either symptomatic (to prove angina), or to reveal 'silent' ischaemia (predictive of reinfarction or death early after MI).

Exercise ECG testing
Ambulatory ECG monitoring

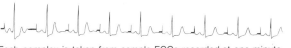

Each complex is taken from sample ECGs recorded at one minute intervals during exercise (top line) and recovery (bottom line). At maximum ST depression, the ST segment is almost horizontal. This is a positive exercise test.

This is an exercise ECG in the same format. It is negative because although the J point is depressed, the ensuing ST segment is steeply upsloping.

Cardiac catheterization

This involves the insertion of a catheter into an artery (to examine the left heart and perform a coronary arteriogram) or vein (for right heart imaging). The catheter is manipulated within the heart and great vessels to measure pressures, sample blood to assess oxygen saturation, and for the injection of radio-opaque contrast medium in conjunction with cine-film recording to look at blood flow and the anatomy of the heart and vessels. Using the catheter, it is now possible to insert angioplasty and valvuloplasty balloons, and to perform cardiac biopsy.

Indications:
- Coronary artery disease: diagnostic (including assessment of bypass graft patency) and therapeutic (angioplasty—balloon or laser).
- Valve disease: assessment of severity and therapeutic valvuloplasty. Aortic valvuloplasty (if too ill or declines valve surgery) may be attempted using a retrograde (from the aorta) or a transseptal approach. Mitral and pulmonary valvuloplasty is also performed.
- Other: Congenital heart disease; cardiomyopathy; pericardial disease; endomyocardial biopsy.

Morbidity: Arrhythmias, thromboembolism (coronary and systemic), infection, trauma to heart and vessels (eg dissection).

Pre-procedure assessment: Check peripheral pulses and general fitness. ECG, CXR, FBC, platelets, U&E, clotting screen, HBsAg, group & save blood.

Post-procedure: Regularly check peripheral pulses beyond catheter site; check site for bleeding.

Normal values of pressures and saturations

Location		Pressure (mmHg)	Saturation (%)
Inferior vena cava			76
Superior vena cava			70
Right atrium:	a	2–10	
	v	2–10	
	mean	0–8	
Right ventricle:	systolic	15–30	74
	end diastolic	0–8	
Pulmonary artery:	systolic	15–30	
	end-diastolic	3–12	
	mean	9–16	
Pulmonary capillary wedge:	a	3–15	
	v	3–12	
	mean	1–10	
Left ventricle:	systolic	100–140	98
	end-diastolic	3–12	
Aorta:	systolic	100–140	
	end-diastolic	60–90	

Gradients (mmHg) across stenotic valves

Valve	Normal gradient	Stenotic gradient		
		Mild	Moderate	Severe
Aortic	0	<30	30–50	>50
Mitral	0	<5	5–15	>15
Prosthetic	5–10			

Intracardiac electrophysiological studies via a cardiac catheter can determine the type and origin of arrhythmias and locate aberrant pathways. Arrhythmias may be induced and the effectiveness of control by various drugs assessed. Aberrant pathways can be electrically ablated with good success.

Coronary artery anatomy

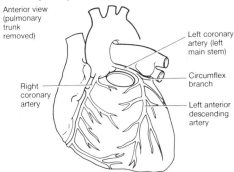

Anterior view (pulmonary trunk removed)

Left coronary artery (left main stem)

Circumflex branch

Right coronary artery

Left anterior descending artery

Angiographic views

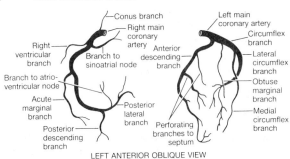

1

RIGHT CORONARY SYSTEM

Conus branch

Right main coronary artery

Right ventricular branch

Branch to sinoatrial node

Branch to atrioventricular node

Acute marginal branch

Posterior lateral branch

Posterior descending branch

LEFT CORONARY SYSTEM

Left main coronary artery

Circumflex branch

Lateral circumflex branch

Obtuse marginal branch

Medial circumflex branch

Anterior descending branch

Perforating branches to septum

LEFT ANTERIOR OBLIQUE VIEW

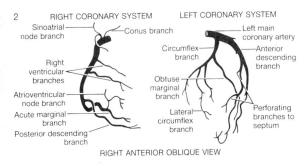

2

RIGHT CORONARY SYSTEM

Sinoatrial node branch

Conus branch

Right ventricular branches

Atrioventricular node branch

Acute marginal branch

Posterior descending branch

LEFT CORONARY SYSTEM

Left main coronary artery

Circumflex branch

Anterior descending branch

Obtuse marginal branch

Lateral circumflex branch

Perforating branches to septum

RIGHT ANTERIOR OBLIQUE VIEW

1. Reproduced with permission from 'Understanding Angina'. © American Heart Association. 2. From M. Sokolow, Clinical Cardiology, Lange, p170, with kind permission.

Echocardiography

This non-invasive technique uses the differing ability of various structures within the heart to reflect ultrasound waves. It not only demonstrates anatomy but provides a continuous display of the functioning heart throughout its cycle.

2D echocardiography produces a picture of a two-dimensional fan-shaped plane. In this way one sees eg all four chambers in a view from the apex or a cross-section of the aortic valve from the parasternal region.

M-mode echocardiography offers a picture of the structures in a single narrow beam beneath the transducer. By holding the transducer still, one can record fine movements of the valves which can be timed and measured more easily than in the 2D mode.

Doppler and colour flow echocardiography will illustrate flow and gradients across valves and septal defects (p320).

Uses of echocardiography

Valve disease: Mitral stenosis is characteristic on M-mode, but aortic stenosis is better demonstrated by 2D. Neither technique is diagnostic in regurgitant lesions which are best evaluated by Doppler. Doppler can also quantify the gradient across a stenosis.

Cardiomegaly: Focal and global hypokinesia, aneurysms, and pericardial effusions are all well demonstrated on 2D.

Vegetations and *mural thrombi* are less reliably shown.

Pericardial effusion is best diagnosed echocardiographically, using either technique.

Nuclear cardiology

Radionuclide ventriculography is a less invasive investigation than cardiac catheterization, and is very effective at assessing gross ventricular anatomy (eg global and focal dyskinesias) and function (eg ejection fraction). Technetium-99 is used to label red cells or albumin.

Myocardial perfusion is assessed using thallium-201 (its distribution mirrors that of K^+) injected just before maximum exercise on a treadmill. Exercise scans are compared with resting views taken $2\frac{1}{2}$–3h later: fixed defects (infarct) or reperfusion (ischaemia) can be seen and the coronary artery involved is reliably predicted.

Myocardial infarction can be labelled by technetium pyrophosphate as it is not taken up by normal tissue. This is valuable if the patient presents when it is too late for enzymes to be of value (but within 10 days of infarction), and/or if the ECG is unhelpful (eg LBBB).

See also positron-emission tomography, p794.

Echocardiography
Nuclear cardiology

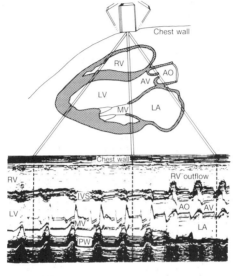

ANT

e

f

POST

Normal mitral valve

CLOSED | OPEN | CLOSED

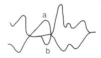

2. Mitral stenosis
• reduced e-f slope

3. Aortic regurgitation
• fluttering of ant. leaflet

a

b

4. (a) Systolic anterior leaflet movement (SAM) in HOCM
(b) Mitral valve prolapse (late systole)

Normal M-mode echocardiogram. (RV = right ventricle; LV = left ventricle; AO = aorta; AV = aortic valve; LA = left atrium; MV = mitral valve; PW = posterior wall of LV; IVS = interventricular septum. After R Hall *Med Intl* (17) 774.

Cardiovascular drugs

Beta-blockers Used for decreasing cardiovascular sympathetic effects. Blockade of β_1 adrenoreceptors is negatively inotropic and chronotropic, and delays AV conduction; blockade of β_2 receptors causes coronary and peripheral vasoconstriction and bronchoconstriction—so blocking agents which are *relatively* specific for β_1 receptors ('cardioselective') have been developed (atenolol, metoprolol, and bisoprolol). β-blockers are used to diminish myocardial oxygen demand in coronary artery disease, and to control hypertension and certain arrhythmias. They are known to have a beneficial effect prophylactically following a myocardial infarction (p284). CI: asthma, obstructive airways disease, claudication (β_2 effect), heart failure/heart block (unhelpful β_1 effects). Use cautiously in unstable diabetes. Other SE: nightmares, depression, $\downarrow$glucose tolerance, lethargy, impotence, headache.

Diuretics are used in hypertension (thiazides only) and heart failure. SE: *frusemide:* $\downarrow$K$^+$, ototoxicity; *thiazides:* $\downarrow$K$^+$, $\downarrow$Mg^{2+}, $\uparrow$glucose, $\downarrow$platelets; *amiloride:* $\uparrow$K$^+$, GI upset.

Vasodilators decrease the work of the heart in heart failure and coronary artery disease, and $\downarrow$BP in hypertension. They may be predominantly venodilators (eg nitrates) which decrease the filling pressure (pre-load), arterial dilators (eg hydralazine) which primarily bring down the systemic blood pressure (after-load), or a mixture (eg prazosin). Captopril and enalapril vasodilate by inhibiting angiotensin converting enzyme (ACE). SE: severe hypotension, so start with low dose (eg captopril 6.25mg PO) and plan its introduction carefully (see p299).

Calcium antagonists These $\downarrow$ cell entry of Ca^{2+} via voltage-sensitive channels on smooth muscle cells and myocytes. Ca^{2+} channels differ in tissues, and Ca^{2+} antagonists vary in selectivity for binding sites.[1]
Indications: Angina ($=$A, below), dysrhythmias^D, hypertension^H.
Typical doses: Phenylalkylamines: Verapamil^{ADH} 40–120mg/8h PO. *Benzithiazepines:* Diltiazem^{AH} (p396). *Dyhydropyridines:* Nifedipine^{AH} 20mg/12h PO pc (slow-release), nicardipine^{AH} 20mg/8h PO, amlodipine^{AH} 5mg/24h PO, isradipine^H 2.5–5mg/12h PO, felodipine^H 5–10mg/24h PO, lacidipine^H 2–6mg/24h PO. Effects include:
• Coronary vasodilatation (diltiazem and verapamil may be better than nifedipine; nifedipine is good if there is coronary artery spasm, p706).
• Slowing of atrioventricular and sinoatrial conduction (antiarrhythmic). NB: diltiazem allows normal pulse increase at peak exercise; nifedipine causes reflex tachycardia (so use a β-blocker). Verapamil must not be given with β-blockers. Amlodipine allows a once-a-day dose régime.
• SE: flushes; headache; oedema (diuretic unresponsive); LV function$\downarrow$.

Digoxin is used to slow ventricular rate in fast AF. It is dangerous in HOCM (p316) and WPW (p288). It has only a very weak +ve inotrope and is generally not now used in heart failure. Dose example: p290. Old people are especially at risk of toxicity: start them on low doses (0.0625mg/24h PO). Toxicity is also more likely with K$^+$$\downarrow$, and in renal failure. Blood levels may be measured. Someone on digoxin should receive less current in cardioversion (start with 5J). SE: any arrhythmia (SVT with AV block is highly suggestive), GI upset, yellow vision, confusion, gynaecomastia. In toxicity stop digoxin, check K$^+$, and treat arrhythmias. Consider digoxin-specific antibodies (Digibind®, p748) to promote excretion.

---

1 *Drug Ther Bul* 1993 **31** 81

Cardiovascular drugs

Anti-arrhythmics may be classified by main site of action:
1 Atrioventricular node: Adenosine, verapamil, β-blockers, digoxin.
2 Ventricles only: Lignocaine, mexiletine, phenytoin, flecainide.
3 Atria, ventricles, and bundle of Kent: Quinidine, procainamide, disopyramide, amiodarone.

Angina pectoris

This is central, crushing chest pain which may radiate to the jaw, neck, or one or both arms. It may only be felt in the jaw or arm. It represents myocardial ischaemia. It is precipitated by exertion (often predictably), anxiety, cold weather and heavy meals, and may be associated with dyspnoea and faintness. It is relieved by rest and nitrates. There are usually no physical signs. Angina remits spontaneously in $1/3$ of patients.

Causes Usually coronary artery disease. Rarely: valve disease eg aortic stenosis; HOCM; hypoperfusion from arrhythmias; arteritis; anaemia.

Incidence 5/1000/yr in males over 40yrs.

Tests ECG: look for ST depression and flat or inverted T-waves, also Q-waves from old MIs. If the resting ECG is normal (it usually is), consider an exercise ECG (p270), or exercise radionuclide scanning (p274). The availability of angiography in the UK is limited. FBC and ESR exclude non-atheromatous causes (eg anaemia, polycythaemia, giant cell arteritis).

Management Optimize weight and fitness. Avoid risk factors (p284).

Drugs: Start with 0.5mg glyceryl trinitrate (GTN) SL or spray (0.4mg/intra-oral puff) PRN, up to every $1/2$h. Pills last ≤8 wks; sprays last years. Work up to maximum tolerated doses of 'triple' therapy—e.g.:
- β-blockers: atenolol 50–100mg/24h PO. CI: asthma, LVF, bradycardias.
- Calcium antagonists: restrict slow channel Ca^{2+} entry into cells, so reducing Ca^{2+} release from sarcoplasmic reticulum with 3 good effects: 1 Myocardial O_2 consumption↓; 2 Dilatation of coronary arteries; 3 Peripheral vasodilatation. Example: *nifedipine* 10–20mg/8h PO; SR form (slow release) 20mg/12h. SE: flushes, headache, ankle swelling (no response to diuretics), headache, gum hyperplasia. CI: fertile women. Reflex tachycardia may be a problem if β-blockers cannot be used—use *diltiazem* here: eg 60mg/8h PO, max 160mg/8h—less if elderly; (SR tabs are 90, 120, 180 or 300mg); CI: pregnancy, pulse↓, heart block.
- Isosorbide dinitrate 5–40mg/6h PO (slow release: 20–60mg/24h, more frequent doses easily cause tolerance), or try adhesive nitrate skin patches (5–10mg/24h, remove for 4–8h/24h) or buccal absorption (eg 1mg, 2mg or 3mg Suscard Buccal® pills). SE: headaches.

Surgery: Coronary artery bypass grafting (CABG) is indicated if drugs fail to control angina or are not tolerated. Surgery is for symptoms; mortality is only reduced in left main stem and possibly triple vessel disease. Mortality is <2% (but operator variable). 70% get total relief and 20% improvement of symptoms—but pain returns to 50% in 5yrs. Aspirin 75–300mg/24h PO ↓risk of graft closure. An alternative is percutaneous transluminal angioplasty. A balloon catheter is inflated at the site of a discrete stenosis, but the risk of complete occlusion demands full surgical back-up. 33% relapse compared with 11% post-CABG at 6 months, and 30% go on to require CABG.[1] Mortality and MI rates are similar.

Unstable angina is angina which is rapidly worsening, eg on minimal or no exertion. Treatment: diltiazem[2] ± β-blocker as above; isosorbide dinitrate 2–10mg/h IVI; aspirin eg 75mg/24h PO; consider heparin (SC or IVI). Monitor pulse and BP. ~15% die in 1yr if untreated, so consider prompt angiography. If pain is uncontrolled, emergency angiography and bypass grafting or angioplasty is needed. Prinzmetal angina: p708.

Mortality 0.5–4%/yr (sudden death, MI, LVF), depending on number of affected vessels. It is doubled if there is left ventricular dysfunction.

1 RITA (Randomized Intervention Treatment of Angina) 1993 *Lancet* 341 573 2 S Spinler 1991 *Clinical Pharmacy* 10 825

Angina pectoris
[*OTM* 13.143]

Myocardial infarction (MI)

This is death of heart muscle. Incidence: 5/1000 per yr (UK). 90% of acute transmural MIs (ECG Q-waves heralded by ST elevation in several leads) are caused by thrombosis, eg following rupture of an atheromatous coronary artery plaque (so amenable to thrombolysis). Non-Q-wave MIs and pre-infarction states may be due to intermittent platelet plugs (amenable to heparin and aspirin treatment), and here the ECG may be normal.

Mortality In one study, of 100 'sudden deaths' and MIs:
   25 died before reaching hospital,
   13 more died before discharge (8 within 1st week),
   7 more died within the 1st year; ie 55 were alive at the end of 1 year.

Symptoms The pain is usually of greater severity and duration (>30mins) than in angina, but similar in nature, and associated with nausea, vomiting, sweating, and extreme distress. Painless ('silent') infarcts are quite common (eg in the elderly).

Signs Distressed, cold, clammy, tachycardia. BP may be up or down. May be cyanosed and have a mild pyrexia (<38.5°C).
Also features of complications—eg LVF: dyspnoea, crepitations.

Tests ECG (p268): ST elevation, T inversion, and Q-waves are typical in the leads adjacent to the infarction. Leads *opposite* the site of the infarct (eg V1–4 in inferior MI; II, III, and avF in anterolateral MI) may show 'reciprocal' ST depression. The absence of Q-waves in a proven infarct is associated with higher risk of subsequent MI.

CXR: Exclude aortic dissection, look for signs of failure, watch for changes in cardiac size. U&E. FBC. ESR can rise to 80mm/h and take some weeks to return to normal. Similarly glucose can be upset, and lipids may be raised from the first 24h post-MI for up to 3 weeks.

Cardiac enzymes:

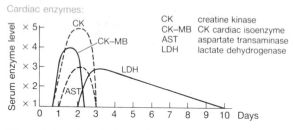

CK	creatine kinase
CK–MB	CK cardiac isoenzyme
AST	aspartate transaminase
LDH	lactate dehydrogenase

Diagnosis is made on the basis of 2 or 3 out of the history, ECG changes, and enzyme changes being suggestive of an MI.

Differential diagnosis: Angina, pericarditis, myocarditis, aortic dissection, massive pulmonary embolism, pleurisy, oesophagitis + spasm, peptic ulcer, pancreatitis, cholecystitis, shingles, nerve root lesion (especially C8), chest wall pathology.

Aortic dissection presents with severe interscapular pain and often with lost peripheral pulses and different BPs in each arm. CXR: widened mediastinum. *Treatment:* bed rest; hypotensives (keep systolic BP <110mmHg). Transfer for urgent arch aortogram.

---

1 KS Channer, 1993 *BMJ* **ii** 1146   **2** AS Wierzbicki 1993 *Lancet* **341** 890

Sequential ECG changes following acute MI

Acute inferolateral infarction with minimal 'reciprocal' ST changes in I, AVL, and V_2–V_3.

Immediate management of MI

▶▶ Greatest risk of death is in the 1st hour; correcting arrhythmias and giving thrombolysis *at home*[1] can lead to a 50% reduction in mortality compared with waiting until hospital admission. See pre-hospital care (*Thrombolysis at home?*, OHCS 4th ed p804).

A patient presenting to his GP days after his MI is at lower risk; manage at home, unless pain relief or complications are problems.

Management in CCU ▶ Your patient is terrified; reassure him—and find time to talk to the relatives.
- 35% oxygen (unless COAD). ● Site an IV cannula for emergencies.
- Diamorphine 5–10mg IV at ≤1mg/min (+ anti-emetic cyclizine 50mg IV) for analgesic, anxiolytic, anti-arrhythmic, and venodilator effects.
- Glyceryl trinitrate (0.5mg SL) for coronary artery vasodilatation.
- If presentation is <~12h after the onset of pain, give streptokinase (SK) 1.5 million units in 100ml 0.9% saline IVI over 1h + aspirin (160mg/day PO for ≥1 month[1]). *Prompt* treatment with a fibrinolytic and an antiplatelet drug reduces mortality by ~43% (from 13% to 8%).[2] Some also give heparin (eg 5000U IV stat, then 1000U/h IVI for 24h).[3] *SK side effects:* Hypotension (use IV fluids and raise legs); anaphylaxis is rare; overall incidence of stroke is not increased. *SK contraindications:* Stroke or active bleeding in past 2 months, BP >200mmHg, surgery or trauma (eg cardiac massage) in past 10 days, bleeding disorder (eg on anticoagulants), pregnancy, menstrual period, proliferative DM retinopathy, previous SK or APSAC treatment in last 5 days to 1yr (use rt-PA here, p593; those studies showing an rt-PA advantage vs SK did *not* use aspirin, and caused excess stroke).[3]
- Heparin 5000U/8h SC until mobilized, as prophylaxis against DVT.
- If pain continues, give isosorbide dinitrate infusion 2–10mg/h IVI. If maximum tolerated dose is inadequate, give diamorphine 5–10mg IV as often as needed. If the pain is still not controlled, refer for angiography with a view to urgent surgery.
- As needed: prochlorperazine 12.5mg IM up to 6-hourly; laxative.
- Discontinue pre-infarction cardiac drugs. ● Prohibit smoking.
- Continuous ECG monitoring. ● 24h bed rest. ● Four-hourly TPR, BP.
- Daily 12-lead ECG, CXR, cardiac enzymes, and U&E for 2–3 days.
- Examine heart, lungs and legs for complications at least daily.

Complications of MI ● Arrhythmias (p286–94)
- Heart failure (p296) or frank cardiogenic shock (p722).
- Hypertension: if BP >200/100mmHg, ensure adequate relief of pain and anxiety. If BP still up, try nifedipine 10mg (bite into capsule and swallow). Do not treat lower BPs.
- DVT and PE; and systemic embolization. ● Dressler's syndrome (p696)
- VSD: presentation is acute cardiac failure; treatment, surgical.
- Papillary muscle rupture: presents with acute pulmonary oedema due to mitral regurgitation; the treatment is urgent surgery.
- Cardiac rupture and subsequent tamponade is similarly devastating and usually irremediable without very rapid surgery.
- LV aneurysm occurs late, presenting as intractable LVF, arrhythmias, or embolization. ECG: persistently raised ST segments. CXR: bulge on L heart border. Anticoagulate. Consider excision.

1 GREAT 1992 *BMJ* ii 548   **2** LIMIT-2 1992 *Lancet* **339** 1553   **3** GUSTO 1993 *NEJM* **329** 673

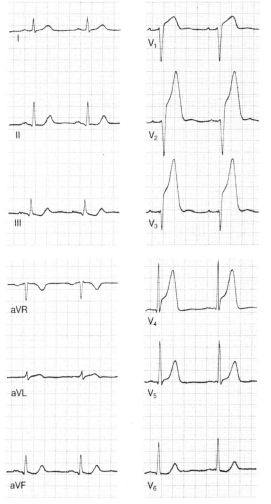

Acute anteroseptal infarction with minimal 'reciprocal' ST changes in III and aVF.

Subsequent management of MI

If uncomplicated, the patient may sit out of bed on day 2, walk freely on day 5 (discontinue heparin now), go home on day 7–10, and return to work after 4–12 weeks, depending on degree of heavy manual labour. He should not drive for 1 month and is responsible for informing the driving licence authorities. They are unlikely to prohibit driving (except heavy goods vehicles). Advise slowly to increase exercise; consider a planned fitness training programme. Specifically mention to him and his partner that sexual activity may be safely resumed. Ask about any fears of returning to a normal life. Discourage risk factors strongly. Arrange for a lipid test if was not done on a sample during the first 24h.

The 5 favourable prognostic signs ● No pre-existing hypertension.
● Normal heart size on CXR. ● No significant arrhythmias after day 1.
● No post-MI pulmonary oedema. ● No post-MI angina. These patients are at very little risk and the best plan is to discharge them not taking any drugs as these encourage doctor-dependency and will have side-effects. Review all other patients at 4–6 weeks for:
● Repeat ECG and CXR (to look for cardiac size and aneurysm).
● If angina recurs, treat it conventionally, perform an exercise test, and consider coronary angiography.
● Start an exercise program *and* a β-blocker (eg atenolol 50mg/24h PO) to improve prognosis.[1] Giving β-blockers to *all* patients—except low-risk patients (above), or if there are contraindications—eg asthma, unstable diabetes (may obviate the need for universal exercise testing followed by selected angiography—a régime which appears to carry less morbidity and mortality than laxer régimes[2]—but probably only because more β-blockers are used).
● Consider aspirin 75mg/24h PO. Warfarin anticoagulation has a very limited effect on mortality, but may reduce re-infarction risk.[3]
● Consider ACE inhibitor (p299) VF (known to ↓ mortality).[4]

Preventing ischaemic heart disease (IHD)

IHD causes 30% male and 22% female deaths in England and Wales. However, the incidence does seem to be falling in UK and USA.

Risk markers Hypercholesterolaemia, smoking, and ↑BP are the principal three. Age is also important. Others include male sex, family history, diabetes, contraceptive pill, alcohol, obesity, geographical factors (eg Scotland high risk, Sweden low), environment (eg climate, soft water), type A personality type, socio-economic class (lower class is higher risk in UK).

Note: the graph of BP against mortality is J-shaped (low-point=75–79mmHg); increased mortality with low BP may be related to high blood viscosity. We know that if viscosity rises, so does BP,[5] and this may be a physiological, to maintain perfusion.

Primary prevention may be possible. Dietary manipulation can lower lipids, and possibly the incidence of IHD. Drug therapy is more controversial (p654). Community health programmes, both general and directed against high risk groups have been shown to have some effects, but their role is not clear. Such changes in life-style may explain the 21% drop in incidence of deaths from IHD in USA from 1968 to 1976. For smoking advice, see OHCS p452. Exercise is also beneficial.[5]

1 G O'Connor 1989 *Circulation* **80** 234–44   **2** J Sanderson 1989 *BMJ* ii 1435   **3** ASPECT 1994 *Lancet* **343** 499   **4** ISIS-4 1995 *Lancet* **345** 969   **5** S N Blair 1995 *JAMA* **272** 1092

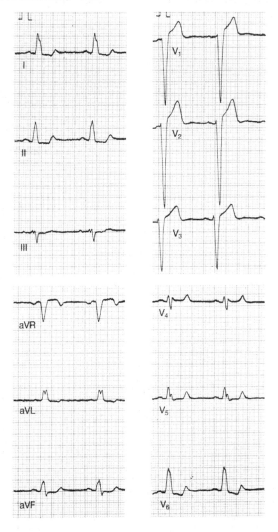

Arrhythmias and conduction disturbances

Causes MI is the commonest setting in which these are seen, usually occurring in the first hour and settling rapidly. They may also occur in the context of chronic ischaemia. (50% of those who die in the 1st month after MI die in the 1st hour; 90% of these are from VF.)

Incidence after acute MI:[1] sinus or nodal bradycardia—38%; atrial flutter or fibrillation—9%; supraventricular tachycardia (SVT)—4%; ventricular extrasystole—57%; ventricular tachycardia (VT)—31%; ventricular fibrillation (VF)—19%.

Conduction disturbances are also common after MI: 25% with inferior MI have 1° or 2° AV blocks which tend to have a benign course. AV block in anterior MI is usually associated with RBBB or LBBB (or hemiblocks). In this situation, asystole is common. Often the AV block resolves, but the intraventricular conduction disturbance persists.

Other causes of arrhythmias and conduction disturbances: Drugs, cardiomyopathy, myocarditis, cardiac dilatation, thyroid disease, electrolyte disturbance. Congenital aberrant pathways cause SVT (p288). RBBB may be a normal finding, LBBB almost never is.

Presentation of arrhythmias Funny turns, collapse, palpitations (ask patient to tap out rhythm to assess regularity and rate). Sometimes they are found incidentally on cardiac monitoring.

Questions ● How do any palpitations start (and stop)? ● Any haemodynamic effects—chest pain, dyspnoea or collapse? ● What drugs is the patient on?

Tests U&E, Ca^{2+}, glucose, thyroid function, FBC, CXR, ECG ± ambulatory ECG (p270). Get a long rhythm strip (P-waves may be best assessed in lead V1; try turning up the 'gain'). Those <50 years old with unexplained arrhythmias need an echocardiogram to exclude eg HOCM (causes ventricular arrhythmias) and mitral stenosis (causes AF). If a ventricular arrhythmia is documented, its origin may be demonstrated by electrophysiological studies at cardiac catheterization. The arrhythmia may then be stimulated and its therapeutic response to various drugs assessed.

Treatment If a normal rhythm is demonstrated during palpitations, reassure the patient. If an arrhythmia is demonstrated, its treatment depends on its type. The patient may be taught autonomic procedures, eg Valsalva manoeuvre for SVT. He may be started on oral preparations, eg verapamil (40–120mg/8h PO) or a β-blocker for SVT. Amiodarone (below) is used for VT. Arrhythmias may be terminated by carotid sinus massage (never do this bilaterally—and have resuscitation facilities to hand) or DC cardioversion. The source of the arrhythmia may be removed, either by transvenous ablation in the catheter lab, or surgically (eg LV aneurysm). Finally, permanent pacing may be used to overdrive tachyarrhythmias, to override bradyarrhythmias, or prophylactically in conduction disturbances (p294). Implanted automatic defibrillators are being evaluated.

Amiodarone: Dose: 200mg/8h PO for 7 days, then 200mg/12h for 7 days. Maintain on 200mg/24h. SE: Corneal deposits, photosensitivity, hepatitis, nightmares, INR↑, hypothyroidism, hyperthyroidism (amiodarone is 37.5% iodine). Monitor LFTs, T_4, T_3. If hyperthyroidism occurs, concurrent antithyroid drugs may be tried.[2] Exogenous thyroxine may be needed later.

1 A Adgey 1971 Lancet ii 501   2 L Reichert 1989 BMJ i 1547

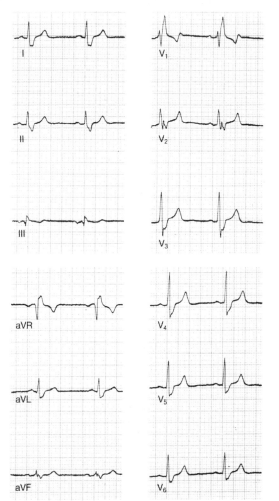

287

Supraventricular tachycardia

The supraventricular tachycardias (SVT) are:
- Sinus tachycardia (p268). • Nodal tachycardia.
- Atrial flutter and fibrillation (p290).
- Paroxysmal atrial tachycardia (PAT) due to:
  (a) re-entry atrioventricular pathway; or
  (b) tachycardia originating from an ectopic atrial focus.

The term SVT has 2 uses: in the acute situation it is used to denote a tachycardia which is not ventricular (VT) (p292). It is relatively easy to define which of the above is the mechanism, and thereafter, SVT is usually reserved for PAT.

Re-entry tachycardias The normal AV node (AVN) has a very slow conduction velocity thus allowing consecutive contraction of atria then ventricles. It also has a long refractory period which prevents a fast atrial rate being conducted to the ventricles, thus allowing adequate diastolic filling time. Some people have a second (accessory) pathway between the atria and ventricles; these may have very different electrophysiological characteristics.

An impulse conducted normally to the ventricles through one AV connection may be reconducted retrogradely up the other, causing a premature atrial contraction. This is then conducted down the first AV connection. On each occasion the ventricle also depolarizes, giving a reentry SVT. It is usually started by an ectopic.

Because of the refractory period of the normal atrio-ventricular node, a ventricular rate of 150 is rarely exceeded. However, if the accessory pathway has a shorter refractory period the ventricular rate may be so fast that the cardiac output falls and the patient collapses. Be alerted to a re-entry SVT by a ventricular rate >200/min.

A special type of re-entrant pathway is the Wolff–Parkinson–White (WPW) syndrome in which the accessory pathway (bundle of Kent) connects to the ventricle some distance from the bundle of His. Thus, after a short PR interval (the bundle of Kent has a rapid conduction velocity), there is a slow depolarization (delta wave) until the depolarization front reaches the bundle of His and rapid ventricular depolarization occurs. The delta wave is only present at rest, and not during the SVT.

Treatment of SVT • Monitor ECG. • Try vagal manoeuvre, eg Valsalva.
- Give adenosine 3mg IV (fast bolus—can give 6mg then 12mg at 2min intervals: safe with β-blockers, and in VT; it may exacerbate bronchospasm in asthmatics).[1] Child's starting dose: 0.0375mg/kg.
- If there is serious haemodynamic compromise, cardiovert (p724).
- If the compromise is moderate, give verapamil 5–10mg slowly IV. ▶ *Do not give IV verapamil to someone who has recently had a β-blocker, and vice versa.*
- Correct any acid–base or biochemical abnormalities (eg K+).
- If there is no compromise, consider starting long-term treatment with digoxin (except in WPW) or amiodarone. One of these agents should be considered also after an acute haemodynamic compromise has been controlled.
- Aberrant pathways may be ablated surgically—or by transcatheter electrical ablation after careful electrophysiological studies.

1 CJ Garrett 1992 BMJ i 3

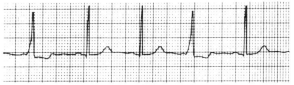

Intermittent Wollf–Parkinson–White syndrome (first and fourth beats). By comparison with the normal beats, it can be seen how the delta wave both broadens the ventricular complex and shorthen the PR interval.

Atrial fibrillation and flutter

Atrial fibrillation is an ineffective, chaotic, irregular and rapid (300–600/min) atrial rhythm. Its discharges approach the AV node from varying angles at varying intervals, thus resulting in a totally irregular ventricular rate. It is common in the elderly (up to 9%) and is often asymptomatic. The chief risk is embolic stroke (~4%/yr); this is much reduced (eg to ~1%/yr[1]) by warfarin anticoagulation.

Common causes: Myocardial infarction and ischaemia, mitral valve disease (commoner in stenosis), hyperthyroidism (may be the only symptom in the elderly), hypertension.

Rare causes: Cardiomyopathy, constrictive pericarditis, sick sinus syndrome, Ca bronchus, atrial myxoma, endocarditis, haemochromatosis, sarcoidosis, any other cause of atrial dilatation. 'Lone' AF means none of the above causes.

Clinical features: Irregularly irregular pulse; first heart sound of variable intensity. The rate at the apex is greater than at radial artery.

Differential diagnosis is multiple ventricular or atrial ectopics, or variable AV block.

Diagnosis is on ECG: Chaotic baseline with no P-waves, and irregularly irregular but normally shaped QRS complexes.

Management: Recent onset AF: Attempt to cardiovert. Warfarinize for one month, then DC shock (200J). If that fails, treat as chronic AF.

Chronic AF: The aim is control of ventricular rate, not sinus rhythm.
- Use digoxin: oral loading dose 0.5mg × 3 doses in 2 days; maintenance 0.25mg/24h PO. (In the elderly, load with 0.75mg in total and use a maintenance dose of 0.0625–0.125mg/24h.)
- If still too fast, check compliance and serum level, then cautiously ↑ dose—± low dose β-blocker (eg propranolol 10–20mg/8h PO).
- Consider low-intensity warfarin anticoagulation (INR 1.4–2.8) unless contraindicated or it is 'lone AF'—when the risk of emboli is small.[1] If contraindicated, or if monitoring is a problem, and in the elderly where bleeding is more likely, consider aspirin 325mg/day PO instead: it is safer and nearly as good,[2] (stroke rate in elderly is 4.6%/yr *vs* 4.3% for warfarin). Warfarin is probably 1st choice if there has been a TIA or minor stroke, or if echocardiography shows abnormalities.

Paroxysmal AF: Quinidine 200–400mg/6–8h PO or amiodarone (load with 200mg/8h PO for 1 week, 200mg/12h for 1 week, then maintenance 100–200mg/24h) may maintain patient in sinus rhythm.

Atrial flutter is a regular circus movement of continuous atrial depolarization with a period of 300/min. The AV node cannot conduct that fast, so there is usually 2:1 or 3:1 block. ▶ Always consider flutter first if a patient has a regular tachycardia of 150/min.

Flutter ECG: Regular sawtooth baseline at a rate of 300/min. This may be difficult to see in 2:1 block, but may be revealed by carotid sinus massage (or adenosine IV, p288) which can increase the block.

Treatment: Treat in the same way as atrial fibrillation.

Some patients seem to vary between flutter and fibrillation.

1 M Ezekowitz 1992 *NEJM* **327** 1406   2 SPAF II 1994 *Lancet* **343** 687

Atrial fibrillation and flutter

[OTM 13.121]

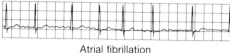

Atrial fibrillation

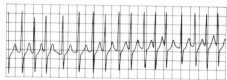

Atrial fibrillation with a rapid ventricular response. Diagnosis is based on the totally irregular ventricular rhythm.

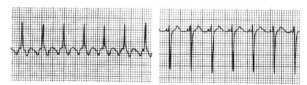

Atrial flutter with 2:1 AV block. Lead aVF (on left) shows the characteristic saw-tooth baseline whereas lead V_1 (on right) shows discrete atrial activity, alternate 'F' waves being superimposed on ventricular T waves.

Ventricular arrhythmias following MI

Ventricular fibrillation (VF) Do immediate cardioversion (p724). Then give prophylactic lignocaine (see below).

Ventricular tachycardia (VT) Defined as ≥3 successive ventricular extra-systoles at a rate >120/min. 2 is a salvo.

ECG: a wide QRS complex (>0.12sec) regular tachycardia.

The main problem may be distinguishing VT from an SVT with aberrant ventricular conduction (fixed or rate-related), but 75% of broad-complex tachycardias are ventricular. ▶ If in doubt, assume VT. VT is certainly commoner after acute MI. In favour of an SVT is:
● The presence of P-waves *associated with* the QRS complex. (There may be dissociated P-waves in VT in about 30%).
● No fusion or capture beats (diagnostic of ventricular origin).
● Classical LBBB or RBBB QRS morphology.
● Same QRS morphology as in sinus rhythm.
● QRS <0.14sec.
● Normal axis (left axis deviation favours VT).

If all complexes have similar polarities in the chest leads ('concordance') suspect VT. ▶ If the diagnosis is uncertain, treat for VT. Even if the patient is conscious, the diagnosis can still be VT.

Management:
If at any stage the patient has impaired consciousness or angina:
● *Synchronous cardioversion*—as for VF (p724–5)+sedation if conscious.
● Correct any acid–base or electrolyte imbalance (eg K⁺).
● In a last ditch attempt, try emergency overdrive pacing.

Otherwise, the indications for treating VT are controversial. Some always treat; some only if there is haemodynamic disturbance or if it persists for >1 minute. One order of drugs used is:
● Adenosine IV (see p288) if the diagnosis VT versus SVT with aberrant conduction is uncertain (VT will continue unaffected).
● Lignocaine 75–100mg IV loading dose over 2mins. If after 10mins it persists then give a further loading dose. If there is still no response move on to the next drug. If it has controlled the VT, begin maintenance therapy: 4mg/min for ½h, 3mg/min for 1h, and 1–2mg/min for 12–24h.
● Procainamide 25–100mg/min IV (ECG monitoring) to maximum of 1g. Stop if BP falls below 100mmHg. Oral maintenance is 250mg/6h.
● Amiodarone 5mg/kg into *central* line over 60mins. See p286.
● If still ineffective, try overdrive pacing.

Remember that all anti-arrhythmics are negatively inotropic. Note: flecainide increases mortality post-MI if given for asymptomatic ventricular ectopics, or asymptomatic non-sustained VT.

Torsade de pointes looks like VF but is in fact VT with a varying axis. It occurs post-MI and in association with congenital, drug-induced or biochemical causes of prolonged Q–T intervals. It may be overdrive paced. Consider stopping anti-arrhythmics; IV Mg²⁺ may help.

Ventricular arrhythmias following MI
[OTM 13.124]

Prevention of recurrent VT: Surgical isolation of the arrhythmogenic area or implantation of tiny automatic defibrillators may help.

Ventricular extrasystoles This is the commonest arrhythmia following infarction. The only proven predictor of VF is the early 'R on T' extrasystole. If these are frequent, (>10/min) treat with lignocaine as above. Some would treat multifocal extrasystoles. Otherwise, most physicians now recommend just observing the patient.

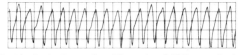

Ventricular tachycardia with a rate of 235 per minute.

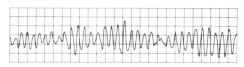

Ventricular fibrillation.

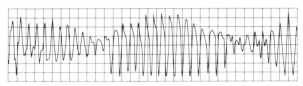

Torsade de pointes ventricular tachycardia.

Bradycardia

If the bradycardia is acute and symptomatic (usually in MI):
- Treat or remove underlying cause, eg β-blockers (remember eye-drops).
- Give atropine 0.3–0.6mg slow IV. Repeat to maximum of 3mg/24h.
- As a second line agent, try isoprenaline 0.5–10μg/min IVI.
- If medical treatment fails, insert a temporary pacemaker.
- If the acute problem does not resolve, pace permanently.

Chronic symptomatic bradycardia should be treated with a permanent pacemaker, if the causes cannot be corrected.

Pacemakers

Pacemakers supply electrical initiation to contraction. They may dramatically benefit even a very old patient. The pacemaker lies subcutaneously where it may be programmed through the skin as needs change—eg for different rates. They last 7–15 years.

Indications for temporary pacemakers
- Symptomatic bradycardia, uncontrolled by drugs.
- Suppression of drug-resistant VT and SVT.
- Acute conduction disturbances:
    After acute *anterior* MI, prophylactic pacing is required in:
        Second or third degree AV block;
        Bi- or tri-fascicular block (p266).
After *inferior* MI, only act on these for symptoms.

Indications for permanent pacemaker
- 2° (Mobitz type II) and 3° heart block.
- Symptomatic bradycardias (eg sick sinus syndrome).
- Occasionally useful in suppressing resistant tachyarrhythmias.
Some say persistent bifascicular block after MI requires a permanent system: this remains controversial.

Insertion is under local anaesthetic. *Pre-operative assessment:* FBC, platelets, clotting screen, HBsAg. Insert IV cannula. Consent. Consider pre-medication. Give antibiotic cover (eg flucloxacillin 500mg IM and benzylpenicillin 600mg IM) 20mins before, and 1 and 6h after. Before discharge, check position (CXR) and function.

ECG of paced rhythm Check each vertical pacing spike is followed immediately by a wide QRS complex. If the system is on 'demand' of 60 beats/min, a pacing spike will only be seen if the natural beats are coming less frequently than 60/min. If it is cutting in on a more rapid natural rate, its sensing mode is malfunctioning. If it is failing to cut in at slower rates, its pacing mode is malfunctioning.

Dual chamber pacemakers pace atria and ventricles sequentially, helping stroke volume—and, via their sensory system, allowing rates to increase with exercise, if sinus node function is normal. They cost twice as much.

Post-operative procedures Inspect wound for haematoma, dehiscence or muscle twitching during the first week. Count the apex rate. If ≥6 beats less than the pacemaker's quoted rate, suspect malfunction. Other problems: lead fracture; pacemaker interference (eg from the patient's own muscles). Driving rules: OHCS p468.

Sick sinus syndrome Common in elderly. Sinus node dysfunction causes sinus bradycardia and arrest, sinoatrial block, and SVT alternating with bradycardia/asystole (tachy–brady syndrome). AF and thrombo-embolism may occur. *Tests:* 24h ECG ± exercise ECG. *Management:* Pace if symptomatic.

Bradycardia
Pacemakers
[*OTM* 13.136]

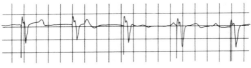

Paced rhythm

Heart failure: basic concepts

Heart failure occurs when the heart fails to maintain sufficient circulation to provide adequate tissue oxygenation in the presence of normal filling pressures. Because the venous (ie filling) pressure is increased, there is pulmonary and peripheral congestion, hence the term congestive cardiac failure (CCF).

Classification Either (i) high-output, low-output, and fluid overload; or (ii) left or right ventricular failure (RVF & LVF).

High-output failure The heart maintains a normal or increased output in the face of grossly elevated requirements. Failure occurs when the heart fails to meet those requirements. It will occur with a normal heart, but even earlier if there is cardiac disease. Causes: anaemia, hyperthyroidism, Paget's disease, arteriovenous malformation, pregnancy. Consequences: features of RVF predominate at first, but later LVF becomes evident.

Low-output failure The heart is either failing to generate adequate output, or can only do so with high filling pressures. Causes:
- *Intrinsic heart muscle disease:* Cardiomyopathy, ischaemia, infarction, myocarditis, Chagas' disease, infiltrative diseases (eg amyloid, eosinophils—p704), beri-beri.
- *Negatively inotropic drugs:* Eg any anti-arrhythmic agent.
- *Chronic excessive afterload:* Eg aortic stenosis, hypertension.
- *Chronic excessive preload:* Eg mitral regurgitation.
- *Restricted filling:* Eg constrictive pericarditis or tamponade, restrictive cardiomyopathy.
- *Inadequate heart rate:* β-blockers, heart block, post infarction.

Fluid overload involves pushing the heart over the hump of the Starling curve, such that increased filling pressures are now associated with *lower* cardiac output. The features are of left ventricular failure. It does not usually occur with normal hearts unless renal excretion is impaired or enormous volumes are involved. It is much more easily precipitated if there is simultaneous compromise of cardiac function and in the elderly.

Left ventricular failure is dominated by *pulmonary oedema*. The *symptoms* are exertional dyspnoea, orthopnoea, paroxysmal nocturnal dyspnoea, wheeze (cardiac asthma), cough (sometimes productive of pink froth), haemoptysis, fatigue. The *signs* are tachypnoea, tachycardia, basal crackles (more widespread in severe LVF), third heart sound, pulsus alternans (alternating large and small pulse pressures), cardiomegaly, peripheral cyanosis, pleural effusion. The peak expiratory flow rate may be reduced, but if it is <150 litres/min suspect COAD or bronchial asthma.

CXR shows prominent upper lobe veins (upper lobe diversion), Kerley B lines (engorged peripheral lymphatics), peribronchial cuffing, fluid in the fissures, diffuse patchy lung shadows, pleural effusion, cardiomegaly. The pulmonary shadows may be perihilar 'bat's wings' or occasionally unilateral (if nursed on one side), or nodular (esp. with pre-existing COAD).

ECG may elucidate cause. *Echocardiography* may also be useful in differentiating focal and diffuse disease, valve disease, and pericardial disease. Occasionally, there may be a place for endomyocardial biopsy.

Right ventricular failure causes peripheral oedema, abdominal discomfort, nausea, fatigue and wasting. *Signs:* raised JVP, hepatomegaly, pitting oedema, peripheral cyanosis.

Heart failure: basic concepts
[*OTM* 13.84]

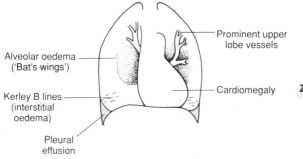

The CXR in left ventricular failure

297

Heart failure: long-term management

- Treat causes if possible, eg valve lesion, hypoxia in cor pulmonale.
- Look for any reversible exacerbating factors (eg anaemia, ↑BP).
- Restrict salt and alcohol intake (if latter is excessive).
- Maintain optimal weight and good nutrition (p104). ● Stop smoking.
- Try to avoid NSAIDs: they cause fluid retention and may interact with diuretics and ACE inhibitors to cause chronic renal failure.

● *Drugs:* 4 types of drugs are used: 1 Diuretics 2 ACE inhibitors 3 Vasodilators such as nitrates 4 Inotropes (digoxin). If symptoms are mild, start with a thiazide diuretic (eg cyclopenthiazide 0.25–0.5 mg/24h PO). Once stabilized, start an ACE inhibitor (p299) if no contraindication, to improve symptoms *and* prognosis (risk of MI[1]); increase its dose as needed (eg captopril 25mg/8h PO).[2] If this is inadequate, add a vasodilator such as isosorbide dinitrate SR 10–20mg/24h PO. If the combination is still not enough, thiazide dose increases will not help (flat dose–response curve). Instead, alter the thiazide to a loop diuretic (eg frusemide 40mg/24h PO, up to 160mg).

Digoxin is out of favour as an inotrope, but meta-analyses suggest 12% of those in heart failure in sinus rhythm benefit from digoxin.[3]

● *Follow up and avoiding diuretic-induced hypokalaemia:* Check body weight, BP, clinical condition, and U&E before and 1 week or so after starting any drugs, and, ideally every 6 months once K^+ is stable. K^+-sparing diuretics are rarely needed (mild hypokalaemia is usually harmless[4]) unless: ● K^+ <3.2mmol/l ● Known predisposition to arrhythmias. ● Concurrent digoxin (its cardio-toxicity is augmented by K^+↓) ● Pre-existing K^+-losing conditions, eg cirrhosis. Dose example: amiloride 5mg/24h PO (=co-amilofruse if combined with 40mg frusemide).

If HCO_3^-↑, ± malaise, thirst or hypotension, suspect over-diuresis.

Intractable heart failure Reassess cause and patient's compliance with treatment. Aim to regain lost ground and improve the patient to his former out-patient-manageable state: so admit to hospital for:
- Strict bed rest. Bedside commode. Avoid straining at stool.
- Use anti-thromboembolic stockings.
- DVT prophylaxis: heparin 5000U/8h SC.
- IV frusemide (<4mg/min—by this route it is a venodilator too).
- Consider metolazone, eg 5–80mg/24h PO. Daily weight, U&E (K^+↓).
- Alter ACE-inhibitor/vasodilator dosage to maximum tolerated.

Sometimes, a compromise must be struck: the patient may have to accept a moderate amount of oedema or limitation of exertion in order to avoid unacceptable symptoms of low output.

● *In extremis*, the patient may benefit from a period on IV inotropes (eg dopamine 2–10μg/kg/min) to tide him over an acute exacerbation. But it may be very difficult to wean him off it.
● Finally, consider the patient for heart transplantation.

Acute heart failure (pulmonary oedema) See p726.

1 S Yusuf 1992 *Lancet* **340** 1173   **2** Consensus 1987 *NEJM* **316** 1429   **3** R Jaeschke 1990 *Am J Med* **88** 279–86   **4** *Drug Ther Bul* 1991 **29** 87

How to start giving an ACE-inhibitor

The danger (rare) is hypotension after the 1st or 2nd dose. Minimize this risk by taking the following precautions before the 1st dose:

- Check BP. If systolic BP <100mmHg, or the patient is prone to fainting, it is prudent to admit to ward before starting therapy.
- Ensure that the patient is not salt or volume depleted (eg D&V).
- Ensure that there is no outflow tract obstruction (eg aortic stenosis). Is the pulse character and volume normal?
- Stop (or minimize) diuretics and hypotensives a few days before.
- Ensure that the patient knows how to get help if he feels faint.
- Ensure the patient is not alone during the hour after the first 2 doses. This does *not* necessarily entail hospital supervision—but suggest that the patient lies down during this time. If symptomatic hypotension *does* occur, elevate the legs. Very rarely, admission is needed for a 0.9% saline infusion (+ atropine if bradycardic).[1]
- Start with captopril 6.25mg PO (shorter half-life than other ACE-inhibitors). Choosing a newer ACE-inhibitor (if needed)—eg enalapril, lisinopril, perindopril, ramipril,[2] can be done later if a once-a-day dose is needed, or troublesome side-effects (eg cough, D&V).
- Check U&E and creatinine 1 week later. Should the patient have renal artery stenosis (p398) worsening renal function will be identified—so ACE inhibitors can be stopped.

Other precautions ● Check U&E and creatinine (do not use ACE-inhibitors routinely in those with pre-existing renal insufficiency—but, a raised urea thought to be due to diuretic use is not a contraindication).
- Weigh the patient (to help monitor response).
- Stop K+ salts and potassium-sparing diuretics (K+↑ is the danger).
- If there is a connective tissue disease, monitor WCC and urine protein.
- Other interactions: Li+ (levels↑), digoxin (levels ↑, possibly), NSAIDs (urea↑, K+↑), anaesthetics (BP↓).

To avoid 1st dose effects the BNF recommends initiation in hospital if:
- ≥70 yrs old*
- Hypovolaemic
- Systolic BP <90mmHg
- Serum creatinine >150μmol/l
- On >80mg frusemide (or similar)/day
- Hyponatraemic (Na+ <130mmol/l)
- Unstable heart failure
- On high-dose vasodilator therapy

Contraindications Symptomatic BP↓; angio-oedema, renal artery or aortic stenosis, pregnancy, breast feeding, collagen disease, porphyria.

▶ Do not be put off by all these caveats: most patients tolerate ACE-inhibitors well, and with all the evidence suggesting that they will live longer and feel better, it is our job to use them when appropriate.

1 Manufacturer's *Data Sheet* 2 AIRE group 1993 *Lancet* ii 821

Hypertension

▶ This is a major risk factor for stroke and myocardial infarction. It is often symptomless, so screening is vital—*before* damage is done. When taking BPs, be sure the cuff is wide enough (>$\frac{1}{2}$ the arm's circumference).

Value judgements There is no BP value above which you must always treat, and below which you can safely leave alone—you must take into account family history, age, and smoking/lipid status—and more complex matters such as the value a patient sets on life. If he is agreeable, bring the spouse or partner in on the decision. No-one remembers to take antihypertensives all the time, and the spouse may be more motivated to get treatment right. If the patient does not set the value of his life at a pin's worth, you may need to spend time treating his depression before talking about antihypertensives. So the first rule is: *know your patient.* Don't think of this as another chore to tick off; dialogue is our reward for all those boring things we do in a day, and helps keep us sane.

Patient's views of the balance of risk to benefit of treatment are critical. Be sure he is well informed. It is unacceptable to cause significant side-effects in people who thought they were healthy before they met you: most of us live in the here-and-now, and relate with difficulty to the stranger who is us in 20 years. The British Hypertension Society suggests drugs are needed if *sustained* diastolic >100mmHg on ⩾3 readings each a week apart. Drugs may also be indicated if diastolic *sustained* at >90mmHg over 6 months, if systolic >160mmHg OR end-organ damage:
●Left ventricular hypertrophy ●PMH myocardial infarct ●PMH stroke/TIA
●Peripheral vascular disease ●PMH angina ●Renal failure
If no end-organ damage and diastolic 90–99mmHg for >6 months, treat if risk factors present, ie ♂; +ve family history; smoker; diabetes; hyperlipidaemia. (Systolic BP ⩾200mmHg also prompts drug treatment even if no risk factors or end-organ damage or raised diastolic BP.) *BP in the elderly (65–80yrs):* 3 trials support treatment eg if systolic >180mmHg.

Causes In 95% the cause is unknown ('essential')—but alcohol and obesity may play an important part. The remaining 5%:
Renal disease: Renal artery stenosis (atheroma or fibromuscular hyperplasia, p398), polycystic kidneys, chronic pyelonephritis, glomerulonephritis, polyarteritis nodosa, systemic sclerosis (p670).
Endocrine disease: Cushing's syndrome (p548) and Conn's syndromes (p554), phaeochromocytoma (p554), acromegaly, hyperparathyroidism, DM.
Others: Coarctation, pre-eclampsia (OHCS p96), (steroids). Hypertension is a natural response to pain and sometimes an analgesic is the best drug.

Presentations Headache, stroke, heart failure, MI, renal failure (eg lethargy with proteinuria), blindness. Always check for retinopathy—either *arteriopathic* (arteriovenous *nipping:* arteries nip veins where they cross as they share the same connective tissue sheath) or *hypertensive* (arteriolar vasoconstriction and leakage producing *hard exudates, macular oedema, haemorrhages,* and, rarely, *papilloedema*—see p684).

Tests Look hard for a cause in the young, severe hypertensive; be more restrained in the older patient (it's not that they matter less, but that diagnostic yield and treatment benefits will be less—and tests may hurt). Start by talking to the patient. Is he or she aware of the hypertension? (Episodic feelings 'as if about to die' suggests phaeochromocytoma). Next do a fullish examination—we religiously listen for abdominal bruits for renal artery stenosis, but have never yet detected it this way. Then do U&E, CA^{2+}, creatinine, random glucose, MSU (for cells, casts, protein, and pathogens)—and then consider renal ultrasound, IVU, renal arteriography, 24 hour urinary VMA $\times$ 3, urinary free cortisol (p550). Assess degree of end-organ damage with ophthalmoscopy, ECG, CXR, U&E.

301

Management of hypertension

▶ Look for and treat any underlying cause (eg pain). See p300.

Non-drug treatment Stop smoking.

- Optimize weight: give the patient realistic aims and a palatable diet sheet; review their current diet.
- Reduce alcohol consumption to ≤1 unit/day.
- Some advise not to add salt to food at the table.[1]
- Consider relaxation therapy (eg slow breathing, muscle relaxation and mental relaxation by imagining pleasant experiences ± cognitive strategies to cope with stress, and knowledge of how to deal with hostile feelings). This may be done in groups with the leader setting regular homework. Typical falls in BP are 6–26mmHg (systolic) and 5–15mmHg (diastolic) in those 50% of trials showing benefit.[2]

Drug therapy Explain that the patient may be on the tablets for life and why he should take them, even though they may not make him feel better. Ask him to see his doctor if he experiences adverse effects, not merely to stop the tablets.

Drug choice has expanded and each physician will have a personal preference. Here is one approach:

- Start with a thiazide (eg bendrofluazide 2.5mg/24h PO—higher doses are unlikely to help). SE: gout, hyperglycaemia, hypokalaemia, hyperlipidaemia, and impotence. Reasonably safe. Avoid in diabetics.
- Next add a β-blocker (eg atenolol 50–100mg/24h PO).
- If still uncontrolled, add a calcium antagonist (eg nifedipine SR 20mg/12h PO). SE: flushing, dizziness, ankle oedema, GI upset.
- If this is inadequate, stop calcium antagonists and add an angiotensin converting enzyme (ACE) inhibitor—eg captopril 6.25–25mg/8h PO or lisinopril 2.5–20mg/24h PO (max 40mg/day) or enalapril (dose as for lisinopril). ▶ See p 299 for starting régime.
- If the hypertension is still not controlled, seek expert help. Agents such as methyldopa, prazosin, minoxidil or hydralazine may be needed.

Adverse effects may dictate choice. With β-blockers and diuretics, 40% adverse reactions occur within 3 months, 80% within 1 year. If a patient fails to tolerate one first-line agent, try another.

Accelerated-phase (malignant) hypertension is pathologically distinct, characterized by fibrinoid necrosis. *Prognosis:* untreated, 90% die in 1 year; treated, 60% 5yr survival. *Detection:* universal screening from aged 20yrs.[3] *Presentation:* heart failure, encephalopathy (seizures, coma), renal failure, headache, or very high BP (diastolic >130mmHg), associated with grade III–IV retinopathy. *Treatment* is strict bed rest, and oral nifedipine 10–20mg/8h. Initially, bite a nifedipine 10mg capsule and swallow. The BP will start to fall gently after ~30mins. Avoid sudden drops in BP as cerebral autoregulation is impaired and there is risk of stroke. If minute-by-minute BP control is essential (eg during seizures), use sodium nitroprusside: 0.3µg/kg/min IVI titrated up to 6µg/kg/min (eg 50mg in 1 litre of dextrose 5%; expect to give 100–200µl/h for a few hours only, to avoid cyanide risk). Do BP every minute until stable, then every 30min. Aim for a diastolic BP of 100–110mmHg. Cover burette with reflective foil to stop photodeactivation. Labetalol is an alternative IV drug (eg 50mg IV over 1min, repeated every 5mins, up to 200mg), as is hydralazine (5–10mg over 20mins, repeated after 30mins if needed).

1 N Boon 1985 *BMJ* i 940   2 *Drug Ther Bul* 1989 27 77   3 J Tudor Hart 1993 *BMJ* i 437

Management of hypertension
[*OTM* 13.371]

Rheumatic fever

This systemic illness is due to cross-reactivity with Group A β-haemolytic streptococci (eg after throat infection). In the susceptible 2% of the population, there may be permanent damage to heart valves—and the risk of subsequent endocarditis (p312) is increased. It is common in the Third World, and rare in the West. Pockets of resurgence are appearing in the USA. It usually follows a latent period of 2–4 weeks. Peak incidence: 5–15yrs.

Diagnosis This is based on the *revised Jones criteria* and can be made in the presence of evidence of previous streptococcal infection plus 2 major criteria or 1 major and 2 minor criteria.

Evidence suggesting recent streptococcal infection: (1 is enough.)
- Recent scarlet fever.
- Positive growth from throat swab.
- Increase in antistreptolysin O titre (ASOT) >200U/ml.

Major criteria:
- Carditis (endo-, myo-, and/or pericarditis) presents as sleeping tachycardia, changing murmurs (most commonly mitral or aortic regurgitation, or Carey Coombs' murmur—p262), pericardial rub, heart failure, cardiomegaly, conduction defects (45–70%).
- Migratory ('flitting') polyarthritis—red and very tender (75%).
- Sydenham's chorea (St Vitus' dance) (10%)—commoner in women.
- Subcutaneous nodules (2–20%).
- Erythema marginatum (2–10%)—trunk, thighs, arms (p686).

Minor criteria:
- Raised ESR or C-reactive protein (CRP).
- Arthralgia (but not if arthritis is one of major criteria).
- Fever.
- History of previous rheumatic fever or rheumatic heart disease.
- Prolonged PR interval (but not if carditis is major criterion).

Management
- Bed rest until CRP normal for 2 weeks—may be 3 months.
- Benzylpenicillin 600mg IM stat then penicillin V 250mg/6h PO.
- Analgesia for carditis and arthritis: aspirin 90mg/kg/day PO (in 4-hourly divided doses; up to 8g) for 2 days then 70mg/kg/day for 6 weeks. Monitor blood levels. Watch for adverse effects: ototoxicity, hyperventilation and metabolic acidosis, GI upset.
- Immobilize joints in severe arthritis.
- Haloperidol (0.5mg/8h PO) for the chorea.

Prognosis The average acute attack lasts 3 months. Recurrence may be precipitated by future streptococcal infection, pregnancy, and the contraceptive pill. 60% who have carditis develop chronic rheumatic heart disease. Of them, disease affecting the mitral, aortic, tricuspid and pulmonary valves occurs in 70%, 40%, 10% and 2% respectively. Incompetent lesions may develop with the acute attack and may regress; stenoses may develop years later.

Secondary prophylaxis Following rheumatic fever, give penicillin V 125mg/12h PO until 25 years old. Thereafter, give antibiotics for dental and other operative procedures for life (p314). Penicillin allergy: use erythromycin.

Rheumatic fever
[*OTM* 13.277]

Mitral valve disease

▶ Defer replacing a mitral valve until symptoms demand.

Mitral stenosis *Cause:* Rheumatic heart disease.

Symptoms: Dyspnoea, palpitations, fatigue; with pulmonary hypertension: haemoptysis, recurrent bronchitis, right heart failure and sometimes chest pain (but suspect other causes of pain).

Signs on *inspection:* Peripheral cyanosis (if on cheeks='malar flush'). *Palpation:* Pulses are of normal character, but irregular (AF, from an enlarged L atrium); left parasternal heave (from pulmonary hypertension causing RVH); undisplaced, tapping apex beat (ie palpable S₁, p260). *Auscultation:* Loud S₁; opening snap (after S₂, from abrupt checking of the valve opening—like the crack of a sail struck by a strong wind) preceding a rumbling mid-diastolic murmur (best heard at the apex), pre-systolic accentuation murmur. Pulmonary hypertension (p366) and Graham Steel murmur (p262) may occur.

Severity: The more severe the stenosis, the worse the dyspnoea; the larger the left atrium, the longer the murmur, and the closer the opening snap is to S_2. Symptoms start if valve area is $<1.25cm^2$.

Tests in mitral stenosis: ECG: P mitrale if in sinus rhythm, but usually in AF, right ventricular hypertrophy. CXR: large left atrium (double shadow behind the heart), splayed carina, pulmonary oedema and later mitral valve calcification. Echocardiography is diagnostic (is the mid-diastolic closure rate of the anterior cusp <50mm/sec?). Significant stenosis if valve orifice $<1cm^2/m^2$ body surface area).

Management: Treat symptoms with diuretics and digoxin (p298 & p276). Anticoagulate (in AF). If this is inadequate, mitral valvotomy or balloon valvuloplasty (if valve is still pliant and not calcified), or replacement. SBE prophylaxis (p314). Regular penicillin to ward off recurrent rheumatic fever if <25yrs (p304). Ensure well nourished (p104).

Complications: Pulmonary hypertension; emboli, from mural thrombus, eg on McCallum's patch, just above the posterior mitral valve cusp.

Mitral regurgitation (incompetence) *Causes:* Rheumatic heart disease, mitral valve prolapse, ruptured chordae tendinae, papillary muscle rupture (after MI) or dysfunction (in ventricular dilatation or dysfunction), endocarditis, cardiomyopathy (hypertrophic and dilated), rheumatoid arthritis, congenital.

Symptoms: Dyspnoea, fatigue.

Signs: Displaced and forceful apex, soft S₁, loud S₃, pansystolic murmur at apex radiating to axilla. Sometimes AF.

Severity: The more severe, the larger the left ventricle. Eventually, left ventricular failure occurs.

Tests: in mitral incompetence: CXR: large heart. Echocardiography may reveal cause and extent of ventricular dilatation *Doppler echocardiography* is semi-quantitative in assessing regurgitation.

Management of mitral incompetence: Anticoagulation in AF. Digoxin, diuretics, and surgery if severe. Antibiotics to prevent endocarditis.

Mitral valve prolapse is common (eg incidental finding in 5%). It may be asymptomatic or be associated with atypical chest pain, palpitations, syncope, and emboli. There may be a mid-systolic click and/or a late systolic murmur. Echocardiography is diagnostic. ECG may show inferior T-wave inversion. There is a risk of SBE, so give antibiotic prophylaxis.

Mitral valve disease
[*OTM* 13.282]

Aortic valve disease

Aortic stenosis Causes: Congenital, rheumatic heart disease, bicuspid valve, degenerative calcification.

Symptoms: Angina, dyspnoea, syncope, dizziness, sudden death.

Signs: Slow rising, small volume pulse, narrow pulse pressure (difference between systolic and diastolic pressures), heaving undisplaced apex beat. In mild stenosis, S_2 is normally split, with A_2 preceding P_2. As the stenosis becomes more severe, A_2 is increasingly delayed, giving first a single S_2 and then reversed splitting. In the presence of aortic valve calcification, A_2 becomes ever softer and may disappear totally. There may be an S_4, ejection click (in pliable valves), and an ejection systolic murmur, loudest in the aortic area (but sometimes elsewhere) which radiates to the carotids and apex.

Tests: ECG: Left ventricular hypertrophy with strain. CXR: Post stenotic dilatation of ascending aorta; calcification on valve (best seen at fluoroscopy). 2D-echo/Doppler is diagnostic, and may estimate the gradient across the valve quite well. If angina is prominent, coronary angiography should be done to show coronary artery disease. Angiography may not be needed if there is no angina.

Management of aortic stenosis: These patients are at great risk of Stokes–Adams attacks and sudden death. Valve replacement is normally required in symptomatic patients. The management of asymptomatic significant AS is an unresolved problem: most cardiologists would advise replacement if the systolic gradient across the valve was >50mmHg. If the patient is not fit for surgery, percutaneous transluminal valvuloplasty may be attempted. Antibiotic prophylaxis against endocarditis is essential (p314) as is good nourishment pre-operatively (p104).

Aortic sclerosis is senile degeneration not associated with left ventricular outflow tract obstruction. There is an ejection systolic murmur, but the character of the pulse remains normal. These patients do not have ejection clicks or abnormal S_2.

Aortic regurgitation Causes: Rheumatic heart disease; endocarditis; hypertension; syphilis; aortic dissection; spondyloarthritides (eg ankylosing spondylitis; Reiter's, p669; Marfan's, p704).

Symptoms: Dyspnoea, palpitations (due to extrasystoles).

Signs of a wide pulse pressure: Collapsing (water-hammer) pulse. Corrigan's sign—visible neck pulsation. De Musset's sign—head nodding. Visible capillary pulsations (eg seen in nail beds—Quincke's sign).

Other signs: Duroziez's sign is a femoral diastolic murmur reflecting flow of blood *backwards* up the aorta in severe aortic regurgitation. To elicit this, make this backflow turbulent by pressing on the femoral artery distal to the point of listening. Hyperdynamic displaced apex beat; high pitched early diastolic murmur; Austin Flint mid-diastolic murmur (due to regurgitant stream vibrating the anterior mitral cusp).

Investigations of AI: CXR: cardiomegaly. Doppler echo is diagnostic. M-mode and 2D show the degree of ventricular dilatation.

Management: Try to replace valve before significant impairment of LV function. SBE prophylaxis.

Aortic valve disease
[*OTM* 13.293]

Right heart valve disease

Tricuspid stenosis is rheumatic. Mitral disease always coexists.
Features: Oedema; giant a wave, small v wave, and slow y descent in JVP; opening snap, early diastolic murmur (left sternal edge in inspiration). Echocardiography with Doppler may be diagnostic.
Treat with diuretics and sometimes surgical repair.

Tricuspid regurgitation usually occurs with RV enlargement (p366). *Other causes:* Endocarditis (IV drug abusers), carcinoid syndrome, rheumatic, congenital (eg Ebstein's anomaly, OHCS p746).

Clinical: Oedema, systolic waves in JVP, pulsatile hepatomegaly (leads to jaundice and hepatic fibrosis), ascites, parasternal RV heave, pansystolic murmur heard best at lower sternal edge in inspiration. Also features of underlying condition (eg dyspnoea).

Management: Diuretics, vasodilators. *Surgical:* annuloplasty, replacement (but 20% operative mortality). Treat underlying cause.

Pulmonary stenosis is usually congenital eg in Fallot's tetralogy (OHCS p748). Other: Rheumatic, carcinoid. *Clinical:* Prominent a waves in JVP, RV heave, S_2 widely split, ejection systolic murmur (loudest to left of upper sternum), radiating to the left shoulder. ECG: RVH. CXR: dilated pulmonary artery.

Pulmonary regurgitation is caused by any cause of pulmonary hypertension (p366). It causes a decrescendo murmur in early diastole at the left sternal edge (the Graham Steel murmur).

Cardiac surgery

Coronary artery bypass grafting An option for angina if drugs fail, or there is L mainstem disease, and angioplasty (p278) is unsuitable. (Type and site of surgery is determined by coronary angiogram.) The heart is stopped and blood pumped artificially by a machine outside the body (cardiac bypass). The patient's own saphenous vein is used as the graft. One end is attached to the aorta, the other to the coronary artery distal to the stenosis. Several grafts may be placed. <50% of vein grafts are patent after 10yrs. Internal mammary artery grafts last longer (but may cause chest wall numbness). *After CABG:* If angina persists, add in drugs, and consider angioplasty. The cause may be poor run off from the graft (distal disease), new atheroma, or graft occlusion. Mood, sex and intellectual problems are common early on, and helped by rehabilitation:

- Exercise: walk→cycle→swim→jog
- Get back to work eg at 3 months
- Drive at 1 month (no need to tell DVLA if non-HGV licences, OHCS p468)
- Attend to: smoking; BP; lipids
- Aspirin 75mg/24h PO for ever

Closed mitral valvotomy is used in stenosis of uncalcified valves (ie loud S_1, prominent opening snap). A finger inserted in the left atrium dilates the valve orifice. Overenthusiasm leads to regurgitation. This procedure may be repeated. *Open mitral valvotomy* is under direct vision at bypass. Balloon valvuloplasty is gaining popularity.

Cardiac surgery
[*OTM* 13.300]

Valve replacements may be pig valves (Carpentier–Edwards; xenografts). They do not need permanent anticoagulation (useful for the elderly and the pregnant), but do not last as long as artificial valves. Two popular artificial valves are the ball and cage valve (Starr–Edwards) and the tilting disc (Björk–Shiley). Each has the risk of thrombosis and embolization, haemolysis, and leaks—so permanent anticoagulation will be needed (p596). These patients need antibiotic SBE prophylaxis.

Cardiac transplantation 5-year survival is now approaching 70%. Main contraindications: significant peripheral vascular or cerebrovascular disease; severe systemic disease; and contraindications of immunosuppression (eg active infection).

Pre-op assessment FBC, platelets, ESR, clotting, U&E, LFTs, HBsAg, CMV serology, CXR, ECG, crossmatch 10 units, MSU, swab nose & groin.

Infective endocarditis (SBE)

▶ *Fever + changing murmur = endocarditis until proven otherwise*.

Classification
1 50% of all endocarditis is on normal valves. It follows a more acute course presenting with acute heart failure.
2 Endocarditis on rheumatic, degenerative or congenitally abnormal valves tends to run a subacute course.
3 Endocarditis on prosthetic valves may be 'early' (acquired at implantation, very bad prognosis) or 'late' (acquired haematogenously). Infected prosthetic valves must usually be replaced.

Pathogenesis Any bacteraemia, eg any dentistry/poor dental hygiene, GU manipulation, or surgery exposes valves to colonization. In 60% no cause is found. The commonest organism, *Strep viridans* (35–50%), more commonly runs a subacute course. Others include other streptococci (mainly *Enterococcus faecalis*) and *Staph aureus* (20% of all cases, but 50% acute endocarditis). Rarely other bacteria, fungi, *Coxiella*, or *Chlamydia*. Non-bacterial causes are SLE (Libman–Sacks endocarditis) and malignancy. Vegetations may embolize. Right heart endocarditis is commonest in IV drug abusers. It may cause pulmonary abscesses. Endocarditis is associated with a vasculitis.

Clinical features of chronic endocarditis a) *Infective:* Fever, weight loss, malaise, night sweats, clubbing, splenomegaly, anaemia, mycotic aneurysms; b) *Heart murmurs* and *failure*; c) *Embolic:* Eg stroke; d) *Vasculitic:* Microscopic haematuria, splinter haemorrhages, Osler's nodes (painful lesions on finger pulps), Janeway lesions (palmar macules), Roth's spots (retinal vasculitis), and even renal failure.

The 4 'best' signs are: fever, haematuria, splenomegaly & murmurs.

Tests Take 3 *blood cultures* at different times and from different veins, preferably when the fever is rising. ESR high. Normochromic, normocytic anaemia. CXR: cardiomegaly. *Echocardiography* may show vegetations, but only if >3mm. *Urinalysis* for haematuria is important.
Strep bovis endocarditis is associated with colonic carcinoma, and will need further appropriate lower GI investigation.

Management
● Antibiotics—for streptococci: benzylpenicillin 1.2g/4h IV + gentamicin 80mg/12h IV (or netilmicin in elderly) for 2 weeks, then amoxicillin 1g/8h PO for 2 weeks. *Staph aureus*: flucloxacillin 2g/6h IV + gentamicin 80mg/12h IV for 2 weeks then flucloxacillin alone for ≥2weeks. For either in penicillin allergy: vancomycin 1g/12h IV slowly +gentamicin. *Staph epidermidis*: vancomycin+rifampicin. Monitor MBCs (minimum bactericidal concentrations) and gentamicin levels (p796–7).
● If blind therapy required: benzylpenicillin eg 1g/4h IV + gentamicin 80mg/12h IV; if acute, add flucloxacillin 2g/6h IV.
● If treatment fails on antibiotics or severe failure supervenes, replace the valve under extra antibiotic cover (eg vancomycin).

Prognosis 30% mortality with staphylococci; 14% with bowel organisms; 6% with sensitive streptococci.

Infective endocarditis (SBE)

[*OTM* 13.314]

Prevention of endocarditis[1]

Any patient with congenital, or rheumatic heart valve disease or prosthetic valves is at risk and should take antibiotics before an elective procedure which may result in bacteraemia. Included in this group are ventricular (but not atrial) septal defect, persistent ductus arteriosus, and mitral valve prolapse, but not aortic sclerosis. The important procedures are dental treatment (including scaling), urogenital procedures (including assisted but not normal deliveries), upper respiratory tract (URT) instrumentation, GI instrumentation, and open surgery.

Recommendations for dental procedures

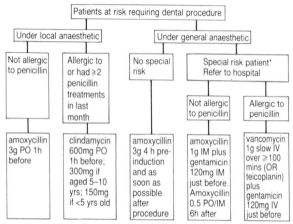

Patients at risk requiring dental procedure

Under local anaesthetic

- Not allergic to penicillin → amoxycillin 3g PO 1h before
- Allergic to or had ≥2 penicillin treatments in last month → clindamycin 600mg PO 1h before; 300mg if aged 5–10 yrs; 150mg if <5 yrs old

Under general anaesthetic

- No special risk → amoxycillin 3g 4 h pre-induction and as soon as possible after procedure
- Special risk patient* Refer to hospital
  - Not allergic to penicillin → amoxycillin 1g IM plus gentamicin 120mg IM just before. Amoxycillin 0.5 PO/IM 6h after
  - Allergic to penicillin → vancomycin 1g slow IV over ≥100 mins (OR teicoplanin) plus gentamicin 120mg IV just before

*patients who have received antibiotics in the preceding month, who have a prosthetic valve, who are allergic to penicillin, or who have previously had endocarditis. Teicoplanin dose: 400mg IV.

CHILDREN: Amoxycillin: ½ dose if 5–10yrs; ¼ dose if <5yrs old. Teicoplanin 6mg/kg; gentamicin 2mg/kg IV. Use the same dose frequency as in adults.

Recommendations for other procedures (doses as above)

Procedure:	Valve type:	Antibiotics:
GU instrumentation	Native damaged or prosthetic	Amoxycillin IM plus gentamicin IM just before induction; and amoxycillin PO/IM 6h post-op
Obstetric & gynaecological procedures (eg coil insertion)	Prosthetic only	
GI endoscopy, Barium enema	Prosthetic only	
Tonsillectomy/adenoidectomy	Prosthetic	
Tonsillectomy/adenoidectomy	Native damaged	Amoxycillin IM pre- and post-op

1 British Society for Antimicrobial Chemotherapy 1992 *Lancet* i 1292

314

Prevention of endocarditis
[*OTM* 13.323]

Diseases of heart muscle

Cardiomyopathies are disorders of heart muscle. There are 3 kinds:

● *Congestive (dilated):*
Essence: Ventricles are dilated and contract poorly.
Prevalence: 0.2%.
Mortality: 40% in 2yrs (eg sudden death, cardiogenic shock).
Typical patient: A young man with RVF, LVF, cardiomegaly, AF ± emboli.
Diagnose by excluding ischaemia, valvular, pericardial and specific heart muscle diseases.
Echocardiography shows a *globally* hypokinetic, dilated heart.
Other tests show no or non-specific abnormalities.
Management: As for heart failure. Consider transplantation.

● *Hypertrophic (obstructive) cardiomyopathy (HOCM):* Hypertrophy of ventricular muscle. Autosomal dominant, but ½ are sporadic. Screening relatives may yield mixed blessings: helpful preventive measures, but possible life insurance refusal. *Symptoms:* Dyspnoea, angina, faints, palpitations, sudden death (risk probably ↑ by strenuous exercise).
Signs: Jerky pulse, double impulse at apex, S_3, S_4, late systolic murmur (outflow tract obstruction ± mitral regurgitation).
ECG: LVH, LBBB. *Echocardiography* is usually diagnostic: Asymmetrical septal hypertrophy, systolic anterior movement of mitral valve, mid-systolic closure of aortic valve. *Cardiac catheterization* shows a small banana-like LV cavity + thickened papillary muscles and trabeculae; cavity obliteration in systole.
Management: β-blockers for angina. Amiodarone for arrhythmias (has been shown to improve prognosis). Try and maintain sinus rhythm, as the loss of atrial kick is serious. Finally consider septal myectomy or myotomy. In paroxysmal AF, anticoagulate.

● *Restrictive cardiomyopathy* is due to endomyocardial stiffening. Clinically, it resembles constrictive pericarditis (p318). It is commonest in the tropics where it is due to idiopathic endomyocardial fibrosis. In UK the commonest cause is amyloid.

Specific heart muscle diseases (cause known) usually behave like dilated cardiomyopathy. But amyloid and carcinoid may be restrictive; amyloid and the cardiac involvement of Friedreich's ataxia may resemble HOCM. *Commonest causes:* Ischaemic heart disease and BP↑ (the patient may present in failure with normal BP: examine fundi to reveal signs of earlier hypertension). *Other causes:* Infections, alcohol, post-partum, smoking, connective tissue diseases, diabetes, hyper- and hypothyroidism, acromegaly, Addison's, phaeochromocytoma, haemochromatosis, sarcoid, Duchenne muscular dystrophy, myotonic dystrophy, irradiation, drugs (eg cytotoxics), storage diseases.

Acute myocarditis is inflammation of the myocardium and may present similarly to myocardial infarction. The commonest causes are viral eg *coxsackie*, but it may also occur with diphtheria, other infections, acute rheumatic fever, and drugs.
Clinical: Faints, pulse↑, angina, dyspnoea, arrhythmia; heart failure.
Investigation: Aim to exclude eg infarction, pericardial effusion. Viral or Chagas' serology rarely may help.
Management: Supportive. These patients may recover spontaneously or develop intractable heart failure.

Diseases of heart muscle

[*OTM* 13.196]

Left atrial myxoma This rare, benign, primary tumour presents with left atrial obstruction (like mitral stenosis), systemic emboli, AF, or fever, weight loss, and ↑ESR. ♀/♂=2. Notable features distinguishing it from mitral stenosis are the occurrence of emboli (eg large enough to obstruct the aortic bifurcation) while the patient is in sinus rhythm. Auscultation may reveal a 'tumour plop', as the pedunculated tumour reaches the end of its tether. Auscultatory signs vary according to posture. Whenever emboli are excised, put them under the microscope to see if myxoma cells are present. *Tests:* Echocardiography is diagnostic. *Treatment:* Excision.

Diseases of the pericardium

Pericarditis

Causes: ● Infections—viruses (commonest cause, especially *coxsackie*); TB (often rapid effusion, look for calcification); other bacteria; parasites. ● Malignant pericarditis. ● Uraemia. ● Myocardial infarction (MI). ● Dressler's syndrome (occurs ~10 days post MI, p696). ● Trauma. ● Radiotherapy. ● Connective tissue disease. ● Hypothyroidism.

Symptoms: A sharp, constant sternal pain, relieved by sitting forward. It may radiate to the left shoulder, and sometimes down the arm or into the abdomen. It may be worsened by lying on the left, inspiration, coughing, and swallowing.

Signs: Pericardial friction rub—a scratchy superficial sound heard best at the left sternal edge.

Tests in pericarditis: ECG: Concave-upwards (saddle-shaped) ST segments in all leads except aVR. Unlike an infarct, there are no reciprocal changes. (This classical pattern is quite rare.)
CXR: Normal unless there is an effusion.

Treatment of pericarditis: Treat the cause. Indomethacin 25–50mg/6h pc for pain. Consider steroids in resistant cases.

Pericardial effusion is accumulation of fluid in the pericardial sac, caused by anything that causes pericarditis.

Clinical features are of left and right heart failure. If the effusion becomes so large as to cause the blood pressure to fall, *cardiac tamponade* has occurred (p722). It is characterized by tachycardia, hypotension, peripheral shutdown, pulsus paradoxus (fall of systolic blood pressure of >10mmHg on inspiration), and a high JVP which paradoxically rises with inspiration (Kussmaul's sign). Beck's triad: rising JVP; falling BP; small, quiet heart.

Differential diagnosis: MI and pulmonary embolism. Although tamponade is much less common, it must be remembered in high-risk situations, eg in the days following cardiac surgery.

CXR shows a large globular heart. ECG shows loss of voltages and alternating QRS morphologies (electrical alternans).

Echocardiography is diagnostic, showing an echo-free zone surrounding the heart. The fluid may not be evenly distributed.

Management: Treat the cause. If there is tamponade the effusion should be drained urgently. Insert a needle upwards and to the left from the xiphisternum. Send fluid for culture, cytology and haematocrit. It may be wise to leave a pericardial drain *in situ*.

Constrictive pericarditis is the encasement of the heart within a non-expansile pericardium. It is usually due to TB, but may follow any cause of pericarditis.

Clinical features: These are mainly of right heart failure with severe ascites, hepatosplenomegaly, a high JVP which rises paradoxically with inspiration, hypotension, pulsus paradoxus and a loud high-pitched S_2 (pericardial knock).
CXR shows a small heart (in 50%) and may show calcification.

Management: This is by surgical excision of the pericardium.

Diseases of the pericardium
[*OTM* 13.304]

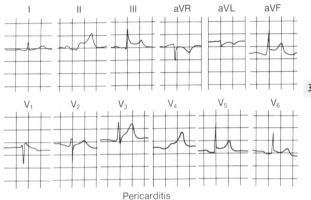

Pericarditis

Atrial septal defects (ASD)

Essence There is a hole connecting the two atria. Often, they are not detected until they cause symptoms in adult life. Ostium secundum holes (high in the septum) are commonest; ostium primum holes (opposing the endocardial cushions) are associated with AV valve anomalies.

Presentation The majority over 40yrs will have symptoms, namely arrhythmias (especially AF) and cardiac failure. Paradoxical emboli (embolus passing from right to left through the defect) are rare.

Signs There is wide and fixed splitting of S_2. There is a pulmonary ejection systolic murmur (increased flow across the valve). Pulmonary hypertension may cause pulmonary regurgitant murmurs (from pulmonary artery dilatation) or tricuspid regurgitant murmurs (both diastolic). ASDs should be suspected whenever murmurs are best heard in the 2nd intercostal space at the left sternal edge.

Complications Shunt reversal (as above). There is *no* significant increased risk of SBE (little turbulence in the low-pressure system).

Investigations CXR: pulmonary plethora (peripheral 'pruning' with onset of pulmonary hypertension); globular heart. ECG: right bundle branch block + right (secundum) or left (primum) axis deviation. 2D-echocardiogram with Doppler colour flow. Cardiac catheter: RA O_2 >IVC O_2.

Treatment best prognosis with operative correction in first 2 decades. Close ASD if physical signs (unless >65 years old, with a small shunt)—or Eisenmenger's syndrome is developing (p698).

Ventricular septal defects

Essence There is a hole connecting the two ventricles. When congenital (the commonest congenital heart defect) this is usually in the membranous part of the septum, rather than the muscular part. Acquired septal rupture occurs after myocardial infarction.

Presentation In adults, this ranges from an incidental finding to advanced pulmonary vascular disease (Eisenmenger's syndrome, p698). AF may occur—and, in some sites, aortic incompetence may be caused.

Signs The characteristic murmur is pansystolic, best heard at the 3rd/4th intercostal space. Small holes give loud murmurs, but are haemodynamically insignificant (*maladie de Roger*). There may be a systolic thrill, ±left parasternal heave. There may pulmonary hypertension signs (loud P_2; pulmonary diastolic incompetent murmur).

Complications Reversal of shunt if a significant defect is left untreated (Eisenmenger's syndrome, p698); infective endocarditis (at point of maximal turbulence where the jet impinges on the right ventricular wall); ventricular ectopics; ventricular tachycardia.

Associations Down's syndrome; Turner's syndrome; Fallot's tetralogy (VSD, pulmonary stenosis, overriding aorta, and right ventricular hypertrophy—OHCS p746).

Investigations CXR: pulmonary plethora. ECG: normal or left axis deviation. 2D-echocardiogram with Doppler colour flow. Cardiac catheter: RV O_2 > vena caval O_2 and right atrial O_2.

Treatment Small defects can close spontaneously, even in adulthood. Otherwise, close if symptomatic, shunt >3:1, or a single attack of endocarditis. SBE prophylaxis is vital for untreated defects (p314).

Cardiovascular medicine

Relevant pages in other sections:
Symptoms and signs: The respiratory examination (p28); chest pain (p256); clubbing (p42); cough (p42); cyanosis (p42); dyspnoea (p44); haemoptysis (p48); stridor (p58).

Others: Acute severe asthma (p730); pulmonary embolism (p728); pulmonary oedema (p726); acid–base balance (p630); tuberculosis (p198–200)

Points on examining the chest

Expansion (on inspiration) of one hemithorax reflects the tidal volume of that lung. Tracheal position indicates the position of the upper mediastinum. Increased tactile vocal fremitus (TVF, p28), increased vocal resonance (VR, p28), bronchial breathing, and whispering pectoriloquy all imply the presence of consolidation or fibrosis stretching from the chest wall to a large conducting airway. The combination of decreased TVF, decreased VR, and reduced breath sounds indicates that transmission of sound from a large airway to the chest wall is impaired eg pleural effusion or bronchial occlusion.

Added sounds
Wheezes (rhonchi) may be monophonic, a single note localized in the chest (signifying a partial obstruction in one airway) or polyphonic, multiple notes generalized throughout the chest (signifying widespread narrowing of airways of differing calibre eg asthma or COAD). In general, the higher the pitch of the wheeze, the narrower the airway it is coming from.

Crepitations (crackles) are fine and high pitched if coming from the distal air spaces (eg pulmonary oedema, fibrosing alveolitis), but coarse and low pitched if they originate more proximally (eg bronchiectasis, in which condition the crepitations may be audible at the mouth and in both inspiration and expiration). NB: if crepitations disappear on coughing, they are of no significance.

Pleural rubs occur with inflammatory conditions of the pleura eg adjacent pneumonia, pulmonary infarction.

The silent chest may represent severe bronchospasm in the pre-terminal asthmatic (or patients with COAD), with almost no air entering to cause wheezing. By now his $PaCO_2$ will be rising. Delay at his peril.

CO_2 retention causes peripheral vasodilation, bounding pulses, flap, confusion, and finally narcosis.

Respiratory distress occurs when high negative intrapleural pressures are needed to generate air entry. It is indicated by nasal flaring, tracheal tug (pulling of thyroid cartilage towards sternal notch in inspiration), the use of accessory muscles of respiration, intercostal, subcostal and sternal recession, and pulsus paradoxus (systolic BP falls in inspiration by >10mmHg).

Points on examining the chest

[*OTM* 15.51]

(There may be bronchial breathing at the top of an effusion.)

PLEURAL EFFUSION

Expansion: ↓
Percussion: Stony dull ↓
Air entry: ↓
Vocal resonance: ↓

Trachea + mediastium central
Expansion ↓
Percussion note ↓
Vocal resonance ↑
Bronchial breathing ± coarse crackles
(with whispering pectoriloquy)

CONSOLIDATION

325

Expansion ↓
Percussion note ↓
Breath sounds ↓

EXTENSIVE COLLAPSE
PNEUMONECTOMY /LOBECTOMY

Expansion ↓
Percussion note ↑
Breath sounds ↓

PNEUMOTHORAX

Expansion ↓
Percussion note ↓
Breath sounds bronchial
± crackles

FIBROSIS

Some physical signs

The chest x-ray

Be systematic in your approach to the chest film. (See p774.)

The heart is normally $<\frac{1}{2}$ the width of the thorax, but may appear smaller in the hyperexpanded chest of COAD. Enlargement of the heart and left ventricular failure results in characteristic x-ray appearances (see p775 and p296).

The mediastinum may be enlarged on a PA CXR due to a number of different disorders. A lateral chest film is essential to localize and help elucidate any masses seen:
- *Any section* —aortic aneurysm, lymph nodes (lymphoma, metastases).
- *Superior mediastinum* —thymoma, retrosternal thyroid, para-oesophageal (Zenker's) diverticulum.
- *Anterior and middle mediastinum* —dermoid cyst, teratoma, bronchogenic cyst, pericardial cyst, Morgagni diaphragmatic hernia.
- *Posterior mediastinum* —neurogenic tumours, any paravertebral mass, Bochdalek diaphragmatic hernia, achalasia, hiatus hernia.

The hila The left should be slightly higher. They can be pulled up or down by fibrosis or collapse (see p775). They may be enlarged by pulmonary arterial hypertension, lymphadenopathy (eg TB, sarcoid, lymphoma, metastases). Hilar calcification suggests TB, silicosis or histoplasmosis.

The diaphragm The right side is usually slightly higher. An apparently elevated hemidiaphragm may be due to loss of lung volume, phrenic nerve palsy, subpulmonic effusion, subphrenic abscess, hepatomegaly, diaphragmatic rupture (eg post traumatic).

Diffuse lung disease produces the most interesting CXRs and often the most difficult to interpret. Try to categorize the pattern into nodular (small or large, equal size or varying), reticular (a network of thin lines) or alveolar ('fluffy') or often a combination of all three.

Nodular shadows
- Viral pneumonia (varicella)
- Granulomas (miliary TB, sarcoid, histoplasma, hydatid, Wegener's)
- Mitral stenosis (microlithiasis pulmonale due to pulmonary haemosiderosis)
- Malignancy (metastases, lymphangitis carcinomatosis, bronchoalveolar Ca)
- Pneumoconioses (except asbestos—linear shadow), Caplan's sy
- Septic emboli

Reticular shadows
- Fibrosis of chronic infections (TB, histoplasma)
- Sarcoid, silicosis, asbestosis
- Early LVF
- Malignancy (lymphangitis carcinomatosa)
- Extrinsic allergic alveolitis
- Cryptogenic fibrosing alveolitis
- Autoimmune diseases (Wegener's, SLE, PAN, CREST (p670), rheumatoid arthritis)

Alveolar shadows
- Pulmonary oedema
- Infection
- Pulmonary haemorrhage
- Smoke inhalation
- Drugs (heroin, cytotoxics)

- ARDS, O_2 toxicity, fat emboli, DIC
- Renal or liver failure
- Head injury, neurosurgery
- Alveolar proteinosis
- Near drowning, heat stroke

Left upper-lobe collapse

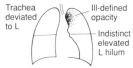

Trachea deviated to L

Ill-defined opacity

Indistinct elevated L hilum

Sharply-defined posterior border due to anterior displacement of oblique fissure

Left lower-lobe collapse

Oblique fissure displaced posteriorly

Triangular opacity visible through the heart with loss of medial end of diaphragm

327

Lingular consolidation

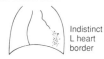

Indistinct L heart border

Right upper lobe collapse

Trachea deviated to R

Horizontal fissure and R hilum displaced upwards

Triangular opacity with well-defined margins

Right middle lobe collapse

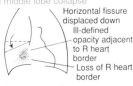

Horizontal fissure displaced down
Ill-defined opacity adjacent to R heart border
Loss of R heart border

Well-defined triangular opacity running from hilum

Right lower-lobe collapse

Horizontal fissure displaced downwards

Oblique fissure and hilum displace posteriorly

Well-defined posterior opacity

Well-defined opacity adjacent to R heart border (R heart border still visible)

Bedside tests in chest medicine

Arterial blood gas analysis gives information about:

- Ventilatory efficiency: $PaCO_2$ >6.5kPa diagnoses *hypoventilation*; $PaCO_2$ <4.5kPa diagnoses *hyperventilation*. Example: in severe asthma, at first the patient hyperventilates to maintain his PaO_2—he has a low $PaCO_2$. As he tires, he hypoventilates—his $PaCO_2$ becomes normal then rises. A normal $PaCO_2$ is thus a danger sign in asthma.

- *Oxygenation:* PaO_2 is principally dependent on the ventilation: perfusion (V/Q) balance which is upset in almost all lung diseases. PaO_2 is also affected by inspired O_2 concentration (F_IO_2), ventilatory efficiency, and the affinity of blood for O_2. A PaO_2 <8kPa defines respiratory failure (see p352).

- An estimate of the *alveolar:arterial O_2 concentration gradient* (A–a gradient) can be made with this formula:
$$F_IO_2 - (PaO_2 + PaCO_2)$$
At 25 years of age the normal range is 0.2–1.5kPa; at 75 years, 1.5–3.0. Larger gradients suggest *impaired gas exchange*. For practical purposes, if the units are kPa, F_IO_2 (partial pressure of inspired oxygen) equals its %: ie 20kPa in air.

- *Acid–base status:* See p630.

Sputum examination This involves inspection, microscopy to look for bacteria, particulate matter, and malignant cells, and culture. *Gram staining:* See p183. Bacteria are only likely to be pathogenic if seen in large numbers, or are intracellular. Make sure the sample you send for analysis is sputum, not saliva. The physiotherapist may help with sputum production, sometimes loosening secretions with a hypertonic saline nebulizer.

Spirometry defines type of respiratory problem by measuring functional lung volumes. This relationship may not be preserved in severe obstructive disease.

Pattern	FEV_1/FVC (normal >70%)
Restrictive	>70%
Obstructive	<70%

(FVC: forced vital capacity; FEV_1: forced expiratory volume in first second)

Peak expiratory flow rate (PEFR) is valuable when measured serially to establish the pattern of obstructive airways disease and monitor its response to treatment, particularly in asthma.

The disposable end tidal carbon dioxide detector (Easy Cap™)[1,2] This uses a chemical pH indicator (metacresol) to detect the presence of CO_2 in expired gases. The colour varies from mauve to yellow with inspiration and expiration (respectively). Mauve indicates a CO_2 of <0.5%; tan≈0.5–2%; yellow≈2–5%. The device's *cyclical* colour change may be used to distinguish oesophageal from tracheal intubation—except that gastric fluid contamination produces a *permanent* orange colour which may be falsely reassuring. Also, if the patient is being resuscitated with lignocaine, or adrenaline given via an endotracheal tube (p724) a permanent yellow colour occurs. The device does not work if wet, or very cold. After ~2h replacement is needed. It is not reliable when used in cardiac arrest cardiopulmonary resuscitation.

1 J Muir 1990 BMJ ii 42 2 M Hayes 1990 BMJ ii 42

Normal Peak Expiratory Flow Rates

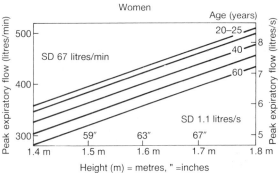

Charts for prediction of peak expiratory flow rates. After J Cotes 1978, *Lung Function* 3 ed, Blackwell, Oxford.

Examples of spirograms

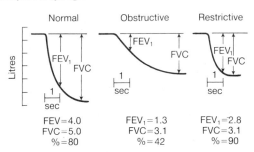

J West 1982; in *Harrison's Principles of Internal Medicine* 11 ed, ed E Brunnward *et al.*, p1055, McGraw-Hill, New York. Reproduced with permission.

Further investigation in chest medicine

Lung function tests FEV$_1$, FVC, PEFR (see p328). *Total lung capacity* (TLC) and, more so, residual volume (RV) are increased in obstruction because of air trapping. Obstruction may be so severe that the spirogram no longer shows the typical pattern. *Flow volume loop* measures peak flows at various lung volumes: flows are most affected at low volumes in distal obstruction (eg asthma) and at high volumes in proximal obstruction (eg trachea). *Transfer factor* (T$_{CO}$; DL$_{CO}$) measures gas transfer by assessing carbon monoxide uptake. Abnormalities are non-specific, but it is a good test of gas exchange in interstitial disease. The effect of *bronchodilators* on obstruction may be assessed: 10% increase in any of FEV$_1$, FVC, or vital capacity (VC) is useful. Similarly >15–20% increase in FEV$_1$/FVC.

Fibre-optic bronchoscopy is done using topical anaesthesia to the naso-pharynx and larynx. It allows examination with a flexible bronchoscope as far as the subsegmental bronchi.

Uses: Direct examination of airways; obtaining biopsies, cytological brushings, traps for bacteriology and cytology; bronchoalveolar lavage; removal of foreign bodies.

Preparation for bronchoscopy: CXR, spirometry, HIV and HBsAg serology (for biopsy: clotting studies, platelets, urea). Blood gases may be needed (bronchoscopy is hazardous in hypoxia; use O$_2$).

Bronchoalveolar lavage is performed by wedging the tip of the bronchoscope into a subsegmental bronchus, then instilling and aspirating a known volume of buffered saline into the distal airway. It may be diagnostic in pulmonary haemosiderosis, eosinophilic granuloma and alveolar proteinosis. It may also help differentiate certain interstitial diseases, eg extrinsic allergic alveolitis and sarcoidosis usually produce a highly lymphocytic lavage. It is used therapeutically to wash out secretions.

Lung biopsy may be obtained in several ways. *Percutaneous needle biopsy* under radiological imaging is useful for discrete peripheral lesions. *Bronchial biopsy* is helpful in diagnosing proximal lesions, usually tumours. *Transbronchial biopsy* at bronchoscopy may help in diagnosis of diffuse lung disease eg sarcoidosis. Surgical *open lung biopsy* may be considered.

Surgical procedures *Rigid bronchoscopy* provides a wide lumen, better for controlling haemorrhage, large biopsies, and removal of viscid secretions (eg proteinosis) and foreign bodies. *Mediastinoscopy* allows examination and biopsy of the right hilum. *Mediastinotomy* allows similar access to the left hilum. *Thoracoscopy* allows examination and biopsy of pleural lesions. *Open lung biopsy* is used for diagnosis and staging of interstitial lung disease.

Radiology CXR p326. *Conventional tomography*, for long used to define hilar and other masses, is being superseded by *computerized tomography* (CT). CT scans can visualize mediastinum, hilum, pleura and lung parenchyma. *Bronchography* is unpleasant for the patient but is useful for defining the extent of bronchiectasis if surgery is contemplated. (See p792.)

Lung volumes: physiological and pathological[1]

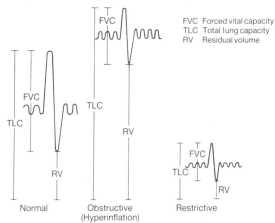

FVC Forced vital capacity
TLC Total lung capacity
RV   Residual volume

Normal   Obstructive (Hyperinflation)   Restrictive

Flow volume loops[2]

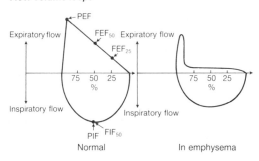

PEF
Expiratory flow
FEF$_{50}$ Expiratory flow
FEF$_{25}$

75 50 25
%

Inspiratory flow

Inspiratory flow

PIF FIF$_{50}$
Normal

In emphysema

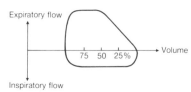

Expiratory flow

75 50 25% Volume

Inspiratory flow

In extrathoracic upper airway narrowing

PEF   Peak expiratory flow
FEF$_{50}$ Forced expiratory flow at 50% total lung capacity
FEF$_{25}$ Forced expiratory flow at 25% total lung capacity

PIF   Peak inspiratory flow
FIF$_{50}$ Forced inspiratory flow at 50% total lung capacity

1 After D Flenley 1987 *Med Intl* 1 (20) 240   2 After B Harrison in *Thoracic Medicine* 1981, ed P Emerson, Butterworths, London; and B Harrison 1971 *Thorax* 26 579

Pneumonia

Symptoms and signs Systemic upset, fever, pleuritic pain, cough, green sputum (may be scanty at first; 'rusty' sputum of pneumococcal pneumonia), haemoptysis, confusion (sometimes the *only* sign in the elderly); the signs of consolidation (see p324) or just localized crepitations. Tachypnoea is a valuable early sign eg in the elderly in whom there should be a high index of suspicion.

Differential diagnosis of unilateral signs Acute bronchitis (usually less systemically unwell), effusion (may coexist, but the signs of effusion are dictated by gravity, not lobar anatomy), pulmonary embolism, carcinoma, pulmonary oedema (occasionally unilateral).

Differential diagnosis of bilateral signs Pulmonary oedema (may coexist and be cause or effect of the pneumonia), acute bronchitis, alveolitis (look for clues in the history; the crepitations are very fine), bronchiectasis (*chronic* sputum production; coarse leathery crepitations). Clinical and radiographic features of pneumonia may develop as a result of atelectasis (collapse of alveolae due to hypoventilation producing a dull percussion note and quiet breath sounds) or pooled secretions in old, immobile, or post-operative patients. Superinfection is common; prevent it with physiotherapy.

Investigations CXR, U&E, FBC (with a differential white cell count), blood cultures, sputum culture and microscopy, serology for 'atypical' organisms: *Legionella*, *Mycoplasma*, *Chlamydia*, and *Coxiella* (send 10ml clotted blood on admission and a convalescent sample 10 days later to demonstrate an antibody rise). If the patient is very ill and has not responded to conventional treatment, microbiological specimens may be obtained with the fibre-optic bronchoscope from bronchial washings or by percutaneous lung aspiration. In some centres *countercurrent immunoelectrophoresis* (CIE) can be performed on blood, urine, sputum, or CSF to identify pneumococcal antigen. Antibiotics do not interfere with CIE. Consider: ECG, cardiac enzymes.

Bacteriology *Primary pneumonia* occurs in previously healthy people (often young). It is often due to pneumococci or 'atypical' organisms. Rarer causes: *Staph aureus* (especially as a complication of a flu epidemic) and viral pneumonias.

Secondary pneumonia occurs in patients with damaged defences, either locally (eg COAD) or systemically (eg malignancy). The most common pathogens are *Pneumococcus* and *Haemophilus*, others include *Staph aureus* (which may cavitate or cause pneumothorax), *Klebsiella* (which may cause lobar 'swelling' on CXR), *Pseudomonas aeruginosa*, TB, anaerobes and Gram negative organisms (both associated with aspiration pneumonia), or opportunists (eg *pneumocystis* in AIDS, p214).

Complications Pleural effusion, empyema, lung abscess, septicaemia, metastatic infections, respiratory failure, jaundice.

Preventing pneumococcal infection Offer pneumococcal vaccine (23 valent Pneumovax II®, 0.5ml SC) to those with: ● Chronic heart or lung conditions ● Cirrhosis ● Nephrosis ● Diabetes mellitus ● Immunosuppression (eg splenectomy, AIDS, or on chemotherapy). CI: pregnancy, lactation, fever. Those with the highest risk of fatal pneumococcal infection (asplenia, sickle-cell disease, nephrosis) should be revaccinated after 6yrs (3–5yrs in children), unless they have had a severe vaccine reaction.

Pneumonia
[*OTM* 15.92]

'Atypical' pneumonias

▶ Suspect if the CXR is much worse than the symptoms. Ask yourself: *Is he immunocompromised?*

Chlamydia pneumoniae The commonest chlamydial infection.[1] Person-to-person spread occurs (no animal needed) causing a biphasic illness: pharyngitis, hoarseness and otitis, then pneumonia. *Tests:*[2] Chlamydia serology (non-specific). *Treatment:* erythromycin 500mg/6h PO for ~10d.

Chlamydia psittaci causes psittacosis (contracted eg from parrots) or ornithosis (turkeys, pigeons). *Incubation:* 4–14 days. *Presentation:* Pneumonia or systemic illness with pneumonitis ± hepatosplenomegaly. CXR: Patchy consolidation. *Diagnosis:* Serology. *Treatment:* Tetracycline 500mg/6h for 21 days. Try to eradicate source. Relapses occur. (Pigeon fancier's lung, an extrinsic allergic alveolitis, is unconnected.)

Mycoplasma pneumoniae Epidemics may occur. *Incubation:* 12–14 days. Often the cough is dry. *Extrapulmonary features:* 50% have cold agglutinins ± haemolytic crises; 50% have eg D&V; 25% have skin signs (eg erythema multiforme); arthralgia; myalgia; ~6% have meningo-encephalitis or myelitis. CXR: lower zone consolidation, usually unilateral. *Tests:* 80% have 4-fold↑ in antibody titre; cold agglutinins suggest recent infection. *Treatment:* Erythromycin as above. Relapse is common.

Legionella pneumophila can spread by vapour or ventilation systems. *Incubation:* 2–10 days. ♂/♀=3. It begins as high fever and myalgia. Nausea and vomiting, abdominal pain, and blood PR may precede chest problems. The patient may be very ill, with haemoptysis, CNS signs (eg delirium), or renal failure. U&E: Na↓ and Ca²⁺↓ are common. Myoglobinuria may be present. *Diagnosis:* serology or immunofluorescence. *Treatment:* erythromycin 1g/6h IVI (1 week) then PO (3 weeks) + rifampicin if severe.

Pneumocystis carinii pneumonia (PCP) This variously classed (fungus / protozoa) organism causes pneumonia in the immunocompromised (eg AIDS). *Presentation:* Tachypnoea, dry cough, respiratory failure, ± fever. *Diagnose* from clinical setting, sputum cytology, bronchoalveolar lavage or lung tissue microscopy. CXR: Normal, or diffuse bilateral alveolar shadows, sparing lower zones—or show ground-glass appearance. *Treatment:* Co-trimoxazole, p180. Peak serum sulphamethoxazole >100µg/ml may be needed. (Prophylaxis dose: 960mg/24h PO.) SE: rash—do not always stop co-trimoxazole. Alternative: pentamidine 4mg/kg/24h IV over 2h for 21d. Consider prednisolone if hypoxic: 40mg/24h PO; lower dose after 5d.

Other causes of atypical pneumonia: TB, influenza (p206), CMV, measles, varicella, *Coxiella* (p236), aspergillosis (p340), actinomycosis.

Lung abscess

Causes 1 Inadequately treated pneumonia (especially Staph, *Klebsiella*). 2 Aspiration: eg alcoholism, bulbar palsy, oesophageal disease (achalasia, obstruction). 3 Bronchial obstruction (tumour, foreign body). 4 Pulmonary infarction. 5 Septic emboli.

Clinical Malaise, weight↓, cough±sputum, fever, clubbing, haemoptysis.

CXR Walled cavity with fluid level (moves in decubitus film).

Treatment Attempt to identify the organism (bronchoscopy may be necessary; and will exclude obstruction). Antibiotics as appropriate (eg in the light of sensitivities found from ultrasound-guided percutaneous sampling—anaerobes are the most important pathogens), usually for several weeks. Postural drainage. Surgery is rarely necessary.

1 SJ Bourke 1993 *BMJ* i 1219   **2** M Sillis 1993 *BMJ* ii 63

Management of pneumonia

- Bed rest. The patient should sit up, rather than lie flat.
- Analgesia for pleurisy (eg NSAID).
- Fluids (IV if necessary) to rectify dehydration and maintain an adequate urine output (>1.5 litres/24h). Remember losses are increased if the patient is febrile. However, free water clearance is impaired, so beware of over-hydration.
- Oxygen—but take care in COAD (see p348).
- Antimicrobials: see below and opposite.
- Treat for septicaemic shock (see p720) if very ill.
- Physiotherapy is no longer recommended in acute lobar pneumonia without copious secretions. It achieves nothing and may worsen gas exchange.

Antimicrobial use: general points

- As the patient is often too ill to wait for the results of cultures, it is common to start IV therapy immediately *after* taking appropriate cultures and advice from a senior colleague, if available. Antibiotic choice for this 'blind' treatment depends on the clinical picture (p337).
- 'Blind' treatment may also be appropriate in out-patients.
- IV therapy is necessary if the patient is very ill, cannot swallow, or if the GI tract is not functioning.
- Always consider consulting with a microbiologist to discuss the sensitivities of local pathogens, prescribing policies, and the significance of culture growths, if equivocal.

If the pneumonia is slow to resolve or recurrent, consider atypical or resistant organisms, a foreign body, carcinoma of the bronchus, or aspiration, eg from a nasal sinus.

'Blind' treatment of pneumonia

Clinical picture	Organisms to be covered	Antibiotic route: PO	IV
Primary pneumonia[1]	*Pneumococcus* If 'atypicals' (p334) likely	Ay add E	A add E
Secondary pneumonia			
Previous lung disease (eg COAD)	*Pneumococcus*, *Haemophilus*	Co	Co+E or C+E
If follows flu, or associated with URTI	Staph is also possible	add F	add F
Aspiration	*Pneumococcus*, anaerobes, *Klebsiella*, Gram negative organisms	x	P+M+G
Immunosuppression* eg leukaemia	*Pseudomonas* as well as usual organisms	x	T+G or Cz+G
Hospital acquired*	Gram negative as well as usual organisms	?x	C+G
Sepsis elsewhere*	Treatment depends on origin of infection		
Severely ill patient*	Widest possible range	x	Ct+E+G

For treatment of a patient with HIV infection, see p214.

337

Key to antimicrobials

A   Ampicillin 500mg/6h IV
Ay   Amoxycillin 250mg/8h PO
C   Cefuroxime 750mg/8h IV
Co   Co-amoxiclav 1 tablet (250/125) PO or 1.2g (1000/200)/8h IV
Ct   Cefotaxime 1g/8h IV
Cz   Ceftazidime 2g/8h IV
E   Erythromycin 250–500mg/6h PO or 500mg/6h slowly IV (IV erythromycin is painful and may cause phlebitis)
F   Flucloxacillin 250mg/6h PO or 250–1000mg/6h IV
G   Gentamicin: dosage nomogram p796; dose example: 120mg IV as loading dose then 2–5mg/kg/day in 2–3 divided doses (eg 80mg/8h); check serum levels within 48h (p757). (IV gentamicin is painful.)
M   Metronidazole 500mg/8h IV (for up to 7 days)
P   Benzylpenicillin 600mg/6h IV (dose may be increased)
T   Ticarcillin 4g/6h IV
x   Oral therapy is inappropriate in this situation

*With these serious infections, it is always wise to discuss the matter with your microbiologist first. There may be hospital policy on the choice of antibiotics: there are many powerful régimes and it is important to minimize the risk of resistance by having a uniform policy. In addition, the microbiologist may know about a similar pathogen that is causing trouble elsewhere in the hospital and your patient may benefit from experience already gathered.

1 M Woodhead 1992 *Br J Hosp Med* **47** 684

Bronchiectasis

▶ Consider this in any recurrent or persistent chest infection.

Pathology Irreversibly dilated bronchi act as sumps for persistently infected mucus which is expectorated daily. *Haemophilus influenzae* and *Pseudomonas aeruginosa* are commonest pathogens.

Causes Congenital: Cystic fibrosis, Kartagener's sy (bronchiectasis, sinusitis, dextrocardia). *Post-infection:* TB, measles, pertussis, pneumonia. *Other:* Bronchial obstruction, ABPA (p340), hypogammaglobulinaemia.

Clinical Chronic bronchial sepsis with exacerbations, persistent sputum, haemoptysis, clubbing, low-pitched 'leathery' inspiratory and expiratory crackles (may be audible at mouth), wheeze.

Tests CXR: cystic shadows with fluid levels, thickened bronchial walls (tramline and ring shadows); sputum culture; spirometry to assess extent of lung damage and reversible airway obstruction, serum immunoglobulins, sweat test, skin-prick test for aspergillus. Bronchography if diagnosis is uncertain or to assess extent of disease prior to surgery. CT may help.

Management 1 Physiotherapy—daily postural drainage; 2 Antibiotics—intermittently if exacerbations are infrequent (eg doxycycline 200mg PO on first day, then 100mg/24h for 7–14 days; or as a 5-day course of high dose amoxycillin 3g/12h), or continuously if recurrent infections are frequent; may be given PO, IV or by nebulizer; 3 Bronchodilators, for airway obstruction; 4 Surgical excision, only for localized disease.

Complications Acute infective exacerbations; haemoptysis (may be life-threatening); pulmonary hypertension; pneumothorax; brain abscess.

Cystic fibrosis

See *OHCS* p192

One of the commonest inherited diseases: 1 in 2000 live births. Inheritance: autosomal recessive; gene now identified (codes for a transmembrane protein) and allows prenatal diagnosis. 1:25 Caucasians are carriers.

Pathology Cells, relatively impermeable to chloride, make salt-rich secretions. Viscid mucus blocks and damages glands.

Clinical features *Neonate:* Meconium ileus. *Child or young adult:* Recurrent chest infections, bronchiectasis, pancreatic insufficiency (steatorrhoea, failure to thrive, good appetite), DM, distal intestinal obstruction syndrome (meconium ileus equivalent). Fertility problems.

Tests Sweat test. CXR: shadowing suggesting bronchiectasis, especially in upper lobes. Malabsorption screen. Glucose tolerance test. Spirometry. Sputum culture. Skin test for *Aspergillus* (20% develop allergic bronchopulmonary aspergillosis—p340).

Diagnosis Sweat sodium >70mmol/l in children.

Management 1 Chest: as for bronchiectasis, but with appropriate antibiotics (Staph is the commonest pathogen in children; *Pseudomonas* becomes increasingly common with age); 2 Pancreatic enzyme and vitamin supplements (A, D, E. K); 3 Diabetes management; 4 In end-stage, some may be considered for heart–lung transplantation.

Prognosis Improving—most reach early adulthood.

Bronchiectasis
Cystic fibrosis
[*OTM* 15.100; 15.103; 15.102]

Fungi and the lung

Fungi may be commensals, allergens, or pathogens in the lung. Fungal infection is much commoner in the immunocompromised.

Aspergillosis *Aspergilloma:* A tangled mass of hyphae, fibrin, and inflammatory cells, which accumulate in a lung cavity, often after TB. Usually asymptomatic, it may cause cough and haemoptysis (which may be torrential). CXR: round opacity, usually in the upper zone, surmounted by a thin dome of air. Organisms may be cultured from sputum, and *Aspergillus* precipitins are present in high titre. Treatment is only required for symptoms (usually severe haemoptysis) and surgical excision is the most effective, although mortality is high. Local instillation of antifungal drugs often fails.

Allergic bronchopulmonary aspergillosis (ABPA): This results from an allergic reaction to *Aspergillus fumigatus*. Early on, the allergic response causes bronchoconstriction, but as the inflammation persists, permanent damage occurs, causing bronchiectasis (classically proximal). The clinical picture is of wheeze, cough (productive of plugs of mucus) and dyspnoea. ABPA should be suspected in asthmatics who have transient shadows on CXR and eosinophilia. The shadowing may be permanent and proximal bronchiectasis is seen. A positive skin-prick test confirms hypersensitivity. Serum precipitins are present but rarely in high titre. Serum IgE is raised, as is the peripheral eosinophil count. It is almost impossible to eradicate the fungus, so prednisolone 30–45mg/24h PO is given to reduce inflammation until clinically and radiographically clear. Some maintain patients on prednisolone 20mg/alternate day. Sometimes bronchoscopic aspiration of mucus plugs is necessary.

Invasive aspergillosis: This only occurs in the immunocompromised, and should be considered in the differential diagnosis of any focal infection or septicaemia in such a patient. Diagnosis may only be made at biopsy—or autopsy. Treat with amphotericin B.

Using amphotericin B: First give a test dose, 1mg in 20ml 5% dextrose IV. If there is no reaction, give 0.3mg/kg IV in 50ml dextrose over 6h on day 1. Thereafter increase to 1mg/kg/24h (given over 6h at a concentration of 10mg/100ml. Max dose: 1.5mg/kg/24h. Do not give any other drug in the same IVI. SE: after rapid infusion, anaphylaxis, arrhythmias, seizures. Nephrotoxicity (common), vomiting, phlebitis, hypokalaemia, anaemia, fevers, and weight loss. Monitor K⁺ and renal function daily. Ambisome® is liposomal amphotericin (less nephrotoxic, fewer side-effects, much more expensive; dose example:[1] start with 1mg/kg/day IVI over 1h, increase by 1mg/kg/day up to 3mg/kg/day or more).

Other fungi

Although aspergillomas are the commonest fungus balls (mycetoma), other fungi (eg Candida) may also be implicated. In the Americas, *Histoplasma capsulatum*, *Coccidioides immitis*, *Blastomyces dermatitidis*, and *Paracoccidioides brasiliensis* may cause acute respiratory infections (with cough, dyspnoea, arthralgia, erythema nodosum and multiforme) and chronic infections (causing cavitation) which may resemble TB and malignancy. See also p244.

Ketoconazole is the drug of choice for some forms of histoplasmosis, blastomycosis, paracoccidioidomycosis (and mucocutaneous but not systemic candidosis).[2] Dose: 200–400mg/24h PO with food. It is ineffective in aspergillosis. SE: LFTs↑ in 5–10%. Liver failure is rarer and not dose-related. CI: pregnancy; porphyria; LFT↑.

1 *Drug Ther Bul* 1993 **31** 93   **2** Br Soc Antimicrob Working Party 1991 *Lancet* **i** 1577

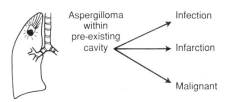

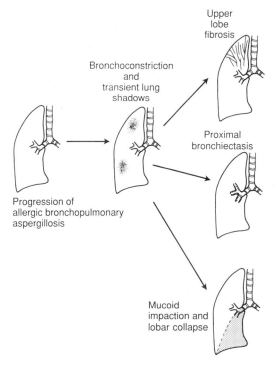

After PJ Kumar and ML Clark 1990 *Clinical Medicine* 2 ed, p688, Bailliere Tindall

Lung malignancy

Bronchial carcinoma[1] ≈19% of all cancers and 27% of cancer deaths. Incidence increasing. Major risk factor: cigarette smoking:

Cigarettes/day	0	1–14	15–24	≥25
Male deaths/yr/1000	0.07	0.78	1.27	2.51

Risk in those who give up declines in time; former smokers of <20 cigarettes/day take 13 years to reach risk of non-smokers. *Adenocarcinoma is not smoking-related.* Other risk factors: age, male sex, asbestos, radio-activity, chromium, iron, iron oxides. Chronic inflammation predisposes to alveolar cell carcinoma.

Histology 4 main types: squamous (35% of cases); adenocarcinoma (20%); large cell (19%); small (oat) cell (25%).

Clinical features Cough 80%, haemoptysis 70%, dyspnoea 60%, chest pain (often mild) 40%, wheeze (often monophonic) 15%. Patients may present with pneumonia (eg recurrent; slow to resolve).

Complications: *Intrathoracic* Pleural effusion, recurrent laryngeal nerve palsy, SVC obstruction, brachial neuritis, Horner's (p54), rib erosion, pericarditis, AF. *Metastatic:* Brain, bone, liver, adrenals.

Non-metastatic: **a.** *Endocrine:* ADH or ACTH secretion, hypercalcaemia. **b.** *Neuromuscular:* Neuropathy, myopathy, (dermato-)myositis, dementia, cerebellar degeneration, and the myasthenic syndrome (p472). **c.** *Other:* Weight loss, anaemia (rarely classical leucoerythroblastic), clubbing, thrombophlebitis migrans, bleeding disorders, and hypertrophic pulmonary osteoarthropathy (HPOA, ie clubbing and painful, swollen wrists and ankles, with typical x-ray).

Investigation 1 *Confirm suspicion:* CXR (±tomograms) (see p326); central lesions may only be diagnosed bronchoscopically. **2** *Determine histology:* Cytology (sputum, bronchoscopic brushings); biopsy (bronchoscopic for proximal lesions, percutaneous biopsy for peripheral); pleural aspiration and biopsy for effusions; lymph node biopsy. **3** *Are there metastases?* Alk phos and other LFTs for liver/bone metastases. Isotope bone scan, mediastinoscopy and mediastinotomy, abdominal ultrasound, CT (thorax, brain, abdomen).

Management[2] *Non-small cell tumours:* If no sign of extrapulmonary or pleural spread (~25%), attempt excision. *Small cell tumours* have almost always disseminated at presentation. Combination chemotherapy prolongs survival. Squamous and small cell tumours are also radiosensitive. For most patients, the main aim is symptom relief: radiotherapy for eg bronchial obstruction causing lung collapse, SVC obstruction, bony pain, distressing haemoptysis, Pancoast's tumour (p706); methadone linctus 2mg/4h for cough; drain symptomatic effusions then consider pleurodesis.

Prognosis *Non-small cell:* 50% 2-yr survival without spread, 10% with spread. *Small cell:* 3 months if untreated, 1yr treated (median survival).

Bronchial 'adenoma' Most are carcinoid tumour; slow growing, metastasize late and are usually successfully excised.

Mesotheliomas are tumours of pleura. If malignant there is usually a remote history of asbestos exposure. They present with severe chest pain, dyspnoea, bloody effusions. Certain diagnosis requires pleural biopsy (open or thoracoscopic). Prognosis very poor. Treat symptoms.

1 R Doll 1976 *BMJ* **ii** 1525   **2** *Drug Ther Bul* 1988 **26** 29

Coin lesions of the lung

Malignancy (1° or 2°)
Abscesses (p334)
Granuloma
Carcinoid tumour
Pulmonary hamartoma

Arteriovenous malformation
Encysted effusion (fluid, blood, pus)
Cyst
Foreign body
Skin tumour

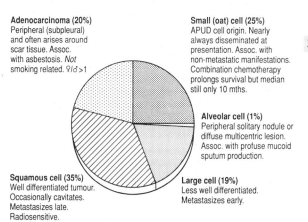

Adenocarcinoma (20%)
Peripheral (subpleural)
and often arises around
scar tissue. Assoc.
with asbestosis. *Not*
smoking related. ♀/♂>1

Small (oat) cell (25%)
APUD cell origin. Nearly
always disseminated at
presentation. Assoc. with
non-metastatic manifestations.
Combination chemotherapy
prolongs survival but median
still only 10 mths.

Alveolar cell (1%)
Peripheral solitary nodule or
diffuse multicentric lesion.
Assoc. with profuse mucoid
sputum production.

Squamous cell (35%)
Well differentiated tumour.
Occasionally cavitates.
Metastasizes late.
Radiosensitive.

Large cell (19%)
Less well differentiated.
Metastasizes early.

Histological types of bronchial carcinoma

Bronchial asthma

▶ Respect this condition: young people are still dying from it and the mortality appears to be rising.

Essence Generalized airways obstruction, which, in the early stages at least, is paroxysmal and reversible. There are 3 separate obstructing factors: mucosal swelling, increased mucus production, and bronchial muscle constriction.

Prevalence Up to 8% in some populations are episodic wheezers.

History Intermittent wheeze, dyspnoea or cough. Nocturnal cough may be the *only* symptom. Ask about: ● *Disturbed sleep* (quantify as nights per week: it is a sign of serious asthma) ● *Exercise* (quantify distance to breathlessness) ● *Days per week off work or school* ● *Diurnal variation* (marked morning dipping of peak flow is common and may be fatal, despite normal peak flow at other times) ● *Precipitants*: emotion, exercise, infection (often rhinovirus or coronavirus URTI), allergens, drugs eg aspirin, β-blockers ● *Other atopic disease* (eczema, hay fever, allergy); family history of asthma ● *Acid reflux* (a known asthma association) ● *Occupation* (find out if symptoms remit at weekends or holidays). Rarely, asthma may be part of the systemic Churg–Strauss syndrome (p674) or be complicated by allergic bronchopulmonary aspergillosis (p340).

Examination The cardinal sign is widespread polyphonic, high-pitched wheezes (airways of varying size, but mostly small calibre, affected). Chronic asthma may give a 'barrel chest' thorax with indrawn costal margins (Harrison's sulci). Acutely ill asthma patients (p730) present with signs of respiratory distress (p324). Air entry may be inadequate to generate any wheeze ('the silent chest'—an ominous sign).

Tests Teach him to do peak expiratory flow rates (PEFR) regularly (eg 4–hourly for a week, if worsening). Diurnal patterns help to plan therapy. Severity markers: CXR (hyperinflation), spirometry ($FEV_1/FVC\downarrow$), residual volume (↑ means marked air trapping), blood gases in acute severe asthma (p730). Look for aspergillus: CXR, FBC (eosinophils), sputum (aspergillus plugs, eosinophils). Do prick tests as appropriate (good for aspergillus, but disappointing in identifying specific allergens amenable to therapy; they may merely confirm atopy). In acute asthma, investigations aim to find a precipitant or complication, eg infection, pneumothorax.

Differential diagnosis Pulmonary oedema ('cardiac asthma'), chronic obstructive airways disease (which very often coexists, or may have a reversible component), large airway obstruction (producing stridor) eg foreign body or tumour, pneumothorax, pulmonary embolism.

Natural history Most childhood asthmatics either grow out of their disease in adolescence, or suffer much less as adults. A significant number of people develop chronic asthma late in life.

Treatment See p346. Emergency treatment (severe asthma): p730.

Mortality Death certificates give a figure of 2000/yr (UK): more careful surveys more than halve this figure.[1] 50% are >65yrs old.

1 W Berrill 1993 *BMJ* ii 193

Bronchial asthma

Examples of serial peak flow charts

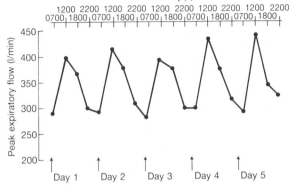

Classical diurnal variation of asthma
Arrows point to morning 'dips'

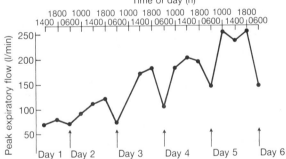

Recovery from severe attack of asthma
Predicted PEFR was 320 l/min
Arrows point to early morning 'dips'

After B Harrison 1981 *Thoracic Medicine*, ed P Emerson, Butterworths, London

Management of bronchial asthma[1]

▶ The aim is full symptom control. Rarely be satisfied with less.

Behaviour Stop smoking. Avoid any *relevant* allergens (p344). Education. Specific advice about what to do in an emergency.

Drugs available β_2 *adrenoreceptor agonists* relax bronchial smooth muscle, acting within minutes. Salbutamol, the most popular, is ideally given by inhalation (aerosol, powder, or nebulizer), but may also be given PO or IV. SE: tachyarrhythmias, ↓K+, tremor, and anxiety. If more than 2 puffs (200μg)/12h are needed, add an inhaled steroid. Salmeterol is a new long-acting inhaled β_2-agonist that may be useful for short-term relief of nocturnal symptoms and for reducing morning dipping.[2] It is *not* a substitute for steroids. SE: paradoxical bronchospasm, tolerance.[3] ▶ Avoid all β-blockers and NSAIDs: they worsen asthma.

Corticosteroids are best inhaled, eg beclomethasone spacer (or powder) but may also be given PO or IV. The keystone of therapy in moderate to severe asthma, they act over a period of days by decreasing bronchial mucosal inflammation. The patient should gargle after inhaled steroids to prevent oral candidiasis. Oral steroids are used acutely (high dose, short courses, eg prednisolone 30–40mg/24h PO for 7 days) and longer-term in lower dose (eg 7.5mg/24h) if not optimally controlled on inhalers.

Aminophylline (metabolized to theophylline) may act by inhibiting phosphodiesterase, thus decreasing bronchoconstriction. It is usually given PO as a prophylactic agent, eg at night to prevent morning dipping. It is also useful as an adjunct if inhaled therapy is inadequate. In acute severe asthma, it may be given IVI. It has a narrow therapeutic ratio, causing arrhythmias, GI upset and fits in the toxic range. Therapy should be controlled by checking blood theophylline levels (p797), and IV therapy should be accompanied by ECG monitoring, if available.

Anticholinergics (eg ipratropium) may ↓muscle spasm synergistically with β_2 agonists. They may be of more benefit in COAD than in asthma. Try each bronchodilator alone, and then together; assess with spirometry.

Cromoglycate, in asthma, can only be inhaled. It is sometimes useful for prophylaxis in mild asthma, and exercise-induced asthma, especially in children. Occasionally, it may precipitate asthma.

British Thoracic Society 1993 guidelines[1] Check inhaler technique; address any fears of the patient. Prescribe peak flow meter. Start at the step most appropriate to severity; move up as needed (or down, if control good for >3 months). Courses of prednisolone may be needed at any time.

Step 1: Try occasional β-agonist inhaler. If needed >daily, add step 2.

Step 2: Add inhaled beclomethasone or budesonide 100–400μg/12h, if nedocromil sodium or cromoglycate do not control symptoms.

Step 3: ↑Inhaled steroid ≤1mg/12h by large volume spacer, ±theophylline, eg 250mg/12h PO (keep to one slow-release brand).

Step 4: Add ipratropium±salmeterol or long-acting β-agonist tablets.

Step 5: Check compliance; add regular prednisolone tabs (1 daily dose).

Pitfalls ● Failing to prescribe steroids soon enough.
● Failing to notice severe morning dips in peak flow. Always ask about nocturnal waking: it is a sign of dangerous asthma.
● Not admitting those with moderately severe attacks to hospital.

Emergency treatment of acute severe asthma See p730.

1 BTS Guidelines *BMJ* 1993 i 776–800   2 MF Fitzpatrick *BMJ* 1990 **301** 1365–8   3 D Cheung 1992 *NEJM* **327** 1198

Doses of some inhaled drugs used in bronchoconstriction

	Inhaled aerosol	Inhaled powder	Nebulized
Salbutamol:			
single dose	100µg	100µg or 200µg	1mg/ml or 5mg/ml
recommended régime	200µg/6h	400µg/6h	2.5mg/4–6h
Terbutaline:			
single dose	250µg	500µg*	2.5mg/ml
recommended régime	≤500µg/4h	500µg/6h	5–10mg/12–6h
Salmeterol			
single dose	25µg	50µg	–
recommended régime	50–100µg/12h	50–100µg/12h	–
Ipratropium bromide:			
single dose	18µg or 36µg	–	250µg/ml
recommended régime	18–72µg/6h	–	100–500µg/6h
Beclomethasone:			
Becotide*			
single dose	–	100µg or 200µg	50µg/ml
recommended régime	–	200µg/6–8h up to 800µg/12h	100µg/12h
Becotide 50*			
single dose	50µg	–	–
recommended régime	100µg/6–8h or 200µg/12h	–	–
Becotide 100*			
single dose	100µg	–	–
recommended régime	100µg/6–8h or 200µg/12h	–	–
Becloforte*			
single dose	250µg	–	–
recommended régime	250µg/6h or 250–1000µg/12h 500µg/12h**	–	–

In the form of a Turbohaler—which is much easier to use than the comparable form of salbutamol (Diskhaler*) and which, unlike inhalers, contains no fluorocarbons.

**The dose may be increased to 1500-2000µg/day—if so, significant systemic absorption occurs, and the patient should carry a steroid card.[1] Note: systemic absorption (via the throat) is minimized if a Nebuhaler* device is used. SE: local (oral) candidiasis (p500).

1 *Drug Ther Bul* 1990 **28** 45

Chronic obstructive airways disease (COAD)

Essence COAD includes the spectrum of chronic bronchitis and emphysema. *Chronic bronchitis* is defined clincally as sputum production on most days for 3 months of 2 successive years. It causes obstruction by narrowing the airway lumen with mucosal thickening and excess mucus. *Emphysema* (defined histologically) is the dilation of the air spaces by destruction of their walls. It causes obstruction by decreasing the lungs' elastic recoil, which normally holds the airways open in expiration. The two coexist in varying proportions in COAD, principally due to smoking. α-1-antitrypsin deficiency is a rare inherited cause of emphysema.

Clinical features *Symptoms:* Cough, sputum, dyspnoea, and wheeze. *Signs:* Hyperinflation, descended trachea (decreased distance between thyroid cartilage and sternal notch), hypertrophied accessory muscles, respiratory distress (p324), quiet breath sounds (quietest over bullae), crepitations, wheezes, cyanosis and cor pulmonale (p366). (Haemoptysis and clubbing are unusual in COAD. If present consider malignancy.)

Pink puffers and **blue bloaters** Some patients with COAD have an increased alveolar ventilation. They have a normal or low $PaCO_2$ and a relatively normal PaO_2. They are breathless but not cyanosed, hence the term 'pink puffers'. They may progress to type I respiratory failure (p352). Other COAD patients have a reduced alveolar ventilation, resulting in a high $PaCO_2$ and low PaO_2. They are cyanotic, and if cor pulmonale develops (p366), become 'blue and bloated'. Their respiratory centres are relatively insensitive to CO_2 and they rely on their hypoxic drive to maintain respiratory effort. It is dangerous to give these patients supplemental oxygen without careful observation, as hypoventilation or apnoea may result. These extremes represent ends of a spectrum; most patients fall in between.

Investigations *CXR:* Hyperinflation (>6 anterior ribs seen above diaphragm in mid-clavicular line), flat hemidiaphragms, reduced peripheral vascular markings, bullae. *Lung function tests:* obstruction with air trapping ($FEV_1/FVC \ll 70\%$); RV (residual volume) and TLC (total lung capacity) high; DL_{CO} (p330) low in emphysema, unlike asthma. FBC: secondary polycythaemia.

Management ● Stop smoking.
● Physiotherapy: breathing exercises and postural drainage.
● Aim for maximum bronchodilatation using drugs listed on p347.
● Treat pulmonary hypertension and polycythaemia (p366). Consider domiciliary O_2 for stable non-smokers with resting PaO_2 <7.3kPa. O_2 concentrators are prescribable by GPs. Some may benefit from overnight NIPPV (see p352).
● Give the patient a supply of antibiotics eg oxytetracycline 250mg/6h PO (1h ac) and annual flu vaccinations.
● Treat acute exacerbations as infections with some reversible obstruction. Out-patient management: give amoxycillin 500mg/8h PO (or as co-amoxiclav) plus a course of prednisolone reducing from 30–40mg PO over 14 days. If severe: admit and give, in addition, nebulized bronchodilators 4-hourly (or more often if effective) in maximal doses and hydrocortisone 200mg/6h IV for 24h.

Chronic obstructive airways disease (COAD)

Self-administration of inhaled drugs This is the cornerstone of therapy for asthma and COAD. Many find it difficult—mostly because they are badly taught.

Shake the inhaler (to mix drug with propellant). The patient should exhale fully. Then, putting the mouthpiece between his teeth, press the canister once just *after* beginning to inhale through the mouth. He should hold his breath in inspiration for 10 seconds if possible. If the patient cannot master the technique alternatives include: a 'spacer' (which is first filled from the canister with the aerosolized bronchodilator and the breath is then taken from it); several breath-activated devices for inhaled powders eg Rotahaler®; the breath-activated aerosol inhaler (avoiding the need to coordinate breathing and hand movements); and the nebulizer in which an aqueous solution of the drug is mixed with saline and aerosolized by passing air through it from an air compressor.

349

Always assess your patient's inhaler technique. Omitting to do so may be as effective as forgetting to write the prescription.

Advise taking a bronchodilator several minutes before an inhaled steroid so that the steroid may have greater access to the airways.

Causes of acute exacerbation of COAD
- Infection
- Pneumothorax
- Left ventricular failure
- Pulmonary embolism
- Sedative drugs
- Inspissated secretions & collapse

Acute respiratory distress syndrome

Synonyms Adult respiratory distress syndrome (ARDS); acute lung injury.

Essence Increased permeability of pulmonary microvasculature causes leakage of proteinaceous fluid across the alveolar-capillary membrane. It is thought to be the pulmonary manifestation of a generalized disruption of endothelial structure and function precipitated most commonly by trauma or sepsis and resulting in hypoxia and multiple organ failure.

Definition 1 Antecedent history of precipitating condition.
2 Refractory hypoxaemia (PaO_2 <7.5kPa, F_IO_2 >0.5; arterial/alveolar oxygen tension ratio <0.25). NB: less strict criteria are a PaO_2 <10kPa with inspired O_2 of 50% ($PaO_2:F_IO_2$ ratio <20kPa).[1]
3 CXR signs of bilateral alveolar oedema.
4 Pulmonary capillary wedge pressure <15–18mmHg; oncotic pressure ↔.
5 Total thoracic compliance <30ml/cmH$_2$O.

Risk factors

Trauma	Near drowning	Drugs: paraquat, heroin, aspirin
Gastric aspiration	Head injury	Multisystem diseases, eg:
Smoke inhalation	Bruising	Vasculitis; Acute pancreatitis
Sepsis	DIC; ICP↑	Acute hepatic failure
Massive transfusion	Burns	Eclampsia; Amniotic fluid embolism
Heart/lung bypass	Fat emboli	Thrombotic thrombocytopenic purpura

The more risks the more the chance of ARDS ($1\approx25\%$, $2\approx42\%$, $3\approx85\%$).

Management[2]
● Treat the cause, and ensure O_2 delivery to all organs.
● Treat any source of sepsis vigorously with broad spectrum antibiotics.
● *Circulatory support:* Dobutamine 2.5–10μg/kg/min may help maintain cardiac output and O_2 delivery. Pulmonary artery catheterization aids haemodynamic monitoring. Avoid fluid overload (but vigorous lung drying out does not work[1]). Haemofiltration may be needed in renal failure.
● Pulmonary hypertension may need treatment with β_2-agonists to limit right ventricular dysfunction and to limit pulmonary oedema formation.
● *Respiratory support:* Those with mild lung injury achieve satisfactory oxygenation with continuous positive airway pressure (CPAP), which re-expands collapsed alveoli, increases functional residual capacity and compliance. Most patients require positive-pressure ventilation and conventional techniques can provide large tidal volumes (10–15ml/kg) to recruit collapsed alveoli. Reduced compliance may lead to high peak airway pressures and increase the risk of pneumothorax. Positive end-expiratory pressure (PEEP) will help oxygenation but at the expense of cardiac output (venous return↓) and reduced perfusion of kidneys and liver. New approaches to ventilation include inverse ratio ventilation (a longer inspiratory time raises inspiratory to expiratory ratio from 1:2 to 4:1), high-frequency jet ventilation, permissive hypercapnia (small tidal volumes to keep peak airway pressure <35cmH$_2$O; oxygenation may be maintained and hypercapnia is well tolerated) and extracorporeal membrane oxygenation (ECMO). Oxygenation devices implanted in the inferior vena cava offer uncertain benefit.[1]
● Enteral feeding carries a lower risk of sepsis than TPN, probably by preserving mucosal integrity, and should be encouraged. Give stress ulcer prophylaxis. Steroids do not improve mortality in the acute stage.

Mortality ~50%. Bad prognosis if elderly or no response to treatment within the first 24h, or if aspiration/infection is the cause.

1 R Beale 1993 *BMJ* ii 1335 **2** PD MacNaughton 1992 *Lancet* **339** 469

Adult respiratory distress syndrome (ARDS)

[*OTM* 15.161]

Respiratory failure

Definition Respiratory failure is present if PaO_2 is <8kPa.
It is subdivided into two types by the $PaCO_2$.

In *type I*, $PaCO_2$ is <6.5kPa and there is pure ventilation–perfusion mismatch: less O_2 is absorbed from the hypoperfused alveoli, but compensation cannot take place in the hyperperfused ones because of the sigmoid shape of the Hb dissociation curve. By contrast, the CO_2 dissociation curve is linear, so this compensation does occur.

In *type II*, $PaCO_2$ is >6.5 kPa and the problem is pure hypoventilation; both gases are affected reciprocally. In practice, causes of types I and II may coexist.

Causes 1 Acute exacerbations of underlying pulmonary or neuromuscular disorders, eg COAD, asthma, pulmonary fibrosis, kyphoscoliosis, respiratory muscle disease, myasthenia gravis. Infection, left ventricular failure, and sedative drugs may all contribute.
2 In previously healthy people (eg ARDS, p350), severe pneumonia, muscle weakness (eg Guillain–Barré syndrome, polio), brainstem disorders.

352

Clinical features Signs of any underlying disease. Hypoxia causes restlessness, confusion and ultimately coma. The patient becomes cyanosed. Hypercapnia causes drowsiness, flapping tremor, bounding pulse, warm peripheries, headaches, and papilloedema.

Management
- Assess previous disability, deterioration, potential reversibility, and current clinical state.
- Do CXR, arterial blood gases, peak expiratory flow rate, ECG, U&E, FBC, sputum and blood culture, save serum (for serology).
- Treat any underlying precipitant—eg infection, LVF, airways obstruction.
- Physiotherapy.
- Use supplemental O_2 to keep PaO_2 >8kPa. Caution is required in patients with COAD (see p348): give O_2 24%, then check blood gases within 30mins. If $PaCO_2$ has risen by >1kPa, consider artificial ventilation (p354) or start doxapram 1–4mg/min IVI and monitor its effects clinically and with frequent ($\frac{1}{2}$-hourly) blood gases. Remember, doxapram increases *rate* but not depth of inspiration. (Note: its adverse effects resemble the clinical features of hypercapnia, eg twitching.)
- If adequate oxygenation is difficult to achieve despite optimum conservative management, consider artificially ventilating the patient, but only if there is a prospect of weaning the patient off the ventilator. An alternative is nasal intermittent positive pressure ventilation (NIPPV). This augments inspiration when triggered by the patient and requires a nasal mask and a simple bedside ventilator.
- Long-term management depends on the underlying cause: see COAD (p348), assisted ventilation (p354), and pulmonary hypertension (p366).

Oxygen dissociation curves

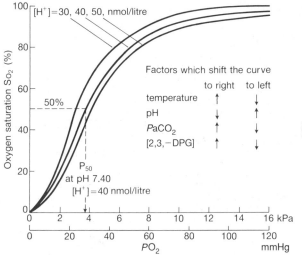

After D Flenley 1982 *Med Intl* 1(20) 940.

353

Assisted ventilation

Indications Deteriorating gas exchange due to a potentially reversible cause of respiratory failure eg pneumonia, pulmonary oedema, exacerbation of COAD, massive atelectasis, exhaustion, weakness of respiratory muscles (acute infective polyneuritis, myaesthenia gravis), head injury, cerebral hypoxia (asphyxiation, cardiac arrest), chest trauma or burns.

Intermittent positive pressure ventilation (IPPV) expands the lungs by actively blowing air into them.

CPAP (continuous positive airways pressure), provides a constant positive pressure throughout the respiratory cycle. This re-expands collapsed alveoli, increasing functional residual capacity (FRC) and compliance, such that the work of breathing is reduced and gas exchange is improved. The patient breathes an O_2–air mixture at high flows, spontaneously through a tightly fitting mask. BIPAP is a similar device that allows one to set a higher pressure for inspiration and a lower pressure for expiration. The ventilator is triggered by the patient initiating inspiration. In conditions such as acute asthma and COAD, where positive expiratory pressure is not desirable, NIPPV (nasal IPPV) delivers a positive pressure, when triggered by the patient initiating inspiration, for a prescribed inspiratory time, allowing the patient to exhale to atmospheric pressure. These 'pressure generating' devices have the advantage that if a leak develops around the mask, the ventilator can continue to provide the patient with the prescribed positive pressure. However, the patient must be able to protect their airway and generate enough effort to trigger the machine.

Ventilators like 'the BIRD' are also triggered by the patient, but in contrast deliver a pre-set tidal volume to augment inspiration. The 'Bromptonpac' triggers via a nasal mask and delivers a pre-set volume, but can be used to provide a number of mandatory breaths/min.

If the patient is unable to trigger these devices, controlled mandatory ventilation (CMV), via an endotracheal tube, with sedation and neuromuscular blockade, may be preferable. Conventional techniques provide the desired volume of gas. The *minute volume* = *tidal volume* × *rate* and these parameters, as well as the inspired O_2 (F_iO_2) are varied to achieve a PaO_2 where tissue oxygenation is adequate whilst avoiding high inspiratory pressures (>40cmH$_2$O). This may include *permissive hypercapnia* (allowing $PaCO_2$ to rise to double figures). Oxygenation may be furthered by PEEP (positive end-expiratory pressure), but at the expense of a decrease of cardiac output and increased risk of barotrauma. The relative proportions of time spent in inspiration and expiration (I:E ratio) is normally set at 1:2, but may be altered, eg in ARDS a prolonged inspiratory time is beneficial (see p350) whereas in acute asthma a long expiratory time is needed.

Ventilation may either be terminated abruptly or by gradual transfer of the ventilatory workload from the machine to the patient. An alternative to cyclical mandatory ventilation is SIMV (synchronized intermittent mandatory ventilation)—the patient initiates a breath and the machine provides a predestined tidal volume and rate. This may be appropriate when weaning the patient whose respiratory muscles have wasted—when ventilation may be prolonged, a tracheostomy is advisable.

Intermittent negative pressure ventilation (INPV) works by 'sucking' out the chest wall and is used in patients with chronic hypoventilation (eg polio, kyphoscoliosis or muscle disease). Some patients may manage for months on their own after a few weeks of INPV. Alternatives to the old 'iron lung' include thoracic cuirasses and other devices which may be custom built.

Assisted ventilation

Pneumothorax

This is accumulation of air in the pleural space.

Causes Often spontaneous (especially in young thin men) due to rupture of a pleural bleb. Other causes include trauma (especially iatrogenic, eg CVP lines), asthma, COAD, positive pressure assisted ventilation, TB, pneumonia and abscess, lung carcinoma, cystic fibrosis, sarcoidosis, fibrosing alveolitis and any diffuse lung disease, and ascent in aeroplanes.

Clinical features There may be no symptoms (especially in fit people with small pneumothoraces) or there may be dyspnoea and pleuritic pain (may be transient). The signs are ipsilateral diminished expansion and breath sounds, and hyperresonant percussion note: they are easily missed. The mediastinum may be shifted away from the side of a tension pneumothorax.

Investigation On the CXR the pneumothorax will show as an area devoid of lung markings peripheral to the edge of the collapsed lung (do not confuse this with the medial edge of the scapula). It may be better demonstrated in an expiratory film when the lung is deflated. It may be missed in a supine film.

Management[1] Small pneumothoraces (ie a small rim of air around a lung) usually need no treatment. If no chronic lung disease, send home with an appointment for the chest clinic in 7–10 days. Advise the patient that flying is dangerous. Admit all others. Moderate (lung collapsed ½-way towards heart border) and complete pneumothoraces (airless lung, separate from diaphragm) need simple aspiration:

- Under 1% lignocaine anaesthesia insert a 16G cannula in 2nd intercostal space in the midclavicular line (or 4–6th space in midaxillary line).
- Enter pleural space; withdraw needle's point. Connect to 50ml syringe with a Luer lock and 3-way tap. Aspirate ≤2.5 litres of air (50ml × 5).
- Stop if resistance is felt, or if the patient coughs excessively.
- Request CXR in inspiration. If the pneumothorax is gone or is small, you have been successful. If you fail after a second attempt, use an underwater drain, p754. (Never clamp these drains, as this impairs successful inflation.)
- If a drain is used, remove 24h after bubbling has stopped. Then do CXR.
- If a drain continues to bubble, it may be dislocated, or air may be leaking around it, or a bronchopleural fistula may exist. Pleurodesis, pleurectomy or surgical closure may be the answer to persistent leaks.

Tension pneumothorax is an emergency. Air drawn into the pleural space with each inspiration has no route of escape during expiration. The mediastinum is pushed over onto the contralateral hemithorax. Unless the air is rapidly removed, cardiorespiratory arrest will occur. Signs: respiratory distress, tachycardia, hypotension, distended neck veins, trachea deviated away from side of pneumothorax. ►► Insert a large bore needle (19G) into the chest on the side of the suspected pneumothorax *before* requesting a CXR, then proceed as above. See p754.

1 British Thoracic Society Guidelines 1993 *BMJ* ii 114

Pneumothorax
[*OTM* 15.54]

Pleural effusion

This is fluid in the pleural space.

Causes are usually divided by the effusion's protein concentration into *transudates* and *exudates*, having <30g/l and >30g/l respectively. In practice, it is common to have borderline values making the distinction less useful. A collection of blood in the pleural space is a *haemothorax*; if there is blood and air, it is a *haemopneumothorax*; if there is pus, it is an *empyema*.

Transudates: May be due to a) increased venous pressure (heart failure, constrictive pericarditis, fluid overload), or b) hypoproteinaemic states causing a reduced capillary oncotic pressure (nephrotic syndrome, liver failure, protein-losing enteropathy). Transudates are usually bilateral, but if unilateral tend to affect the right. They may also be found in hypothyroidism, Meigs' syndrome (p704), and the yellow nail syndrome.

Exudates: May be due to increased leakiness of the pleural capillaries. They are usually unilateral in focal diseases (pneumonia, empyema, TB, pulmonary infarction, lung carcinoma, mesothelioma, lymphoma, subphrenic abscess, pancreatitis) and bilateral in systemic and diffuse lung diseases (disseminated malignancy, multiple pulmonary infarcts, collagen vascular diseases eg RA, SLE, Dressler's syndrome, asbestos exposure). In *lymphangitis carcinomatosa* malignant cells, especially from breast cancer, block the pleural capillaries causing effusions.

Symptoms Dyspnoea varies with size of effusion; dull chest pain; symptoms of underlying cause. May be asymptomatic.

Signs Ipsilateral diminished expansion, stony dull percussion note, absent breath sounds with an area of bronchial breathing and bleating vocal resonance ('aegophony') at the effusion's top. The mediastinum may be pushed away from very large effusions.

Investigations CXR: Effusion is seen as a water-dense shadow with a concave upwards upper border. A completely horizontal upper border implies that there is also a pneumothorax.
Diagnostic aspiration: Draw off 30ml using a 21G needle inserted just above the upper border of an appropriate rib (be guided by the CXR); this avoids the neurovascular bundle. Send fluid for protein, glucose, Gram and auramine (or Ziehl–Neelsen) stain, culture (including for TB), cytology, rheumatoid factor, ANA, and complement. NB: TB and malignancy cannot be ruled out by microscopy and cytology. It is often wise to perform a parietal pleural biopsy with an Abrams' needle at the same time; this has a high yield in TB, particularly if biopsy is cultured.

Management is of the underlying cause. If the effusion is symptomatic, it should be drained, if necessary repeatedly. The fluid is best removed slowly (≤2 litres/24h). It may be aspirated as in a diagnostic tap, or an indwelling chest drain may be inserted (see p754). But unlike in pneumothorax, a chest drain for an effusion must be able to drain the base of the pleural space, not the apex. Chemical pleurodesis (tetracycline, bleomycin, talc) should be considered for malignant effusions.

Pleural effusion
[*OTM* 15.55]

Sarcoidosis

Sarcoidosis is a disease of unknown cause characterized by non-caseating granulomata. It may affect any organ and occur at any age, but most commonly affects chests of young adults. Incidence: 1–64/100,000. Blacks are more commonly and more severely affected, often by extrathoracic disease.

Thoracic sarcoid classically presents as bilateral hilar lymphadenopathy (BHL) on CXR. It is asymptomatic (30%) or associated with cough, fever, arthralgia, malaise or erythema nodosum (Löfgren's syndrome) in a third. Parenchymal lung disease may also be asymptomatic and without signs, although dyspnoea and fine crepitations may develop. CXR shows diffuse bilateral mottled shadows (2–5mm diameter) ± BHL. Reticular shadows occur when fibrosis develops, typically in the upper lobes. Rarely a restrictive ventilatory defect develops.

Extrathoracic sarcoidosis ● *Skin:* Erythema nodosum, nodules, lupus pernio (blue-red rash on nose, cheeks, digits), scar infiltration.
● *Eye:* Lacrimal gland enlargement, anterior or posterior uveitis. The latter causes most of the visual loss induced by sarcoid, and consists in perivascular cuffing of retinal equatorial veins which look like '*tache de bougie*' (candle wax dripping).
● *Bones:* Arthralgia, dactylitis, osteopenic 'punched out' lesions.
● *Heart:* Sudden death, ventricular tachycardia, complete heart block, cardiomyopathy, pericardial effusion.
● *CNS:* Cranial and peripheral nerve palsies (especially VII), space occupying lesions (± seizures) or diffuse CNS disease, granulomatous meningitis, pituitary (eg diabetes insipidus).
● *Kidney:* Hypercalcaemic nephropathy, renal calculi.
● *General:* Lymphadenopathy, hepatosplenomegaly, fever, malaise.

Diagnosis Characteristic histology, eg in bronchial or transbronchial lymph node biopsy. The Kveim test is an intradermal injection of sarcoid spleen tissue. If +ve, non-caseating granulomata are seen at the injection site in 4–6 weeks. A +ve test is highly specific for sarcoid, but if −ve, it is not excluded. Tuberculin tests are −ve. Other signs: Ca^{2+}↑; hypercalciuria; vital capacity↓.

Treatment[1] ► In active disease prevent blindness, nephropathy and lung fibrosis with prednisolone (eg 20–40mg/24h PO for 1–3 months, then tailing off; ±cyclopentolate and steroid eye-drops—fluorometholone is least likely to precipitate glaucoma). Ca^{2+}↑ is also an indication for steroids. Markers of activity are Ca^{2+}↑, angiotensin converting enzyme↑, macrophages and CD4 cells in bronchoalveolar lavage, and activated macrophages in granulomas (shown on gallium scans). Methotrexate (eg 10mg/wk for 3 months; monitor LFT) and hydroxychloroquine (eg 200mg alternate days for 9 months) may help skin and lung fibrosis. Effervescent phosphate (eg 500mg/12–24h PO) may promote normocalcaemia. Azathioprine 50mg/8h PO may allow steroid reduction. Most patients with BHL and half with BHL and pulmonary shadowing resolve without treatment. If shadowing lasts for >2yrs, risk of fibrosis increases: give prednisolone, monitor with CXR and RFTs.

Other causes of BHL: TB, lymphoma, carcinoma, fungi.

Other granulomatous diseases: TB; brucella; Hodgkin's, Wegener's, and Crohn's diseases; primary biliary cirrhosis; allergic alveolitis; drugs; leprosy; mycoses; toxoplasmosis; leishmaniasis; berylliosis.

1 G James 1992 *Prescr J* **32** 9–14

Sarcoidosis

Obstructive sleep apnoea

(OSAS, or Pickwickian syndrome)

Essence This is a disorder characterized by intermittent and recurrent closure of the pharyngeal airway, with partial or complete cessation of breathing during sleep. This apnoea is terminated only by partial arousal. The typical patient is a middle-aged fat man who presents because of snoring or daytime somnolence. The spouse or partner usually describes the patient stopping breathing frequently during sleep.

Mechanism The normal decrease in tone of the oropharyngeal musculature during sleep, in association with a small starting size or external loading by neck obesity, results in the upper airway (at the level of the pharynx) collapsing. Partial narrowing causes snoring. Arousal is provoked mainly by increased respiratory effort to overcome the blockage, and thus heavy snoring *without* apnoea can disturb sleep as well.

In children narrowing of the airway by enlarged tonsils may cause apnoea.

Clinical features
- Loud snoring
- Daytime somnolence
- Poor sleep quality
- Morning headache
- Decreased libido
- Cognitive performance

Features of type II respiratory failure (p352) may be evident, if there is associated chronic obstructive pulmonary disease. There is also an increased risk of cardiovascular diseases, and road traffic accidents.

Tests Polysomnography (monitoring oxygen saturation, airflow at the nose and mouth, ECG, EMG chest and abdominal wall movement during sleep). Simpler studies (oximetry and video recordings) will often be all that is required for diagnosis. The occurrence of 15 or more episodes of apnoea or hypopnoea during an hour of sleep is the conventional indicator of significant sleep apnoea, but severe symptoms occur with lower or normal apnoea rates when snoring alone is provoking sleep disruption.

Management ● Continuous positive airway pressure (CPAP) via a nasal mask during sleep is the best treatment for symptomatic patients willing to use this cumbersome device. ● Weight reduction also helps. ● Tonsillectomy, uvulopalatopharyngoplasty and tracheostomy are very rarely needed.

Extrinsic allergic alveolitis

Essence Various inhaled antigens provoke a hypersensitivity reaction in the lungs of sensitized individuals.

Causes Allergens are usually spores of microorganisms (eg *Micropolyspora faeni* in farmer's lung) or avian proteins (eg droppings or feathers in bird fancier's lung or pigeon breeders' disease).

Clinical *Acute:* 4–8h following exposure systemic illness of fever, malaise, dry cough, and dyspnoea. *Chronic:* Gradually increasing malaise, weight loss, exertional dyspnoea, and fine crepitations. Chronic disease may follow or be independent of acute episodes, which themselves may have no chronic sequelae.

Pathology *Acute:* Predominantly type III (immune complex) reaction producing alveolitis. *Chronic:* Predominantly type IV (cell mediated) reaction with lymphocyte infiltration, granulomata, giant cells, endobronchial exudates and bronchiolitis obliterans.

1 TD Bradley 1993 *BMJ* **306** 1260

Obstructive sleep apnoea
[OTM 15.158]
Extrinsic allergic alveolitis
[OTM 15.119]

Investigations *Acute:* Neutrophilia (eosinophils usually normal); ESR↑; serum precipitins detectable (ELISA tests quantify for example IsG antibody to pigeon gammaglobulin); CXR: normal *or* widespread small nodules *or* ground-glass appearance; respiratory function tests (RFTs): slight and transient reduction in lung volumes and transfer factor. *Chronic:* Serum precipitins detectable; CXR: mainly upper zone shadowing due to fibrosis; RFTs: more marked restrictive changes and reduced gas transfer. Bronchoalveolar lavage fluid in acute or chronic disease shows a predominance of lymphocytes and mast cells.

Management Avoid exposure to allergens, failing which wear a very fine mesh face mask or positive pressure helmet. Acute symptoms may be controlled by prednisolone 40mg/24h PO; when asymptomatic, reduce dose rapidly. In chronic disease, longer term steroids often achieve radiological and physiological improvement. In practice, in the face of overzealous advice about eliminating allergens, many patients with bird fancier's lung end up avoiding their doctors, not their birds—and it has to be said that frequently symptoms wane in spite of repeated exposure.[1]

Note: Psittacosis and ornithosis are infections (zoonoses). They are not a cause of allergic alveolitis. See p334.

1 S Bourke 1989 *Thorax* **44** 415–8

Fibrosing alveolitis[1]

Essence This is a disease of unknown cause ('cryptogenic'), characterized by increased numbers of inflammatory cells in the alveoli and interstitium, together with variable degrees of pulmonary fibrosis. A third are associated with connective tissue diseases (RA—10% of all fibrosing alveolitis, SLE, systemic sclerosis, mixed connective tissue disease, Sjögren's, poly/dermatomyositis)—or chronic active hepatitis, renal tubular acidosis, autoimmune thyroid disease, ulcerative colitis. The rest are 'cryptogenic'.

Clinical features Progressive exertional dyspnoea, dry cough, general malaise, weight loss, arthralgia; finger clubbing (65%), fine end-inspiratory crepitations, cyanosis.

Investigations CXR: Bilateral diffuse nodular or reticulonodular shadowing favouring the lung bases. Honeycomb shadows develop as the disease progresses and lung volume is lost. Hypergammaglobulinaemia. ESR raised. ANA and rheumatoid factor are present in 35% of cases. Respiratory function tests: Reduced lung volumes (restrictive pattern) and reduced transfer factor. Bronchoalveolar lavage: Increased total cell count with increased proportion of neutrophils and/or eosinophils. Lymphocytes are occasionally raised and may indicate a favourable response to treatment. Open lung biopsy may be needed for diagnosis and staging.

Management 35% respond to immunosuppression, eg with prednisolone 60mg/24h PO or cyclophosphamide 100–120mg/24h PO with prednisolone 20mg PO on alternate days. Monitor response with CXR and lung function tests. Response to steroids may take 1–2 months, after which the dose may be cautiously reduced. Responding to cyclophosphamide may take longer.

Survival 50% die in 4–5yrs, but the range is wide (1–20yrs).

Other causes of pulmonary fibrosis Drugs (bleomycin, busulphan, amiodarone, nitrofurantoin), adult respiratory distress syndrome (p350), chronic extrinsic allergic alveolitis (p362), eosinophilic granuloma, radiation, uraemia.

Mineral fibrosis

Simple coal worker's pneumoconiosis (CWP) results from pulmonary coal dust deposition. It is asymptomatic and causes no signs. CXR: multiple rounded opacities (1–5cm), predominantly in the upper zones. It may proceed to progressive massive fibrosis (PMF), which causes dyspnoea and a restrictive defect. In PMF, CXR opacities are larger (1–10mm).

Silicosis is similar radiographically, but is associated with egg-shell calcification of hilar nodes. Unlike CWP, silicosis is a risk factor for TB.

Asbestosis is similarly a progressive fibrotic lung parenchymal disease, clinically and radiographically resembling cryptogenic fibrosing alveolitis. This is distinct from the other asbestos related chest conditions: pleural plaques (which may calcify), malignant mesothelioma, carcinoma of the bronchus.

1 R Stein 1987 Med Intl 2 1554

Fibrosing alveolitis
Mineral fibrosis
[*OTM* 15.123]

Cor pulmonale

Cor pulmonale is right heart failure due to chronic pulmonary hypertension.

Clinical features are ankle oedema, hepatic and GI engorgement (which may cause anorexia and nausea), ascites, raised JVP, and tricuspid regurgitation.

CXR: Prominent right atrium and ventricle and pulmonary artery.

ECG: P pulmonale, right axis deviation, and right ventricular hypertrophy with inverted T-waves in V1–4 ('strain pattern').

FBC: There may be a secondary polycythaemia.

Causes of pulmonary hypertension

Chronic pulmonary venous hypertension: Chronic LVF, long-standing mitral stenosis.

Increased pulmonary vascular resistance: Pulmonary embolic disease, many lung diseases (eg COAD, severe pulmonary fibrosis of any cause, pneumonectomy and multiple lobectomies, bronchiectasis and cystic fibrosis), primary pulmonary hypertension, schistosomiasis, any cause of chronic hypoxia (eg Pickwickian syndrome, p362).

Increased pulmonary blood flow: Congenital shunts: atrial and ventricular septal defects, patent ductus arteriosus. (As increased flow causes progressively increasing right-sided pressures, so the shunt decreases. Eventually, the right-sided pressures rise above those in the left heart. At this point the left-to-right shunt reverses to a right-to-left shunt, and the patient becomes cyanosed. This is Eisenmenger's syndrome and is irreversible.)

Clinical features of pulmonary hypertension
Hypoxia and cor pulmonale—and the features of the underlying cause.

Management of pulmonary hypertension[1]
If the underlying cause is irreversible, or pulmonary hypertension has developed despite the cause having been treated, there is likely to be a steady decline towards cor pulmonale and death. This can be slowed by minimizing pulmonary vascular resistance. Just as the commonest cause of pulmonary vasoconstriction is hypoxia, so the most potent therapeutic pulmonary vasodilator is oxygen. Continuous oxygen therapy for at least 15h/day improves survival in cor pulmonale secondary to COAD. This will reverse hypoxia and increase right ventricular function. Home oxygen concentrators are now available. Oxygen may be piped to several rooms in the house. O_2 concentrators are prescribable by GPs. Other exacerbating factors include smoking, sedatives, obesity, infection, and stroke. Fluid overload is treated with frusemide (eg 40–160mg/24h PO) and appropriate potassium supplements. Venesection may be performed for polycythaemia. This will decrease the risk of thrombosis, and also decrease the work of the heart. Oral β-agonists may be useful. Selected patients should be considered for heart–lung transplantation.

5-year survival 40%.

1 J Rees 1984 *BMJ* **ii** 1398

Cor pulmonale

9 Renal medicine

Relevant pages in other chapters:

Symptoms and signs: Frequency (p46); loin pain (p50); oedema (p52); oliguria (p52); polyuria (p52).

Other pages: Urinary retention (p134); catheters (p134); the biochemistry of kidney function (p632); electrolyte physiology (p636); sodium (p636–8); potassium (p640); calcium (p642); uric acid and the kidney (p634); osteomalacia (p648); urinary incontinence (p74).

In OHCS: Gynaecological urology (OHCS p70); bacteriuria and pyelonephritis in pregnancy (OHCS p160); obstetric causes of acute tubular necrosis (OHCS p160); chronic renal disease in pregnancy (OHCS p160); UTI in children (OHCS p188); urethral valves (OHCS p220); horseshoe kidney (OHCS p220); ectopic kidney (OHCS p220); hypospadias (OHCS p220); Wilms' nephroblastoma (p220); acute and chronic renal failure in children (OHCS p280); nephritis and nephrosis in children (OHCS p282); Potter's syndrome (OHCS p120); the haemolytic uraemic syndrome (OHCS p280).

Introduction to nephrology

Nephrology is a branch of medicine which is often given less attention than it deserves. The complications of primary renal disease and the nephrology of systemic disease make it a multisystem specialty.

One of the main problems in approaching nephrology is that there is no strict correlation between the clinical syndrome, histological classification and cause. The reason for this is that the kidney has a strictly limited number of ways of responding to a multitude of pathogenic stimuli. Consequently the diagnosis and classification of renal disease must be considered at each of these 3 different levels.

Classification by clinical presentation
  Proteinuria and the nephrotic syndrome
    (the latter is also termed nephrosis)
  Haematuria and the nephritic syndrome ('nephritic' signifies haematuria from glomerular bleeding in glomerulonephritis)
  Acute renal failure
  Chronic renal failure
  Hypertension
  Urinary tract infection

Classification by histology
  Glomerulonephritis
  Tubulointerstitial disease:
    tubular necrosis
    acute interstitial nephritis
    chronic interstitial nephritis
  Vasculitis
  Atherosclerosis

Classification by cause
  Primary renal disease:
    Mostly associated with immune complex deposition, or, rarely (but importantly), to antiglomerular basement membrane antibodies.
  Secondary to systemic disease:
    Particularly diabetes mellitus, hypertension, drugs, amyloid, systemic vasculitis (eg polyarteritis nodosa and systemic lupus erythematosus).

Urine

▶ If you suspect renal disease always obtain some fresh midstream urine for: 1 'dipstick' testing, 2 microscopy, and 3 culture.

Causes of abnormalities on ward test (eg Multistix®)

● **Haematuria** Causes of urinary red blood cells may be divided into:
Renal: Glomerulonephritis (p380); vasculitis (eg endocarditis, SLE, and other connective tissue disease); interstitial nephritis; carcinoma (renal cell, transitional); cystic disease (polycystic disease, medullary sponge kidney); trauma; vascular malformations; emboli.
Extra-renal: Calculi; infection; neoplasm; prostatitis; trauma; urethritis; bladder-catheterization; post-cyclophosphamide.
Coagulation disorders: (or anticoagulants; if INR is in therapeutic range, it is wise to consider other causes). *Others:* Sickle trait/disease; vasculitis.
Work-up: MSU, FBC, ESR, U&E, creatinine clearance, plain films + ultrasound, ?renal biopsy before cystoscopy in those <45yrs (p373). NB: free Hb and myoglobin make stix +ve in the absence of red cells. Discoloured urine is also seen with porphyria, rifampicin or beetroot ingestion.

● **Glycosuria** Causes: diabetes mellitus, pregnancy, sepsis, tubular damage (p400) or low renal threshold (check blood glucose).

● **Proteinuria** UTI, vaginal mucus, diabetes, glomerulonephritis, nephrosis, pyrexia, CCF, pregnancy, postural proteinuria (2–5% of adolescents, rare if >30yrs)—protein excretion (generally <1g/24h) when upright, and normal in recumbancy. Uncommon causes: haemolytic-uraemic syndrome, hypertension, SLE, myeloma, amyloid (p398).
Microalbuminuria: See p650 and p396. Causes: DM; acute illness.

● **Bile** See p484.

● **Nitrite** Suggests UTI (or a high-protein meal).

Microscopy

1 Find a scrupulously clean slide. It need not be new.
2 Almost close the iris diaphragm of a light microscope.
3 Put a tiny drop of urine on a slide; cover urine with coverslip.
4 Open iris diaphragm a little.
5 Use the ×40 objective.
6 Look for blood, pus, crystals, casts, cocci, and rods. A counting chamber helps to quantify bacteria: 1+ means a few bacteria in every square; 2+ means many bacteria in every square, 3+ means innumerable bacteria. UTI is very likely if 3+, and unlikely if <1+. If 1+ or 2+ bacteria, then UTI is very likely if there are >400 WCC/µl.[1]

RBCs: See *haematuria*, above; >2/mm³ is considered abnormal.

WBCs: Bacterial or chemical cystitis, urethritis, prostatitis, pyelitis, TB, analgesic nephropathy, stones. ≤15/mm³ may be normal.

Casts are best seen in counting chambers. The person who meticulously examines urines may do so for months and see no other variety than insignificant, transparent, jelly-baby-like *hyaline casts*, or *granular* **casts** (signifying a wide range of renal diseases, eg acute tubular necrosis or rapidly progressive glomerulonephritis, or no disease at all—simply the result of exercise). *Red cell casts* mean glomerulonephritis; they look greenish. *WBC casts* mean pyelonephritis.

Crystals: Combine to form stones. To find a few crystals is common, eg in old or cold urine—when they need not signify any pathology.

24-h urine quantification Na⁺, K⁺, Ca²⁺, urea, creatinine, protein. Do a simultaneous plasma creatinine for creatinine clearance (p632).

1 F Culclasure 1994 *Arch Int Med* **154** 6490

Urine microscopy

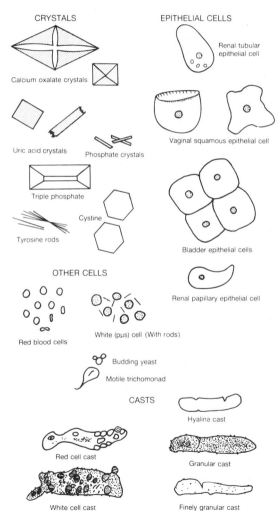

CRYSTALS

Calcium oxalate crystals

Uric acid crystals

Phosphate crystals

Triple phosphate

Cystine

Tyrosine rods

EPITHELIAL CELLS

Renal tubular epithelial cell

Vaginal squamous epithelial cell

Bladder epithelial cells

Renal papillary epithelial cell

OTHER CELLS

Red blood cells

White (pus) cell (With rods)

Budding yeast

Motile trichomonad

CASTS

Hyaline cast

Red cell cast

Granular cast

White cell cast

Finely granular cast

Principal source: M Longmore 1986 *An Atlas of Bedside Microscopy*
Royal College of General Practitioners occasional paper No. 32

Radiology of the urinary tract

Plain abdominal film Look for: 1 Renal outlines—normal size 10–12cm and situated at T12–L2. 2 Abnormal calcification—90% of renal stones are radio-opaque; tumour; TB; nephrocalcinosis.

Ultrasound This is a key investigation in acute renal failure, in particular to exclude urinary tract obstruction, and to show renal size—small kidneys are typical of chronic renal failure. Also used to guide biopsy. Doppler ultrasound may be used to look for renal vein thrombosis. With plain x-ray, this is at least as good as IVU for investigation of haematuria.[1]

Intravenous urography (IVU or IVP) Preparation: laxatives; nil by mouth (unless uraemic or myeloma when dehydration is contraindicated) for 12h. Always look at the plain film for renal size and calcification.

The nephrogram: (The image of contrast diffusing through the kidney—usually *immediate* and *brief*). It is *intense* in obstruction and glomerulonephritis (GN); *prolonged* in obstruction, ATN (acute tubular necrosis), chronic GN and renal vein thrombosis; *absent* in a non-functioning kidney (infarction, severe GN); *delayed* in renal artery stenosis.

- Absent kidney: obscured by bowel gas or perinephric abscess, renal agenesis, nephrectomy (look for amputated 12th rib).
- Small kidney: (<12cm) chronic pyelonephritis, renal artery stenosis. Bilateral in chronic renal failure (chronic GN or pyelonephritis).
- Large kidney: (>14cm) hydronephrosis, tumour, renal vein thrombosis, post-contralateral nephrectomy. Bilateral in polycystic kidneys, amyloidosis, myeloma, lymphoma, bilateral obstruction.
- Scarred outline: pyelonephritis, ischaemia, TB.
- Low kidney: (normal T12–L2) hepatomegaly, transplants, congenital.
- Pelvicalyceal filling defects: one calyx not seen in tumour, TB, partial nephrectomy. Irregular filling defect—clot or tumour. Smooth defect—papilloma, renal artery aneurysm, pressure from vessel.
- Some patterns: *Obstruction:* prolonged and dense nephrogram, clubbed calyces, mega-ureter. *Chronic pyelonephritis:* small scarred kidneys, thin cortex, clubbed calyces. *Papillary necrosis:* linear breaks at papillary bases. *Ureteric obstruction.*

Micturating cystourethrogram The bladder is catheterized and filled with contrast. Used to show ureteric reflux during micturition.

Retrograde pyelogram The ureters are catheterized and contrast injected. Used for non-functioning kidneys and to define the site of obstruction eg papillary necrosis and non-opaque stones.

Renal arteriography via femoral arteries. Used for renal artery stenosis or embolus, tumour circulation, renal trauma.

Isotope renal scans (DMSA and DTPA scans) See p790.

1 J Spender 1990 *BMJ* i, 1074–6

Renal biopsy

Renal biopsy should be performed where a knowledge of the histology will influence management. Small kidneys are difficult to biopsy and results usually unhelpful.

- Before biopsy: check FBC and clotting. Group and save serum.
- Check size and position of the kidneys by IVU or ultrasound—IVU will also ensure that there are 2 functioning kidneys.
- Biopsy is performed under ultrasound guidance with the patient in a prone position and breath held.
- Send samples for normal stains, immunofluorescence and electron microscopy. Special stains (eg Congo Red if indicated).
- After biopsy the patient should rest for 24h. Monitor pulse, BP, symptoms and urine colour. The major complication is bleeding.

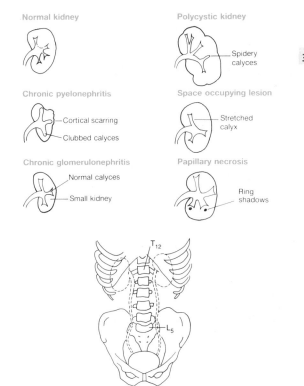

Normal kidney

Polycystic kidney
— Spidery calyces

Chronic pyelonephritis
— Cortical scarring
— Clubbed calyces

Space occupying lesion
— Stretched calyx

Chronic glomerulonephritis
— Normal calyces
— Small kidney

Papillary necrosis
— Ring shadows

T₁₂

L₅

The course of the ureters

Urinary tract infection (UTI)

▶ UTI is common; most women will have ≥1 UTI in their lives. Men have longer urethras, and fewer UTIs. Risk factors: DM; pregnancy; impaired voiding (causes of obstruction, p378); GU malformations.

Definitions *Bacteriuria:* any bacteria in urine uncontaminated by urethral flora. It may be asymptomatic (*covert*). If there are >10⁸ organisms/litre (of a pure culture) bacteriuria is *significant*. UTI denotes bacteriuria, with features of GU inflammation at particular sites—eg kidney (*pyelonephritis*), bladder (*cystitis*), prostate (*prostatitis*) or urethra (*urethritis*—non-specific urethritis is the commonest cause of male GU infection, p220). Acute and chronic pyelonephritis are separate entities: *acute pyelonephritis* is acute kidney infection with symptoms of fever, chills, rigors and flank pain with tenderness. *Chronic pyelonephritis* presents as chronic renal failure or one of its complications, and probably arises from childhood UTIs associated with reflux and renal scarring (OHCS p188). IVU (p372) is the key to its diagnosis.

Urethral syndrome: (Women only.) Symptoms of urgency, frequency and dysuria (ie painful micturition) in the absence of readily identifiable bacteriuria. (As there is no urethral disease, the term *abacterial cystitis* is preferred.) It is induced by cold, stress, intercourse or nylon underwear. Cause: unknown.

Presentation of UTI

Frequency/dysuria/haematuria	Retention of urine	*Pains:*
Incontinence	Fever with D&V	RIF; loin
	Urgency Strangury (p376)	Suprapubic

Examine for: large prostate, renal mass, meatal ulcer, vaginal discharge, loin tenderness, hypertension, signs of CRF (p388).

Bacteria *E coli* >70%—but ≤41% in hospital UTIs; Staphs, Streps, *Pseudomonas*, *Proteus*. *Klebsiella* is rare.

Tests If the urine *looks* clear, it probably *is* clear ($p \geq 0.97$)—true only because most MSUs are uninfected; of *infected* MSUs, ⅔ look clear.[1]
The midstream urine (MSU): Get a clean catch, and perform microscopy (p370). Test cultured organisms for sensitivity to a range of antibiotics. MSUs must be fresh. If there are >10⁸ organisms/l the result is significant. If there are <10⁸/l ± pyuria (eg >20 WBCs/mm³) the result may still be significant. Repeat the test.

Consider: U&E, creatinine, blood culture. IVU or cystoscopy if:
● Recurrent UTI (or 1st UTI in men)   ● Overt haematuria   ● Pyelonephritis
● Persistent microscopic haematuria   ● Unusual organism   ● Fever persists

Ultrasound or IVU?[2,3] Ultrasound may miss stones, papillary necrosis, small carcinomas, and clubbed calyces—but is radiation-free.

Treatment of simple UTIs No 'holding on'; urinate often. Drink plenty: 2 cups/h (even though this reduces antibiotic levels in urine). Double voiding (going again after 5mins) and voiding after intercourse may prevent reinfection. Specific therapy: eg trimethoprim 200mg/12h PO twice or amoxycillin 250mg/8h PO or one 3g sachet, with a second dose 12h later PO. Give longer courses in males, pregnancy, GU malformations, immunosuppression, pyelonephritis, *relapses* (2nd UTI, same organism), and *re-infections* (2nd UTI, different organism).

Urinary tract infection (UTI)

[*OTM* 18.67]

Causes of sterile pyuria
- Inadequately treated UTI
- Infection: TB (or atypical Streptococci, Corynebacteria, and other organisms which are fastidious in their culture requirements)
- Calculi
- Bladder-tumour, chemical cystitis eg from cytotoxics.
- Prostatitis
- Papillary necrosis (eg analgesic abuse, p394)
- Interstitial nephritis or polycystic kidneys
- Appendicitis

▶ If persistent, do 3 early morning urine samples (TB culture+ZN stain).

375

1 G Phillips 1993 *BMJ* i 653   2 J Spencer 1993 *BMJ* i 211   3 M Wilkie 1993 *BMJ* i 211

Renal stones and renal colic

▶ *Beware infection above the stone.*

The patient Surprisingly large stones may be symptomless and tiny stones may cause excruciating ureteric spasm. Stones in the *kidney* cause loin pain. Stones in the *ureter* cause renal colic which often radiates from loin to groin but may only be felt in the tip of the penis. *Bladder stones* cause strangury (the distressing desire to pass something that will not pass) and *urethral stones* may cause interruption of the urine flow. Nausea and vomiting may accompany colic. Pain is often eased by curling up and patients classically cannot lie still. *Other presentations*: UTI, haematuria, renal failure. *Signs*: Little abdominal tenderness; loin tenderness common. Haematuria (macro- or microscopic).

Tests Ask yourself: Is there an obvious stone? Is there infection? Why has this patient got a stone? Test urine for blood; culture and microscopy for infection. Abnormal crystals usually occur in acid urine. Crystals found in alkaline urine may be of no significance because phosphates crystallize out. Plain abdominal films ('KUB'—kidneys, ureters and bladder). 90% of renal stones are radio-opaque. Look along the line of the ureters for calcium that may represent a stone.

Blood tests: U&E, uric acid, Ca^{2+}, bicarbonate. Consider IVU and 24h urine for Ca^{2+}, PO_4^{3-}, oxalate, urate, creatinine clearance.

Management of acute renal colic
- Pain relief: Diclofenac sodium 75mg IM or 100mg PR or pethidine 100mg IM. (NSAIDs are very effective and avoid problems with potential opiate misusers, but they can decrease renal blood flow; monitor renal function whilst on these drugs).
- Increase fluid intake.
- Sieve all urine to 'catch' the stone for analysis—or use a chamber pot, being sure to prepare the patient's mind to hear 'the tinkle in the pot like far-away chimes'[1]—the triumphant vindication of your diagnosis.

Most stones will pass spontaneously. If there is any hint of obstruction or infection (fever, pyuria) seek urological help urgently. Procedures include:
- Antegrade nephrostomy (p372) and percutaneous nephrolithotomy (for stones in the kidney or upper half of ureters). These may be combined with lithotripsy (ultrasonic dissolution) if the stone is not too close to a bone. • Surgical removal is only occasionally required.

Prevention Drink more (>3 litres-24h), especially in summer. Reduce milk intake. For *oxalate* stones: ↓ oxalate intake (no chocolate, tea, rhubarb, spinach) and ↓ Ca^{2+} excretion (bendrofluazide 5mg/24h PO); pyridoxine if hyperoxaluria. For *calcium phosphate* stones: thiazides, low Ca^{2+} diet. For *triple phosphate* (*calcium, magnesium, ammonium*) 'staghorn' stones: antibiotics. For *urate* stones: allopurinol, urinary alkalinization (pH >6.5). For *cystine* stones: vigorous hydration, D-penicillamine, urinary alkalinization.

1 Graham Greene 1947 & 1995 *Under the Garden* ISBN 0-146-00057-9

Renal stones and renal colic
[*OTM* 18.87]

Composition of stones and causes

Calcium stones are the commonest (calcium with phosphates/oxalates, 43%; calcium oxalate alone, 35%; calcium phosphate, 3%).

Radio-opaque
- Calcium oxalate (49%)—hyperoxaluria (p396), hypercalciuria, hypercalcaemia, dehydration, renal tubular acidosis (p400), medullary sponge kidney (p396).
- Calcium phosphate (25%)—as above and UTI.
- Struvite (15%)—magnesium ammonium phosphate—UTI (*Proteus*) especially if accompanied by high urine pH or stasis.

Semi-opaque
- Cystine (5%)—cystinuria—one of the aminoacidurias; urine contains excess cystine, ornithine, arginine and lysine.

Lucent
- Urate (8%), xanthine.

Urinary tract obstruction

▶ *Is the bladder palpable?* (See urinary retention, p134.)
▶ *Think of urinary tract obstruction whenever renal function worsens.*
Always exclude by ultrasound. Beware infection *above* obstruction.

Obstruction may be bilateral if below the level of the ureter or unilateral if
ureteric. The latter may remain clinically silent for a long time as long as
the other kidney is functioning.

Bilateral obstruction *Causes:* Urethral stricture, urethral valves, prostatic hypertrophy, bladder tumours, blood clot, calculi, pelvic malignancy.
Presentation: Lower abdominal pain (eg with retention), renal impairment, urinary infection, overflow incontinence.

Unilateral obstruction *Causes:* Calculi, tumours, pelvi-ureteric junction (PUJ) obstruction, retroperitoneal fibrosis, compression by lymph nodes, impaction of sloughed papilla.
Presentation: Loin ache, worsened by drinking; a mass; UTI.

Tests Check U&E and creatinine and urine pH (p400). Ultrasound is the
best. It assesses bladder and kidney size and any hydronephrosis (renal
pelvis dilation). IVU is also useful (p372) with plain films for calculi.
Anterograde pyelography assesses the site of obstruction and allows
sampling of urine and insertion of nephrostomy tubes. Retrograde pyelograms are performed at cystoscopy (risks of introducing infection).

Treatment Relieve obstruction (eg by catheter or nephrostomy tube).
Treat the cause. Beware a large diuresis after relief of obstruction—a temporary salt-losing nephropathy may occur involving litres a day. Dehydration and hypokalaemia may occur. Careful monitoring of urine output,
weight and regular electrolytes is necessary.

Retroperitoneal fibrosis (RPF)

Also known as chronic periaortitis, the ureters are imbedded in a dense
fibrous plaque. Thought to be an IgG-mediated reaction to insoluble lipid
(ceroid) leaking from aortas in which there are severe atherosclerotic
plaques with medial thinning.[1]

Clinical features Middle-age males, backache, fever, malaise, sweating,
leg oedema, hypertension, palpable mass, pyuria and chronic or acute
obstructive renal failure. Associated with drugs (methysergide), carcinoma, Crohn's, connective tissue diseases, Raynaud's and fibrotic
diseases (eg mediastinitis, alveolitis).

Diagnosis IVU or contrast CT—dilated ureters with medial deviation of
the ureters. Biopsy—periaortic inflammation with lymphocyte and
plasma cell infiltrate.

Treatment Percutaneous nephrostomies to relieve obstruction. Surgery
(wrap the ureters in omentum). Consider steroids.

1 N Bullock 1988 *BMJ* ii 240

Urinary tract obstruction
Retroperitoneal fibrosis (RPF)

Glomerular disease

Renal disease may be classified by clinical syndromes, pathology, or by cause. The pathology of glomerulonephritis (GN) is complex. Important terms are *focal* (some of the glomeruli affected) versus *diffuse* (all affected), and *segmental* (part of each glomerulus affected) versus *global*. Analysis involves light microscopy (LM), electron microscopy (EM) and immunofluorescence (IF) for immunoglobulin (Ig) and complement.

1 Minimal change GN Typically a child presenting with nephrotic syndrome. *Histology:* LM normal; IF negative; EM fused podocytes. Associated with Hodgkin's disease. Usually steroid responsive.

2 Membranoproliferative GN (mesangiocapillary) 50% present as the nephrotic syndrome. Most have low C3. *Histology:* LM—cellular expansion of the mesangium. The basement membrane (BM) appears split, producing 'tram-tracks'; EM—two types: *1* subendothelial deposits and *2* deposits within the basement membrane. Associations include shunt nephritis, SBE, mixed essential cryoglobulinaemia, partial lipodystrophy, C3–nephritic factor (a circulating autoantibody), and measles.

3 Membranous GN Accounts for 50% of adult nephrotic syndrome. *Histology:* LM—thickened BM; IF—IgG and C3; EM—subepithelial deposits. There may be an underlying malignancy in up to 10% of adults. *Prognosis:* 25% enter remission but ~50% progress to CRF in 10 years.

4 Proliferative GN Usually follows 10–14 days after streptococcal infection. Low C3 common. *Histology:* LM—proliferation of endothelial and mesangial cells with neutrophils; IF—IgG and C3; EM—subepithelial deposits. *Prognosis:* excellent.

5 Focal segmental GN A non-specific response to variety of insults. *Histology:* LM—proliferative or sclerotic lesions; IF and EM variable. *Associations:* SLE, SBE, PAN, Wegener's, deep visceral infection.

6 IgA disease (Berger's disease) common cause of recurrent haematuria in young men. *Histology:* IF:IgA and C3. Similar picture seen in Henoch–Schönlein purpura (HSP). 10% progress to CRF. See p694.

7 Rapidly progressive GN—RPGN (≡crescentic GN) *Presentation:* haematuria, oliguria, hypertension. *Histology:* hypercellularity with crescents in >60% of glomeruli; EM and IF variable but typically subepithelial IgG and C3. Maximal treatment may save renal function. Associations: anti-GBM antibodies, Wegener's, microscopic polyarteritis, Henoch–Schönlein purpura, and occasionally post-streptococcal infection.

8 Focal segmental glomerulosclerosis *Presentation:* proteinuria or nephrotic syndrome. *Histology:* focal sclerosis with hyaline deposits; other glomeruli may be normal; IF—IgM and C3. Seen with heroin abuse. *Prognosis:* >50% progress to CRF. Usually steroid resistant.

Investigations for glomerulonephritis

- *Urine microscopy:* RBC casts and dysmorphic RBC.
- *Urine:* 24h protein excretion and creatinine clearance.
- *Serum:* U&E, FBC, ESR, CRP, albumin, ANA, antineutrophil cytoplasmic antibody (ANCA), DNA-binding, complement (C3, C4), antistreptolysin O (ASO) titre, HBsAg.
- *Culture:* blood, throat, ear if otitis, skin if cellulitis.
- *Radiology:* CXR, renal ultrasound, IVU.
- *Renal biopsy.*

Glomerular disease
[*OTM* 18.36]

The nephrotic syndrome

Essence Proteinuria (>3.6g/24h) sufficient to cause hypoalbuminaemia and oedema—often with hypercholesterolaemia. It may be due to loss of negative charge of the basement membrane±increased pore size.

Presentation Insidious swelling of eyelids, ascites and peripheral oedema. The urine may be frothy due to protein.

Causes 1 Minimal change GN (90% of children and 30% of adults). 2 Membranous GN. 3 Membranoproliferative GN. 4 Focal segmental glomerulosclerosis. 5 Others: DM, amyloid, neoplasia (<10% of adults with membranous GN) especially lymphoma or carcinoma, SBE, PAN, SLE, sickle-cell, malaria, drugs (penicillamine, gold).

Investigations *Urine:* 24h specimen for creatinine clearance and protein. Microscopy for RBCs and casts. *Blood:* U&E, albumin, cholesterol, FBC, ESR, CRP, Ig level, protein strip, ANA, DNA-binding, C4, C3, ANCA. *Renal biopsy:* not indicated in a child unless complicated (BP↑ or haematuria) or failure to respond to treatment.

Protein selectivity refers to the ratio of high molecular weight (MW) to low MW protein excreted in the urine (ie IgG:transferrin). Low values suggest steroid responsiveness in children with minimal change GN.

Treatment Bed rest (and a warm bath) to stimulate diuresis. High protein diet (though no proven benefit). Frusemide 40–80mg PO with amiloride as K⁺-sparing agent (or spironolactone 100mg/24h); diuretics relieve oedema but do not treat the underlying disorder. Beware IV diuretics as patients are often intravascularly depleted and may precipitate pre-renal failure. In general give IV diuretics only with IV plasma protein (salt poor), at least until diuresis is established. Monitor U&E and weight daily (aim to lose 1kg/day). Consider anticoagulation (see below).

Complications 1 Venous thromboses and pulmonary embolism (urinary loss of antithrombin III, low plasma volume, ↑clotting factors II, V, VII, VIII and X). 2 Infection (pneumococcal peritonitis—consider pneumococcal vaccination if nephrotic state persists). 3 Hypercholesterolaemia: (atherosclerosis, xanthomata—consider treatment, p654). 4 Hypovolaemia and renal failure. 5 Loss of specific binding proteins (eg transferrin, presenting as iron-resistant hypochromic anaemia).

Renal vein thrombosis This occurs in 15–20% of those with nephrosis, 30% of patients with membranous GN, and children with acute dehydration.

Presentation: There may be no symptoms or mild abdominal or back pain. Others develop severe pain and loin tenderness. Suspect if there is unexplained loss of renal function with RBC in the urine.

Diagnosis: This is by renal venogram or Doppler ultrasound.

Treatment: Anticoagulate. Streptokinase may be used to lyse acute thromboses. Pulmonary thromboembolism may occur.

The nephrotic syndrome
[*OTM* 18.58]

Acute renal failure (ARF):
1. Diagnosis

Essence A rapidly rising serum urea, creatinine and K$^+$, usually (but not invariably) with anuria or oliguria (<15ml/h). A careful history may provide the diagnosis: enquire about recent surgery or radiographic procedures, trauma, drugs (past and present), allergies, recent infections, underlying renal disease, family history of renal disease.

Signs Oliguria, vomiting, confusion, bruising, GI bleeding. There may be signs of *volume overload* (pulmonary oedema, BP↑) or *volume depletion* (postural hypotension, dehydration) secondary to vomiting.

Tests ● Urgent urine microscopy: RBC or RBC casts suggest GN (refer to nephrologist); haem +ve but no RBC suggests myoglobinuria (p394) or haemoglobinuria; lack of casts, cells and minimal protein suggests pre-renal. ● Urine electrolytes (especially urea). ● Blood: U&E, blood cultures, blood gases, FBC; when diagnosis uncertain consider eg C3/C4, ANA, DNA, ANCA, ASO titre, Bence Jones protein (p614) in urine. ● ECG. ● CXR. ● Renal ultrasound. ● Daily weight.

Biochemical changes 1 *Rise in urea and creatinine:* In catabolic state urea may rise by up to 10mmol/l/day and creatinine by 100μmol/l/day.
2 *Hyperkalaemia:* Especially if there is ischaemic tissue.
3 *Hypocalcaemia & hyperphosphataemia:* More marked in tissue necrosis.
4 *Hyponatraemia:* Due to relative water overload.
5 *Hyperuricaemia:* Not excreted. ↑ in catabolic state and tissue necrosis.
6 *Acidosis:* Usually with an increased anion gap.
7 *Urinary biochemistry:* See p385.

Causes
Pre-renal: ● Shock, burns, dehydration, sepsis, low output cardiac failure.
● Renal artery stenosis with dehydration or ACE inhibitors.
● Renal artery emboli, trauma, hepato-renal syndrome (p110).

Renal ● Acute tubular necrosis (ATN) secondary to ischaemia, septicaemia, tubular toxins (myoglobin, Bence Jones protein, contrast), drugs (eg gentamicin), acute pancreatitis.
● Acute cortical necrosis after prolonged hypotension or pregnancy related (abruptio placentae).
● Acute interstitial nephritis: drugs and infections (p394).
● Acute glomerulonephritis.
● Vascular: BP↑, emboli (thrombi, cholesterol), vasculitis (Wegener's, PAN), *in situ* thrombi (thrombotic thrombocytopenic purpura (p398), DIC, post-partum, eclampsia), renal vein thrombosis.
● Acute pyelonephritis rare unless in pregnancy or DM.
● Myeloma kidney.

Post-renal ● Intra-tubular obstructive lesions (eg uric acid crystals).
● Urinary tract obstruction (p378)—typically treatable prostatic disease (all too easy to miss unless one has a rule to always examine the abdomen for a large bladder, and to examine the prostate *per rectum*[1]). In women, do a vaginal examination (is there cervical carcinoma?—OHCS p34).

---

1 T Feest 1993 *BMJ* i 481

Acute renal failure (ARF) 1. Diagnosis
[*OTM* 18.124]

Urinary indices useful in distinguishing pre-renal from renal causes
of oliguria

	Pre-renal	Renal (acute tubular necrosis)
Urine osmolality	>500	<350
Urine sodium	<20	>40
Urine/serum creatinine	>40	<20
Urine/serum osmolarity	>1.2	<1.2
Fractional excreted Na*	<1	>1
Renal failure index (RFI)**	<1	>1

* Fractional excreted sodium $= \dfrac{[\text{Urine Na/Serum Na}]}{[\text{Urine Cr/Serum Cr}]} \times 100$

** Renal failure index $= \dfrac{[\text{Urine Na} \times \text{serum Na}]}{\text{Urine Cr}}$

Acute renal failure: 2. Management

▶ Seek specialist help earlier rather than later.

Is there need for urgent treatment?
Hyperkalaemia (K+ >6.5mmol/l±ECG change: peaked T, wide QRS).
Give:
- 10–30ml calcium gluconate (10%) IV slowly, for cardio-protection.
- 15 units soluble insulin with 50g glucose 50% IV. Continuous infusion thereafter of dextrose 50% at 50ml/h and insulin according to frequent blood glucose measurements.
- Polystyrene sulphonate resin 15g/6h PO or as enema (30g).
- Bicarbonate (100ml of a 4.2% solution by IVI to correct acidosis).

These are only temporary: dialysis may be the only long-term remedy.

Refractory pulmonary oedema, pericarditis, tamponade (▶ p722), arrange urgent dialysis.

Septicaemia, and *volume depletion:* treat appropriately.

Optimize fluid balance
- Replacement if depleted, diuresis if overloaded (frusemide IV 40mg–2g slowly). May need central venous pressure monitoring.
- If oliguric, 24h fluid maintenance = 24h urine output + allowance for diarrhoea/vomiting + 500ml insensible losses (more if febrile).
- If anuric/oliguric, try frusemide 40mg–1g IV (slowly, as can cause deafness) + renal dose dopamine (2–5μg/kg/min).
- Catheterize initially to assess urine output hourly. If anuric/oliguric remove catheter as source of sepsis.
- If in pulmonary oedema and no diuresis, think of removing a unit of blood as a temporary measure before dialysis commences.

Sepsis treat with appropriate antibiotics only after cultures are taken. There is no indication to treat prophylactically. Aim to remove all unwanted routes of infection (eg lines).

Further management • Has obstruction been excluded? ▶ Examine for masses PR and *per vaginam*; arrange urgent ultrasound; is the bladder palpable? Bilateral nephrostomies relieve obstruction, provide urine for culture and allow anterograde pyelography to determine the site of obstruction. • H_2 antagonists (eg ranitidine 150mg/12h PO) as high risk of GI bleeding. Avoid nephrotoxic drugs (check all drugs in *BNF*). • If rapidly deteriorating renal function but dialysis independent, consider renal biopsy. • Diet: high in calories (2000–4000kJ/day) with adequate high quality protein. Consider NG feeding or parenteral route if too ill.

Prognosis Depends on underlying disease; (ATN mortality: surgery/trauma—60%, medical illness—30%, pregnancy—10%). Oliguric acute renal failure worse than non-oliguric—more GI bleeds, sepsis, acidosis and higher mortality.

Indications for urgent dialysis
- K+ persistently high (>6.0mmol/l). • Acidosis (pH <7.2).
- Pulmonary oedema and no substantial diuresis.
- Pericarditis. • Cardiac tamponade (▶ p722).
- High catabolic state with rapidly progressive renal failure.

Acute renal failure (ARF) 2. Management
[*OTM* 18.124]

Chronic renal failure (CRF)

▶ Why has the patient presented now? Is there anything reversible?

Common causes DM, hypertension (malignant), chronic pyelonephritis, obstruction, glomerulonephritis, polycystic kidneys. Others: amyloid, myeloma, SLE, PAN, gout, $Ca^{2+}\uparrow$, interstitial nephritis. *Ask about:* family history (polycystic kidneys), UTIs, drugs especially analgesics.

Symptoms Nausea, lethargy, itch, nocturia, impotence. *Later*—oedema, dyspnoea, vomiting, confusion, seizures, hiccups, neuropathy, coma.

Signs Pallor, 'lemon-tinge' to skin (rarely uraemic 'frost' on skin), pulmonary or peripheral oedema, pericarditis (± tamponade), pleural effusions, hypertension, retinopathy (DM, BP↑), metastatic calcification.

Investigations ● *Urine* microscopy and culture, urinary electrolytes and osmolality, 24h urinary protein and creatinine clearance (p632).
● *Blood:* U&E, glucose, $Ca^{2+}\downarrow$, $PO_4^{3-}\uparrow$, urate↑ (may be primary or secondary to CRF), protein, FBC, ESR, protein electrophoresis.
● *Radiology:* renal ultrasound for renal size (typically small; large if obstructed, DM, polycystic, amyloid or myeloma) and to rule out obstruction. x-ray bones, especially hands, for renal osteodystrophy. Consider IVU, DMSA, DTPA scans (p790) and retrograde pyelography (p372).
● *Renal biopsy:* consider this if renal size is normal.

Treatment Refer to renal unit for early assessment. Dialysis may be needed for symptomatic relief when GFR <5% of normal (p390).
● Treat any reversible cause: relieve obstruction, avoid nephrotoxic drugs (aminoglycosides, cephalosporin with frusemide, NSAIDs), treat ↑Ca^{2+} and ↑uric acid, ?improve diabetic control (not formally proven).
● Monitor and treat ↑BP which may slow decline in renal function.
● Dietary advice: adequate calories, vitamins and iron. Salt restrict if ↑BP. Protein restriction: 0.5g/kg/day if symptomatic uraemia.
● Preparation of access for dialysis, eg Cimino fistulae or insertion of Tenchkoff catheter, before dialysis required.
● Anaemia, if due to CRF, may respond to erythropoetin (p574). Consider other causes of anaemia (p572).
● Renal bone disease: treat early, as soon as PTH raised. Lower PO_4^{3-} (aim 0.6–1.4mmol/l) with phosphate binders (eg calcium carbonate 300–1200mg/8h PO) and ↓dietary intake. (Aluminium-binders may cause aluminium accumulation with encephalopathy and osteomalacia-type bone disease.) Use vitamin D analogues (alfacalcidol 0.25–1µg/24h PO) and Ca^{2+} supplements *after* PO_4^{3-} lowered to reduce risk of metastatic calcification.
● Pre-transplant work-up: blood group and tissue type. Is there a related donor? Consider pre-transplant immunization.
● Miscellaneous: greatest mortality in CRF is from cardiovascular disease. Consider monitoring and treating hyperlipidaemia (p654). Hiccups: try chlorpromazine 25mg/8h PO. Oedema may require high doses of diuretics (frusemide 250mg–2g/day + metolazone 5–10mg/day).

Chronic renal failure (CRF)
[*OTM* 18.134]

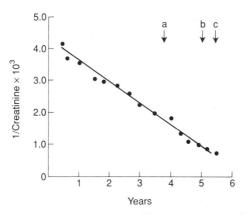

Plot of reciprocal plasma creatine (μmol/l) against time in adult polycystic disease patient: (a) work promotion, (b) arterio-venous fistula, (c) haemodialysis.

The majority of patients with CRF lose their renal function (ie GFR) at a constant rate. Creatinine is produced at a fairly constant rate and rises on a hyperbolic curve as renal function declines. The reciprocal creatinine plot is thus a straight line, parallel to the fall in GFR. This is widely used to monitor renal function and to predict when end-stage renal failure requiring dialysis will occur. Rapid decline in renal function greater than that expected may be due to: infection, dehydration, uncontrolled ↑BP, metabolic disturbance (eg Ca^{2+}↑), obstruction, nephrotoxins (injudicious use of drugs). Careful investigation and treatment at this point may delay end-stage renal failure.

389

Dialysis and transplantation

(See p386 for indications for urgent dialysis.)

Haemodialysis (HD) Blood flows opposite dialysis fluid and substances cleared down concentration gradient across a semipermeable membrane. *Problems:* ● Infection (HIV, hepatitis, bacteria). ● U&E imbalance ● BP↓. ● Vascular access (bleeding, thrombi, infection, vascular insufficiency). ● Dialysis arthropathy (amyloid formation especially in shoulders and wrists—due to accumulation of β_2-microglobulin). ● Aluminium toxicity (dementia ± a bone disease similar to osteomalacia), so use aluminium-free dialysis fluid and avoid aluminium-containing phosphate binders. NB: expense is also a frequent problem. *Out-patient follow-up:* Check dietary compliance, including vitamins (iron, B complex, C), general health. Examine dialysis site, BP, JVP, weight, oedema. Check: FBC, U&E, creatinine, Ca^{2+}, albumin, PO_4^{3-}, lipids.

Haemofiltration Used continuously eg in ITU for treatment of acute renal failure. Blood is filtered across a semipermeable membrane allowing removal of small molecules. Easy to use and causes smaller fluid shifts so fewer hypotensive episodes. Achieves good clearance values.

Intermittent peritoneal dialysis (IPD) Dialysate is introduced into the peritoneal cavity via a catheter for 10–30mins then withdrawn. Mainly used for ARF in hospitals without access to haemodialysis.

Continuous ambulatory peritoneal dialysis (CAPD) A permanent catheter is inserted into the peritoneum via a subcutaneous tunnel. Up to 2 litres of dialysate is introduced and exchanged up to 5 times a day. The patient is not tied to a machine and uraemic anaemia is less than with HD. *Contraindications:* Abdominal sepsis, anticoagulants, respiratory failure, previous abdominal surgery. *Problems:* Peritonitis (usually Staph, Strep, or coliforms—add antibiotics to the dialysate), catheter blockage (adhesions and fibrin—try urokinase into the catheter), weight gain, poor diabetic control and lactic acidosis (dialysate high in sugar and lactate), lactic acidosis, pleural effusion, leakage.

Renal transplantation Usually sited in an iliac fossa and so easy to biopsy. 1 year graft survival: HLA identical: 95%, 1 mismatch: 90–95%, complete mismatch 75–80%. *Acute graft rejection* (ie in first 3 months) often responds to increased immunosuppression. *Chronic rejection* presents as gradual decrease in graft function; it responds poorly to treatment. *Immunosuppressive drugs:* ● Cyclosporin, 10mg/kg reducing to 4mg/kg after some weeks; monitor plasma levels. ● Azathioprine 2mg/kg/day. ● Steroids. ● Antilymphocyte globulin.

Complications: ● Rejection ● Obstruction at ureteric anastamosis. ● Persistent ↑BP. ● Cyclosporin-induced nephrotoxicity. ● Atherosclerosis (renal disease and steroids). ● Side-effects of immunosuppression—opportunistic infection, eg *Pneumocystis carinii* (give prophylactic co-trimoxazole for 6 months), CMV, fungi, bacteria. ● Skin malignancy (squamous cell cancer) and lymphoma.

Dialysis and renal transplantation
[*OTM* 18.134]

Urinary tract malignancies

Hypernephroma (Clear cell or renal adenocarcinoma)
Typical patient: Male (the ♂/♀ ratio is 2:1), 40–60 years old.
Pathology: Renal tubular epithelium cell origin. Cords of tumour may invade the inferior vena cava via the renal veins.
Spread: Direct to nearby tissues and haematogenous to bone, liver and lung where 'cannon ball' secondaries may be seen on chest x-ray.

Clinical features: Classical triad of haematuria, loin pain and abdominal mass. Other presentations: PUO (night sweats), polycythaemia in 2% (the tumour secretes erythropoietin), anaemia, hypercalcaemia and rarely left varicocele as left renal vein compressed.

Investigations: Uroscopy for RBCs; CXR; ultrasound; IVU with tomography; angiogram; CT scan.

Treatment: Nephrectomy. Progesterone has been tried in some patients. Embolism may help symptoms. Metastatic disease is probably one of the few indications[1] for immunotherapy with interleukin-2 (chemotherapy and radiotherapy are disappointing); SE include rigors (may need IV pethidine), myalgia, renal failure and the vascular leak syndrome which may require colloids and dopamine (interleukin-2 stimulates the release of tumour necrosis factor and cytokines which make vessels leak macromolecules). Overall 5yr survival is 30–50%.

Other renal tumours Transitional cell (renal pelvis) 10%, lymphoma, secondaries, Wilms' nephroblastoma (*OHCS* p220).

Bladder tumours See p136.

Prostatic carcinoma In 10% this malignancy is so indolent that discovery is incidental. The patient is usually elderly but young men may have a fast-growing anaplastic variety. It is the third most common malignancy of men. Previous vasectomy may be associated with increased risk.
Spread: Bone (sclerotic or lytic lesions); direct to bladder, rectum.
Symptoms: Prostatism (hesitancy, frequency, dribbling) indicates obstruction—but unreliably;[2] bone pain indicates metastases.
Signs: On rectal exam: no median sulcus between lobes of prostate; nodules; mucosa stuck to prostate. Systemic: bone tenderness.
Risk: Testosterone↑, +ve family history; *not* benign hypertrophy.
Tests: FBC, ESR, U&E, creatinine, Ca^{2+}, prostate specific antigen >~10nmol/ml (reflects disease activity; NB: 25% large benign prostates have PSA up to 10), transrectal ultrasound morphology, bone x-rays and scintiscan. Outflow obstruction tests: bladder ultrasound (residual urine volume); urine flow rates; IVU.
Treatment: TURP (transurethral resection of the prostate) if there is obstruction. If there is no obstruction, treatment is controversial. Hormonal therapy should probably be reserved until metastases cause symptoms. Testosterone determines prostatic growth, so orchidectomy may slow tumour growth (if unacceptable offer gonadotrophin-releasing analogues, eg monthly goserelin injections, which first stimulate, and then inhibit pituitary gonadotrophin output). Cyproterone acetate 100mg/8h PO, stilboestrol 1mg/24h PO, and local radiotherapy are alternatives. Warn the patient that he may become impotent. See p135.
Symptomatic relief: Analgesia and radiotherapy to painful metastases.
Prognosis: Very variable. 10% die in 6 months, but 10% live >10yrs.
Screening: PR; PSA (but ↔ in 30% of small cancers); transrectal USS.

1 T Hamblin 1990 *BMJ* i 275 & 745 **2** M Stott 1988 *BMJ* i 1470

Urinary tract malignancies
[*OTM* 18.166]

Acute nephritic syndrome

Presentation Classically follows 1–3 weeks after a throat, ear or skin infection with Lancefield Group A β-haemolytic streptococcus. There is haematuria (RBC±cellular casts) with an acute fall in GFR leading to fluid retention, oliguria, hypertension and variable uraemia. Proteinuria is usually present. Common findings are: low C3, circulating mixed cryo-globulins, hypergammaglobulinaemia.

Management Risks are from hypertensive encephalopathy, pulmonary oedema and acute renal failure. Strict bed rest is unnecessary unless severely hypertensive or gross pulmonary oedema. Restrict dietary salt and fluid intake if oliguric. Give diuretics (frusemide PO or IV) and appropriate antihypertensive treatment (β-blockers may precipitate pulmonary oedema). Give a course of penicillin to eradicate residual infection. Acute renal failure may require dialysis. Steroids and cytotoxic drugs not indicated. Prognosis is excellent in children. In adults, moderate proteinuria ± urinary sediments may persist. Developing rapidly-progressive glomerulonephritis and chronic renal failure in weeks/months is rare.

Interstitial nephritis

This is an important cause of acute and chronic renal failure. It is associated with inflammatory cell infiltration mostly affecting the renal interstitium and tubules. The cause is often unknown, but it may follow drug hypersensitivity, analgesic nephropathy, reflux nephropathy, DM, sickle-cell disease, graft rejection or Alport's syndrome (p400).
Acute interstitial nephritis is often due to drugs (eg antibiotics, NSAIDs, diuretics) or infections (Staphs, Streps, brucella, leptospira).
Presentation: Renal failure, T°↑, arthalgia, eosinophiluria/eosinophilia.
Diagnosis: This is confirmed by renal biopsy.
Treatment: Régime on p386, and corticosteroids. *Prognosis:* Favourable.
(NB: if frank haematuria persists, suspect carcinoma of the renal pelvis.)

Myoglobinuria

Myoglobin (a muscle protein) can appear in the urine if there is muscle injury or necrosis (rhabdomyolysis), eg crush injury, exercise, trauma, burns, electric shock, coma, sepsis, fits or with low K^+. Myoglobinuria is often symptomless but can cause acute renal failure (ARF), possibly by tubular damage and/or blockage. *Diagnosis:* The urine is dark brown. Ward urine tests are +ve for blood but there are no RBC on microscopy. Urinary myoglobin must be looked for in the first 48h. Serum CK and aldolase are high. If ARF develops there will be expected biochemical profile (p384) but K^+ and PO_4^{3-} may be markedly ↑ if there is massive tissue breakdown. *Treatment:* Mannitol diuresis (12.5g loading dose then 2g/h IV—but do a smaller test dose first: 200mg/kg slowly IV), alkalinization of urine with IV bicarbonate. Avoid frusemide as it may acidify the urine and precipitate myoglobin. Dialysis or haemofiltration may be required.

Analgesic nephropathy

Prolonged use of NSAIDs (including aspirin and ibuprofen) may give an interstitial nephritis-like picture. In addition there is an increased incidence of UTI. Frank haematuria should lead one to suspect carcinoma of the renal pelvis, a known association.

Renal medicine

Diabetic nephropathy

DM is a common cause of renal failure and ~25% of diabetics diagnosed before age 30yrs develop chronic renal failure. Diabetics are prone to atherosclerosis, UTIs, and papillary necrosis, but glomerular lesions cause most of the problems. Typically there is thickening of the glomerular basement membrane and glomerulosclerosis which may be diffuse or nodular (Kimmelstiel–Wilson lesion). Early on there is an elevated glomerular filtration rate with enlarged kidneys, but the principal feature of diabetic nephropathy is proteinuria. This develops insidiously—initially intermittent microalbuminuria (p370), progressing to constant proteinuria and occasionally nephrotic syndrome. Once proteinuria is established, there is usually a delay before gradual, irreversible decline in GFR.

Treatment Scrupulous metabolic control may reduce elevated GFR and microalbuminuria but has little effect on established proteinuria. A low protein diet may slow the declining GFR. BP↑ worsens diabetic glomerulopathy, and its treatment reduces microalbuminuria, and delays the onset of renal failure[1]—renal failure is predicted by microalbuminuria—provided the duration of DM is <15yrs (if >15yrs, microalbuminuria is not predictive).[2] ACE inhibitors (eg enalapril 20mg/day PO, p299) reduce microalbuminuria and may slow progression to CRF even in normotensive subjects.[3] Many will come to dialysis or transplantation and the prognosis is now quite good, but limited by extensive vascular disease elsewhere.

Adult polycystic kidney disease

This is an autosomal dominant condition (gene on chromosome 16) and is a common cause of CRF. Cysts develop anywhere in the kidney and cause gradual decline in renal function. *Presentation:* Haematuria, UTI, abdominal mass, lumbar and abdominal pain, hypertension. Aneurysms of intracerebral arteries and subarachnoid haemorrhage are associated. Examination reveals bilateral, irregular, abdominal masses (~30% may have cysts in the liver or pancreas) and BP↑.

Tests: Uraemia, polycythaemia. Diagnosis is confirmed by ultrasound or IVU which shows large kidneys with long 'spidery' calyces.

Treatment is supportive. Control infection and BP. Some will need dialysis and transplantation. Check family members, although cysts may not be identified with certainty in patients <30yrs.

Infantile polycystic kidney disease (autosomal recessive): OHCS p220.

Medullary sponge kidney

Essence Dilatation of renal collecting tubules.

Presentation Urinary infections, nephrolithiasis, gross or microscopic haematuria. There is often a family history.

Diagnosis is confirmed by IVU which demonstrates the dilated tubules (1–7.5mm diameter) with contrast medium appearing in the renal papillae. Associated stones may be seen in the renal collecting system but are frequently present within the deformed ducts. Association with nephrolithiasis: Medullary sponge kidney has been found to be present in 12% of men and 19% of women presenting with calcium renal stones who do not have predisposing metabolic abnormalities.

1 E Mathiesen 1991 *BMJ* ii 81   2 C Forsblom 1992 *BMJ* ii 1051   3 M Hallab 1993 *BMJ* i 175

Nephrocalcinosis

This is the deposition of calcium in the kidney and is seen as calcification on plain x-rays. It is usually medullary and may cause symptoms of UTI or stone disease (p376). Causes of *medullary nephrocalcinosis:* hyperparathyroidism, and distal renal tubular acidosis comprise 60%. Also medullary sponge kidney, idiopathic calciuria, papillary necrosis, oxalosis. *Cortical calcification* is rare (<5%) and follows serious renal disease eg cortical necrosis or chronic glomerulonephritis.

Nephronophthisis

This is an inherited medullary cystic disease, which accounts for 20% of childhood renal failure. The kidneys are small and fibrosed. *Synonyms and related diseases:* juvenile nephronophthisis, medullary cystic disease, Senior Loken syndrome. It may exist in association with retinal degeneration, optic atrophy, retinitis pigmentosa (giving tunnel vision) and congenital hepatic fibrosis. *Treatment* is of the associated renal failure (*OHCS* p160).

Vascular disease and renal failure

Renal artery stenosis may be due to fibromuscular hyperplasia or atherosclerosis. It is associated with hypertension (sometimes malignant). If bilateral or extensive, dehydration, hypotension or ACE-inhibitor therapy may precipitate renal failure. In renal artery stenosis, GFR is maintained by angiotension II, and renal scanning after captopril shows that GFR is lowered markedly. *Diagnostic pointers:* Vascular disease elsewhere; severe or drug-resistant hypertension; abdominal bruit, urea↑, proteinuria. *Treatment:* Percutaneous renal angioplasty or bypass surgery.[1]

Malignant hypertension is associated with rapid renal failure. There is low C3 and microangiopathic haemolytic anaemia. Urinary sediment usually shows dysmorphic RBC and red cell casts. Usually no cause is found but it is important to exclude renal artery stenosis, Conn's syndrome, phaeochromocytoma and vasculitis. *Treatment:* See p302.

Haemolytic–uraemic syndrome (HUS) and thrombotic thrombocytopenic purpura (TTP). These may be two ends of a spectrum of the same disorder characterized by microangiopathic haemolytic anaemia, decreased platelets, and renal failure without the features of DIC (ie no clotting abnormality). In TTP neurological features are prominent including confusion and fits and skin lesions are seen. Histology shows fibrin and platelets in arterial and glomerular lesions with no immune deposits. In childhood, HUS is associated with bacterial and viral gastroenteritis, and treatment is supportive, often including dialysis. Mortality is <5% and haematological and renal features tend to resolve in 2 weeks. *Management:* Treatment of adult HUS is often problematic and up to 80% may need long-term dialysis. TTP is more serious with a very high mortality. Treatment must be prompt and may involve steroids and plasma exchange. It may relapse.[2]

Pre-eclampsia is the association of hypertension, proteinuria and oedema in pregnancy. The kidney shows glomerular endotheliosis with narrowing or obliteration of the glomeruli and deposition of fibrin and platelets. With severe disease there is headache, epigastric pain, visual disturbances and hyperreflexia. Investigation shows **h**aemolysis, **e**levated **l**iver enzymes, **l**ow **p**latelet counts (=HELLP syndrome). *Treatment:* Bed rest, antihypertensives and induction of labour if severe (see *OHCS* p96).

Cholesterol emboli Cholesterol emboli are seen in association with widespread atherosclerosis and raised cholesterol. The clinical syndrome consists of livedo reticularis, skin purpura, GI bleeding, renal failure, myalgia, cyanotic peripheries with intact distal pulses. It often follows arterial instrumentation, but may not be seen for weeks or months. It is also seen with thrombolysis, abdominal trauma and spontaneously. Investigation shows raised ESR, eosinophilia, mild proteinuria, inactive urinary sediment, raised urea and creatinine, C3 and C4 normal and ANCA negative. Diagnosis is confirmed by the finding of cholesterol clefts in renal biopsy. The course is usually progressive and fatal, but mortality may be due to underlying atherosclerosis. A few cases have regained renal function after dialysis.[3,4]

1 T Kremer Hovinga 1990 *Nephrol Dial Transplant* **5** 481–8 **2** J Moake 1991 *NEJM* **325** 426 **3** R Scully 1991 *NEJM* **325** 563 **4** R Scully 1991 *NEJM* **324** 113

Vascular disease and renal failure

Renal tubular disease

Renal tubular acidosis (RTA) The tubules either fail to create an acid urine (type 1 or 'distal' RTA) or there is bicarbonate wasting and serum bicarbonate falls until the 'filtered load' equals the reduced resorbtive capacity and the urine may become acid again (type 2 or 'proximal' RTA): usually occurs with other tubular disorders (Fanconi syndrome, below).

Type 1 ('distal') RTA: Presentation: in childhood with polyuria, failure to thrive and bone pain; in adults with weakness ($\downarrow$K$^+$), bone pain (osteomalacia), constipation, renal calculi (calcium phosphate, struvite, nephrocalcinosis) or renal failure. *Diagnosis:* urine pH >6 with spontaneous, or ammonium chloride (100mg/kg) induced, acidaemia. Normally urine pH falls below 5.3. (Do not acid load if plasma bicarbonate <19mmol/l). *Treatment:* bicarbonate replacement (1–3mmol/kg/day) and K$^+$. Hypokalaemia must be corrected before acidosis to prevent further fall in K$^+$ and the risk of cardiac arrest. *Prognosis* is variable.

Type 2 ('proximal') RTA: Presentation: in infancy with growth failure, hyperchloraemic acidosis and alkaline or slightly acid urine. It may be secondary to cystinosis, Fanconi's syndrome, or after renal transplant. There is no nephrocalcinosis and the prognosis is good. *Treatment:* Bicarbonate 10mmol/kg/day. (There is no 'type 3' RTA.)

Type 4 RTA: Occurs in disease states associated with aldosterone deficiency, or associated with distal tubular damage. It is seen in Addison's disease, DM (hyporeninaemic hypoaldosteronism), amyloidosis, obstructive uropathy and drugs (amiloride). If aldosterone deficient treat with fludrocortisone or a loop diuretic.

When to suspect RTA: ● High urine pH and negative urine culture.
● Hyperchloraemic hypokalaemic acidosis. ● Normal anion gap.
● Hypouricaemia and hypophosphataemia (type 2 RTA). ● Failure of urinary acidification in acidaemia. ● Hyperkalaemia and acid urine, eg in elderly diabetics or obstructive uropathy (type 4 RTA).

Fanconi's syndrome is a disorder with multiple proximal tubular defects causing glycosuria, aminoaciduria, phosphaturia and RTA.

● *Cystinosis* (autosomal recessive) Cystine crystals accumulate in lysosomes. Presents in first year of life as failure to thrive, polyuria, polydipsia and rickets. Renal failure develops <10yrs age. Treatment is with HCO$_3$$^{1-}$, K$^+$, PO$_4$$^{3-}$ and ergocalciferol. Cysteamine may halt progression.
● *Idiopathic adult Fanconi's syndrome* often transient and presents with bone pain, weakness, polyuria and polydipsia. Treat with replacement of excreted ions and vitamin D. ● *Secondary Fanconi's* occurs in myeloma, Wilson's disease, lead toxicity and post-renal transplantation.

Alport's syndrome (autosomal dominant) Nephritis and sensorineural deafness (OHCS p742). Associated with lens abnormalities, platelet dysfunction and hyperproteinaemia. Does not recur after transplantation.

Hyperoxaluria Autosomal recessive. May be primary, or secondary to gut resection or malabsorption. There are two types of primary oxalosis. Type 1 is more common and calcium oxalate stones are widely distributed throughout the body. It presents as renal stones and nephrocalcinosis in children and 80% are in renal failure by age 20yrs. Pyridoxine may help. Type 2 is more benign with only nephrocalcinosis and no renal failure.

Cystinuria The commonest aminoaciduria. In urine: cystine↑, ornithine↑, arginine↑, lysine↑. Cystine stones form. *Treatment:* ↑fluid intake and alkalize urine. (NB: in cells most cystine is reduced to form cysteine.)

Renal tubular disease

Willis and neurology Thomas Willis (1621–75) is one of those happy Oxford heroes belonging to Christ Church College who hold a bogus DM degree—awarded in 1646 for his Royalist sympathies. He had a busy life inventing terms such as 'neurology' and 'reflex'. Not only has his name been given to his famous circle, but he was the first to describe myasthenia gravis, whooping cough, the sweet taste of diabetic urine, and a variety of small nerves. He was the first person (and few have followed him) who knew the course of the spinal accessory nerve—which he discovered. He is unusual among Oxford neurologists in that, at various times of his life, he developed the practice of giving his lunch away to the poor. He also developed the practice of iatrochemistry: a theory of medicine according to which all morbid conditions of the body can be explained by disturbances in the fermentations and effervescences of its humours.

403

Further reading in neurology: IM Wilkinson 1993 *Essential Neurology*, 2ed, Blackwell.
C Warlow 1991 *Handbook of Neurology*, Blackwell.
1 J Lourie 1982 *Medical Eponyms: Who was Coudé?*, Pitman

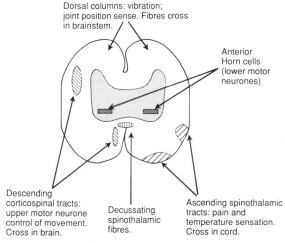

Dorsal columns: vibration; joint position sense. Fibres cross in brainstem.

Anterior Horn cells (lower motor neurones)

Descending corticospinal tracts: upper motor neurone control of movement. Cross in brain.

Decussating spinothalamic fibres.

Ascending spinothalamic tracts: pain and temperature sensation. Cross in cord.

Transverse section of spinal cord showing position of some of pathways

A neuroanatomy revision

Cerebral arteries

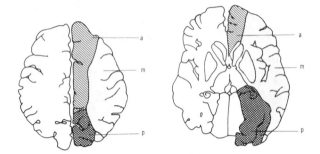

The distribution of supply of the anterior (a), middle (m), and posterior (p) cerebral arteries.

Attempt first to localize the lesion, then to identify the pathology. Is the main problem in: muscle; nerve; neuromuscular junction; spinal cord; brainstem; cerebellum; basal ganglia; cerebral cortex?

Anterior cerebral artery: Supplies the medial aspect of each hemisphere. Occlusion may cause: a weak, numb contralateral leg with milder disturbance in the arm.

Middle cerebral artery (and internal carotid): Supplies the external aspect of each hemisphere (concavity). Occlusion may cause: contralateral hemiplegia and sensory loss mainly of face and arm; dysphasia (dominant hemisphere); dyspraxias (non-dominant hemisphere); sometimes contralateral homonymous hemianopia.

Posterior cerebral artery: Supplies the occipital lobe. Occlusion may cause contralateral homonymous hemianopia.

Vertebrobasilar circulation: Supplies the cerebellum and brainstem (and also posterior cerebral artery). Occlusion may cause: hemianopia; cortical blindness; diplopia; vertigo; nystagmus; hemi- or quadriplegia; sensory symptoms on one or both sides; cerebellar symptoms; drop attacks; coma.

Brainstem vascular syndromes: Infarction of arterial territories in the brainstem can produce a number of syndromes. One such is the *lateral medullary syndrome*, due to occlusion of posterior inferior cerebellar artery, or one vertebral artery and leading to infarction of lateral medulla and inferior surface of cerebellum. This causes: severe vertigo with vomiting; dysphagia; nystagmus on looking to side of lesion; ipsilateral hypotonia and ataxia; ipsilateral paralysis of soft palate; ipsilateral Horner's syndrome; dissociated sensory loss (analgesia to pin-prick on ipsilateral face and contralateral trunk and limbs).

Subclavian steal syndrome Proximal stenosis of subclavian artery may cause blood to be *stolen* from ipsilateral vertebral artery. Brainstem ischaemia follows exertion of arm. A difference in BP of >20mmHg between the arms supports the diagnosis.

A neuroanatomy revision

Spinal cord level and the vertebral column The spinal cord does not occupy the full length of the the vertebral column. In adults, it ends at the level of the L1,2 disc.

Upper motor neurone lesions The following signs result from lesions in the motor system proximal to (above) the α-motor neurone. The signs are: *spasticity* (hypertonia predominant usually in flexors of arms and extensors of legs*; the hypertonia is suddenly reduced on overcoming the initial resistance to straightening the patient's limb—hence the term *clasp-knife spasticity*); *paralysis* or *weakness* predominantly of extensors in the arm and flexors in the leg*; *hyperreflexia*; *extensor (upgoing) plantars*; *clonus* (rhythmic muscular beats, eg on suddenly flexing the patient's foot at the ankle) and *Hoffman's reflex* (flicking one finger causes neighbouring digit(s) to flex). *Common causes are: stroke*; any *space-occupying lesion* (p456) inside the cranium; *cord pathology*.

Lower motor neurone lesions The following signs result from lesions to the motor nerves which directly innervate the muscle (the α-motor neurones). The lesion may be to the peripheral nerves themselves, or to the cell bodies (within the anterior horn of the spinal cord). *The signs are:* **405** muscle wasting, muscle fasciculation, weakness, decreased or absent reflexes. *Common causes are:* root lesions; neuropathies. *Note:* Diseases of muscle (p470) may give similar clinical picture, although reflexes are commonly preserved.

Reflexes and their spinal cord level See p32. **Roots of muscle groups** See p408.

*These patterns are known as the pyramidal distribution.

Drugs and the nervous system

The brain is a gland which secretes both thoughts and molecules: both products are amenable to drug therapy via neurotransmitter systems, eg:

[1] Precursor of the transmitter (eg L-dopa).
[2] Interference with the storage of transmitter in vesicles within the pre-synaptic neurone (eg tetrabenazine).
[3] Binding to the post-synaptic receptor site (bromocriptine).
[4] Binding to receptor-modulating site (benzodiazepines).
[5] Interference with the breakdown of neurotransmitter within the synaptic cleft (monoamine oxidase inhibitors—MAOIs).
[6] Interference with re-uptake of transmitter from synaptic cleft into pre-synaptic cell (tricyclic antidepressants).
[7] Binding to pre-synaptic 'autoreceptors'—involved in feedback loops, often reducing neurotransmitter release.

The proven neurotransmitters include:

Amino acids Glutamate and aspartate acting as excitatory transmitters on NMDA and non-NMDA receptors. These are relevant to epilepsy and ischaemic brain damage, but drugs acting on them are not in general use. γ-aminobutyric acid (GABA) is an inhibitory neurotransmitter. *Drugs enhancing GABA activity are used in:* Epilepsy (phenobarbitone, benzodiazepines, vigabatrin); spasticity (baclofen, benzodiazepines).

Peptides Opioids and substance P.

406 Histamine and Purines (such as ATP) Clinical relevance is not clear.

Dopamine (DA) *Drugs enhancing DA activity used in:* Parkinson's disease; hyperprolactinaemia; acromegaly. SE: nausea; vomiting; BP↓; chorea; dystonia; hallucinations and delusions; (reduced prolactin). *Drugs reducing DA activity used in:* Schizophrenia; delusions; chorea; tics; nausea; vertigo. SE: parkinsonism; dystonias; akathisia; (prolactin↑).

Noradrenaline (NAd) and adrenaline (Ad) 4 receptor types: α_1, α_2, β_1, β_2. Noradrenaline is more specific for α receptors but both transmitters affect all receptors. In the periphery: *α-receptor stimulation* leads to arteriolar vasoconstriction and pupillary dilatation; *β_1 stimulation* to increase in heart rate and contractility; *β_2 stimulation* to bronchodilation, uterine relaxation and arteriolar vasodilation. *Drugs enhancing adrenergic activity are used in:* Asthma (β_2); anaphylactic shock (adrenaline); heart failure (dobutamine); depression (tricyclics, MAOIs may act through increasing synaptic noradrenaline centrally). *Drugs reducing adrenergic activity are used in:* Angina, BP↑, some arrhythmias, symptoms of thyrotoxicosis and anxiety (β_1); hypertension due to phaeochromocytoma (α).

5-Hydroxytryptamine (Serotonin) 4 main receptors: 5-HT_{1-4}. 5-HT_1 has 5 subtypes (5-HT_{1A-E}). *Agonists:* Lithium$_{1A}$; sumatriptan$_{1D}$. *Partial agonists:* Buspirone$_{1A}$; LSD$_2$. *Antagonists:* Mianserin$_{1C}$; ondansetron$_3$; pizotifen$_{1\&2}$; methysergide$_{1\&2}$. *Re-uptake inhibitors:* Fluoxetine. Fenfluramine and MDMA (p184) ↑nerve terminal 5-HT release.

Acetylcholine (Muscarinic and nicotinic receptors)
Centrally acting anticholinergic drugs are used in: Parkinsonism, dystonias, travel sickness. Central toxic effects: confusion, delusions.
Peripherally acting antimuscarinic drugs are used in: Asthma (ipratropium); incontinence; to dry secretions pre-operatively; to dilate pupils; to increase heart rate (atropine).
Peripherally acting cholinergic agonists use: Glaucoma (pilocarpine); myasthenia (anticholinesterases). SE: dry mouth, colic, accommodation↓.

Drugs and the nervous system

Some drugs acting on neurotransmitter systems

	Enhancing *	*Decreasing* *
Dopamine	L-dopa (1) Bromocriptine [D$_2$-agonist] (3) Selegiliine [MAO-B inhibitor] (5) Amantadine (6)	Major tranquillizers (3) Some anti-emetics (3)
Noradrenaline/ adrenaline	Salbutamol (β$_2$) (3) Adrenaline ? Tricyclic antidepressants (6) Monoamine oxidase (MAO) inhibitors (5)	Propranolol (β) (3) Atenolol (β$_1$) (3) Clonidine (α$_2$ agonist) Phentolamine (α) (3)
(7)		
5-HT	LSD and other hallucinogens Sumatriptan (3) Some tricyclic antidepressants (eg trazodone) (6)	Cyproheptadine Pizotifen
Acetylcholine	Carbachol (3) Pilocarpine (3) Anticholinesterases (eg neostigmine) (5)	Atropine (3) Scopolamine (3) Ipratropium (?3) Benzhexol Orphenadrine (?3) Procyclidine
GABA	Baclofen (GABA$_B$) (3) Benzodiazepines (4) Vigabatrin Barbiturates	

* Numbers in parentheses, eg (3), refer to probable mechanism — see page opposite

Testing peripheral nerves[1]

Nerve root:	Muscle:	Test—by asking the patient to:
C3,4	trapezius	Shrug shoulder, adduct scapula
C4,5	rhomboids	Brace shoulder back
C5,6,7	serratus anterior	Push forward against resistance
C5,6,7,8	pectoralis major (clavicular head)	Adduct arm from above horizontal and forward
C6,7,8 and T1	pectoralis major (sternocostal head)	Adduct arm below horizontal
C5	supraspinatus	Abduct arm the first 15°
C5,6	infraspinatus	Externally rotate arm, elbow at side
C6,7,8	latissimus dorsi	Adduct horizontal and lateral arm
C5,6	biceps	Flex supinated forearm
C5,6	deltoid	Abduct arm between 15° and 90°

The radial nerve

C7,8	triceps	Extend elbow against resistance
C5,6	brachioradialis	Flex elbow with forearm half-way between pronation and supination
C6,7	extensor carpi radialis longus	Extend wrist to radial side with fingers extended
C5,6	supinator	Arm by side, resist hand pronation
C7,8	extensor digitorum	Keep fingers extended at MCP joint
C7,8	extensor carpi ulnaris	Extend wrist to ulnar side
C7,8	abductor pollicis longus	Abduct thumb at 90° to palm
C7,8	extensor pollicis brevis	Extend thumb at MCP joint
C7,8	extensor pollicis longus	Resist thumb flexion at IP joint

Median nerve

C6,7	pronator teres	Keep arm pronated against resistance
C6,7,8	flexor carpi radialis	Flex wrist towards radial side
C7,8 T1	flexor digitorum sublimis	Resist extension at PIP joint (while you fix his proximal phalanx)
C8 T1	flexor digitorum profundus I & II	Resist extension at the DIP joint
C8 T1	flexor pollicis longus	Resist thumb extension at interphalangeal joint (fix proximal phalanx)
C8 T1	abductor pollicis brevis	Abduct thumb (nail at 90° to palm)
C8 T1	opponens pollicis	Thumb touches 5th finger-tip (nail parallel to palm)
C8 T1	1st & 2nd lumbricals	Extent PIP joint against resistance with MCP joint held hyper-extended

Ulnar nerve

C7,8	flexor carpi ulnaris	Abducting little finger, see tendon when all fingers extended
C8 T1	flexor digitorum pro-fundus III & IV	Fix middle phalanx of little finger, resisting extension of distal phalanx
C8 T1	dorsal interossei	Abduct fingers (use index finger)
C8 T1	palmar interossei	Adduct fingers (use index finger)
C8 T1	adductor pollicis	Adduct thumb (nail at 90° to palm)
C8 T1	abductor digiti minimi	Abduct little finger
C8 T1	opponens digiti minimi	With fingers extended, carry little finger in front of other fingers
L4,5 S1	gluteus medius & minimus (superior gluteal nerve)	Internal rotation at hip, hip abduction.
L5 S1,2	gluteus maximus (inferior gluteal nerve)	Extension at hip (lie prone)
L2,3,4	adductors (obturator nerve)	Adduct leg against resistance

409

Femoral nerve

L1,2,3	ilio-psoas	Flex hip with knee flexed and lower leg supported (patient lies on back)
L2,3	sartorius	Flex knee with hip externally rotated
L2,3,4	quadriceps femoris	Extend knee against resistance

Sciatic nerve

L4,5 S1,2	hamstrings	Flex knee against resistance
L4,5	tibialis posterior	Invert plantar-flexed foot
L4,5	tibialis anterior	Dorsiflex ankle
L5 S1	extensor digitorum longus	Dorsiflex toes against resistance
L5 S1	extensor hallucis longus	Dorsiflex hallux against resistance
L5 S1	peroneus longus & brevis	Evert foot against resistance
S1	extensor digitorum brevis	Dorsiflex hallux (muscle of foot)
S1,2	gastrocnemius	plantarflex ankle joint
S1,2	flexor digitorum longus	Flex terminal joints of toes
S1,2	small muscles of foot	Make sole of foot into a cup

1 *Aids to the Examination of the Peripheral Nervous System* HMSO 1976

Dermatomes

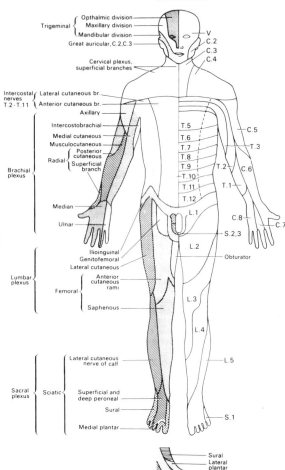

Trigeminal
- Opthalmic division
- Maxillary division
- Mandibular division
- Great auricular, C.2, C.3

V
C.2
C.3
C.4

Cervical plexus, superficial branches

Intercostal nerves T.2–T.11
- Lateral cutaneous br.
- Anterior cutaneous br.
- Axillary
- Intercostobrachial
- Medial cutaneous
- Musculocutaneous
- Radial { Posterior cutaneous / Superficial branch }

Brachial plexus

- Median
- Ulnar

T.5
T.6
T.7
T.8
T.9
T.10
T.11
T.12

C.5
T.3
T.2
C.6
T.1
C.8
C.7

L.1
S.2,3
L.2

Lumbar plexus
- Ilioinguinal
- Genitofemoral
- Lateral cutaneous
- Femoral { Anterior cutaneous rami / Saphenous }

Obturator

L.3
L.4
L.5

Sacral plexus
- Sciatic { Lateral cutaneous nerve of calf / Superficial and deep peroneal / Sural / Medial plantar }

S.1

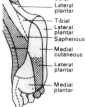

- Sural
- Lateral plantar
- Tibial
- Lateral plantar
- Saphenous
- Medial cutaneous
- Lateral plantar
- Medial plantar

ANTERIOR ASPECT

410

Dermatomes

Ophthalmic division ⎱
Maxillary division ⎬ Trigeminal
Mandibular division ⎰

Mastoid branch, C.2, C.3 ⎱ Superficial
Great auricular branch, C.2, C.3 ⎰ cervical plexus

Occipital, C.2
Occipital, C.3 ⎱
Occipital, C.4 ⎬ Dorsal
Occipital, C.5–C.8 ⎰ branches

Supraclavicular, C.3, C.4

Dorsal rami of thoracic nerves

Cutaneous branch of axillary

Lateral cutaneous branches ⎱
of intercostal nerves ⎰

Medial and lateral cutaneous br. of radial

Intercostobrachial

Medial cutaneous

Musculocutaneous

Anterior branch of radial

Median

Dorsal cutaneous branch of ulnar

Gluteal branch of 12th intercostal
Lateral cutaneous br. of iliohypogastric

Lateral branches of dorsal ⎱
rami of lumbar and sacral ⎰

Medial branches of dorsal rami, L.1 – S.6

Perforating branch of ⎱ Pudendal plexus
Posterior cutaneous ⎰

Lateral cutaneous
Obturator
Medial cutaneous ⎱ Femoral ⎱ Lumbar plexus
Saphenous ⎰

Posterior cutaneous

Superficial peroneal ⎱ Common
Sural ⎰ peroneal ⎱ Sacral plexus
Tibial
Lateral plantar ⎰

C.2
C.2
C.3
C.4
C.5
T.3
T.2
T.1
C.6
C.8
C.7
T.4
T.5
T.6
T.7
T.8
T.9
T.10
T.11
T.12
L.1
L.2
L.3
S.3
S.2
L.3
L.5
L.4
S.1

L.5

S.1
L.4
L.5

411

POSTERIOR ASPECT

Lumbar puncture (LP)

▶ If suspicous of raised intracranial pressure (falling level of consciousness with falling pulse, rising BP, vomiting, focal signs, papilloedema) discuss immediately with relevant clinician with a view to CT scan. Lumber puncture is normally contraindicated.

Method Explain everything fully to the patient. Emphasize that he will be able to tell you how he feels throughout.

Place the patient on his left side, his back on edge of bed, fully flexed (knees to chin).

- Landmarks: plane of iliac crests through L4. In adults, the spinal cord ends at L1,2 disc. Mark L4,5 or L3,4 intervertebral space, eg by a *gentle* indentation of a thumb-nail on the overlying skin (better than a ballpoint pen mark, which might be erased by the sterilizing fluid).
- Wash hands. Don a mask and sterile gloves.
- Sterilize the back with tincture of iodine unless allergic.
- Open the spinal pack. Check manometer fittings. Have 3 plain sterile tubes and 1 fluoride (for glucose) tube ready.
- Inject 0.25–0.5ml 1% lignocaine under skin at marked site.
- Wait 3mins; then insert spinal needle (stilette in place) through the mark aiming towards umbilicus. You will feel the resistance of the spinal ligaments, and then the dura, followed by a 'give' as the needle enters the subarachnoid space.
- Withdraw stilette. Wait for CSF.
- Measure CSF pressure with manometer.
- Catch fluid in 3 sequentially numbered bottles (<3ml).
- Remove needle. Send specimens promptly for microscopy, culture, protein, glucose (send a simultaneous plasma sample too), and, if applicable: virology, syphilis serology, cytology, AFB, oligoclonal bands if multiple sclerosis suspected.
- Patient should lie flat for >4h (some advise 24h) and have hourly neurological observation and BP measurement. Encourage fluids to lessen post-LP headache.

CSF composition *Normal values:* Lymphocytes <4/mm³; polymorphs 0; protein <0.4g/l; glucose >2.2mmol/l (or about 70% plasma level); pressure <200mm CSF. *In meningitis:* See p442. *In multiple sclerosis:* see p454.

Bloody tap ie artefact due to piercing blood vessel is indicated by fewer red cells in successive bottles, no yellowing of CSF. To estimate how many white cells (w) were in the CSF before the blood was added, use the following:

$$w = \text{CSF WCC} - [\text{blood WCC} \times \text{CSF RBC} \div \text{blood RBC}]$$

If the patient's blood count is normal, the rule of thumb is to subtract from the total CSF WCC (per µl) one white cell for every 1000 RBCs. To estimate the true protein level subtract 10mg/l for every 1000 RBCs/mm³ (be sure to do the count and protein estimation on the same bottle).

Subarachnoid haemorrhage: Yellow CSF (xanthochromia). Red cells in equal numbers in all bottles. The RBCs will excite an inflammatory response (eg CSF raised) most marked after 48h.

Very raised CSF protein: Acoustic and spinal tumours; Guillain–Barré. *Raised:* meningitis; MS; hypothyroidism; DM.

Post-LP headache This may be partly preventable by reducing post-LP CSF leak—by using finer needles.

Lumbar puncture (LP)
[*OTM* 21.10]

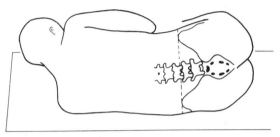

Method of defining the interspace between the 3rd and 4th lumbar vertebrae.

After Vakil and Udwadia *Diagnosis and Management of Medical Emergencies* 2 ed, OUP, Delhi.

Headache

Every day, *thousands* of patients visit their doctors for headache; these consultations can be rewarding as the chief skill is in the history—not in *taking* the history, so much as in *allowing* it to unfold. Let the patient tell you about the headache's causes and antecedent events: even, on occasion, *who* their headache is. Stress is the usual cause, and referral to a neurologist, or lab tests are rarely needed. But in those over 55yrs with recent-onset headache, always consider giant cell arteritis ▶ *and do an ESR.*

● Acute single episode If bilateral with drowsiness, neck stiffness then *meningitis, encephalitis* and *subarachnoid haemorrhage* must be excluded, eg by LP. If its onset is like a sudden blow to the head a *subarachnoid haemorrhage* is likely even if no neck stiffness.

After *head injury* headache is common. Sinister signs are drowsiness, focal signs, and very severe pain.

Sinusitis may present with a constant, unilateral frontal pain or a boring pain in the centre of the forehead, with local tenderness. Confirm by x-ray.

● Acute recurrent attacks Migraine (p416) is the commonest cause. *Cluster headache (migrainous neuralgia):* Commoner in men than women, begins at any age. It is unilateral and always on the same side. The pain becomes rapidly very severe around one eye which may become watery, bloodshot, with lid swelling, lacrimation, and rhinorrhoea. The pain often occurs at night. The patient may be unable to control his posture (eg driven to stand on one leg), or to bang his head against something solid. A common action is to drive the finger nails into the palm. It lasts 20–60mins and may recur several times every 24h for several weeks. There is then a pain-free period for many months before another cluster. Clusters last 4–12 weeks. It has been suggested that these headaches are caused by local release of 5-HT in vicinity of the superficial temporal artery.[1] Treat with 1 ergotamine suppository (2mg) 1h before expected attack. Stop medication one day a week to see if headaches return. Consider sumatriptan by injection (p416). Methysergide (a 5-HT antagonist) is a second line drug (seek expert help).

● Subacute onset If patient over 55yrs exclude *giant cell arteritis* (p674). Look for tender, thickened, pulseless temporal arteries and ESR >40mm/h. Ask about jaw claudication.

● Chronic headache **Tension headache:** Typically pain is present all day, worse in evening, usually bitemporal, occipital or like a tight band. The scalp may be tender. Treatment aimed at the stresses in the patient's life. Amitriptyline 25–50mg/12h PO may be valuable (even if no depressive illness). Analgesics and tranquillizers are of limited value and there is danger of dependency.

The headache of *raised intracranial pressure*, *severe hypertension* and *benign intracranial hypertension* (p458) tends to be present on waking often accompanied by vomiting but persists less during the day than tension headache. It may be made worse by coughing or bending forward and be accompanied by blurred vision.

Analgesic rebound headache: This is a persistent headache associated with long-term overuse of analgesics, hypnotics and tranquillizers.[2]

Examine for fever, neck stiffness, hypertension, papilloedema, other neurological signs. Also examine temporal arteries for signs of arteritis.

Tests Consider ESR. CT scan if space-occupying lesion suspected.

Headache in children See OHCS p257.

1 P Aubineau 1992 *Lancet* i 1294   **2** *Southern Medical Journal* 1993 **86** 1202–5

Trigeminal neuralgia

Brief (seconds) intense stab of pain in distribution of one or more divisions of trigeminal nerve (eg cheek or jaw). Usually (96%) unilateral. Face may screw up with pain (hence *tic doloureux*). Pain may be precipitated by touching a trigger point on face, or by eating or talking. More common in women than men, and in those over 50yrs. Cause usually unknown. In young patient consider MS. If untreated, usually progresses with shorter and shorter periods of remission.

Treatment Aim for pain relief despite high doses. *Batting order:* carbamazepine 100-400mg/8h PO; phenytoin 200-400mg/24h PO; clonazepam 2–6mg/24h PO; baclofen 5–25mg/8h PO. If these fail, surgical treatment (section of surgical root or thermocoagulation of ganglion) may be necessary, but may cause permanent anaesthesia.

415

Migraine

Migraine is a common cause of recurrent misery. Its prevalence is 8%, and it costs the UK economy £200 million a year in lost production.[1]

Symptoms The classical symptoms of migraine are of visual (or other) aura followed after 10–60mins by a unilateral, throbbing headache. Two other symptoms pattern are: aura with no headache; and, episodic (often pre-menstrual) severe headache (often unilateral), with nausea, vomiting, and sometimes photophobia but with no aura ('common migraine').

Pathogenesis Reduction in cerebral blood flow, possibly due to vaso-constriction, leading to the aura followed by increased cerebral and extracranial blood flow leading to the headache. Attacks often associated with changes in plasma 5-hydroxytryptamine levels.

Triggers to migraine include cheese, chocolate, contraceptive steroids, anxiety, exercise, travel. In many cases (~50%) no trigger found, and in only a few does stopping triggers prevent attacks.

Differential diagnosis Cluster headaches, tension headaches, cervical spondylosis; BP↑; depression; sinusitis; subdural; caries; otitis; TIA.

Prophylaxis of migraine (There are no panaceas.)

Useful if frequency > twice a month. If one drug doesn't work after adequate trial (2–3 months) try another. 60% of patients can expect some benefit. A reasonable *batting order* (but note contraindications):
- Propranolol 40–80mg/12h PO.
- Pizotifen 500µg/8h PO; or 1.5mg PO at night (a 5-HT *antagonist*).
- Amitriptyline 25–75mg/8h PO (even without depression)
- Methysergide (under specialist supervision)

Pre-menstrual migraine may respond to diuretics (eg acetazolamide 250mg/12h PO) or depot oestrogens.

Treatment of migraine attack *batting order* as follows:
- Paramax® (each tablet contains paracetamol 500mg and metoclopramide 5mg). 2 tabs at onset, then every 4h as necessary to maximum of 6 in 24 hours.
- Ergotamine (a 5-HT agonist) 1mg PO as attack begins, repeat at 30mins to maximum of 3mg in one day, and 6mg in one week; or, better as suppository (2mg ergotamine and 100mg caffeine) to maximum of 2 in 24h or 4 in one week. Emphasize the danger of ergotamine (gangrene, permanent vascular damage). Contraindications: peripheral vascular disease; ischaemic heart disease; pregnancy; breast-feeding; hemiplegic migraine; Raynaud's; liver or renal impairment; hypertension.
- Sumatriptan is a new 5-HT_1 (serotonin-1) *agonist*. Its rôle in migraine treatment is unclear. It may precipitate angina or arrhythmias and rarely myocardial infarction—even if no pre-existing heart disease.[2]

NB: rebreathing into paper bag (raising $PaCO_2$) may abort attack. Some patients find warm or cold packs to the head ease the pain.

1 J Blau 1991 *Migraine*, Office of Health Economics, 12 Whitehall London SW1A 2DY
2 J Ottervanger 1993 Lancet **341** 861

Blackouts

Take a detailed history from patient and a witness.

What does the patient mean by 'blackout'? Loss of consciousness, a fall to the ground without loss of consciousness (*drop attack*, p426), a clouding of the visual field, diplopia, or vertigo?

10 points in the history ● Timing: frequency; duration; time of day.
● Is there a warning? Eg an aura preceding attacks.
● What are the exact circumstances in which they occur?
● What does the patient look like during an attack (colour, movement)?
● What happens during an attack (eg injury, incontinence)?
● What is the pulse during the attack?
● Was the patient confused or sleepy after the attack?
● How much does the patient remember?
● Were there accompanying symptoms?
● Can the patient prevent attack?

Principal diagnoses *Epilepsy* (p448) Can occur any time (including at night); often has aura; may be precipitated by TV; often, during attack, altered breathing with cyanosis; various movements common during attack; sometimes incontinent (including faecally); may bite tongue (virtually diagnostic if it occurs); usually period of drowsiness after attack; often prolonged (>30mins) period for which patient is subsequently amnesic; occasionally residual paralysis.

Syncope (vaso-vagal): Rare at night; usually preceded by brief period of nausea and closing in of visual field; can be precipitated by anxiety or pain; cannot occur if lying down; usually pale during attack; occasionally incontinent of urine, never of faeces; may jerk limbs but no tonic–clonic sequence; no prolonged subsequent confusion nor amnesic period; crouching may abort attack. *Stokes–Adams attack* (Transient cardiac arrhythmia causing decrease in cardiac output and thus loss of consciousness.) Often no warning (except palpitations); patient goes deathly pale (may flush on recovery); pulse irregular or slow *during* attack. Similar attacks may result from aortic stenosis or intracardiac thrombus/tumour. Injury during an attack is a cardinal sign of intermittent arrhythmia.

Other diagnoses ● *Postural hypotension* on sudden standing.
● *Carotid-sinus syndrome* in those >50yrs. Suggested if blackout only on turning head. Massage one carotid and feel for extreme bradycardia.
● *Vertebro-basilar ischaemia* (on turning head).
● *Micturition syncope* (on urinating; seen in middle-aged men; benign).
● *Cough syncope* (consider airways obstruction).
● *Hypoglycaemia.* ● *Intracranial pressure* ↑ (headache & vomiting).
● *Anxiety with hyperventilation* (suggested if patient conscious at time of generalized paraesthesiae or stiffening, or known to hyperventilate).
● *Alcohol/drug ingestion* ● *Simulated blackout*

Examination Cardiovascular, neurological. BP lying and standing.

Tests Unless diagnosis very clear, thorough investigation is needed. ECG (heart block, arrhythmias, long Q–T interval), U&E, FBC. Consider EEG, 24h EEG, 24h ECG, echocardiogram, CT scan, 2mins hyperventilation.

Advice about jobs (if epilepsy) Restrict life as little as possible. Don't bath infants alone, or swim alone. Don't work with heavy moving machinery or at heights. *Driving:* Law varies. In UK need *either* to be fit-free for 2yrs; *or* only fits whilst asleep for last 3yrs.

For Heavy Goods Vehicle Licence: no fit since age 5yrs. See: *OHCS* p468.

Blackouts
[OTM 21.51]

Vertigo

This is the illusion of movement, usually rotatory. It indicates disease of the labyrinth (80% of cases) or its central connections.

Examine for deafness and nystagmus: ears, cranial nerves.

Investigations Audiometry. Caloric tests. Brainstem auditory evoked responses (BAER)

Pointers to labyrinthine damage (eg brainstem) Nystagmus: multi-directional, vertical, permanent. Other cranial nerve lesions. NB: conductive deafness suggests wax or middle-ear disease.

Causes of acute episode with labyrinthine damage
Acute vestibular failure (vestibular neuronitis): Abrupt onset, patient feels very ill, lies still in bed and may vomit. No deafness or tinnitus. Probably caused by virus in young and vascular lesion in elderly. Gets better over 3–4 weeks. Reassure. Sedate.
Drugs (eg aminoglycosides) and Ramsay Hunt syndrome (OHCS p756) may cause acute vertigo.

Causes of acute episode with central damage
MS, brainstem infarct, drugs: barbiturates, phenytoin, alcohol (NB: Wernicke's encephalopathy, p712, needs urgent treatment), cerebellar damage.

Causes of recurrent attacks with labyrinthine damage
Benign postural vertigo: Patient gets vertigo whenever making a particular movement (eg turning in bed). Usually follows acute vestibular failure or head injury. Improves slowly.

Ménière's disease: A syndrome of vertigo, tinnitus, deafness. Attacks last minutes to hours and may include vomiting, nystagmus, and past pointing. Treat with prochlorperazine. There is no evidence that cinnarizine or betahistine are specific for Ménière's.

Hypoglycaemia: For causes see p536.

Causes of recurrent central attacks
Migraine, epilepsy; vertebro-basilar ischaemia.

Dizziness

Dizziness is a less specific symptom than vertigo. It may be caused by any of the above. Other causes of *episodic dizziness* are: migraine; drugs; postural hypotension; hypoglycaemia; arrhythmias; phaeochromocytoma; MS; epilepsy.

Other causes of persisting dizziness are drugs; brainstem infarct; MS; acoustic neuroma; posterior fossa tumour; otosclerosis; anaemia.

Vertigo
Dizziness
[*OTM* 21.91]

Nystagmus

This is oscillation of the eyes on attempted fixation. It is defined by the direction of the fast phase, but it is the slow phase which is pathological: it represents a drift away from fixation, the fast phase attempts to correct that drift. Ask the patient to fix on your finger (more than two feet away, to avoid the interference of convergence). Move finger to the 12, 3, 6, & 9 o'clock position; wait 5sec for nystagmus. Do not move further than 45° from central, beyond that nystagmus may be physiological. Nystagmus must be sustained for more than a few beats to be significant. It is caused by lesions of inner ear, cerebellum and brainstem.

Vestibular nystagmus occurs only in one direction of gaze (away from the side of the lesion) and is made worse by gaze in that direction (Alexander's law). It may be horizontal or rotational, but not vertical. If visual fixation is removed (eg with Frenzel's glasses or electronystagmography), the nystagmus is worse. With time the nystagmus may disappear in peripheral vestibular disease. This is in contrast with central disease. Associated tinnitus and deafness suggests Ménière's (p420).

Positional nystagmus occurs in benign positional vertigo (BPV) and is prompted by rapid head movements or on rapidly lying supine. It may follow viral labyrinthitis or head injury. It is diagnosed by Hallpike's test: turn head to one side and then lie back; support head and allow it to fall below the horizontal plane off the top of the bed. In BPV, the nystagmus will follow a delay of a few seconds and will fatigue after 10–20sec. It will be associated with vertigo but if the test is oft repeated, the response will adapt. It only occurs in one direction. The unidirectionality, latency, fatiguability, and adaptation do not occur with central lesions.

Central nystagmus results from disease of the brainstem eg MS, vertebrobasilar ischaemia, Wernicke's encephalopathy, phenytoin toxicity. It may cause nystagmus in any direction including vertical. Up-beat nystagmus is associated with lesions in the floor of the fourth ventricle, pontine tegmentum, and perhaps anterior cerebellum. Down-beat nystagmus is associated with extrinsic compressive lesion of the foramen magnum.

Cerebellar nystagmus is towards the side of the lesion.

Ataxic nystagmus occurs in internuclear ophthalmoplegia. Although the individual movements of each eye are intact, there is a gaze palsy: on looking to one side the adducting eye does not move and there is nystagmus of the abducting eye. This is caused by damage to the medial longitudinal bundle and is most common in MS. Also seen in brainstem infarction, pontine glioma, Wernicke's encephalopathy, encephalitis, and phenytoin toxicity.

See-saw nystagmus One eye rises and turns in; the other falls and turns out. Seen in parasellar tumours.

Pendular nystagmus There is no rapid phase. It is congenital and caused by poor visual acuity. It is asymptomatic.

Nystagmus
[*OTM* 21.88]

Deafness

Simple hearing tests To establish deafness the examiner should say a number increasingly loudly in one ear whilst blocking the other ear with a finger. The patient is asked to repeat the number. Make sure that failure is not from misunderstanding.

Rinne's test Place the base of a vibrating 256 (or 512) Hz tuning fork on the mastoid process. At the moment when the patient can no longer hear it, move the fork so that the prongs are 4cm from the external acoustic meatus. If it is not heard better there, bone conduction (BC) is better than air conduction (AC) implying conductive deafness. If AC is better than BC, this is either normal or represents sensorineural deafness.

Weber's test Place a vibrating tuning fork in the middle of the forehead. Ask which side the patient localizes the sound to—or is it heard in the middle? In unilateral sensorineural deafness the sound is located to the good side. In conduction deafness the sound is located to the bad side (as if the sensitivity of the nerve has been turned up to allow for poor air-conduction). Neither of these tests is completely reliable.

Conduction deafness Usually due to wax, which should be removed by syringing with warm water after being softened by drops (olive oil); or removed under direct vision. Also: otitis media; otosclerosis.

Causes of sensorineural deafness Ménière's disease; acute labyrinthitis; head injury; acoustic neuroma; central causes including MS; Paget's disease; excessive noise; aminoglycosides; frusemide; lead; several rare congenital syndromes; maternal infections during pregnancy.

Tinnitus

(See *OHCS* p544.)

This is ringing or buzzing in the ears. It is a common phenomenon.

Causes Unknown; hearing loss (20%); wax; viral; presbyacusis; noise (eg gunfire); head injury; suppurative otitis media; post-stapedectomy; Ménière's; head injury; anaemia; hypertension; impacted wisdom teeth.
▶ Investigate unilateral tinnitus fully to exclude an acoustic neurinoma.

Drugs: Aspirin; loop diuretics; aminoglycosides (eg gentamicin).

Objectively detectable tinnitus: Palatal myoclonus; temporomandibular joint problems; AV fistulae; bruits; glomus jugulare tumours (*OHCS* p544).

Psychological associations: Redundancy, divorce, retirement.

Treatment *Psychological support* is very important (eg from a hearing therapist). Exclude serious causes; reassure that tinnitus does not mean madness or serious disease and that it often improves in time. Cognitive therapy is recommended.[1] Patient support groups can help greatly.[2]

Drugs are disappointing. Avoid tranquillizers, particularly if the patient is depressed (use nortryptiline here),[1] but hypnotics at night may be helpful. Carbamazepine has been disappointing; betahistine only helps in some of those in whom the cause is Ménière's disease.

Masking may give relief. White noise is given via a noise generator worn like a post-aural hearing aid. Hearing aids may help those with hearing loss by amplifying desirable sound. Section of the cochlear nerve relieves disabling tinnitus in 25% of patients (but deafness results).

1 L Luxon 1993 *BMJ* i 1490   **2** British Tinnitus Assocn, 105 Gower Street, London W1E 6AH

Deafness
Tinnitus

Weak legs

▶ Cord compression is a medical emergency. It must be considered when there is a rapid progression of leg weakness and/or sphincter failure.

There are many causes of weak legs but only a few cardinal questions: was the onset gradual or sudden? Is the tone spastic or flaccid? Is there sensory loss, in particular a 'sensory level' (a strong clue to spinal cord disease)? Is there any loss of sphincter control (bowels, bladder)?

Sudden weak legs with spasticity
Cord compression—a neurosurgical emergency (disc prolapse at L1–2 or higher, tumours, fractures, epidural abscess, spinal TB, Hodgkin's, myeloma, Paget's disease).
Other causes: MS; post-infectious myelitis; carcinomatous meningitis; cord infarction (due to any vasculitis, eg PAN, syphilis, thrombosis of anterior spinal artery, trauma or compression, dissection of aortic aneurysm, surgery).

Sudden weak legs with flaccidity
Cauda equina compression: This is a neurosurgical emergency. Causes are: tumour; prolapsed disc, canal stenosis; TB. Acute cord trauma; acute cord infarction; myelitis (in early stages with back pain, fever, double incontinence, sensory loss at a defined level surmounted by a zone of hyperalgesia); Guillain–Barré syndrome (p700); lumbosacral nerve lesion; hypokalaemia.

Chronic spastic paraparesis
MS; cord compression (eg cervical spondylosis); syringomyelia; motor neurone disease (MND); subacute combined degeneration of the cord (vitamin B_{12} deficiency) p582.

Chronic flaccid paraparesis
Tabes dorsalis; peripheral neuropathies (p462); myopathies (p470); nerve trauma.

Weak legs with no sensory loss Motor neurone disease

Absent knee jerks and extensor plantars
Friedreich's ataxia; taboparesis (syphilis); MND; subacute combined degeneration of the cord.

Unilateral foot drop
DM; stroke; prolapsed disc; MS; palsy of common peroneal nerve.

Investigations Spinal x-rays. MRI, if available, is a valuable alternative to myelogram. Where cord compression is suspected speedy investigation and treatment are of the essence: myelography, biopsy, then surgery or radiotherapy. FBC, ESR, serum B_{12}, folate, syphilis serology, U&E, LFT, acid phosphatase (prostate carcinoma), protein electrophoresis (myeloma), CXR (TB, Ca bronchus), LP (p412).

Management Depends on cause. Turn totally paralysed patients every 2h; devoted bladder care (OHCS p730).

Drop attacks
Sudden falling to knees without loss of consciousness and without warning. Immediate recovery. Usually stop occurring after a year or two. Cause unknown and probably not a TIA (p436).

Weak legs

Abnormal involuntary movements (dyskinesia)

Tremor (see p58) Chorea (see p40)

Hemiballismus Hemichorea (but often very violent), due usually to stroke (subthalamic lesion) in elderly diabetic patients. Usually recovers over months. Treat as for chorea.

Tics Brief, repeated and stereotyped movements. Often no known cause; may be secondary to drugs; post encephalitis lethargica; post-trauma. Occasionally part of *Gilles de la Tourette's syndrome* (p700). Treat severe tics as for Tourette's syndrome.

Myoclonus Sudden, involuntary jerks. Generalized myoclonus occurs in: (1) certain encephalopathies (mostly progressive due to a variety of metabolic disorders, eg lysosomal storage enzyme defects); (2) epilepsy; (3) *benign essential myoclonus* (often autosomal dominant with onset in childhood). Focal myoclonus may result from viral myelitis. *Treatment:* if epilepsy, use anticonvulsants (p450) (in particular sodium valproate and clonazepam).

Dystonia Prolonged spasms of muscle contraction. Occasionally symptom of neurological disease (eg Wilson's) but usually of unknown cause and without other neurological problems (*idiopathic torsion dystonias*), eg *spasmodic torticollis* in which the head is pulled and held turned to left or right by the contraction of one sternomastoid. Blepharospasm (involuntary contraction of orbicularis oculi) is one of the focal dystonias and is described in detail in OHCS p522. Drug treatment is often ineffective but an oral anticholinergic eg benzhexol, may be tried. An *acute dystonia*, with head pulled back, eyes drawn upward and mouth open (trismus) may occur (particularly in young men) within a few days of starting neuroleptic medication. *Treatment* is with an anticholinergic (eg procyclidine 5–10mg IV).

Oro-facial dyskinesia (tardive dyskinesia) involuntary chewing and grimacing due usually to several years of neuroleptic ingestion especially in elderly. The problem is that the dyskinesia may remain (and indeed worsen) on withdrawal of drug. Long-term neuroleptics (which include the anti-emetics metoclopramide and prochlorperazine) should be avoided where possible. *Treatment:* withdraw neuroleptic and wait 3–6 months. If dyskinesia remains consider tetrabenazine 25–50mg/8h PO.

Akathisia (restless legs) Unpleasant sense of motor restlessness and uncontrollable desire to move). Usually caused by neuroleptics. *Treatment:* stop neuroleptic if possible. Propranolol 40mg/8h PO. Diazepam 2–10mg/24h PO.

Athetosis Slow, smooth, writhing movements. Usually congenital but may result from focal lesions to corpus striatum.

Note: The Akinetic-rigid syndrome is synonymous with 'parkinsonism' except that tremor may be absent. The causes and treatment are as for parkinsonism (p452). This is a *movement disorder* but not a dyskinesia as there are no involuntary movements.

Principal source: D Parkes 1987 *Neurological Disorders*, Springer-Verlag

428

Abnormal involuntary movements (dyskinesia)

[*OTM* 21.218]

Dysarthria, dysphasia, and dyspraxia

Dysarthria Difficulty with articulation due to incoordination or weakness of the articulatory apparatus. Language is normal (see below). *Assessment:* Ask the patient to repeat 'British Constitution', 'baby hippopotamus'. In cerebellar disease the speech is slurred, staccato and scanning in quality. Extra-pyramidal disease results in soft, monotonous voice with slurring of the syllables together. Pseudo-bulbar palsy results in 'Donald Duck' speech (p466). In myasthenia gravis dysarthria may worsen as patient continues to speak.

Dysphasia is a disorder of language causing impaired formulation or comprehension of speech. Assessment: 1 Spontaneous speech: Listen to the patient speak. Does he make sensible and appropriate responses? 2 Comprehension: 'Touch your ear.' Ask for tasks within your patient's capability. Give no non–verbal clues. 3 Repetition: 'Please repeat: "no ifs, ands or buts"'. 4 Writing: if normal, not aphasic (may be mute). 5 Reading. *Classify:* (p431): In an *expressive dysphasia* (lesion in frontal lobe, Broca's area), fluency of speech is lost; it is usually hesitant, circumlocutious, poorly articulated and accompanied by considerable effort and frustration as the patient tries to express himself. Speech may be incorrectly structured with vague words ('the thing') or *paraphasias* (eg 'brush' for 'comb' or 'spoot' for 'spoon'). Reading and writing are usually also impaired but comprehension is relatively intact so that the patient will understand commands and questions and attempt to convey meaningful answers. In a *receptive dysphasia* (superior temporal or parietal lobe lesion), speech may be fluent, well-articulated and grammatically correct but the content is oddly empty. The patient does not appear to attempt to answer questions appropriately and does not understand his own errors of speech. In *conduction dysphasia* the site of the lesion is the structural link between auditory cortex in the superior temporal gyrus and Broca's area, so there is very poor repetition though comprehension and fluency are relatively preserved. *Treatment:* Speech therapy is of some use in Broca's and conduction dysphasias but less so in stroke. Following a stroke, speech may continue to improve long after motor disabilities cease to change.

Dyspraxia is the impairment of performance of complex movements despite intact individual movements, sensation and coordination—eg *dressing dyspraxia* (especially with lesions in the non-dominant hemisphere); *constructional dyspraxia* (difficulty in assembling or drawing—lesion in non-dominant hemisphere but also seen in hepatic encephalopathy—have the patient draw a 5-pointed star); *gait dyspraxia* (especially with bilateral frontal lesions, lesions in the posterior temporal region and in hydrocephalus. Common in the elderly and may be the explanation of a gait disorder even though the individual components involved in walking is unimpaired).

Dysarthria, dysphasia, and dyspraxia
[OTM 21.34]

Classification of dysphasias

Type	Lesion	Spontaneous speech	Compre-hension	Repeti-tion	Common causes
Broca ('expressive')	Broca's area (lateral frontal lobe)	non-fluent	+	±	stroke tumour
Global	Huge lesion (Broca's and Wernicke's)	non-fluent	−	−	stroke tumour
Wernicke ('receptive')	Wernicke's area (posterior superior temporal lobe)	fluent	−	−	stroke tumour
Conduc-tion	parietal lobe (near Sylvian fissure)	fluent	+	−	stroke
Anomic (affecting naming)	angular gyrus	fluent	+	+	dementia head injury ICP↑

NB: left hemisphere dominant (ie controls language) in 99% of right-handed and 60% of left-handed people.

Principal source: N Geschwind 1971 *NEJM* **284** 654

Stroke: clinical features and investigations

A stroke results either from ischaemic infarction of part of the brain, or from intracerebral haemorrhage. It is manifest by rapid onset (minutes) of signs and symptoms of focal brain damage. It is the major neurological disease of our times. The annual incidence is 1.5/1000 overall, rising rapidly with age to 10/1000 at age 75yrs. Men are at only slightly greater risk than women. The *immediate causes* are: athero-thrombo-embolism (from artery); emboli from heart; intracerebral haemorrhage (rupture of aneurysm); occasionally hypotension (<50mmHg); vasculitis; thrombophlebitis.

Risk factors Notably hypertension. Also: smoking; DM; heart and peripheral vascular disease; AF; previous TIA; PCV↑; carotid bruits; contraceptive steroids; obesity; hyperlipidaemia; high plasma fibrinogen; excessive alcohol intake.

Clinical features Usually sudden onset of signs/symptoms. Sometimes step-wise progression over hours, even days. In theory, focal damage relates to distribution of affected artery (p404), but collateral supplies cloud the issue. *Cerebral hemisphere infarcts* (50%) may cause any of: contralateral hemiplegia (usually begins as flaccid—the limb falls as a dead weight if raised and then released—before spasticity sets in); sensory loss; homonymous hemianopia; dysphasia. Motor signs are upper motor neurone. *Lacunar syndromes:* (25%) small infarcts around basal ganglia, thalamus and pons. The patient is conscious. May cause pure motor, pure sensory, mixed motor and sensory signs, or ataxia. *Brainstem infarction:* (25%) wide range of effects which include quadriplegia, disturbances of gaze and vision, locked-in syndrome (aware, but unable to respond).

Tests Prompt investigation may prevent future stroke, or wrong diagnosis but consider carefully how results will affect management. Look for:

- *Hypoglycaemia* and *hyperglycaemia.*
- *Atrial fibrillation* (p290): If present, the stroke may have been caused by emboli from the heart. Look for a large left atrium on CXR. Consider echocardiography. Consider anticoagulation (but first exclude cerebral haemorrhage with CT scan).
- *Hypertension:* Look for hypertensive retinopathy (p684) and large heart on CXR. NB: raised BP may be acute result of stroke.
- *Giant cell arteritis* (p675) is suggested by raised ESR, and history of temporal headaches or tenderness. Give steroids promptly (p675).
- *Syphilis:* Look for active disease only (p218).
- *Thrombocytopenia* and other bleeding disorders.
- *Polycythaemia* (p612).
- *Embolic stroke* (other than AF), eg after myocardial infarct (mural thrombus) or other cardiac disease (eg diseased valves). Consider expert cardiological assessment ± anticoagulation (after CT scan).
- *Carotid artery stenosis:* In carotid territory (p404) stroke/TIA, 2 large randomized trials[1,2] have clearly established the benefit of carotid surgery, so expert bodies now affirm that >70% stenoses (on Doppler screening) merit angiography ± surgery—in fit patients.[3]

Stroke: clinical features and investigations
[OTM 21.164]

The minimum investigation to exclude preventable causes is:

Pulse and BP	FBC, platelets	U&E
Listen for carotid bruit	ESR	Blood glucose
ECG	Sickling tests in blacks	Syphilis serology
CXR	Carotid Doppler	

Then consider: CT scan; echocardiogram; cerebral angiogram; clotting screen (include antithrombin III deficiency, protein C or S deficiency, lupus anticoagulant); SLE serology.

433

1 MRC European trial 1991 *Lancet* **337** 1235   2 North American Trial 1991 *NEJM* **325** 445   3 Association of British Neurologists (M Brown) 1992 *BMJ* ii 1071

Prevention and management of stroke

Some indications for MRI or CT scan
- Uncertain diagnosis—if onset slow or not known. Consider especially cerebral tumour and subdural haematoma.
- To distinguish between haemorrhage and ischaemic infarction (do scan within 2 weeks of stroke), eg if considering anticoagulation.
- Uncharacteristic deterioration after 24h.
- Cerebellar stroke—cerebellar haematoma needs urgent surgery.

Differential diagnosis Cerebral tumour; subdural haemorrhage (p440), Todd's palsy (p710). Consider drug overdose if unconscious. Pointers to haemorrhagic stroke are: history of hypertension, meningism, severe headache, and coma within hours. A pointer to ischaemic infarction is previous TIA. However, these two types of stroke are usually clinically indistinguishable.

Management of strokes (see p432 for *tests* and p68) Explain to patient and relatives what has happened. Consider the kindest level of intervention taking into account the quality of life, coexisting conditions and prognosis. No drug is of proven value. The only thing known to save lives is admission to a stroke unit for skilled nursing and physiotherapy.[1] Decide how far to go down this list.
- Maintain hydration, but take care not to overhydrate (cerebral oedema).
- Nil by mouth if there is no gag reflex.
- Turn regularly and keep dry (consider catheter) to stop bed sores.
- Monitor BP but treat only if high (diastolic >130mmHg consistently) and even then do not lower to below 110mmHg diastolic in first 48h.
- If conscious level impaired consider treating cerebral oedema with glycerol 10% 500ml IVI (infuse over 4h) daily for 6 days.
- If cerebral haemorrhage suspected consider immediate referral for evacuation (familiarize yourself with local current management).
- Consider naftidrofuryl 100mg/6h IV then PO for 6–12 months.[2]
- Consider heparin infusion (see p92) for 'stroke in evolution' (eg hemiplegia extending from arm to leg).

Secondary prevention (ie to prevent further strokes.) Control risk factors. Consider aspirin (p436) or warfarin if stroke embolic or patient in chronic AF.

Prognosis Mortality: 20% at one month, 5–10% per annum thereafter. 40% complete recovery. Drowsiness, head and eye deviation and hemiplegia betoken poor prognosis.

Complications Pneumonia; depression; contractures; constipation; bedsores.

Primary prevention Good medical care must be directed at preventing strokes. This means controlling risk factors (p432) and in particular hypertension, smoking, DM and hyperlipidaemia. Lifelong anticoagulant treatment for those with rheumatic or prosthetic heart valves on left side. Consider anticoagulation in patients with chronic AF.[3] For prevention following TIA see p436.

---

1 GA Donnan 1993 *Lancet* ii 384   **2** T Steiner 1986 *Neuroepidemiol* **5** 121   **3** P Peterson 1989 *Lancet* i 175

Prevention and management of stroke

[*OTM* 21.164]

Transient ischaemic attacks (TIA)

These are focal CNS disturbances caused by vascular events such as micro-emboli and occlusion leading to ischaemia. By definition, symptoms last <24h and there are no CNS sequelae. They are harbingers of strokes and myocardial infarcts.

Clinical Abrupt onset of focal CNS features fading over minutes or hours to give full recovery in 24h. Ischaemia in *carotid territory* may cause: contralateral weakness/numbness/paraesthesiae; dysphasia; dysarthria; homonymous hemianopia; amaurosis fugax (p684) in ipsilateral eye lasting few seconds. Ischaemia in *vertebrobasilar territory* may cause: hemiparesis; hemisensory loss; bilateral weakness or sensory loss; diplopia; cortical blindness; vertigo; deafness; tinnitus; vomiting; ataxia; drop attacks.

Causes Athero-thrombo-embolism (from carotid or heart). Arteritis (eg giant cell arteritis, PAN, SLE, syphilis); arterial trauma; haematological causes (eg PRV, sickle-cell).

Differential diagnosis of brief focal neurological symptoms *Migraine* (symptoms spread whilst slowly intensifying, often include visual scintillations followed by headache); *epilepsy* (symptoms usually spread and often include twitching and jerking); *malignant hypertension*; *hypoglycaemia*; MS; *intracranial lesions*; *hysteria*.

Investigations Aimed at: finding cause of TIA; excluding other causes of symptoms; defining risk factors for vascular disease. See also p434. FBC, ESR, platelets, U&E, blood glucose, CXR, ECG, syphilis serology. Consider further cardiological investigations, carotid Doppler and angiography, CT scan. *Note:* echocardiography rarely helps in revealing cardiac causes without suggestive signs.[1]

Treatment
- Control risk factors for stroke (p432), cautiously reduce BP to 100mmHg diastolic.
- Control risk factors for MI (p284); this is the commonest cause of death following TIA.
- Consider antiplatelet drugs, especially aspirin 300mg daily (many experts believe 75mg daily is adequate) for life (not if peptic ulceration). If indigestion, reduce dose (but not below 75mg). The use of aspirin and its dosage is controversial, but evidence suggests it reduces non-fatal stroke and myocardial infarction by 25%, and vascular death by 15%.
- Anticoagulant drugs have a limited role to play. Consider particularly in some heart disease (eg mitral stenosis, recent major septal myocardial infarct) and frequent TIAs not controlled on antiplatelet drugs.
- Carotid endarterectomy should only be considered in patients with TIA in carotid distribution (p404) and where there is stenosis at the origin of the internal carotid artery. If internal carotid stenosis >70%, endarterectomy is probably worthwhile.[2]

Prognosis The commonest cause of death following TIA is myocardial infarction (MI). The risk of stroke and of MI are about 5% per yr each. Mortality is about three times that of a TIA-free matched population.

1 Canadian Cooperative Study *NEJM* 1978 **299** 53; C Warlow 1987 *OTM* 2 ed; D Parkes 1987 *Neurological Disorders*, Springer-Verlag 2 North American Trial *NEJM* 1991 **325** 445

Transient ischaemic attacks (TIA)
[*OTM* 21.157]

Subarachnoid haemorrhage

This is bleeding into the subarachnoid space.

Causes *Ruptured aneurysms* 80% (usually congenital Berry aneurysms, occasionally mycotic aneurysms from endocarditis). *Arteriovenous malformation* 5%. *Bleeding tendency* (eg anticoagulation) very rare. *Trauma. No cause found* 15% (good prognosis).

Berry aneurysms 15% are multiple. Common sites are: the junction of posterior communicating artery with internal carotid; the anterior communicating artery; and, the middle cerebral artery. Genetic influence suggested. Skin biopsy may help identify relatives at risk by demonstrating type III collagen deficiency.[1] Associated with polycystic kidneys, Ehlers–Danlos syndrome, coarctation of aorta.

Clinical features There is a sudden severe headache ('my worst headache ever' or 'as if I had been hit on the head'). Faintness and vomiting common at onset.

Signs	Mortality
None	0%
Neck stiffness and cranial nerve palsies	11%
Drowsiness	37%
Drowsy with hemiplegia	71%
Prolonged coma	100%

Almost all the mortality occurs in the first month. Of those who survive the first month 90% survive a year or more.

Warning leaks These are fairly common (seen before about 20% of major bleeds), and it would be most helpful if they could be identified, so that suitable patients could be operated on while in good condition (prognosis in serious bleeds is much worse if there has been a warning bleed). So let the diagnosis enter your mind in a sudden hemicranial, hemifacial or periorbital headache, particularly if associated with neck or back pain.[2] In some headaches precipitated by sexual intercourse some neurologists recommend a CT scan, and, if normal, an LP >12h after the event with a spectrophotometric search for xanthochromia, to exclude subarachnoid haemorrhage.[3]

Rebleeding is a common mode of death in those who have had a subarachnoid. Rebleeding occurs in 30% of patients, often in the first few days—or around the 12th day. Vascular spasm follows a bleed, often causing ischaemia and permanent neurological deficit. If this happens surgery is not helpful at the time but may be so later.

Investigations The knowledge that CT scans may miss small bleeds has now restored LP to its position as an important investigation (but it will usually be wise to do a CT scan first, to rule out space-occupying lesions). If CSF is yellow or red (after an atraumatic tap), do angiography promptly, if surgery is a possibility.[2] NB: ECG changes can occur (ST elevation; T-wave inversion; prominent V-waves; ventricular and atrial arrythmias).

Subarachnoid haemorrhage
[OTM 21.168]

Treatment Early consultation with neurosurgical unit important. If patient conscious and little deficit, angiography (to show aneurysm) followed by surgery (to clip aneurysm) reduces risk of rebleeding. In very sick patients CT scan may reveal intracerebral haematoma—consider evacuation. Medical treatment involves cautious control of severe hypertension, analgesia for headaches, sedation and bed rest for about 4 weeks. Nimodipine (60mg/4h PO eg for 2–3 weeks, or 1mg/h IVI), a calcium antagonist, is widely used to counter vascular spasm.[4] Management varies considerably between neurosurgical centres. Early surgery runs risk of spasm; but a fatal bleed may occur if surgery delayed.

1 M Venning 1981 *BMJ* **ii** 824   **2** J Ostergaard 1990 *BMJ* **ii** 190   **3** S Ellis 1991 *BMJ* **ii** 421
4 J Pickard 1989 *BMJ* **298** 636

Subdural haemorrhage

▶ *Consider in all whose conscious level fluctuates, and those who seem to have a slowly evolving stroke.*

A subdural haemorrhage may be insidious. Bleeding is from bridging veins between cortex and venous sinuses. The elderly are particularly susceptible because brain shrinkage makes these veins vulnerable. The trauma may not be recalled. Others at risk are those with epilepsy, and those who drink excess alcohol.

Symptoms Be alerted by a fluctuating level of consciousness (present in 35% of cases), and seek history of trauma (50% of cases). Also: headache; localizing neurological symptoms; personality change.

Signs Localizing neurological signs (p456); spasticity and hyperreflexia; signs of raised intracranial pressure (p734); muscle weakness; unequal pupils.

NB: focal signs may develop long after the injury (63 days average).

Investigations CT scan.

Differential diagnosis Evolving stroke; cerebral tumour; dementia.

Treatment Clot evacuation may lead to complete recovery.

Extradural haemorrhage

This results from arterial bleeding between the bone and the dura following head injury and is usually at the site of a skull fracture. It is particularly common following trauma to the temples just lateral to the eyes. Suspect if following head injury conscious level deteriorates and pupil becomes dilated and unresponsive to light (ipsilateral to haemorrhage). Immediate evacuation of the clot through burr holes to skull (on the side of the dilated pupil) may be necessary.

Encephalitis

▶ *Herpes simplex encephalitis needs urgent treatment* (p203).

This can present in many ways, depending partly on the organism. There is no absolute distinction between meningitis and encephalitis, the difference being the relative involvement of inflammation of the meninges and the brain itself. Typically there are fever, symptoms of meningitis (p442), and signs of raised ICP (p734), together with altered consciousness, convulsions and sometimes focal neurological signs and psychiatric symptoms. Viruses are the commonest cause including *Herpes simplex encephalitis* (HSE) (p203); *post-infectious encephalitides* (occurring up to 14 days after infection with measles, varicella zoster, rubella, mumps, influenza) with involuntary movements, cranial nerve lesions, nystagmus and ataxia common. Measles may also cause *subacute sclerosing pan-encephalitis* (SSPE) about 6yrs after original infection with slow deterioration leading usually to death. Diagnosis of encephalitis is made from knowledge of epidemics, specific clinical features (eg temporal lobe features such as lobe swelling seen on CT scan and periodic complexes on EEG in HSE), associated infection (eg mumps); demonstration of virus in CSF, throat swabs, stool culture; MRI or CT scans. Note: *HSE has a specific treatment; if the diagnosis is in doubt treat for HSE, p203.*

Neurology

Subdural haemorrhage
Extradural haemorrhage
Encephalitis
[*OTM* 21.184]

Other causes: Polio; rabies; HIV. Bacterial causes (notably *Staph aureus*), usually secondary to pus elsewhere, lead to *brain abscesses* which are space-occupying lesions. *Neurosyphilis* is a non-pyogenic bacterial encephalitis.

441

Meningitis

▶ Meningitis is an infection of the pia mater and arachnoid. It is one of the great killers, and it kills quickly. If your patient could have meningitis, you must do a lumbar puncture (p412) to make an accurate diagnosis. But time is not on your side: so if you think your patient is already dying from meningitis *nothing* should delay prompt 'blind' treatment (p444). Delays from ward rounds, waiting for expert advice, or for the result of investigations may well be fatal. ▶▶ *Act now*—and don't worry about future criticism from purists. You may even end up with a live patient to take your side.

Clinical features Rapid onset (<48h) of:
- *Meningeal signs:* Headache (eg accentuated by jolting); stiff neck on passively moving chin toward chest; Kernig's sign may be +ve: pain and resistance on passive knee extension when hips fully flexed. Brudzinski's sign (hips flex on bending head forward); opisthotonus.
- *ICP ↑ signs:* Headache, irritable, drowsy, vomiting, fits, pulse↓, consciousness↓, coma, irregular respirations. Papilloedema (late).
- *Septic signs:* Fever, arthritis, odd behaviour, rashes—any, petechiae suggest meningococcus, shock, DIC, pulse↑, tachypnoea, WCC↑.

Predisposing factors *Crowding*, poverty and malnutrition. *Head injury* (especially basal skull fractures), usually occurring within 2 weeks. *Sinusitis*; *mastoiditis*; *otitis media. Immunosuppression* (eg carcinoma; AIDS; splenectomy; sickle-cell disease). DM. *Pneumonia. Alcoholism.* CSF *shunts.*

Differential diagnosis Acute infection of any kind (LP will distinguish). Acute encephalitis (but signs of neurological damage more marked than meningism). Subarachnoid haemorrhage (but sudden onset, blood in CSF).

Tests ● LP *is the key test*, and should be done at once unless signs of ICP↑ (p734), or focal CNS signs (in which case do CT scan 1st), or suspected meningococcaemia (epidemic area, petechial rash, hypotension, rapid progression) when benzylpenicillin must be given urgently. Note that an early LP in pyogenic meningitis may be normal. Do not hesitate to repeat LP if signs persist. Send 3 bottles of CSF (4 drops in each) for: *urgent* Gram stain; culture; virology; glucose. Normal CSF values: p412 & p801.
- CT scan, if diagnosis is in doubt and raised intracranial pressure is suspected, or there are focal neurological signs. In these cases CT scan should *precede* LP as the point is to rule out space-occupying lesions.
- Blood for culture. ● Plasma glucose.
- FBC; U&E. ● Skull x-ray if history of head injury.
- Culture urine, nose swabs (and stool, for virology)

CSF in meningitis	Pyogenic	Tuberculous	Aseptic
Appearance	often turbid	often fibrin web	usually clear
Predominant cell	polymorphs eg 1000/mm³	mononuclear eg 10–350/mm³	mononuclear 50–1500/mm³
Glucose	<²⁄₃ plasma level	<²⁄₃ plasma level	>²⁄₃ plasma level
Protein (g/l)	1–5	1–5	<1.5
Organisms	in smear and culture	often absent in smear	not in smear or culture

Meningitis

Meningitis: organisms and treatment

▶ *Prompt antibiotics save life.* Give penicillin (1.2g IV) *before* **admission. If, on LP, pus emerges from the needle, start a broad-spectrum agent (eg cefotaxime, below) at once while awaiting its analysis.** ▶ *Liaise early with your senior and a microbiologist, but don't wait for sensitivities to be known.*

Causative organisms The most likely organisms causing a pyogenic meningitis is influenced by age: *Neonates: E coli;* β-haemolytic *streps. Children: Haemophilus influenzae* if <4yrs old and unvaccinated; *Meningococcus* (ie *Neisseria menigitidis*); TB. *Young adults: Meningococcus. Older adults: Pneumococcus (Streptococcus pneumoniae) Elderly (or immunocompromised): Pneumococcus; Listeria;* TB; Gram negative organisms, *Cryptococcus.*

The antibiotics of choice[1] (alternatives in brackets).
Pneumococcus: Cefotaxime (or ceftriaxone) ± vancomycin.
Meningococcus: Benzylpenicillin (or cefotaxime or ceftriaxone).
H influenzae: Cefotaxime (or ceftriaxone—or chloramphenicol, but in some areas >50% of isolates are chloramphenicol-resistant); give rifampicin for 4 days before discharge in type b infections.
E coli: Cefotaxime (or ceftriaxone). *Staph aureus:* Cloxacillin (p.177).
Listeria monocytogenes: Gentamicin + ampicillin (or penicillin G; see p222.)
M tuberculosis: Rifampicin *and* ethambutol *and* isoniazid *and* pyrazinamide. NB: the first few LPs may be normal. See OHCS p261.
Cryptococcus neoformans: Amphotericin B + flucytosine (p340, p244).

Treating pyogenic meningitis[1] (See OHCS for children's doses.) Send blood and CSF for culture before treating unless clinical evidence of meningococcal meningitis—or LP is contraindicated by CT (do CT if: consciousness↓, focal signs, headache is *very* bad, or frequent seizures[2]).
● Set up IVI. Dose examples of the drugs above: Ampicillin 1–2g/4h IV. Ceftriaxone 2–4g/day by IV (over >4min, 1 daily dose, p178). Cefotaxime 2–4g/8h IVI; ↓doses in renal failure; see p178 + Datasheet. Benzylpenicillin 2.4g/4h slowly IV.
● Treat shock with plasma IVI until BP >80mmHg systolic and urine flows (p720)—as guided by CVP on ITU. If ICP↑, ask a neurosurgeon's advice: is dexamethasone indicated?[1] Diamorphine may be needed for *severe* pain (eg 5mg/4h IM); for nausea, try domperidone (eg 60mg/8h PR).
● Once organism is known, use most effective antibiotic (discuss with microbiologist). At what dilution does patient's blood (which contains the active antibiotic) inhibit organism's growth *in vitro*?
● Continue parenteral antibiotics for 10 days. Follow with rifampicin to eliminate nasal carriage (dose as for contacts, below).
● Look for complications: cerebral oedema; cranial nerve lesions; deafness; cerebral venous sinus thrombosis (p477); mental retardation.

Meningococcal prophylaxis Offer to: ● Household/nursery contacts (ie within droplet range). ● Those who have kissed the patient's mouth. ● Yourself (possibly: risk is remote). Give rifampicin (600mg/12h PO for 2 days; children >1yr 10mg/kg/12h; <1yr 5mg/kg/12h) or ciprofloxacin (500mg PO, one dose—adults only). Neither drug is guaranteed in pregnancy, but such a short course is unlikely to harm and is recommended.[3] Vaccination (Mengivac®) protects from group A and C strains and *may* be used for school contacts (if >2yrs) or for travellers to endemic areas.

'Aseptic' meningitis *Infective causes:* viruses (eg echovirus, mumps, coxsackie, herpes simplex and zoster, HIV, measles, 'flu); atypical TB; syphilis; Lyme disease; parameningeal infection; leptospirosis; listeria; brucella; partially treated bacterial meningitis; fungi. *Non-infective causes:* SLE; sarcoid; Behçet's. *Pointers to non-viral cause:* ↓CSF glucose. Treatment of viral meningitis: supportive (but see also *Herpes* infections, p202–3).

1 A Tunkel 1995 *Lancet* **346** 1675   2 *JNNP* 1993 **56** 1243–58   3 *Drug Ther Bul* 1990 **28** 34

Acute confusional state

▶ *Any unexplained behaviour change in a hospital patient is likely to be part of an acute confusional state.*

Clinical features The central clinical feature is impaired conscious level with onset over hours or days. Impairment of consciousness is difficult to describe; take any opportunity to be shown patients who demonstrate it. Essentially you have the sense when trying to talk to the patient that he or she is not really with you. The patient is likely to be disoriented in time (does not know the day or year—or the time of day) and, with greater impairment, in place. The clinical presentation is varied. Sometimes the patient becomes unusually quiet, and is found to be drowsy; sometimes s/he appears agitated, and you are called because s/he is disrupting the ward. On other occasions the patient appears deluded (for example accusing staff of plotting against him/her) or to be hallucinating (either seeing things or hearing voices which are not there). In the latter two cases the question of a primary mental illness (eg paranoid state, schizophrenia) is raised. However, if there is no past psychiatric history, and in the setting of a general hospital (ie the patient is physically ill or has had recent surgery), mental illness is rare and acute confusional state is common. A confusional state is particularly likely if the symptoms are worse in the late afternoon and at night.

Differential diagnosis If agitated consider an anxiety state (usually readily distinguished by talking to the patient). If the onset is uncertain consider dementia.

Causes Virtually any physical disease and any drug. However, by considering the following general categories you will find most causes:
- Infection (chest; UTI; surgical wound; IVI sites)
- Drugs (especially analgesics and sedatives)
- Alcohol withdrawal (2–5 days after coming into hospital; LFT, history suggestive)
- Metabolic disorder (check U&E and glucose; common if IVI)
- Hypoxia (consider measuring arterial blood gases)
- Vascular (stroke, myocardial infarction)
- Consider head injury (especially subdural haematoma)

Management
1 Find the cause. Review notes; examine patient with above causes in mind; do necessary investigations. Start relevant treatment. *Note:* if hypoxic give oxygen.
2 If agitated and disruptive, sedation may be necessary, sometimes before examination and investigations are possible. Use major tranquillizers (eg haloperidol 5–10mg IM/PO; or, chlorpromazine 50–100mg IM/PO). Wait 20mins to judge effect. Give further dose as necessary. *Note:* in alcohol withdrawal use chlormethiazole (p682).
3 Nurse ideally in *moderately lit* quiet room with same staff in attendance (to minimize confusion). Repeated reassurance and orientation. In practice a compromise between a quiet room, and a place where staff can keep under close surveillance has to be made.

Acute confusional state

[*OTM* 21.48; 25.13]

447

Epilepsy: diagnosis

Epilepsy is a tendency to spontaneous, intermittent, abnormal electrical activity in part of the brain, manifest as seizures. They may take many forms: for a given patient they tend to be stereotyped.

Clinical There may be a *prodromal* period lasting hours or days preceding the seizure. This is not part of the seizure itself. The patient or friends may notice a change in mood or behaviour. The presence of a prodromal period adds weight to a diagnosis of epilepsy, but does not help distinguish the type. An *aura*, which is itself part of the seizure, may precede its other manifestations. The aura may be a strange feeling in the gut, or a sensation or an experience such as *déjà vu* (a disturbing sense of familiarity). It implies a partial seizure, often, but not necessarily, temporal lobe epilepsy (TLE). After a partial seizure involving the motor cortex (Jacksonian convulsion) there may be temporary weakness of the affected limb(s) (Todd's palsy).

Diagnosis Decide first: is this epilepsy (cf p418)? A detailed description from a witness of the fit and its setting is vital. Decide next what type of epilepsy. The onset of the attack is the key to the major question: partial or generalized? If the fit begins with focal feature, it is a partial seizure however rapidly it is generalized. Ask next: what if anything brings it on (eg flickering light (TV) and alcohol)? Can this be avoided (TV-induced seizures—almost always generalized—rarely require drugs)? Then decide what the appropriate drug is. Start at low dose. Gradually (eg over 8 weeks) build up until fits controlled. Persist until drug at maximum dose before considering a change of therapy (p450).

Classification

1 *Partial seizures:* (Signs and symptoms referrable to a part of one hemisphere suggesting structural disease). (a) *Elementary symptoms* (consciousness not impaired eg focal motor seizures); (b) *Complex symptoms* (consciousness impaired eg olfactory aura followed by automatism). Usually TLE; (c) *Becoming generalized:* evidence of focal origin before generalized seizure.

2 *Generalized seizures:* (No features referrable to only one hemisphere). (a) *Absences (petit mal):* brief (10sec) pauses (eg stops talking mid sentence, carries on where left off). Sudden onset and termination; (b) *Tonic–clonic* (classical *grand mal*). Sudden onset, loss of consciousness, limbs stiffen (tonic) then jerk (clonic). Drowsy afterwards. (c) *Myoclonic jerk* (eg thrown suddenly to ground). (d) *Clonic* (alone). (e) *Tonic* (alone). (f) *Atonic* (becomes flaccid). (g) *Akinetic*. Note also *infantile* spasms.

3 *Unilateral seizures* (tonic, clonic or tonic–clonic).

Causes (Often none is found) Trauma; tumour; alcohol withdrawal. *Physical:* Space occupying lesions, stroke, raised BP, tuberous sclerosis, SLE, PAN, sarcoid, and vascular malformations. *Metabolic causes:* Hypoglycaemia, hyperglycaemia, hypoxia, uraemia, hypo- and hypernatraemia, hypocalcaemia, liver disease, and drugs (eg phenothiazines, tricyclics, cocaine or withdrawal of benzodiazepines). *Infections:* encephalitis, syphilis, cysticercosis, HIV.

Epilepsy: diagnosis
[*OTM* 21.60]

Epilepsy: investigation and treatment

▶ *Exclude cardiac causes of unexplained neurological events.*

Investigation of patient presenting with seizure for first time This is controversial. A seizure with focal onset, or with focal signs on examination warrants investigation with CT scan and, possibly, EEG. Childhood seizures warrant EEG (to identify 3/second spike and wave activity—this is common and is treated with sodium valproate). If no suggestion of focus, opinions differ as to how far to take investigations. Consider: EEG, FBC, ESR, U&E, Ca^{2+}, LFT, CXR, plasma glucose, syphilis serology, ECG.

Therapy Treat with one drug (and by one doctor) only. Slowly build up dose (over 2–3 months) until seizures controlled, or toxic effects manifest, or maximum plasma level (p797) or drug dosage reached. Beware drug interactions (consult formulary). Most specialists would not recommend treatment after one fit but would start treatment after two.

Carbamazepine, sodium valproate and phenytoin are first-line treatment. All are equally effective for partial seizures. Sodium valproate is drug of choice for generalized seizures.

Commonly used drugs
Carbamazepine: Start 100mg/12h PO. A slow release form is available,[1] which is useful if intermittent side-effects experienced when dose peaks. Toxic effects: rash, nausea, diplopia, dizziness, fluid retention, hyponatraemia, blood dyscrasias.
Phenytoin: Start 200 mg/24h PO. Toxic effects: rash, sedation, memory and behaviour disturbance, gum hypertrophy, acne, facial coarsening, blood dyscrasias, folate deficiency, cerebellar syndrome.
Sodium valproate: Start 200mg/12h PO. Max dose 30mg/kg/24h. Monitor liver function. Give after food. Toxic effects: sedation, tremor, weight gain, hair thinning, ankle swelling, hyperammonaemia (causing encephalopathy), liver failure.

Other drugs
Ethosuximide (for absences, first choice in infants); vigabatrin, or lamotrigine[2] (as second-line treatment for partial seizures); phenobarbitone; benzodiazepine.

Changing from one drug to another Indications: on inappropriate drug; side-effects unacceptable; treatment failure. Introduce new drug at its starting dose; slowly increase to middle of its therapeutic range. Only then start to withdraw old drug over about six weeks.

Management of uncontrolled epilepsy Review diagnosis. If epilepsy, decide which type. What is appropriate drug? Has it been tried to its maximum dose? If not introduce appropriate drug slowly and remove other drugs slowly (as above). The patient should be on only one drug. If seizures not controlled change to second most appropriate drug (as above). Only consider maintenance on two drugs if all appropriate drugs tried singly at maximum dose. Consider one of the new anticonvulsants, vigabatrin or lamotrigine. Other refractory epilepsies sometimes amenable to surgery.

Driving and jobs (p418)

1 S Ryan 1990 *Arch Dis Chi* 65 930–5   2 M Brodie 1992 *Prescri J* 32 21–6

Stopping anticonvulsants

80% of patients who develop epilepsy will enter prolonged remission with drug therapy—and there is no sure way of finding out for how long treatment is needed. Unnecessary use of anticonvulsants is a problem because of their many side-effects. Impaired cognition and memory are very real problems[1]—but so is the risk to employment and driving prospects if seizures return. A key question, therefore, is when to try without drugs.

Discuss the risks and benefits with patients, to enable them to make informed choices.

At 2-year follow-up in the large (*n*=1013) Medical Research Council (UK) randomized study of withdrawal of anticonvulsants in those who had had no fits for 2 years, 59% of those who stopped remained seizure-free, compared with 78% in those who continued.[2] In this study, the 3 main predictors of remission were: ● Longer seizure-free periods. ● Previous use of only one antiepileptic drug. ● Absence of tonic–clonic seizures. In general, prognosis for remission from seizures is better in patients with less severe epilepsy who lack evidence of structural CNS abnormalities on examination or investigation (EEG and CT scan).

The best way to withdraw anticonvulsants is unknown, but the Medical Research Council used decrements every 4 weeks as follows:

	Adults	*Children*
Phenobarbitone	30mg	1mg/kg
Phenytoin	50mg	1.5mg/kg
Carbamazepine	100mg	3mg/kg
Valproate	200mg	6mg/kg
Primidone	125mg	4mg/kg
Ethosuximide	250mg	4mg/kg

1 JS Duncan 1990 *Epilepsia* **31** 584–91 2 *Lancet* 1991 i 1176

Parkinsonism & Parkinson's disease

Parkinsonism is a syndrome of *tremor*, *rigidity* and *bradykinesis*.

Tremor: 4–6Hz. Most marked at rest; coarser than cerebellar tremor. Seen as 'pill rolling' of thumb over fingers.

Rigidity: Limbs resist passive extension throughout movement (*lead-pipe rigidity*, not spasticity which is clasp-knife rigidity). *Cogwheel rigidity* means combined rigidity and tremor, best elicited by extending the flexed forearm, or by pronation/supination movements.

Bradykinesis: Slow in initiating and executing movement and speech. Expressionless, mask-like face. The gait is characterized by a shuffling forwards with flexed trunk as though forever a step behind one's centre of gravity (*festinant gait*).

Parkinson's disease (PD) is one cause of parkinsonism. Symptoms usually start between 60 and 70 years old, are progressive and may lead to dementia. PD is due to degeneration of dopaminergic neurones in the substantia nigra. The cause of the degeneration is unknown. There is current interest in toxins: injecting a synthetic substitute for meperidine known as MPTP (related to heroin) can cause parkinsonism.

Management: Assess disability and cognition regularly and objectively (eg time how long to walk 20 yards; can patient dress alone; can he or she turn over in bed?). Choose lowest dose of drug to give adequate symptom relief without troublesome side-effects. Avoid Zimmer frames (flow of movement is spoilt) unless fitted with wheels and handbreak.[1]

Drugs:

1 L-dopa 100mg/12h PO, increasing to 100mg/8h after few days. Increase dose slowly to ≤1000mg/24h. Ensure adequate dose of peripheral dopa-decarboxylase inhibitor (≥25mg/100mg L-dopa). Compromise between L-dopa's side-effects (unwanted movements, nausea) and improved mobility.

2 Selegiline 10mg/24h PO may be given from the outset in the hope of slowing disease progression.[2]

3 Over the years therapy may become less effective and patients tend to switch between periods of immobility and periods of exaggerated involuntary movements ('on–off effect'). This problem may be ameliorated by: slow-release L-dopa; frequent (every 2h) small doses of L-dopa; and dopaminergic agonists such as bromocriptine.

4 Be alert to depression in patient and stress on carers.

5 Anticholinergic drugs (below) may be useful adjunct if tremor.

6 Subcutaneous apomorphine (an injectible dopamine agonist) may help patients with severe on–off effect.[3]

7 Neural transplants may be a drug delivery system in the future.

Causes of parkinsonism *Drugs* (neuroleptics including metoclopramide and prochlorperazine); *arteriosclerosis*. Rarely *postencephalitis*; *supranuclear palsy* (Steel–Richardson–Olszewski syndrome with absent vertical (both upward and downward) gaze and dementia); *Shy–Drager syndrome* (orthostatic hypotension, atonic bladder); *carbon monoxide poisoning*; *Wilson's disease*; *communicating hydrocephalus*

Treatment of drug-induced parkinsonism Preferably by reducing or stopping the drug. If this is not possible (eg schizophrenia needing drugs) add anticholinergic drug (eg procyclidine 2.5mg/8h PO).

Parkinsonism & Parkinson's disease

Drugs combining L-dopa and dopa-decarboxylase inhibitors

Trade name	L-dopa	Benserazide	Carbidopa
Madopar 62.5®	50mg	12.5mg	
Madopar 125®†	100mg	25mg	
Madopar 250®	200mg	50mg	
*Sinemet-110®	100mg		10mg
*Sinemet-275®	250mg		25mg
Sinemet-plus®	100mg		25mg

*The proportion of carbidopa may be suboptimal.
†This proportion of drugs is available in a slow-release formulation (Madopar CR®).

The generic equivalent of Madopar® is co-beneldopa (1 part benserazide to 4 parts levodopa). The generic equivalent of Sinemet® is co-careldopa x/y (where x and y are the strengths (mg) of carbidopa and levodopa respectively).

453

1 JM Pearce 1993 *Parkinson's Disease and its Management*, OUP   2 J Tetrud 1989 *Science* 245 519   3 C Stibe 1988 *Lancet* i 403

Multiple sclerosis (MS)

This is a chronic or relapsing disorder caused by episodic demyelination and formation of many small plaques throughout the CNS. Peripheral nerves are unaffected. The pathogenesis involves focal disruption of the blood–brain barrier with associated immune response and myelin damage. Relapses may be triggered by otherwise innocuous infections.

Epidemiology MS is more common in temperate zones and there is considerable local variation; prevalence in England is 40/100,000 and 120/100,000 in Orkney and Shetland. Lifetime risk in UK is about 1:1000. Adult travellers take their risk with them, whereas children acquire the risk of the natives they settle among.

Clinical Usually presents in young adult either with single focal lesion, especially unilateral optic neuritis (which causes rapid deterioration in central vision), isolated numbness, or with weakness of leg(s). Other features are: vertigo; nystagmus; double vision; pain; incontinence; cerebellar signs; Lhermitte's sign (parasthesiae in arms or legs on flexing neck. This is also rarely seen in myelopathy from cervical spondylosis or B_{12} deficiency). Less commonly: facial palsy; epilepsy; aphasia; euphoria; dementia. The usual course involves initial remissions and relapses with later progressive accumulation of disability. Steady progression from outset sometimes occurs.

Examination Look carefully for signs of a CNS lesion other than the presenting problem. Do not forget the abdominal reflexes may be lost.

454

Diagnosis is based on history of recurrent episodes of neurological disturbance implicating more than one part of CNS, where no other explanation found. Isolated neurological features are never diagnostic, but may become so if a careful history reveals previous neurological features (such as unexplained blindness for a day or two), or investigations reveal involvement of distant parts of the CNS.

Tests CSF: pleocytosis, slightly raised protein (<1g/l), and IgG (in 70%[1]). CSF electrophoresis reveals oligoclonal bands. Visual, auditory and somatosensory evoked potentials prolonged. Magnetic resonance imaging is very sensitive for MS lesions. No test is pathognomonic.

What to tell patients Discuss this with the GP. If the diagnosis is in doubt, harm is done by sharing uncertainties with some patients (terms like 'neuritis' are often tried). Many patients cannot stand this sort of paternalism, and cope better with uncertainty than some of their doctors.

Treatment There is no cure. Dietary supplements with polyunsaturated fats may help. Optic neuritis treated with methylprednisolone for 3 days (250mg/6h IV), then oral prednisolone (1mg/kg/day) for 11 days appears to reduce the risk of developing MS in the next 2yrs, from ~15% to 7%.[2] Seek expert advice. ACTH or methylprednisolone may also be used to shorten relapses. β-interferon 1b appears to reduce relapse rate.[3] Treat spasticity with baclofen 15–100mg/day PO or diazepam 2–15mg/24h PO (but beware of addiction). Offer incontinence aids (p74). Help patient to live with disability (p66). Many aids are available. Advice available from Disabled Living Foundation and MS society (p804).

Prognosis After 5yrs 70% are still employed (35% at 20yrs), and 20% are dead from complications. Relapses last a few months, and remissions may last many years, especially if the 1st sign is optic neuritis, when the remission may outlive the patient (45–80% go on to MS after 15yrs).

1 D McFarlin 1982 *NEJM* **307** 1183   2 RW Beck 1993 *NEJM* **329** 1764   3 IFNS MS Study Group 1993 *Neurology* **43** 653 & 662

Multiple sclerosis (MS)
[*OTM* 21.211]

Space-occupying lesions

May present with signs of raised intracranial pressure (p734), focal brain damage, seizures, or false localizing signs. Ask yourself first *where* the lesion is, and then *what* it is.

False localizing signs Cerebellar signs and vith nerve signs (p32). The vith nerve is at risk because of its long intracranial course.

Localizing signs *Temporal lobe:* Hallucinations of smell, taste, sight, sound. Dysphasia (p430). Memory defects. *Déjà vu*. Contralateral upper quadrantic field defect.

Frontal lobe: Hemiparesis. Seizures (contralateral body movement). Personality changes (indecent, indolent, indiscreet). Grasp reflex (fingers drawn across palm are grasped) which is significant only if unilateral. Dysphasia (Broca's area p430). Loss of smell unilaterally (one nostril).

Parietal lobe: Hemisensory loss. Decreased two-point discrimination. Decreased stereognosis (ability to recognize object in hand). Stroke both patient's hands simultaneously and ask: 'which am I touching?'. Patient may systematically ignore one side (sensory inattention). In drawing clock face, omits half contralateral to lesion. Dysphasia (p430). Gerstmann's syndrome (p700).

Occipital lobe: Contralateral visual field defects (hemianopia).

Cerebellum: Intention tremor (eg worse when picking up a cup of tea). Past-pointing (ie if patient is asked to first touch her nose with her index finger, and then your index finger, held about a yard away, she tends to point beyond your finger). Furthermore, the tremor is worst at the end of each movement. Dysdiadochokinesis (slow and uneven at fast hand movements, eg tap left hand fast on back of right hand). Nystagmus (p422). Truncal ataxia (if worse when eyes closed then lesion is of dorsal columns, not cerebellum). Causes of cerebellar signs: acoustic neuroma; Friedreich's ataxia; stroke; haemangioma; tumours; MS; alcohol excess; abscess.

Cerebellopontine angle (usually acoustic neuroma): Ipsilateral deafness. Nystagmus. Reduced corneal reflex. Facial and vth nerve palsies. Ipsilateral cerebellar signs.

Corpus callosum: Usually severe rapid intellectual deterioration with focal signs of adjacent lobes. Signs of loss of communication between lobes (eg left hand unable to carry out verbal commands).

Midbrain: Unequal pupils. Inability to direct eyes up or down. Amnesia for recent events with confabulation. Somnolence.

Causes of space-occupying lesion Tumour: aneurysm; abscess (25% multiple); haematoma; granuloma; tuberculoma; cyst (eg cysticercosis). *Other causes of focal CNS signs:* Stroke; head injury; vasculitis (SLE, syphilis, PAN, giant cell); MS; encephalitis; post-ictal (Todd's p710).

Tests SXR, CT scan, MRI. Consider biopsy if abscess suspected.

Histology 30% are secondaries (breast, lung, melanoma; 50% multiple). Primaries: astrocytomas, glioblastoma multiforme, oligodendrogliomas, ependymomas (all with <50% 5yr survival); cerebellar haemangioblastomas (40% 20yrs survival); meningiomas (♀/♂ = ~2; complete recovery if removed).

Management Excision if possible. Treat headache with codeine phosphate 60mg/4h PO (adult dose). Dexamethasone 4mg/8h PO (adult dose) for cerebral oedema. Mannitol for raised ICP (p734 for adult dose).

Space-occupying lesions
[*OTM* 21.170]

Cranial nerve lesions

(See p32 for examination of cranial nerves.)

The causes of cranial nerve lesions are as follows:

Any cranial nerve lesion DM; MS; sarcoid; syphilis; tumour.

I Trauma; frontal lobe tumour; meningitis.

II *Optic neuritis* (features are: pain on moving eyes, loss of central vision, afferent pupillary defect, disc swelling): demyelination; rarely sinusitis, syphilis, collagen vascular disorders.

Optic atrophy (pale optic discs and reduced acuity): MS, frontal tumours; Friedreich's ataxia; retinitis pigmentosa; syphilis; glaucoma; Leber's optic atrophy; optic nerve compression.

Papilloedema (swollen discs): (1) ICP↑ (tumour, abscess, encephalitis, hydrocephalus, benign intracranial hypertension—see below); (2) retro-orbital lesion (eg cavernous sinus thrombosis); (3) inflammation (eg optic neuritis); (4) ischaemia, including hypertension.

Field defects Bitemporal hemianopia (optic chiasm compression eg from pituitary); homonymous hemianopia (affects the same half of visual field through each eye) is caused by lesion beyond optic chiasm opposite to the field defect, due to, eg stroke, tumour.

Scotomas (small blind patches): MS; alcohol; tobacco; drugs.

III alone DM; giant cell arteritis; aneurysm of posterior communicating artery; idiopathic; pressure from uncus (temporal lobe) as it herniates through the tentorium from ICP↑.

Nystagmus: p422.

III, IV, VI Stroke; tumours; Wernicke's encephalopathy; aneurysms; MS; myasthenia gravis; cavernous sinus thrombosis.

V *Sensory* (corneal reflex is affected first): *Herpes zoster*; nasopharyngeal carcinoma. *Motor:* Bulbar palsy; acoustic neuroma.

VII *Lower motor neurone* (all of one-half of face affected): Bell's palsy; *Herpes zoster* (Ramsay Hunt syndrome, OHCS p756); polio; otitis media; skull fracture; cerebello-pontine angle tumours; parotid tumours.

NB: if both lacrimation and taste (on anterior $^2/_3$ tongue) are unimpaired, the lesion is distal to stylomastoid foramen. Hyperacusis (excessive sensitivity to sound) suggests lesion proximal to stapedial branch of VII (in temporal branch).

Upper motor neurone (forehead spared): stroke.

VIII (Sensorineural deafness): noise; Paget's disease; Ménière's disease; *Herpes zoster*; neurofibroma; lead; aminoglycosides; frusemide; aspirin.

IX, X Trauma; brainstem lesions; neck tumours.

XI Polio; syringomyelia; tumours near jugular foramen.

XII Stroke; bulbar palsy; polio; trauma; TB.

Benign intracranial hypertension (pseudotumour cerebri)

A condition typically affecting young obese women who present with ICP↑ but no focal neurological signs or symptoms. Consciousness and intellectual function are preserved. Headache (of ICP↑) and papilloedema are the principal features, VI nerve palsy (diplopia) and enlargement of blind spot may occur. *Cause:* not usually found. *Treatment:* repeated LP; dexamethasone; consider shunt. *Prognosis:* usually resolves spontaneously. Permanent significant visual loss in 10%. Recurrence in 10%.

Cranial nerve lesions
[*OTM* 21.80]

Bell's palsy

Essence This was considered to be an idiopathic paralysis of the seventh cranial nerve. Many now accept that it is a viral polyneuropathy with inflammatory demyelination extending from the brainstem to the periphery.[1] Other nerves involved include V, IX, X, and C2,[1] although most authorities reserve the term Bell's palsy only for seventh nerve paralysis. The onset is usually abrupt and may be associated with much pain.

Symptoms due to VII paralysis Mouth sags; dribbling; taste impaired; eye waters (or may be dry). These symptoms are usually unilateral.

Symptoms due to other nerve palsies (not considered Bell's palsy by some) Scalp dysaesthesia (C2); numb face (V); absent gag reflex (IX); palatal weakness (X).

Signs The patient cannot: wrinkle his forehead; close his eyes forcefully; whistle or blow out his cheek. Note also: wide palpebral fissure.

Natural history All who are going to recover completely show a return of movement in the first four weeks. If there is complete axonal degeneration recovery usually begins after three months, and may be incomplete.

Causes of VII nerve palsy *Intracranial:* Brainstem tumour, stroke, MS, polio, cerebello-pontine angle lesions (acoustic neuroma, meningitis).
Within temporal bone: Otitis media, Ramsay Hunt syndrome (*OHCS* p756), cholesteatoma.
Distal: Parotid tumours, trauma.
Other: DM, sarcoid, Guillain–Barré, leprosy, Lyme disease.

Treatment Protect the eye with dark glasses. Instil artificial tears at the slightest evidence of drying.[1] Prednisolone is of unproven value. It is given early by some, to aid recovery (0.5mg/kg/12h PO for 5 days).

Crocodile tears In the month following a Bell's palsy patients may lacrimate unilaterally when they eat.[2] It is held that degeneration of the greater superficial petrosal nerve (serves the lacrimal gland) excites sprouting from the lesser superficial petrosal nerve (serves the parotid) at the point where they meet. These sprouts then innervate the lacrimal gland where they cause the flow of tears, not saliva. Cutting the tympanic branch of IX negates the effect of this misconnection.

1 K Adour 1982 *NEJM* **307** 348   **2** P Golding-Wood 1963 *BMJ* i 1518

Bell's palsy
[*OTM* 21.94]

Mononeuropathies

These are lesions of individual nerve bundles.

Causes Trauma; compression; leprosy; diabetes. If several peripheral nerves affected the term *mononeuritis multiplex* is used. *Causes:* DM; leprosy; sarcoid; amyloid; PAN; cancer.

Median nerve C5–T1 This is the nerve of grasp. Lesions above the cubital fossa:
- Inability to flex the interphalangeal joints of the index finger when hands clasp each other (Ochner's test).
- Inability to flex terminal phalanx of the thumb.
- Loss of sensation over the radial palm. If the lesion is at the wrist the only muscle reliably affected is the abductor pollicis brevis (p464). The area of sensory loss is less than occurs with higher lesions.

Ulnar nerve C8–T1 Abduction of the little finger precedes the claw hand deformity (which only occurs in lesions at the wrist, leaving digitorum profundus intact). There is sensory loss over the little finger and a variable area of the ring finger.

Radial nerve C5–T1 This nerve opens the fist. Test for wrist drop with elbow flexed and arm pronated. Sensory loss is variable, but always includes the dorsal aspect of the root of the thumb.

Sciatic nerve L4–S2 Lesions affect all muscles below the knee, and sensation below the knee laterally.

Lateral popliteal nerve L4–S2 Lesions lead to equinovarus with inability to dorsiflex foot. Sensory loss over dorsum of foot.

Tibial nerve S1–3 Lesions lead to calcaneovalgus with inability to stand on tiptoe or invert foot. Sensory loss over the sole.

Polyneuropathies

These are generalized derangements of peripheral nerves which usually begin distally and spread proximally giving rise to 'glove and stocking' anaesthesia. Distal motor weakness may occur also, and reflexes are reduced. Different patterns are discernible with different causes. Pain and tenderness in affected muscles is notable in neuropathies of diabetes and alcoholics.

Guillain–Barré syndrome This can be a neurological emergency (p700).

Other cause of polyneuropathy
Metabolic: DM, renal failure, liver failure, alcohol, hypothyroidism, porphyria, hypoglycaemia, xanthomatosis.
Connective tissue disease: eg PAN.
Malignancy, including polycythaemia rubra vera.
Infections, especially leprosy, TB. Also syphilis.
Vitamin deficiencies: B_1, B_6, B_{12}.
Genetic causes, notably Refsum's syndrome (p708); Charcot–Marie–Tooth syndrome (p696).
Toxins: Lead, metronidazole; phenytoin; vincristine; nitrofurantoin.

Investigations FBC, ESR, plasma glucose, U&E, thyroid function tests, plasma B_{12}, protein electrophoresis, syphilis serology, ANA, CXR. Consider porphyria screen nerve-conduction studies and electromyography.

Treatment Directed at cause.

Autonomic neuropathy

This may occur in diabetes mellitus, and causes nocturnal diarrhoea, urinary retention, falls from postural hypotension, and dizziness. If hypotension is a real problem support stockings to aid venous return may help, as may fludrocortisone.

Tests Function:	Clinical test:	Symptoms and signs:
Blood pressure	Postural drop >15mmHg	Faints (no vagal pulse↓) scotoma, tunnel vision
Pupils	Instil 2.5% methacholine; constriction '=' para-sympathetic lesion	Horner's syndrome (p700) No dilatation in the dark
Urinary bladder	Pressure studies	Incontinence (atonic or neurogenic bladder)
Sweating	Raise body temperature	Overheating; no sweat
Heart rate	If deep breaths give sinus arrhythmia >9/min then vagal efferent fibres OK	
GI motility		Dysphagia, diarrhoea, constipation
Sexual function		Impotence

Causes[1] *Central, primary:* Parkinson's; progressive autonomic failure.
Central, secondary: Craniopharyngioma or vascular hypothalamic lesion; cord lesion; tabes dorsalis; Chagas' disease; familial dysautonomia.
Distal: DM; amyloid; porphyria; Fabray's disease; autoimmunity eg B$_{12}$↓, hypothyroidism, Guillain–Barré, myasthenia, rheumatoid arthritis.
Drugs and toxins: Alcohol; β-blockers; methyldopa; captopril; anti-malarials; vincristine; phenothiazines; barbiturates; antidepressants; prazosin; hydralazine; ganglion-blocking drugs.

Progressive autonomic failure: ♂/♀≈3; HLA AW32–associated. This is an insidious condition, progressing over 10–15yrs, sometimes with multi-system atrophy. It may present with drop attacks (eg preceded by pain in the back of the neck), and go on to rigidity (tremor and akinesia are rarer), or urinary incontinence, muscle wasting, and unequal pupils. Stridor and apnoea are late signs (may need tracheostomy).

Treatment Ahmad & Watson régime: ● Diagnose any treatable cause.
● If there is symptomatic postural hypotension, start non-pharmacological measures: counsel to stand up slowly in stages. Try compression stockings (OHCS p598). Reserve fluid-retaining drugs (eg indomethacin ≤75mg/24h PO or fludrocortisone 0.1mg/24h PO increasing as needed) in those severely affected when other measures fail.
● If there is pure autonomic nervous system dysfunction, consider a trial of partial β-agonist (pindolol) unless there is cardiac or airways dysfunction (here xamoterol 200mg/12–24h PO has been tried).
● If CNS involvement, desmopressin, or ergotamine have been tried.
● If complete denervation, clonidine may help.
● If persistent post-prandial symptoms occur, caffeine or dihydroergotamine may be tried.
● L-Dopa rarely helps those with extrapyramidal effects from progressive autonomic failure.

▶ *Liaise with a specialist to establish which, if any, of the above may best help your patient.*

463

1 R Bannister 1987 in *OTM* 2ed, OUP, 21.76. **Acknowledgement:** this page owes much to Eric Meyer

Carpal tunnel syndrome

Essence Compression of the median nerve within the wrist, causing wasting, tingling, and pain in the hand.

Causes Any cause of soft tissue swelling can cause nerve compression within the unyielding carpus.

Pregnancy, menopause	Gout, TB	Acromegaly
Rheumatoid arthritis	Myxoedema	Amyloidosis

Signs and symptoms Pain and tingling in the night, not always confined to the thumb, index, middle and radial half of the ring finger. The pain (a burning sensation) is relieved by placing the hand in cold water. When pain at its worst, the patient characteristically shakes the wrist to bring about relief. Light touch, two-point discrimination, and sweating may be impaired.

How to test the muscles involved[1,2]
Abductor pollicis brevis (C8; T1): the thumb is placed so that the nail is in a plane at right angles to the palm, which should face upwards. The patient tries to abduct (ie raise) the thumb against resistance. Opponens pollicis (C8; T1): the patient tries, against resistance to touch the tip of the little finger with the thumb, whilst the thumb nail remains in a plane parallel to that of the palm.

Causes of wasting of the small muscles of the hand

In carpal tunnel syndrome, the wasting is most pronounced in the outer half of the thenar eminence (abductor pollicis brevis). Generalized wasting may be caused by arthritis, carcinomatosis, stroke, cervical cord pathology (esp. syringomyelia), trauma to the brachial plexus and its nerves, motor neurone disease, neuropathies, and, very rarely, myopathies.

Special tests The pain can be brought on by palmar flexion of the wrist maximally for 1min (*Phalen's test*). *Tinel's test:* Tapping over the median nerve in the forearm induces tingling in the hand. Electrophysiology to find level of median nerve lesion.

Treatment Diuretics, splint, local hydrocortisone injection (OHCS p658) and surgical decompression may be all effective.

1 *Aids to the Examination of the Peripheral Nervous System*, 1976 HMSO
2 W Pryse-Phillips 1984 *J Neurol Neurosurg and Psychiatry* **47** 873

Carpal tunnel syndrome
[*OTM* 21.121]

Bulbar palsy

This is palsy of the tongue, muscles of mastication, muscles of deglutition, and facial muscles due to loss of function of brainstem motor nuclei. The signs are those of a *lower motor neurone lesion*. The tongue is flaccid and shows fasciculation (p33), the jaw jerk is normal or absent, speech is quiet, hoarse, or nasal.

Causes: Motor neurone disease; Guillain–Barré; polio; syringobulbia; brainstem tumours. Also as part of *central pontine myelinolysis* seen occasionally in malnourished alcoholics and those with carcinoma or severe hyponatraemia. Demyelination in pons causes progressive quadriparesis and bulbar palsy, often fatal.

Pseudobulbar palsy An *upper motor neurone lesion* involving the muscles of eating, swallowing, and talking due to bilateral lesions above the midpons. The tongue is spastic and small, the jaw jerk increased, the speech like Donald Duck. Sometimes the patient's emotions are labile.

Causes: Pseudobulbar palsy is more common than bulbar palsy and caused usually by strokes affecting the corticobulbar pathways bilaterally. MS and MND are rare causes.

Motor neurone disease (MND)

466

MND is a disease of unknown cause which leads to a degeneration of Betz cells, pyramidal fibres, cranial nerve nuclei, and anterior horn cells. Both upper and lower motor neurones may be affected but there is never any sensory abnormality. This absence of sensory abnormality distinguishes MND from MS and polyneuropathies. MND never affects the external ocular movements (cranial nerves III, IV, and VI) which distinguishes it from myasthenia gravis (p472). Several clinical patterns of MND are distinguished.

- *Amyotrophic lateral sclerosis* (50% of patients) Features: combined LMN wasting and UMN hyperreflexia, both contributing to weakness.
- *Progressive muscular atrophy* (25%) Anterior horn cell lesion affecting distal muscles before proximal. Better prognosis than the above.
- *Bulbar palsy* Accounts for about 25% of patients.

Prevalence 7 per 100,000. Men affected more than women.

Prognosis Always fatal, eg within 3–5yrs. Median age at death: 60yrs.

Treatment Kindness is more helpful than expert neurological advice. Aim to help patient overcome difficulties (p66), and to reduce symptoms:
Spasticity: As for MS.
Drooling: Propantheline 15–30mg/8h PO or amitriptyline 25–50mg/8h PO.
Dysphagia: Blend food; would patient like NG tube, or percutaneous catheter gastrostomy?—or would this prolong death?
Joint pains and distress: Consider large doses of narcotics.
Respiratory failure: Tracheostomy and ventilation may not be kind or wise, but can allow some patients to go home.

Bulbar palsy
Motor neurone disease (MND)
[*OTM* 22.115; 21.167; 21.42]

467

Cervical spondylosis

Essence This is a degenerative condition of the synovial intervertebral joints and ligaments. It causes narrowing of the disc space and the formation of osteophytes, which cause pressure on nerve roots and the cord itself. The nucleus pulposus also degenerates.

Signs and symptoms result from a combination of causes:
- *Degenerative symptoms:* Painful, tender spine with reduced neck mobility.
- *Root pathology:* Pain in arms and fingers and diminished reflexes, dermatomal sensory loss, lower motor neurone weakness.

The intervertebral joints involved (in decreasing order of frequency) are C5/C6 (thumb sensations, biceps muscle); C7/C8/T1 (little finger sensation, flexor carpi ulnaris); C6/C7 (middle finger sensation, latissimus dorsi, triceps reflex); C4/C5 (elbow sensation, supraspinatus).

NB: there are eight cervical roots and only seven cervical vertebrae. Cervical roots exit above, thoracic roots below their vertebra, and C8 root leaves below C7 vertebra.

Test of biceps:[1] Patient resists forearm extension.
Flexor carpi ulnaris (ulnar nerve): Patient resists wrist extension.
Latissimus dorsi: Patient resists arm abduction.
Supraspinatus: Patient tries to abduct the fully adducted arm.

Special tests
- The *inverted supinator reflex:* The fingers flex on eliciting the supinator reflex, but there is no other movement. NB: this may also result from other causes of C5/C6 root and cord compression (eg meningioma, neurofibroma).

Unusual features Bladder involvement. Proprioception is lost less often than vibration and cutaneous sensation.

Investigations Neck x-rays (include flexion, extension and oblique views). Myelography may demonstrate cord or root compression, but is being replaced by MRI.

Differential diagnosis Multiple sclerosis, neurofibroma of the nerve root, subacute combined degeneration of the cord.

Treatment Neck collar. Surgery only if conservative management failed and objective evidence of root lesion or myelopathy. Beware of dismissing a patient with chronic root pain in the arm as suffering simply from 'wear and tear' spondylosis in the neck—you may be denying them relief from surgery (compare with protracted sciatic nerve root irritation from prolapsed lumbar intervertebral disc).

---

1 *Aids to the Examination of the Peripheral Nervous System* 1976 HMSO

Cervical spondylosis
[*OTM* 21.112]

Primary disorders of muscle

Signs and symptoms: Muscle weakness usually (but not in polymyositis) with reduction in relevant tendon reflex (contrast UMN disease). Rapid onset suggests inflammatory or metabolic cause. Fatiguability (weakness increases with exercise) suggests myasthenia (p472). Myotonia (delayed muscular relaxation after contraction, eg on shaking hands) is characteristic of 'myotonic disorders'. Spontaneous pain at rest occurs in inflammatory disease as does local tenderness. Pain on exercise suggests ischaemia. Oddly firm muscles (due to infiltrations with fat or connective tissue) suggest 'pseudohypertrophic' muscular dystrophies. Muscle tumours are rare; common causes of lumps are herniation of muscle through fascia; haematoma, and tendon rupture. Fasciculation (spontaneous, irregular and brief contractions of part of a muscle) suggest LMN disease. Look carefully for evidence of systemic disease.

Investigations: Consider electromyography and muscle biopsy; and investigations relevant to systemic causes (eg T_3).

1 *Muscular dystrophies* are a group of genetically determined diseases with a progressive degeneration of some muscle groups. The primary abnormality may be in the muscle membrane. The secondary effects are marked variation in the sizes of individual fibres and the deposition of fat and connective tissue. The main symptom is progressive weakness of the muscle groups involved. The commonest type is Duchenne's (pseudo-hypertrophic), inherited as sex-linked recessive (30% due to spontaneous mutation) and is (almost always) confined to boys. The Duchenne gene is on the short arm of the X chromosome (Xp21), and its product, dystrophin, is absent (or present in only very low levels) in Duchenne's muscular dystrophy.[1] The serum creatine kinase is raised at least 40-fold.[1] It presents at about age 4yrs with increasingly clumsy walking progressing to difficulty in standing up. Few survive to age 20 years. There is no specific treatment (p66). Genetic counselling is important.

2 *Myotonic disorders* are characterized by myotonia (tonic spasm of muscle). Muscle histology shows long chains of central nuclei within the fibres. The most important of these disorders is *Dystrophia myotonica* which is inherited as autosomal dominant. The onset is usually 20–30yrs with weakness (hands and legs) and myotonia. Muscle wasting and weakness in the face gives a long, haggard appearance. Additional features include: cataract; frontal baldness (in men); atrophy of testes or ovaries; cardiomyopathy; mild endocrine abnormalities (eg DM); and mental impairment. Most patients die in middle age of intercurrent illness. Procainamide hydrochloride may help the myotonia. Genetic counselling is important.

3 *Acquired myopathies of late onset* are often a manifestation of systemic disease. Look carefully for evidence of: carcinoma; thyroid disease (especially hyperthyroidism); Cushing's disease.

4 *Polymyositis* (p670).

5 *Myasthenia gravis* (p472).

1 L Kunel 1988 *NEJM* **318** 1363

Primary disorders of muscle

Myasthenia gravis (MG)

Essence An antibody-mediated autoimmune disease with too few functioning muscle acetylcholine receptors, leading to muscle weakness. Anti-acetylcholine receptor antibodies are detectable in 90% of patients, and cause depletion of functioning post-synaptic receptor sites.

Presentation Usually in young adults with easy muscle fatiguability. It may progress to permanent weakness. Muscle groups commonly affected (most likely 1st): extraocular; bulbar; neck; limb girdle; distal limbs; trunk. Look especially for: ptosis; diplopia; 'myasthenic snarl' (on smiling). On counting aloud to 50 the voice may weaken. Reflexes are often brisk. Weakness may be exacerbated by pregnancy, K+↓, infection, overtreatment, change of climate, emotion, exercise, drugs (aminoglycosides, opiates, tetracycline, quinine, quinidine, procainamide, β-blockers). ♀/♂= 2.

Association Thymic tumour; hyperthyroidism; rheumatoid arthritis; SLE.

Tests 1 'Tensilon® test' (only do if resuscitation facilities+atropine to hand). Prepare 2 syringes—1 with 10mg edrophonium and 1 with 0.9% saline. Give 1st 20% of each separately IV as test dose. Ask independent observer to comment on effect of each. Wait 30sec before giving rest of each syringe. The test is +ve if edrophonium injection improves muscle power. This test is not always as dramatic as is often stated.
2 Anti-acetylcholine receptor antibody (raised in MG).
3 Neurophysiology if limb muscles involved.

Treatment options 1 Symptomatic control with anticholinesterase, eg *pyridostigmine* 60–600mg/24h PO taken through the day; SE: diarrhoea, colic, controllable by propantheline 15mg/8h PO; signs of cholinergic toxicity: hypersalivation, lachrymation, sweats, vomiting, meiosis.
2 Immunosuppression with *prednisolone*, using a single-dose alternate day régime. Starting dose 5mg, increased by 5mg/week up to 1mg/kg on each treatment day.[1] Remission (may take months) prompts a dose reduction. SE: muscle weakness (hence the low starting dose). If no remission, and weakness is severe, try *azathioprine* 2.5mg/kg/day (do FBC and LFT weekly for 8 weeks, then every 3 months) or weekly methotrexate.
3 Thymectomy (if thymoma and if early onset with troublesome symptoms. This achieves remission in ~30%, and worthwhile benefit in another 40%.[1]
4 Plasmapheresis (emergency or pre-thymectomy) gives ~4 weeks benefit.

Prognosis Relapsing or slow progression. If thymoma, 5-yr survival 30%.

Myasthenic syndrome

This typically occurs in association with bronchial small cell carcinoma (=Eaton–Lambert syndrome) or rarely with other autoimmune diseases. Unlike true MG it: ● Affects especially proximal limbs and trunk ● Ocular and bulbar muscles are rarely affected ● There is *hypo*reflexia ● Only a slight response to edrophonium ● Repeated muscle contraction may lead to increased (not decreased) muscle strength ● It is the presynaptic membrane which is affected (the carcinoma appears to provoke production of antibody to calcium channels). Treatment (by expert only): guanidine hydrochloride, to ↑ acetylcholine release from nerve terminals. ► Do regular CXR as symptoms may predate lung cancer by >4yrs.

Other causes of muscle fatigability Polymyositis; SLE; botulism.

Myasthenia gravis (MG)
Myasthenic syndrome
[*OTM* 22.16]

473

Neurofibromatosis (Von Recklinghausen's disease)

Essence Peripheral nerves become lumpy and expanded by an overgrowth of Schwann cells and fibroblasts. The autonomic and central nervous systems are also involved. Those who grow acoustic neuromas are unlikely to exhibit cutaneous signs.[1] Indeed there may be two types of neurofibromatosis: *type 1* being Von Recklinghausen's (defective gene on chromosome 17) and *type 2* being bilateral acoustic neurofibromatosis (autosomal dominant, chromosome 22, with 95% penetrance).[2,3]

Inheritance Autosomal dominant.

Prevalence 1 in 3000 births. Neurofibromata may occur singly or as a part of a generalized disease characterized by:

(1) Dermatological and pulmonary manifestations
- *Café au lait spots:* Light brown macules. 5 spots (each >1.5cm diameter) or more suggest a diagnosis of neurofibromatosis. Some are present at birth. Others may appear during adult life.
- *Axillary freckling.*
- *Violaceous dermal neurofibromata* (like nipples).
- *Subcutaneous nodules* felt along the course of a nerve. They may be very large.
- *Honeycomb lung:* Thickening of alveolar septa producing cystic space, giving honeycomb appearance on CXR. Clinical features: progressive dyspnoea on exercise; spontaneous pneumothorax; rarely clubbing. Other causes of honeycomb lung: histiocytosis X; tuberous sclerosis; lymphangioleiomyomatosis (a rare condition causing chylous ascites and pleural effusions).

(2) Some manifestations around the eye
- Pulsating exophthalmos (from encephalocoele).
- Retinal white dots (astrocytic hamartomas).
- Melanocytic hamartoma of iris (Lisch nodule).

(3) Local effects of neurofibromata These can be endlessly elaborated as any site may be affected, eg nerve roots—causing compression; gut—causing chronic GI bleeds; intra-abdominal—causing obstruction; bone—causing cystic lesions, kyphoscoliosis, and pseudoarthrosis; CNS—causing spinal cord compression and epileptic fits.

Associations (Without evidence of local tumour growth.) Radicular pain syndrome; spina bifida; diffuse spinal canal stenosis;[4] renal vascular lesions.

Associations with other tumours Optic nerve glioma; acoustic neuroma; other cranial nerve tumours; astrocytomas; glioblastomas; ependymomas; meningiomas; phaeochromocytoma; medullary thyroid carcinoma.

Treatment Local excision.

Complications Sarcomatous degeneration.

1 R Martuza 1984 *NEJM* **311** 520   **2** V Riccardi 1981 *NEJM* **305** 1617   **3** R Martuza 1988 *NEJM* **318** 684   **4** P Vancoillie 1979 *Lancet* **ii** 1246

Neurofibromatosis (Von Recklinghausen's disease)
[*OTM* 22.170]

Syringomyelia

Syrinx was one of those versatile virgins of Arcadia who, on being pursued by Pan to the banks of the river Ladon, conveniently turned herself into a reed—from which Pan made his pipes, and, in so doing, she gave her name to all manner of tubular structures, eg syringes—and syrinxes, which are tubular or slit-like cavities which form for unknown reasons in the grey commissure behind the central canal of the cervical spinal cord.

Pathogenesis As the syrinx enlarges it may extend into the dorsal horns and into the white matter. The lateral and dorsal white columns are especially affected. As the pressure builds up in the fluid within the syrinx pressure atrophy obliterates grey and white matter nearby. Extension to the midbrain is called *syringobulbia*.

Aetiology Obscure. That a developmental anomaly may be the cause is suggested by an association with the Arnold–Chiari (p694) malformation in which the cerebellum extends through the foramen magnum and down the spinal cord.

Signs and symptoms Cardinal signs and symptoms are wasting and weakness of the hands and arms, with loss of pain and temperature sensation ('dissociated' sensory loss) over the trunk and arms, classically in a 'cape' distribution ('suspended' sensory loss). These features reflect early involvement of fibres conveying pain and temperature sensation which decussate anteriorly in the cord, and of cervical anterior horn cells. The sensory loss may lead to painless burns and Charcot's joints. Other features may include Horner's syndrome (cervical sympathetics) and upper motor neurone signs in the legs.

Investigations CT or MRI scan.

Natural history The symptoms may be mild and not change for years. Then there may be a rapid deterioration.

Treatment Surgical decompression at foramen magnum may be attempted if there is a Chiari malformation. Surgery aims at promoting free flow of CSF through the foramen magnum, preventing dilation of the syrinx. These procedures may relieve pain, and prevent progression of symptoms.

Charcot's joints These are joints destroyed by too great a range of movement which occurs when the normal sensation to the joint is lost. They are swollen and mobile. They occur in syringomyelia (eg shoulder), tabes dorsalis (eg knee), diabetes, and leprosy.

Cavernous sinus thrombosis

This thrombosis of the intracranial venous sinuses is usually secondary to either trauma (eg causing fracture to the skull) or infection (eg from nasal sinuses or skin of face).

Signs and symptoms Pain in eye, forehead; exophthalmos; paresis of cranial nerves: III, IV, VI.

Investigations Blood culture; CSF examination (increased pressure, protein may be raised, leucocytosis may be present); CT scan (to distinguish from cerebral abscess); carotid angiography (sinus with thrombus fails to fill).

Treatment Treat infection.

Prognosis Depends on extent of thrombus. Cerebral damage may be permanent.

Notes

Notes

480

Relevant pages in other chapters:
Signs and symptoms: Abdominal distension (p38); epigastric pain (p46); faecal incontinence (p46); flatulence (p46); guarding (p46); heartburn (p48); hepatomegaly (p48); LIF and LUQ pain (p50); palmar erythema (p52); rebound tenderness (p56); regurgitation (p56); RIF pain (p56); RUQ pain (p56); skin discoloration (p56); splenomegaly (p56); tenesmus (p58); vomiting (p60); waterbrash (p60); weight loss (p60).

Surgical topics: The contents page to *Surgery* (p80)

Other topics: Alcohol (p682).

Healthy eating

There are no good or bad *foods*, only good or bad *diets*. Eating should be a pleasure; it can also be healthy.

A manifesto for the nation[1] Evidence concerning diet and disease is generally weaker than that relating smoking to, for example, lung cancer. Dietary advice has changed dramatically over recent years. The following advice embodies the main principles consistent with current ideas. Perhaps even this advice needs to be taken with a (small) pinch of salt.

1 Aim, if possible, for body mass index within the range 20–25 (p502).

2 **Reduce fat** (particularly saturated fats; it is unclear whether mono- and poly-unsaturated fats should also be reduced[2]). Grill, don't fry. Select lean meat; remove obvious fat and the skin of poultry. Eat fish (30g/day or 1–2 fish dishes/week is probably enough to protect against heart disease[3]), pulses, nuts. Choose low-fat dairy products (including low-fat cheese such as Edam and cottage cheese) and unsaturated fats for cooking (olive, rape seed and sunflower oils rather than butter). Avoid high-fat foods (eg shop-bought pastries, cakes, most biscuits, potato crisps, sausages, salami, meat pies, taramasalata, chips, chocolate). No more than 4 eggs per week. Eating oily fish rich in omega-3 fatty acid is particularly valuable in those with hyperlipidaemia. Such fish are: mackerel, herring, pilchards and salmon. If tinned fish avoid those in oil. Nuts are also valuable here: walnuts lower total cholesterol and have one of the highest ratios of polyunsaturates to saturates (7:1).[4]

3 **Reduce total sugar** Use fruit rather than sugar to add sweetness. Choose low sugar drinks. Avoid adding sugar to drinks and cereals.

4 **Increase dietary fibre** (The term *fibre* is not precise and is going out of fashion. Most fibre is **non-starch polysaccharides—NSP** which is now the preferred expression.) Eat fresh fruit (preferably with skins) and lightly cooked vegetables, wholemeal bread, potatoes, pasta, rice, porridge oats and higher fibre breakfast cereals.
NB More fluid is needed with high fibre diet. Aim for 8 cups daily (almost 3 pints), and warn patients about bulky stools.

5 **Reduce salt** Use spices (and herbs) to add variety to life.

6 **Enjoy a moderate alcohol intake** ♀: <15U/week; ♂: <20U/week— eg as wine taken regularly with food, not in binges. This may be the key to the 'French paradox'—that while the French have a lipid profile similar to their neighbours, and eat *more* dairy fat, their death rate from heart disease is ⅓ that of their neighbours. We know that alcohol inhibits platelet aggregation: this may be why, taken in the right dose, alcohol is one of the best drugs protecting us from coronary heart disease.[4] But too much alcohol causes problems! (p682).

This diet is inappropriate if: Less than 5yrs old; need for low residue diet (eg Crohn's, ulcerative colitis—p516); need for special diets (eg coeliac disease, p522). See also: hyperlipidaemia (p654); diabetes (p530); obesity (p502); constipation (p488); liver failure (p504); cirrhosis (p506); chronic pancreatitis (p522) and renal failure (p388).

1 Committee on medical aspects of food policy report 1991 HMSO *Report on Health and Social Subjects* **41**   **2** J Mann 1990 *BMJ* i 1297   **3** D Kromhout 1993 *Proc Nutrit Soc* **52** 437 and 1985 *NEJM* **312** 1205   **4** J Sabate 1993 *NEJM* **328** 603 **5** S Renaud 1992 *Lancet* i 1523

Healthy eating

Jaundice

Jaundice (=icterus) refers to the yellow pigmentation of skin, sclerae and mucosae due to a raised plasma bilirubin (eg >35µmol/l). Examine the sclerae in a good light (they are the most sensitive indicator).

Biochemistry Bilirubin is formed from the breakdown of haemoglobin. It is usually all conjugated by hepatocytes. If there is excess bilirubin (eg in haemolysis) or an inborn failure of uptake (Gilbert's syndrome) or conjugation (Crigler–Najjar syndrome), some bilirubin is unconjugated, and remains in the circulation. Being water insoluble, it does not appear in the urine. This is *pre-hepatic jaundice* ('acholuric'). Normally, conjugated bilirubin flows out into the gut, where intestinal bacteria convert it to urobilinogen, some of which is reabsorbed and appears in the urine. The rest is converted to stercobilin, which colours faeces brown. If the common bile duct is blocked, plasma conjugated bilirubin rises. Being water soluble, some is excreted in the urine, making it dark. By contrast, less bilirubin passes into the bowel; the faeces now have less stercobilin and so are pale. This is *post-hepatic* ('*obstructive*' or '*cholestatic*') jaundice. Obstruction is rarely complete. *Hepatocellular jaundice* implies diminished hepatocyte function (usually with varying degrees of cholestasis) and conjugated *and* unconjugated hyperbilirubinaemia. Ward urine tests are a sensitive marker of acholuric jaundice (no urinary bilirubin) and total biliary obstruction (no urinary urobilinogen). In hepatocellular disease, clotting times are the best marker (also AST↑ and ALT↑, p652). In obstructive jaundice, the best markers are the canalicular enzymes (plasma GGT and alk phos↑).

Causes Pre-hepatic: Haemolysis; ineffective erythropoiesis; Crigler–Najjar and Gilbert's syndromes (see above).

Hepatocellular: Viruses (hepatitis A, B, C, E, Epstein–Barr virus); leptospirosis; autoimmune diseases (eg chronic hepatitis, p512); drugs (eg halothane, paracetamol, methyldopa, barbiturates) and rare syndromes (Rotor, Dubin–Johnson, Wilson—p708, p696, p712), cirrhosis.

Cholestatic: Intrahepatic—primary biliary cirrhosis; drugs (chlorpromazine, isoniazid); part of hepatocellular disease. Extrahepatic—gallstones, bile duct or head of pancreas carcinoma; enlarged nodes in the porta hepatis (eg lymphoma).

At the bedside ask about *all* drugs (tablets, injections, anaesthetics, recreational; find all drug charts); alcohol; jaundiced contacts; travel; family history; sexual activity. Look for hepatomegaly (and splenomegaly, signifying infection or portal hypertension), signs of chronic liver disease (p687). ► Examine the urine and stools.

Tests Urine for bilirubin and urobilinogen (see above). Blood (wear gloves—hepatitis risk): U&E + creatinine; LFTs + serum proteins + GGT + glucose + ethanol; FBC + reticulocytes; INR; EBV + CMV serology + hepatitis A IgM+HBsAg. If a rare cause of cirrhosis is suspected: serum copper, caeruloplasmin, α_1-antitrypsin, iron, total iron binding capacity.

The key question is: *Are the bile ducts dilated?* This signifies extrahepatic cholestasis, but dilatation may not occur for some days. Ultrasound is the first step. It is non-invasive, and may also demonstrate the obstructing pathology. If the bile ducts are dilated, consider percutaneous transhepatic cholangiography (PTC). It has a mortality of <1%. SE: bile peritonitis; haemorrhage; cholangitis. An alternative is ERCP (p498), CT scan—or laparotomy. If the bile ducts are not dilated, the next step is liver biopsy (p506).

Jaundice
[*OTM* 12.206]

Diarrhoea and rectal bleeding

Diarrhoea is the passage of >300ml of liquid faeces/24h. In determining the cause, there are three major questions to ask:

- *Is the diarrhoea acute or chronic?* Infections are often acute (has the patient been abroad?); chronic diarrhoea alternating with constipation suggests irritable bowel syndrome (p518); medication abuse eg antacids.
- *Is the large or small bowel to blame?* In the former stools are watery with mucus or blood and there is lower abdominal pain with tenesmus and urgency; in the latter, any pain is often periumbilical (or in the RIF), and stools are bulky and stink.
- *Is there a non-GI cause?* (Eg thyrotoxicosis, anxiety, or autonomic neuropathy from DM—the latter is suggested by nocturnal diarrhoea.) Ask about drugs, eg antacids, cimetidine, digoxin, antibiotics, thiazide diuretics and alcohol.

Osmotic causes of diarrhoea (Fluid is drawn into the GI tract.) Laxatives, eg lactulose; magnesium sulphate (Epsom salts).

Secretory causes Infections: Bacteria (*Campylobacter*, *Vibrio cholerae*, *Staphylococcus*, *E coli*, *Salmonella*, *Shigella*, *Clostridium difficile*, *Yersinia enterocolitica*); giardiasis; rotavirus; amoebiasis; *Cryptosporidium* (a fungus needing special stains).
Inflammatory bowel disease (ulcerative colitis, Crohn's disease).
Laxative abuse; bile salts (eg in malabsorption, p522).

Increased motility Irritable bowel syndrome; thyrotoxicosis.

Causes of bloody diarrhoea 'Dysentery' (ie *Campylobacter*—which may also cause peritonism, *Shigella*, *Salmonella*, and *E coli* infections); amoebiasis; ulcerative colitis; Crohn's disease; colorectal cancer; pseudomembranous and ischaemic colitis (p126).

Causes of rectal bleeding (± diarrhoea): Diverticulitis; colonic cancer; polyps; haemorrhoids; radiation proctitis; trauma; *fissure-in-ano*; angiodysplasia (this common cause of bleeding in the elderly is due to arteriovenous malformation).

Pseudomembranous colitis is caused by overgrowth of *Clostridium difficile*, following oral or IV antibiotic therapy (any, including penicillins[1]). *Treatment:* vancomycin 125mg/6h PO or metronidazole 400mg/8h PO (cheaper; more palatable). Liaise with a microbiologist.

Functional diarrhoea This is chronic diarrhoea, without any readily identifiable cause. It may be a variant of the irritable bowel syndrome, or be caused by malabsorption of bile acids.

Investigating diarrhoea ▶ Do a PR to exclude 'overflow' diarrhoea. Large bowel diarrhoea: examine a fresh stool for pathogens, cysts and ova (repeated tests may be needed). Sigmoidoscopy, barium enema ± colonoscopy if prolonged.
If a small bowel cause is suggested, is there malabsorption? Do a 3-day collection for faecal fats, and measure serum folate and iron. Consider a small bowel barium meal and biopsy (p522).

Management of diarrhoea Treat the cause, if possible. Give clear fluids PO. If dehydrated or elderly, check U&E. If IV fluid is needed, give 0.9% saline (eg with ≥20mmol K+/l). If it is necessary to reduce symptoms, try codeine phosphate 30mg/6h PO.

1 *Lancet* Ed 1982 ii 476

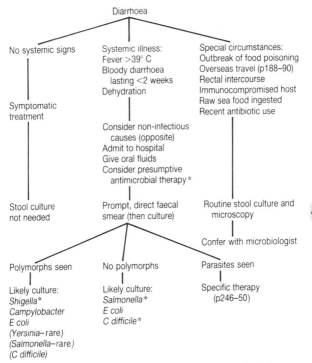

Diarrhoea

No systemic signs

Symptomatic treatment

Stool culture not needed

Systemic illness:
Fever >39° C
Bloody diarrhoea
lasting <2 weeks
Dehydration

Consider non-infectious
causes (opposite)
Admit to hospital
Give oral fluids
Consider presumptive
antimicrobial therapy *

Prompt, direct faecal
smear (then culture)

Special circumstances:
Outbreak of food poisoning
Overseas travel (p188–90)
Rectal intercourse
Immunocompromised host
Raw sea food ingested
Recent antibiotic use

Routine stool culture and
microscopy

Confer with microbiologist

Polymorphs seen

Likely culture:
*Shigella**
Campylobacter
E coli
(*Yersinia*–rare)
(*Salmonella*–rare)
(*C difficile*)

No polymorphs

Likely culture:
*Salmonella**
E coli
*C difficile**

Parasites seen

Specific therapy
(p246–50)

*Prompt, specific treatment (p238 & p486) may be needed before sensitivities are known. Be guided by likely diagnosis following the direct smear.

Principal source: S Gorbach 1987 *Lancet* ii 1378

487

Constipation

▶ Always ask a patient what he means by constipation.

Patients use constipation to mean that:
- Their faeces are too hard
- Defecation hurts
- They do not defecate often enough
- They have diarrhoea

'Medical' causes:

Diet deficient in roughage
Disobeying the call to stool
Myxoedema
Irritable bowel syndrome
Hypercalcaemia
Drugs:
    atropine
    codeine phosphate
    morphine
    tricyclics
    disopyramide

'Surgical' causes:

Anal fissure
Carcinoma of the rectum
Carcinoma of the sigmoid colon
Foreign body
Post-gastrectomy (NB: diarrhoea
   is more common)
Pelvic mass (fibroids, fetus)
Post-operative pain on straining
Bed rest (post-operatively)
Developmental disorders
Any GI obstruction

Treatment Treat any underlying condition. Mobilize the patient. Encourage diet rich in fibre with adequate fluid intake (p482). Consider drugs only if above measures fail. ▶ Try to use drugs for short durations only.

Bulk producers: (To retain water and increase microbial growth.)
Bran: 10–30g/day of bran doubles faecal output, increases frequency of defecation and speeds gut transit.[1] SE: uncooked bran has a high phytate content (so calcium, iron and zinc availability↓[2]).

Mucilaginous polysaccharides: these arise in seeds and gums (eg sterculia 62% granules in 5ml sachets, one sachet/12h PO after food, taken unchewed with plenty of water).

Ispaghula granules, 10ml in plenty of water once or twice daily with meals (halve the dose in children).

Methylcellulose (3–6 500mg tablets/day with plenty of water).

Stimulants: Senna (15–60mg/day); bisacodyl tablets (5–10mg at night) or suppositories (5 or 10mg in the mornings—use the lower dose in children <10yrs); danthron (*co-danthrusate* capsules contain 50mg danthron and 60mg ducosate). *Note:* danthron has been associated with colonic and liver tumours in animals—so reserve its use for the very elderly and terminally ill.

Osmotic agents: ▶ All should be taken with plenty of water.
Epsom salts; magnesium hydroxide mixture (25–50ml/24h PO); lactulose syrup (3.35g/5ml), adult dose: 15ml/12h PO adjusted to produce 2–3 soft stools/day. Children <1yr, 2.5ml/12h; 1–4yrs, 5ml/12h; 5–10yrs, 10ml/12h.

Stool softeners: Liquid paraffin—do not use regularly[2] (SE: anal seepage, lipoid pneumonia, malabsorption of fat-soluble vitamins).

Enemas (sodium phosphate) and glycerine suppositories may be useful additional agents. Excessive use of soapy tap water enemas may lead to water intoxication.

Cost Cheap: senna, bran, co-danthrusate, bisacodyl. ***Moderate:*** magnesium hydroxide, methylcellulose, ispagula granules, sterculia. *Expensive:* lactulose.

---

1 *Drug Ther Bul* 1981 **19** 29 **2** *Drug Ther Bul* 1988 **26** 53

Constipation
[*OTM* 12.20]

Peptic ulcers & hiatus hernia

Ulcer sites oesophagus, stomach, duodenum, Meckel's diverticulum. They are common: men 20%; women 10% (at some time).

Possible exacerbating factors: Stress, smoking, alcohol, NSAIDs, steroids—but only if on NSAIDs,[1] *Helicobacter pylori* on gastric mucus-secreting cells.

Clinical features—duodenal ulcer: Epigastric pain, radiating to the back; eating relieves the pain (so weight↑); worse at night ('I get up and drink some milk'); waterbrash (the mouth fills with saliva). With gastric ulcers the pain may be worse by day.

The chief dangers are GI bleeding and perforation (p112). The history and examination are of little diagnostic help (see p44 for causes of 'indigestion'). Endoscopy and barium studies are needed. Both have a pick-up rate of ~90%, but endoscopy allows biopsy. Consider rebiopsying all gastric ulcers after 4 weeks of treatment, to exclude malignancy.

A common batting order for peptic ulcer treatment is:

1 Remove exacerbating factors (eg smoking), use antacids to relieve pain (until definitive treatment working), and give histamine (H_2) antagonist for 8 weeks, eg cimetidine (800mg PO at night); or ranitidine (300mg PO at night). 90% of ulcers will heal, but relapse on stopping drugs is very common (about 80%). If relapse, consider further courses of histamine antagonist or prolonged maintenance.

2 60–95% of those relapsing on stopping H_2 antagonists have infection with *Helicobacter pylori*. This is best treated with a 'triple' régime, eg 1 week of bismuth chelate (120mg/6h PO 1h ac) + amoxicillin 500mg/6h PO, + 5 daily doses of metronidazole 400mg PO from days 5–7.[2]

3 Surgery (p164–6).

▶ Because of the high relapse rate with H_2 anatagonists and the high prevalence of *Helicobacter pylori*, an argument can be made for 'triple' therapy as first-line treatment in duodenal ulcers.[3]

Other drug treatments are: ● Omeprazole 20mg/24h PO, for 4 weeks (8 weeks for gastric ulcers)—may heal resistant ulcers and works by blocking the 'proton pump' of gastric parietal cells. ● Sucralfate 1g/6h PO 1h ac (it may protect mucosa from acid-pepsin attack)—avoid antacids $\frac{1}{2}$h either side of doses. It is the drug of choice for preventing stress ulcers on ITU. ● Misoprostol 200µg/6h PO (for ulcer healing and preventing NSAID-associated ulcers).

Side-effects Cimetidine: gynaecomastia ± impotence; important interactions, p800. Bismuth: stool blackening; CI: renal diseases, pregnancy. Sucralfate: pruritus, back pain, vertigo, dry mouth, insomnia.

Hiatus hernia (HH) Part of the stomach herniates through diaphragm into thorax. This may occur by the stomach pushing up beside the oesophagus ('paraoesophageal') or (in 80%) it may herniate directly ('sliding' hernias). It is debatable if this leads to oesophagitis by making the cardiac sphincter incompetent. In most cases of reflux, hiatus hernia is not present. Surgery may be indicated (eg Nissen's fundoplication).

1 M Guslandi 1992 *BMJ* i 655   2 RP Logan 1994 *Gut* **35** 15   3 B Marshall 1988 *Lancet* ii 1437

Peptic ulcers & hiatus hernia
Reflux oesophagitis
[*OTM* 12.64; 12.44]

Reflux oesophagitis

This causes a pain which often radiates to the back, worsening with stooping and hot drinks. *Causes:* intra-abdominal pressure↑ (tight clothes, obesity, big meals); faulty lower oesophageal sphincter (smoking, fatty meals, pregnancy, post cardia surgery in achalasia, hiatus hernia); drugs (eg tricyclics, anticholinergics).

Tests: Endoscopy; barium studies; fluid level on CXR (if HH).

Complications: Anaemia, stricture. See Barrett's ulcer (p694).

Treatment: Minimize the above causes. Raise the bed head. Try antacids (magnesium trisilicate mixture, 10ml/6h PO), or coating the oesophagus with alginates (Gaviscon®, 10ml/8h PO). H_2 antagonists, and strengthening the cardiac sphincter with metoclopramide (10mg/8h PO; 5mg/8h if <60kg, or <19yrs old) may help, but are no substitute for healthier living. If resistant or severe, consider omeprazole 20–40mg/24h PO.

Dysphagia

Dysphagia is difficulty in swallowing food or liquid. Unless it is associated with a transitory sore throat, it is a serious symptom, and merits further investigation, usually by endoscopy, to exclude neoplasia. If the experience is one of a lump in the throat *at times when the patient is not swallowing*, the diagnosis is likely to be anxiety (globus hystericus).

Causes

Malignant:	Neurological causes:	Others:
Oesophageal cancer	Bulbar palsy (p466)	Benign strictures
Gastric cancer	Lateral medullary	Pharyngeal pouch
Pharyngeal cancer	syndrome	Achalasia (below)
Extrinsic pressure	Myasthenia gravis	Systemic sclerosis
(eg lung cancer)	Syringomyelia (p476)	Oesophagitis
		Iron-deficiency anaemia

Differential diagnosis There are 4 key questions to ask:

1 Can fluid be drunk as fast as usual, except if food is stuck?
Yes: Suspect a stricture (benign or malignant).
No: Think of motility disorders (achalasia, neurological causes).

2 Is it difficult to make the swallowing movement?
Yes: Suspect bulbar palsy, especially if he coughs on swallowing.

3 Is the dysphagia constant and painful?
Yes (either feature): Suspect a malignant stricture.

4 Does the neck bulge or gurgle on drinking?
Yes: Suspect a pharyngeal pouch (food may be regurgitated).

Investigations FBC; ESR; barium swallow; endoscopy with biopsy; oesophageal motility studies (this requires swallowing a catheter containing a pressure transducer).

Nutrition Dysphagia may lead to malnutrition. Nutritional support often needed prior to treatment.

Oesophageal carcinoma This is associated with achalasia, Barrett's ulcer (p694), tylosis (an hereditary condition in which there is hyperkeratosis of the palms), Plummer–Vinson syndrome (below) and smoking. Survival after resection is rare after 5 years (p164).

Benign oesophageal stricture Causes: oesophageal reflux; swallowing corrosives; foreign body; trauma. Treatment: dilatation (endoscopic or with bougies under anaesthesia).

Barrett's ulcer See p694.

Achalasia There is failure of oesophageal peristalsis and failure of relaxation of the lower oesophageal sphincter. Liquids and solids are swallowed only slowly. CXR: air/fluid level behind the heart, and double right heart border produced by a grossly expanded oesophagus.

Treatment: Myomectomy cures ~75% of patients. Pneumatic dilatation may also help.

Plummer–Vinson (Paterson–Brown–Kelly) syndrome There is an oesophageal web associated with iron-deficient anaemia, and possibly post-cricoid cancer.

Dysphagia
[*OTM* 12.15; 12.37]

▶▶ Upper gastrointestinal bleeding

Haematemesis is the vomiting of blood. *Melaena* is altered blood (black) passed PR. It implies bleeding proximal to the splenic flexure of the colon.

Common causes Peptic ulcer (p492) ● Gastritis (mucosal inflammation and erosion caused eg by alcohol) ● Mallory–Weiss tear (an oesophageal tear caused by vomiting). ● NSAIDs, steroids, thrombolytics, anticoagulants.

Others ● Oesophageal varices (only 5% of patients, but 80% of mortality) ● Nose bleeds (swallowed blood); oesophageal or gastric malignancy; haemobilia (triad of haematemesis, jaundice and biliary colic) ● Oesophagitis ● Angiodysplasia (a GI vascular malformation, seen commonly in the elderly) ● Haemangiomas ● Ehlers–Danlos or Peutz–Jeghers' syndrome (p706) ● Bleeding disorders ● Aorto-enteric fistula (think of this in anyone with an aortic graft: do immediate endoscopy even for small 'herald' bleeds,[1] to prevent impending catastrophe; get expert help).

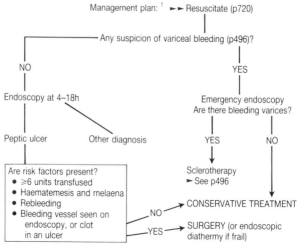

Management plan: [1]   ▶▶ Resuscitate (p720)

Any suspicion of variceal bleeding (p496)?

NO — Endoscopy at 4–18h

Peptic ulcer   Other diagnosis

Are risk factors present?
● ≥6 units transfused
● Haematemesis and melaena
● Rebleeding
● Bleeding vessel seen on endoscopy, or clot in an ulcer

YES — Emergency endoscopy
Are there bleeding varices?

YES — Sclerotherapy ▶ See p496

NO —

NO — CONSERVATIVE TREATMENT

YES — SURGERY (or endoscopic diathermy if frail)

'Conservative' treatment means: bed rest; nil by mouth; have 6 units of crossmatched blood available; IVI 0.9% saline and blood transfusion as needed. Monitor pulse, BP, CVP (may give early warning of a repeat bleed), ECG and fluid balance. Do U&E, LFTs, FBC, platelets (often raised), INR. A normal Hb does not exclude a large bleed (haemodilution takes hours to develop).

There is little evidence that H$_2$ antagonists such as cimetidine stop bleeding or save lives. Mortality: 10%, rising to 30% for rebleeding. It is higher in the elderly, especially in the presence of other disease.

If endoscopy is not diagnostic, and bleeding continues, consider laparotomy or angiography. ▶ *Liaise with surgeons early.*

1 M Woodley 1992 *Manual of Medical Therapeutics* 27ed, p. 288, Little, Brown & Co, Boston (ISBN 0–316–92420–2)

494

Upper gastrointestinal bleeding

[*OTM* 12.21]

▶▶ **Bleeding varices**

In the presence of portal hypertension blood from the gut can reach the right heart, bypassing liver via collateral veins (around the oesophago-gastric junction). With pressures >12mmHg, these veins dilate forming varices, which may bleed catastrophically (eg if clotting factors↓, in cirrhosis). In the UK, 90% of varices are caused by cirrhosis—but world-wide, hepatic fibrosis from schistosomiasis is the major cause.

Classification of portal hypertension *Obstruction before the hepatic sinuses (presinusoidal):* Portal vein thrombosis; pancreatic tumours or pseudocysts; myelofibrosis; Hodgkin's; schistosomiasis; sarcoid. *Post-sinusoidal:* Cirrhosis; the Budd–Chiari syndrome (hepatic vein thrombosis, eg in pregnancy or with steroid contraceptive use).

Assessment Is he shocked? Look for the signs of chronic liver disease—jaundice; spider naevi; palmar erythema; gynaecomastia; reduced sexual hair; testicular atrophy; ascites; encephalopathy.

If the patient is shocked
- Set up several fast plasma IVIs.
- Crossmatch 6 units of blood. (Give ORh−ve blood until ready.)
- Search for, and correct any hypoglycaemia (50g glucose by IVI).
- Set up CVP line to guide fluid replacement. Aim for >5cmH₂O.
- Urgent endoscopy to establish the diagnosis.
- Try vasopressin 20u in 200ml 5% dextrose IV over 15mins with GTN infusion for coronary vasospasm (octreotide may be an alternative).
- Prevent encephalopathy with neomycin and lactulose (p504).
- If bleeding continues, pass a modified Sengstaken tube, with balloons to compress stomach and oesophageal balloons. Before insertion, inflate the balloons with a measured volume (eg 120–300ml) of air giving pressures of 60mmHg (check with sphygmomanometer). Deflate and clamp exits. Pass the lubricated tube (try to avoid sedation) and inflate the gastric balloon with the predetermined volume of air. Inflate the oesophageal balloon. Check pressures. They should be 20–30mmHg greater than on the trial run. Tape to the patient's forehead to ensure the gastric balloon impacts gently on the gastro-oesophageal junction. Place the oesophageal aspiration channel on continuous low suction and the gastric channel drains freely. Leave *in situ* until bleeding stops. Remove after <24h.

Emergency halting of bleeding ● Endoscopic sclerotherapy.
- Oesophageal transection with anastomosis (with a stapling gun).
- Transthoracic transoesophageal ligation of varices.
- The Tanner operation: gastric transection with reanastomosis.

Elective treatments to lower portal pressure Oral β-blockade (propranolol) significantly reduces 1° & 2° bleeding and mortality from large varices;[1] intensive sclerotherapy régimes can also substantially reduce the risk of rebleeding, but have not consistently improved survival (SE: mucosal necrosis, ulceration, stricture and dysphagia).[2]

Shunts (eg portal vein to vena cava—or splenorenal) do not save lives.[3]

Prognosis 40–70% of those bleeding from varices for 1st time will die. Unfavourable signs: jaundice; ascites; hypoalbuminaemia; encephalopathy.

1 C Hayes 1990 *Lancet* **336** 1533   2 Copenhagen Varices Project 1984 *NEJM* **311** 1594
3 N Bass 1987 *NEJM* **317** 893

Bleeding varices
[*OTM* 12.21; 12.231; 12.242]

Child's grading of severity of liver disease in portal hypertension

Grade	Serum bilirubin	Serum albumin	Ascites or encephalopathy	Operative mortality
A	Normal	≥35g/l	none	2%
B	20–50μmol/l	30–35g/l	mild	10%
C	>50μmol/l	<30g/l	severe uncontrolled	50%

Endoscopy

Endoscopy is either diagnostic or therapeutic (marked ● below).

Oesophago-gastro-duodenoscopy involves swallowing a flexible endoscope after sedation, eg diazepam 0.2–0.3mg/kg IV. (Use Diazemuls® emulsion, which is less toxic to veins; monitor with pulse oximetry.) *Pre-op care:* Fast from midnight; consent; warn not to drive till 12–24h after endoscopy; arrange follow-up. *Mortality:* 1:2000. *Morbidity:* 1:200.

Diagnostic indications:	Dysphagia	Dyspepsia
to confirm or exclude:	Haematemesis	Melaena
peptic ulcers	Ingested foreign body	Obscure pain
upper GI carcinoma	Biopsy after a	Pyloric
Post-gastrectomy symptoms	suspicious Ba meal	obstruction

Endoscopic ultrasonography uses a transducer on the tip of the endoscope so allowing better resolution (<1mm) compared with ordinary ultrasound of organs such as the heart or lungs.[1]

Therapeutic indications: ● Diathermy of bleeding ulcers.
● The sclerosis of varices using ethanolamine oleate injections.
● Dilating benign strictures.
● Intubation of obstructing growths using phlanged prostheses (eg Atkinson tubes).
● Lasers for palliation of oesophageal carcinoma or treatment of upper GI bleeding from angiodysplasia or vascular anomalies.[2]

Endoscopic retrograde cholangiopancreatography (ERCP) uses a side-viewing duodenoscope. Mortality is <0.2% overall, but risk rises to 0.4–0.6% with endoscopic stone removal. A catheter is advanced from the endoscope through the ampulla into the bile ducts. X-rays after injection of contrast medium show lesions within the biliary tree and pancreatic ducts. Use antibiotic prophylaxis (eg ciprofloxacin 750mg PO 2h pre-op).

Therapeutic manoeuvres: ● Sphincterotomy with gallstone removal.
● Dilatation of benign biliary strictures.
● Palliation of malignant bile duct obstruction with stents.

Complications: (~1%) Pancreatitis; haemorrhage; ascending cholangitis; perforation. *Pre-op care:* as for liver biopsy (p506). *Pre-med:* pethidine 50mg IM 1h pre-op. Ensure surgical team knows of the patient.

Sigmoidoscopy is useful in many rectal conditions. It precedes barium enemas if cancer is suspected. Flexible instruments gain better access than rigid ones. Do a biopsy, even if appearances are normal. Inflammatory bowel disease, carcinoma, or amyloid may be diagnosed.

Proctoscopy is useful for showing piles in the lower rectum.

Colonoscopy Indications: symptoms suggestive of colonic malignancy, with normal radiology; for biopsy when radiology suggests malignancy or Crohn's; to estimate extent of UC; to do UC surveillance (p514); to look for malignant change in surgical resection lines; to investigate GI bleeding; *S bovis* SBE, p312. Preparation: 'instant' preparation may be tried with 2 hypertonic phosphate enemas (CI: severe constipation, diarrhoea, diverticular disease)—or sodium picosulphate (Picolax®) 1 sachet am & afternoon the day before; or cleansing solutions (GoLytely®) 3–4 litres PO/NGT. Therapeutic indications: ● Polypectomy, using a diathermy snare.
● Diathermy to angiodysplasia. ● Volvulus treatment.

1 P Cotton 1985 *BMJ* i 1373   2 S Bown 1985 *Gut* **26** 1338

Gastroenterology

Endoscopy
[*OTM* 12.1]

499

The mouth

▶ The diagnosis will often come out of your patient's mouth: therefore open it, and inspect it with a bright light.

Aphthous ulcers These are painful white ulcers which occur singly or in crops on the tongue or buccal mucosa. They are associated with Crohn's disease, coeliac disease and Behçet's syndrome (p694), but they also occur in 20% of the normal population. Viruses are a common cause (herpes simplex). Treatment is difficult, and often not needed. Hydrocortisone 2.5mg lozenges held on the ulcer may help, as may tetracycline mouthbath (125mg/5ml), 10ml/8h held in the mouth for 3mins, for ≤3 days.

Other causes of oral ulcers: Trauma (eg false teeth), lichen planus, erythema multiforme, pemphigus, pemphigoid, infections (viral, syphilis, Vincent's angina, p710). ▶ Exclude malignancy in any ulcer which does not heal after 3 weeks (eg by biopsy) as typical features of a rolled edge and induration may be absent.[1]

Oral candidiasis (thrush) may cause small white mucosal flecks, surrounded by a small margin of erythema. Those at risk: the very young and old, DM, the immunosuppressed (eg cytotoxics; steroids; AIDS). ▶ The finding of oral candidiasis in an apparently fit young man suggests AIDS. Differential diagnosis: lichen planus (here the white 'spider's web' patches cannot be wiped off, unlike candida patches). Treatment: slowly suck a nystatin pastille (100,000u) every 6h PO pc, or amphotericin lozenges (10mg) sucked every 3–6h. Systemic candidaemia is a rare complication. Treat with IV amphotericin (p244).

Leucoplakia may follow trauma. It is a white patch, stuck to tongue or oral mucosa. ▶ Frank malignancy may coexist with leucoplakia. *Hairy leucoplakia* is a shaggy white patch on the side of the tongue. Associations: HIV; opportunistic infections. It is not premalignant.[2]

Pigmented lesions in the mouth A bluish tinge opposite the molars suggests Addison's; a dark line below the gingival lining suggests heavy metal poisoning. Brown spots on the lips suggest Peutz–Jeghers' syndrome (p706). Tiny yellow spots may be *Fordyce spots* (sebaceous glands). Telangiectasia may signify Osler–Weber–Rendu syndrome (p706).

Gingivitis Gum inflammation may occur with phenytoin; cyclosporin or nifedipine treatment; scurvy (p520); acute myeloid leukaemia, typically M4—(p602); Vincent's angina (p710).

The tongue may be *furred and dry* (in dehydration, or Sjögren's syndrome), or *black and hairy* (due to overgrowth of papillae and *Aspergillus niger*, eg after antibiotics). Causes of smooth tongues (*glossitis*): iron or vitamin deficiency. Loss of patches of papillae gives rise to the harmless *geographic tongue*. Its outline often changes.

Carcinoma of the tongue typically appears on the tongue's edge as a raised ulcer with firm edges and indurated environs. Examine underneath of tongue and ask patient to deviate his extended tongue sideways. Spread: the anterior ⅓ of the tongue drains to the submental nodes; the middle ⅓ to the submandibular nodes, and the posterior ⅓ to the deep cervical nodes. Treatment: surgery or radiotherapy.

A *ranula* is a bluish salivary retention cyst to one side of the frenulum, named after the bulging vocal pouch of frogs' throats (genus *Rana*).

1 C Grattan 1986 BMJ i 1093   2 *Lancet* Ed 1989 ii 1194

The mouth

Obesity

Those who are thin are often aggressively punishing to the obese. The thin doctor is in danger of abusing his authority by using it as a weapon with which to attack the already vulnerable obese patient. Because if we starve we lose weight, we falsely suppose that our weight is simply under our control. It is not. The control is most complex and includes genetic and physiological factors. Attempts to reduce weight may be resisted by the body's altered metabolism, which, like a thermostat, varies in such a way as to maintain weight around a 'set-point'. An obese person who diets to maintain an average weight experiences the same unpleasantness as does a thin person who is starving himself. So be cautious in telling your obese patient that he cannot have his gall bladder removed unless he first loses 4 stone in weight.

Only occasionally is obesity caused by disease, and obese people do not need costly tests to rule out such diseases as Cushing's, hypothyroidism or hypothalamic disorders, unless there are other signs of these conditions. But there is no doubt that obesity does cause medical complications. Surgery is more dangerous, hypertension more common. There is an excess mortality from diabetes mellitus, heart disease, stroke, pneumonia and accidents.

Classification[1] A simple classification is based on the *body mass index* (BMI) which is: weight/(height)^2, weight in kg, height in metres. The normal range is 20–25. Grade 1 obesity is defined as a BMI=25–30. Grade 2 as BMI=30–40, and grade 3 as BMI >40.

Management[1,2] Weight loss for all grades of obesity should be slow. An average of 0.5–1.0kg weight loss per week is desirable. Rapid weight loss due to a large calorie deficit results in loss of water, glycogen and lean body mass rather than loss of fat. The result is rapid weight gain following cessation of the ferocious diet.

Grade 1 obesity is best managed through healthy diet (p482) and exercise. Aim to promote a healthy life-style for long-term management.

Grade 2 obesity Individual behaviour modification together with an appropriately reduced calorie diet (under medical supervision) will be needed. Psychotherapy may help to enable the patient to understand and eliminate problematic eating habits.

Grade 3 obesity is associated with considerable morbidity and the patient should be informed of this. Dietary management is the same as for grade II obesity, although it is probable that psychotherapy will be needed for long-term success.

Self-help groups can be used for all grades of obesity.

Appetite suppressants (eg fenfluramine) have a limited use. However, in patients reporting failure of compliance due to hunger, they may be of help. They should be used under expert guidance only.

Gastric restriction surgery for grade 3 obesity is an option if all the above methods have failed.

1 J Garrow 1981 *Treat Obesity Seriously*, Churchill Livingstone, Edinburgh   2 K Brownell 1985 in D Barlow (Ed) *Clinical Handbook of Psychological Disorders*, Guildford Press, New York   3 D Murphy 1986 *BMJ* i 992

Obesity

Acute liver failure

Liver failure typically presents the following problems: *encephalopathy*, *hypoglycaemia*, *bleeding*, *overwhelming infection*, and *ascites*. It may occur suddenly in people who have been previously healthy, or, in cirrhotics, it may be precipitated by: infection (always consider spontaneous bacterial peritonitis); sedation; GI bleeds; diuretics and U&E imbalance—they have signs of chronic liver disease (spider naevi, liver palms, gynaecomastia, small testes, jaundice, ascites and oedema). If liver failure occurs within 1 week of the onset of jaundice, it is called *hyperacute* (the group with the best prognosis, even though cerebral oedema is common) if from 1–4 weeks it is *acute*, and it is termed *subacute* if the interval is 5–12 weeks.

Causes Any viral hepatitis (*delta virus co-infection with hepatitis B increases risk); drugs* (paracetamol excess; halothane; isoniazid; methyldopa); fulminant Budd–Chiari (p696); acute fatty liver of pregnancy; carbon tetrachloride; Weil's/Wilson's diseases (* =poor prognosis).

Encephalopathy, cerebral oedema and coma The cause is thought to be excess ammonia, mercaptans and false transmitters. *Early signs:* Being slovenly, lethargy, delirious, forgetful, or psychotic ± coarse tremor (liver flap), incontinence, ophthalmoplegia, extrapyramidal signs, confusion. There may be no jaundice or hepatic foetor. Diagnosis is clinical, although EEGs will often show high voltage slow waveforms. Blood ammonium (uncuffed) may be helpful in monitoring progress. A good sign of early encephalopathy is difficulty in drawing 5-pointed stars (constructional apraxia). Serial attempts are easily compared. *Management:*
- Place in the coma position. Take to ITU. Seek expert help.
- Caution with secretions and blood if hepatitis B is suspected.
- NGT to aspirate blood from stomach (bleeding varices, p496). It is wise to protect the airway with an endotracheal tube.
- Monitor TPR; BP; pupils; urine output; blood glucose; INR; U&E; LFTs; EEG. Perform daily weight and blood culture.
- Prevent worsening coma by emptying the bowel with magnesium sulphate enemas. Aim for 2 soft stools/day (and no diarrhoea).
- Give neomycin 1g/6h PO to reduce numbers of bowel organisms.
- Consider haemodialysis if water overload develops.
- Give cimetidine to keep the gastric pH >5, eg 200mg/8h IV.
- Contact the nearest liver centre, and ask about elective ventilation, prophylactic antibiotics/antifungals, inotropic support, monitoring of intracranial pressure, renal support. Is urgent transfer needed?
- Avoid: sedatives (but diazepam is useful for seizures), drugs with hepatic metabolism. If protein withheld initially give 50% dextrose IV for calories. After 48h protein will be required (obtain expert advice).
NB: attempts to correct acid–base balance are often harmful.

Bleeding ● Vitamin K 10mg/day IV for 2–3 days.
- Platelets (p610) and fresh frozen plasma and blood as needed.

Serious infection Do appropriate cultures. Until sensitivities are known, give gentamicin 1mg/kg/8h + benzylpenicillin 1.2g/8h + metronidazole, eg 500mg/24h (usual *tds* dose is probably too high), all slowly IV.

Ascites The management is described on p506.

Hypoglycaemia Measure blood glucose (eg every 4h). Give IV glucose (eg 50g) if levels fall below 2mmol/l. Watch plasma K+.

Liver transplantation (eg partial orthotopic 'auxiliary')—consider for poor prognosis groups (marked * above), *before* irreversible CNS damage occurs. Immunosuppression with tacrolimus is improving results.

Pain control For mild analgesia, oral paracetamol in normal doses is safe. Avoid NSAID, if possible, because of danger of GI bleed. Opiates are valuable powerful analgesics, but begin at low doses and well-spaced (eg every 12–24h instead of every 4h) until the response of patient is known.

Cirrhosis

Essence The architecture of the liver is irreversibly destroyed by fibrosis and regenerating nodules of hepatocytes.

Causes Unknown (30%), alcohol (25%), post hepatitis B (or C). Others: chronic active hepatitis (p512), Wilson's disease (p712), the Budd–Chiari syndrome (p696), haemochromatosis (p507), and α_1-antitrypsin deficiency (rare autosomal recessive disease often accompanied by emphysema, and caused by lack of the enzyme which counters effects of trypsin; levels can be measured in the blood—normal range 1.3–3.28g/l) p650.

Clinical features These are secondary to hepatocyte failure and portal hypertension. There may be none, depending on the degree of 'compensation', or there may be jaundice, palmar erythema, leuconychia, Dupuytren's contracture (p696), gynaecomastia and small testes (oestrogen metabolism is slow), clubbing or signs of encephalopathy such as liver flap (a slow hand tremor worsened by wrist extension). Bleeding varices, splenomegaly and ascites signify portal hypertension. Secondary hyperaldosteronism and hypoproteinaemia add to the ascites. Altered vitamin D metabolism may lead to osteomalacia.

Spider naevi: These comprise a central arteriole with leg-like branches. They blanch on pressure and are usually (but not always) confined to skin areas drained by the superior vena cava.

Plasma biochemistry is variable, eg bilirubin↑, AST↑, INR↑ (synthesis of clotting factors↓), glucose↓, Na⁺↓, albumin↓.

Diagnosis This is by liver biopsy. Ask for the patient's consent, and arrange FBC, platelets, INR and LFTs. Group and save serum. Biopsy is not safe if the INR is prolonged. After biopsy the patient should lie on the right side for 2h (24h in bed) and have vital signs monitored (eg every ¼h at first). Next day check for signs of pneumothorax and blood or bile leaks. Avoid strenuous exercise for the next week. Arrange follow-up.

Management Avoid (or treat) precipitants of liver failure: alcohol; dehydration; infection; GI bleeds; hepatotoxic drugs. Nutritional support if malnourished (p104).

Ascites: Fluid cytology to exclude other causes (malignancy, TB). Start a low salt diet (40–60mmol/24h if ascites moderate to severe; 80–100mmol/24h if ascites mild), with fluid restriction (<1 litre/day). Check U&E, plasma creatinine, daily weight and urine volume. Give spironolactone 100mg/24h PO, increasing the dose every 2 days up to 600mg/24h. Aim for a daily weight loss of ≤½kg/day (to prevent uraemia). To achieve this loss it may be necessary to use frusemide, up to 120mg/24h PO. Beware hypokalaemia. Stop diuretics if encephalopathy occurs.

Encephalopathy: See p504. *Bleeding varices:* See p496.

Pruritus: Cholestyramine 4g/8h PO, taken 1h after other drugs.

Bad prognostic signs: Encephalopathy, hypoalbuminaemia, plasma Na⁺ <110mmol/l.

Primary haemochromatosis

A not uncommon, and a *treatable* condition: so think of it!

Essence An autosomal recessive condition. Failure of regulation of iron absorption from the small bowel leads to overload, with deposition in, and damage to cells of the liver, heart, pancreas, and pituitary. 11% of the Caucasian population are heterozygous 'carriers' for the gene, which exists on chromosome 6, closely associated with HLA A3. Women (protected by menstrual loss) present 10yrs later than men. Early features: fatigue, arthralgia, arthritis, gonadal failure, hepatomegaly. Chronic liver disease, diabetes, cardiomyopathy develop later. Remember skin pigmentation.

Assessment It may be very difficult to differentiate primary iron overload from overload secondary to alcoholism, or chronic haemolysis, even at biopsy. Take a careful drug and family history.

Tests: LFTs: often normal, even in established disease. Do serum iron and total iron binding capacity (saturation >70%); ferritin (>1000µg/l). Proceed to liver biopsy to define tissue damage, and % iron concentration by weight (>2%). Look for evidence of damage to other organs.

Management Venesection has markedly reduced mortality, although symptom relief is variable. Venesect until iron deficient: this may require months. Monitor iron, TIBC and LFTs.

● Screen family (ferritin). ● Beware hepatoma as a complication.

507

Primary biliary cirrhosis (PBC)

This is a slowly progressive non-suppurative cholangiohepatitis, with destruction of the small interlobular bile ducts. The cause is unknown (putatively autoimmune). 90% are women; peak age: 45yrs. The spectrum of disease activity is very variable. In some surveys, 50% of newly diagnosed patients are symptom-free.[1]

Clinical features *Major:* Pruritus, hepatosplenomegaly, cholestasis (so pale stools and dark urine), melanotic skin pigmentation.

Others: Clubbing, xanthomata, xanthelasmata, arthralgia, fatigue, osteoporosis, osteomalacia, hirsutism, portal hypertension.

Associated diseases (typically in glandular organs) Thyroid disease, Sjögren's syndrome (eg in 70%), pancreatic hyposecretion, CREST syndrome (p670).

Tests At diagnosis plasma bilirubin is rarely raised. Alk phos↑ (but normal in 14%). AST may be slightly raised. Liver biopsy confirms the diagnosis (granulomata imply a good prognosis). ERCP (p498) excludes extrahepatic cholestasis. Do T₄ (often↓).

Immunology: Antimitochondrial antibody is found in 95% of patients, and titres >1:80 make the diagnosis of PBC particularly likely. The antibody may rarely be found in normal people, or those with thyroid disease (and normal liver biopsy). IgM↑.[2]

Staging I Destruction of interlobular ducts; II Ductal proliferation; III Fibrosis; and IV Frank cirrhosis.

508

Treatment This is purely symptomatic (penicillamine has not been found to be helpful[3] and steroids are not indicated)—so no treatment is needed in the asymptomatic. Variceal sclerotherapy may be needed if there is portal hypertension. Cholestyramine 4g/6–12h PO may relieve pruritus, but may worsen osteomalacia by reducing the absorption of vitamin D. If pruritus does not respond, consider ultraviolet light treatment or plasmapheresis (which also benefits xanthomata and xanthomatous neuropathy). Malabsorption prompts the offering of a low-fat diet (restrict according to tolerance) and means that vitamins may be needed: vitamin D as four 'calciferol high strength tablets' (10,000 units each) may be needed for osteomalacia. Also give vitamin A 100,000 units/month IM and vitamin K 10mg/month. If a low fat diet is required then carbohydrate and medium chain triglycerides (MCT) supplements will be needed to maintain or increase energy intake.

Azathioprine has been tried with some success.[3] Liver transplantation is offering encouraging results, both in terms of longevity and quality of life.[3]

Prognosis With severe disease death mostly occurs after 5–7yrs.

1 D Triger 1982 *BMJ* i 1898   **2** E Dickson 1985 *NEJM* **312** 1011   **3** N Finlayson 1987 *BMJ* ii 867

Primary biliary cirrhosis (PBC)
[*OTM* 12.233]

Neoplasia and the liver

Not infrequently, the first place that malignancy declares itself is in the liver, as metastases. These have a characteristic knobbly feeling, and are often found in association with cachexia. Frank jaundice is much less common.

Primary site	Male:	Females:	Rarer malignancies:
	Stomach	Breast	Pancreas
	Lung	Colon	Leukaemia
	Colon	Stomach	Lymphoma
		Uterus	Carcinoid tumours

Differential diagnosis of the knobbly liver Secondaries are by far the commonest cause. Other causes: cirrhosis (the liver is usually shrunken), syphilitic gumma, cysts, abscesses, and primary tumours. The last four are all rare in the UK. Hepatocellular carcinoma is a major cause of malignancy in Africa, South-east Asia, Taiwan, Japan, Greece and Italy.

Investigations Skilful ultrasound is the most helpful initial test for determining the cause of focal liver lesions. Radionuclide scans have poorer resolution.[1] CT also has a rôle.

Cystic lesions: Think of congenital liver cysts, polycystic liver, or a hydatid cyst (p248). If the patient is 'septic' (eg fever with a high wcc) consider pyogenic or amoebic abscesses.

Solid lesions: Suspect malignancy. If the ultrasound scan clearly shows that this is the case further tests are likely to be pointless, as the disease will be too far advanced to treat. Otherwise, tests to consider include cxr, mammography, upper and lower gi endoscopy and bone scans to find the primary malignancy.

If the lesion is a solitary secondary, consider resection. If there are no identified metastases, angiography may be indicated to exclude an haemangioma. Laparoscopy and laparotomy have a place in identifying focal liver lesions.

▶ *If the patient is terminally ill, omit all these tests.*

Primary hepatocellular carcinoma *Incidence:* 1 million deaths/yr, world-wide. *Causes:* hepatitis B virus (the cause in 80%), cirrhosis (co-existing in 50%), aflatoxin. Contraceptive steroids used for >8yrs may increase the risk ~4-fold.[2]

Symptoms: Right upper quadrant pain; weakness and weight loss.

Signs: Hepatomegaly (eg stony hard, with an arterial bruit); ascites; jaundice; signs of cirrhosis. Intraperitoneal bleeds may be devastating. Hypoglycaemia and feminization are rare.

Tests: cxr—raised R hemidiaphragm; lung metastases. Technetium scans, ultrasound and ct scans are useful in diagnosis and in finding a good site for a biopsy. Plasma α-fetoprotein often ↑.

Hepatoma treatment: Chemotherapy, radiotherapy, and transplantation are disappointing. Surgical resection of tumours <3cm diameter gives a survival rate of 59% (baseline 3-yr survival rate is 13%).[3]

Prognosis: Death is usually within six months.

Prevention: Use hepatitis B vaccine (p512).

1 J Thompson 1985 *bmj* i 1643   2 R Doll 1986 *bmj* i 1355 & 1357   3 T Ezaki 1992 *bmj* i 196

Neoplasia and the liver

[*OTM* 12.256]

511

Chronic hepatitis

Chronic persistent hepatitis is an inflammatory reaction which has persisted for >6 months. It is benign, and will remit.

Pathology: Portal fibrosis, but no piecemeal or bridging necrosis.

Causes: Viral hepatitis, alcohol, isoniazid, methyldopa, cytotoxics.

Signs: There may be antecedent acute hepatitis. The liver may be large and tender. Signs of chronic liver disease are absent. **Tests:** Serum transaminase↑ 5–10-fold. IgG ↔. Do a liver biopsy to exclude CAH.

Chronic active hepatitis (CAH) is a slowly progressive condition, and may lead to cirrhosis. There are 2 types—patients may have either:

1 *Lupoid CAH* (often young women). ANF +ve. IgG markedly raised.

2 *Type B CAH* Causes: alcohol, hepatitis C, hepatitis B—particularly if there is superinfection with hepatitis D virus (δ Ag or anti-δ IgM in serum),[1] which is common in IV drug abusers.

About 30% of hepatitis B carriers have CAH, and their mortality is 25–50%. Progression is variable, and remission may occur if HBeAg disappears from the serum and antibody (anti-HBe) appears. There is a strong link between liver cancer and chronic hepatitis B infection.

Pathology: Piecemeal necrosis is the most important process to recognize (death of hepatocytes at an interface between parenchyma and connective tissue, with infiltration of plasma cells and lymphocytes). Other features: fibrosis and bridging necrosis (between central vein and portal tract) with or without the changes of acute hepatitis or cirrhosis.

Clinical features: (variable) Jaundice, exhaustion, spider naevi, hepatosplenomegaly. CAH may present incidentally, eg in blood donors. Lupoid CAH may present with rashes, Cushingoid features, arthropathy and uveitis.

Treatment (for HBsAg −ve patients): Prednisolone 30mg/24h PO for 1 week, reducing to 10–20mg/24h for 6 months. If repeat biopsy shows remission, gradually withdraw steroids. If there is no remission, consider adding azathioprine 50mg/24h PO. Check LFTs monthly (but transaminase levels do not always reflect disease activity).[2] Bed rest is not required. Avoid aspirin and alcohol.[3] Mild forms may not require steroids.[2]

For HBsAg +ve patients steroids are not indicated in the viral replication phase (while HBeAg is present). Interferon given with acyclovir may reduce viral replication and is under trial.[4]

Prevention (of hepatitis B and hepatitis-B-associated cirrhosis, chronic hepatitis and hepatic neoplasia) Use hepatitis B vaccine (Engerix B®) Dose: 1ml into deltoid, repeated at 1 and 6 months (child: 0.5ml ×3—into anterolateral thigh). *Indications:* either everyone (WHO recommendation) or all at-risk populations—health workers (including GPs), dentists, homosexuals, IV drug abusers, those on haemodialysis and the sexual partners of known hepatitis B e-antigen positive carriers. Immunocompromised patients may need further doses. CI: any severe febrile infection. Antibody production may be checked one month after the last dose. Levels eventually begin to wane, so consider revaccination after 3–5 years. Note: protective immunity begins about 6 weeks after the first immunizing dose, so it is inappropriate for those who have had recent exposure. For such people, specific antihepatitis B immunoglobulin is the treatment of choice if they have not previously been immunized.

1 A Shattock 1985 *BMJ* i 1377   **2** J Hegarty 1985 *BMJ* i 877   **3** *Drug Ther Bul* 1983 **21** 89
4 G Alexander 1986 *BMJ* i 915

Chronic hepatitis
[*OTM* 12.221]

Ulcerative colitis (uc)

▶ Suspect uc whenever bloody diarrhoea lasts more than 7 days.

Essence uc is an idiopathic recurrent inflammatory disease of the large bowel—always involving the rectum. Outside the tropics it is the commonest cause of prolonged bloody diarrhoea.

Symptoms Blood and mucus passed PR with a formed stool suggests that disease is limited to the rectum; if accompanied by diarrhoea, suspect more extensive disease. Other features:

Fevers ('uc may present as a PUO)	Abdominal pain
Aphthous ulcers	Weight loss
Faecal urgency with incontinence	Constipation

Signs (of extensive, active disease) Fever; arthritis; iritis
>12 stools/day; GI distension — Anaemia (eg PCV <75% of usual)
Pyoderma gangrenosum (p686) — Erythema nodosum (p686)

Diagnosis *Sigmoidoscopy:* Red, raw mucosa; contact bleeding; inflammatory 'pseudopolyps'; discrete ulcers rare (unlike Crohn's).
Biopsy: Inflammatory infiltrate; mucosal ulcers; crypt abscesses.
Abdominal x-ray: May show colonic dilatation (± perforation).
Barium enema: Loss of haustra, mucosal distortion and colonic shortening are chronic changes. 'Backwash' terminal ileitis.
▶ Exclude amoebiasis—by 'hot' stool examination and serology.

5-day régime (for severe disease) ● Get expert help. Inform surgeons.
● Nil by mouth. Set up IVI (eg 1 litre 0.9% saline + 2 litres dextrose-saline/24h, + 20mmol K+/litre—less if elderly). Chart: TPR, BP.
● Twice daily physical examination. Record stool frequency & character.
● Daily: FBC, U&E, plain films, abdominal girth.
● Hydrocortisone 100mg/6h IV plus two 125mg hydrocortisone acetate foam enemas/day (~50% will be absorbed systemically). Reduce the dose after a week, according to the response.
● Twice daily physical examination. Inform surgeons of progress.
● Consider the need for IV nutrition. Give IM vitamins (p682).
● Colectomy indications: deteriorating colitis after 5 days; 'toxic' dilatation of colon (megacolon); perforation. 'Toxic' means the patient is worsening. Total surgical mortality: 2–7%.

In the less severely ill, steroids may be given PO or PR. Avoid antidiarrhoea drugs: they may promote colonic dilatation.

Maintaining remission Sulphasalazine is sulphapyridine (carrying agent) + 5-amino salicylic acid (5-ASA, the active agent). 1g/12h PO reduces the relapse rate by 65%. SE: rash, infertility. Mesalazine (5-ASA alone) 400–800mg/8h PO is as effective and without the sulphonamide SEs. Long term (eg 6 month) azathioprine may reduce need for steroids. Monitor FBC.

Poor prognosis Early severe illness; extensive disease.

Associations Sacroiliitis, ankylosing spondylitis, cholangitis, hepatitis, amyloid and colonic carcinoma (whose risk depends on the duration and extent of uc; overall risk 11% after 26yrs).

Surveillance Colonoscopy and multiple biopsy is better than sigmoidoscopy. For mild dysplasia, this only needs to be done every 1–2yrs (more frequently if the dysplasia is high grade).

Surgery Proctocolectomy and ileostomy, or ileoanal anastomosis.

Nutrition As for Crohn's disease (p516).

1 J Kirsner 1982 *NEJM* **306** 775, 837   **2** *Lancet* Ed 1986 ii 197

Ulcerative colitis (uc)
[*OTM* 12.126]

Crohn's disease

This chronic inflammatory disorder may affect any part of the gut, especially terminal ileum, colon and the anorectum, with ulcers, fistulae and granulomata. The inflammation is focal, with microerosions, fissuring ulcers, lymphoid aggregates and neutrophil infiltrates. Unlike ulcerative colitis, the whole thickness of the bowel wall may be affected. The cause is unknown, but one model postulates immune hyperreactivity of the gut (lymphocyte activity↑) with excessive GI vulnerability to antigens.[1]

Clinical features Fever, diarrhoea, cramping abdominal pains, subacute GI obstruction, weight loss and slowing of growth in children. Rectal involvement and rectal bleeding is less common than in ulcerative colitis (UC), fever and abdominal pain more so. Anal and perianal lesions (pendulous skin tags, abscesses and fistulae) are characteristic. Granulomata may occur in skin, epiglottis, mouth, vocal cords, liver, nodes, mesentery, peritoneum, even bones, joints, muscle or kidney. Differentiating UC from Crohn's is possible in 80% of patients.

Malignancy Predisposing risks to large and small bowel cancer: prolonged illness (15–20yrs); chronic GI obstruction; blind loops (after surgery). Lymphoma of the colon may also occur.

Barium studies Strictures, 'rose thorn' ulcers, 'cobblestone' mucosal surfaces. ▶ Look for an associated carcinoma. Note: plain abdominal films of good quality may obviate the need for barium studies—which may occasionally be dangerous.

Treatment Drugs used are as for UC, but they are less effective.

Prednisolone relieves symptoms, but there is no evidence that it alters prognosis. A suitable starting dose for symptom control is 20–30mg/day PO. Rectal disease may benefit from hydrocortisone acetate foam enemas, 1 applicator (125mg)/12–24h PR, for 1–2 weeks, and then on alternate days. Metronidazole 400mg/12h PO may help non-responders with perianal or colonic disease (?by reducing antigen load), but courses should last ≤3 months because of the risk of neuropathies. Azathioprine (2mg/kg/day PO) may allow withdrawal of steroid therapy in patients with chronic symptoms (SE: marrow suppression). Sulphasalazine may be tried, but does not help ileal disease[2] and has no role in maintenance therapy. If bile salt malabsorption causes diarrhoea, cholestyramine 4–8g/8h PO may be helpful. Antidiarrhoeals must be stopped if obstructive symptoms occur. If disease activity continues despite intensive treatment, seek expert advice. Low-dose cyclosporin (2.5–4mg/kg/12h PO for 3 months) has been tried to good effect—without inducing too much nephropathy.[3] ▶ Drugs must not be used in an attempt to delay appropriate surgery (eg for obstruction, fistulae and abscesses).

Pregnancy Steroids and sulphasalazine are permitted. Avoid metronidazole and azathioprine.

Nutrition There is evidence of food intolerance (eg cereals, milk, yeasts), and elemental diets may be as good as prednisolone in active disease. Reintroduction of foods one at a time maintains improvement without being too boring.[4,5] If steatorrhoea then give low-fat diet. If lactose intolerant, give lactose-free diet. Only consider total parenteral nutrition (TPN) if enteral route cannot be used (p106).

Surgery[1] Bowel resections; stricture relief. Surgery is never curative (unlike in UC). The aim is to control symptoms.

1 G Sachar NEJM 1990 **321** 894   2 Drug Ther Bul 1986 **24** 13   3 J Brynskov 1989 NEJM **321** 845   4 DA Gorard 1993 Gut **34** 1198   5 AM Riordan 1993 Lancet **ii** 1131

Crohn's disease
[*OTM* 12.126]

The irritable bowel syndrome (IBS)

Essence Intermittent diarrhoea, abdominal colic *relieved by bowel action*, and 'bloated' feelings—due to altered gut motility. The disease is one of negatives: there is no accepted definition; no structural pathology; no test is diagnostic; its identification depends on excluding other disease—and there is no cure. The most positive thing to be said about IBS is that it is the commonest diagnosis made in GI clinics and ⩾20% of the general population will be affected at some time (luckily most visit no doctor). ♀/♂= 1.38.[1]

The typical patient is an adult with intermittent abdominal pain relieved by defecation, abdominal distension, and both more frequent and looser motions when the pain comes on (this may alternate with constipation).[1] Less discriminating features: defecation may be sensed as being incomplete ('rectal dissatisfaction'); rectal mucus. Symptoms are chronic, with remissions and relapse—which may be brought on by stress, food poisoning or changes in bowel flora produced by antibiotics.

Causes Psychological factors have been thought to be important, and more recently specific food intolerances have been implicated. IBS may be part of a generalized smooth muscle/autonomic disorder as there may be detrusor irritability, asthma, migraine, and flushing too.[1]

Differential diagnosis Colonic cancer; inflammatory bowel disease.

Tests No tests are in routine use to confirm the diagnosis. Radiotelemetry may demonstrate abnormal motility patterns (throughout the bowel, so demonstrating that the illness is not confined to the colon). It is wise to do a sigmoidoscopy, FBC, ESR and tests for faecal occult blood before the diagnosis is made. A double contrast barium enema is not required except in the elderly when malignancy must be excluded. Some doctors carry out many tests (eg barium studies, jejunal biopsy, colonoscopy with multiple biopsy, pancreatic scans, cholecystograms, culture of jejunal aspirates and stool, estimation of faecal water and fat, lactose tolerance tests, and laparotomy). These are rarely needed.

Treatment Reassurance, explanation and dietary fibre have been the mainstays of treatment. If food intolerance suspected, dietary supervision may help. Results are often disappointing and this may lead to the trial of many other remedies of uncertain value, such as mebeverine 135mg/8h PO ac (which acts on smooth muscle); anticholinergics such as dicyclomine (10mg/8h ac) and propantheline (15mg/8h ac with 30mg at night) and tranquillizers such as diazepam 2–5mg/8h PO.

Prognosis >50% will continue to have symptoms after 5 years.

1 R Jones 1992 *BMJ* **i** 87

Carcinoma of the pancreas

This accounts for 1–2% of all malignancies. The incidence is rising (more in America than in the UK). Risk factors: smoking—and possibly diabetes. There is no evidence that acute or chronic pancreatitis are risk factors.

Pathology 60% of tumours involve the head of the pancreas, 15% the tail, and 25% the ampulla. Usually enough β cells survive to maintain normoglycaemia. Pancreatic duct obstruction may cause pancreatitis.

Clinical features Abdominal pain (worse at night; radiates to back); weight↓; dyspepsia and pruritus. *Signs:* cachexia, fever, jaundice, and an enlarged gall bladder (suggesting gallstones, until Courvoisier's 'law' is remembered: *If in a case of painless jaundice, the gall bladder is palpable, the cause will not be gallstones*—the reason is that stones excite a fibrotic reaction in the gall bladder, which prevents later distension). Hepatomegaly is from biliary obstruction or metastases. Thrombophlebitis migrans (seen in 10%) and marantic endocarditis may both cause embolism.

Tests Ultrasound; CT; ERCP (p498); needle biopsy (eg under CT control).

Treatment Only 13% are suitable for Whipple's operation (partial gastrectomy + cholecystectomy + distal choledochectomy + partial pancreatectomy).

Survival Operative mortality (Whipple's): 20%. Often, palliative bypass surgery is all that can be done. Median survival after this is 24 weeks (after no treatment: 9 weeks; post-Whipple's: 40 weeks).

519

Carcinoid tumours

These tumours of argentaffin cell origin often appear in the appendix (25%), ileum or other GI site (rarely gall bladder, bronchus or gonads).

Histology Tumours are submucosal, rarely causing ulcers. They consist of small groups of indolent epithelial cells with malignant potential. 80% of large tumours (>2cm across) produce metastases (eg to liver).

Signs Before metastases these may be few. Tumours can cause: appendicitis, intussusception, GI obstruction and (in 5%) carcinoid syndrome.

The carcinoid syndrome This occurs only if secondaries in the liver. Features are: flushing (after alcohol, coffee, various foods or drugs), diarrhoea, oedema and asthma, associated with tricuspid incompetence or pulmonary stenosis (ie right heart)—caused by the release of pharmacologically active mediators (5-hydroxytryptamine, prostaglandins, kinins and polypeptides). These may rarely cause additional features such as Cushing's syndrome. Pellagra (below) may occur because the tumour consumes tryptophan, which is nicotinamide's precursor. There are 4 kinds of flushes: 1 Paroxysmal erythema; 2 Telangiectasia with a violaceous flush; 3 Prolonged flushes, eg for 48h (suggests bronchial carcinoid); 4 White and red patches (suggests gastric carcinoid).

Diagnosis 24h urinary excretion of >15mg of 5-hydroxyindoleacetic acid (5-HIAA) is suggestive. CT and laparotomy may be needed for localization. A gulp of whisky may help diagnostically by inducing the flush.

Treatment Surgery may cure. If metastases (do liver scan) palliate by avoiding precipitating foods. Methysergide 2mg/8h PO blocks 5-HT, and with cyproheptadine 4mg/8h PO may help diarrhoea. Phenoxybenzamine 20–200mg/24h PO can help flushing. Other procedures: enucleating liver metastases; hepatic artery ligation, embolization, 5-fluorouracil injection.

Prognosis Median survival 5–8yrs; 38 months if metastases are present.

Specific nutritional disorders

Scurvy This is due to lack of vitamin C in the diet. ► *Is the patient poor, pregnant, or on a peculiar diet?* Signs: 1 Listlessness, anorexia, cachexia. 2 Gingivitis, loose teeth and foul-smelling breath (halitosis). 3 Bleeding from gums, nose, hair follicles, or into joints, bladder, gut. *Diagnosis:* No test is completely satisfactory. WBC ascorbic acid ↓. *Treatment:* Dietary education, ascorbic acid 250mg/24h PO.

Beri-beri There is heart failure with generalized oedema (wet beri-beri) or neuropathies (dry beriberi) due to lack of vitamin B_1 (thiamine). For treatment and diagnostic tests, see Wernicke's encephalopathy (p712).

Pellagra This is due to lack of nicotinic acid. Other nutritional deficiencies are likely to be present. The classical triad of signs is: diarrhoea, dementia and dermatitis. Others: neuropathy, depression, tremor, rigidity, ataxia, fits. Pellagra may occur in the carcinoid syndrome. *Treatment:* education, electrolyte replacement, nicotinamide 500mg/24h PO.

Xerophthalmia This vitamin A deficiency is a major cause of blindness in the tropics. Conjunctivae become dry and develop oval or triangular spots (Bitôt's spots). Corneas become cloudy and soft (keratomalacia). *Treatment:* a good diet, vitamin A (retinol) 25,000U/24h PO.

Lathyrism This is an acute spastic paralysis occurring in lathyrus pea eaters (due to a toxin). Treatment is unsatisfactory.

Favism There is sudden severe haemolysis in those with certain types of G6PD deficiency on eating broad beans or inhaling their pollen.

Carcinoid tumours
Specific nutritional disorders
[*OTM* 12.1250; 8.12; 8.51]

521

Gastrointestinal malabsorption

Clinical features Diarrhoea; weight↓; fatty stools (steatorrhoea). Signs of deficiencies: anaemia (Fe↓, B_{12}↓, folate↓), bleeding (vit. K↓) oedema (protein↓). The common UK causes are coeliac disease, Crohn's and chronic pancreatitis.

Malabsorption due to mucosal causes *Coeliac damage (below).*
Tropical sprue: Villous atrophy and malabsorption of fats and vitamins. It occurs in the Far and Middle East and the Caribbean, but not in Africa. Cause: unknown. Tetracycline 250mg/6h PO and folic acid 15mg/24h PO may help. *Others:* Crohn's; amyloid; Whipple's disease (p712).

Intraluminal causes (non-structural, biopsy −ve)
Intestinal hurry: Gastrectomy; post vagotomy; gastrojejunostomy.
Infections: Giardiasis; diphyllobothriasis (B_{12} malabsorption); ankylostomiasis; strongyloidiasis.
Defective secretions of: Bile salts (eg obstructive jaundice); enzymes (pancreatic disease).
Bacterial overgrowth: Occurs spontaneously or in diverticula or postoperative blind loops. The condition may respond to intermittent courses of metronidazole (400mg/8h PO) or oxytetracycline (250mg/6h PO).

Tests *Haematology:* FBC (iron-deficient or macrocytosis).
Biochemistry: Ca^{2+}↓; PO_4^{3-}↓; Fe↓; RBC folate↓; INR↑ (vit. K↓).
Faecal fats: Collect stool for 3–5 days (normal fat <6g/24h).
Barium meal may show diverticula.
Breath hydrogen tests: Lactulose is ingested and if the malabsorption is due to bacterial overgrowth there is an increase in exhaled hydrogen 1h after ingestion (better than tests using radioactive ^{14}C bile salts).
Small bowel biopsy with the Crosby capsule: This is swallowed on the end of a tube, and is monitored by x-ray screening until it reaches the jejunum. It is fired by suction, and a biopsy is caught in its jaws.

Coeliac disease There is a permanent intolerance to gluten (a wheat protein) leading to villous atrophy and GI malabsorption. Associations: HLA DR3 (in 90%) and dermatitis herpetiformis.

Symptoms: As above. The stools stink, and are hard to flush away. Aphthous ulcers. Growth retardation, failure to thrive and osteomalacia may occur. Coeliac disease occurs at any age.

Complications: Anaemia; GI lymphoma (suspect if rapidly worsening); gastric/oesophageal cancer; myopathies; neuropathies; hyposplenism.

Tests As above. Reticulin antibodies may be found.

Diagnosis: Villous atrophy must reverse on a gluten-free diet, and to recur after gluten challenge (unlike the villous atrophy of sprue).

Treatment: Permanent gluten-free diet: no wheat, barley, rye, oats (usually), or any food containing them (eg bread, cake, pies). Rice, maize, soya, potatoes, sugar, jam, golden syrup and treacle are allowed. Gluten-free biscuits, flour, bread and pasta are prescribable in the UK.

Chronic pancreatitis This is characterized by prolonged ill health, recurrent abdominal pain, steatorrhoea (from malabsorption), weight loss, and diabetes mellitus. The pain characteristically radiates to the back.

Causes: Chronic alcohol abuse is the chief cause. Others: cystic fibrosis, haemochromatosis, obstructed pancreatic or biliary ducts (eg gallstones).

Gastrointestinal malabsorption
[*OTM* 12.98]

Tests: Plain abdominal x-rays: speckled pancreatic calcification. Plasma glucose↑. Other tests to consider: ultrasound, CT scan, ERCP (p498).

Treatment: Pancreatic extracts (pancreatin) by mouth preferably as enteric-coated capsule (eg Nutrizym GR® 1–4 caps with meals). A low-fat diet and avoiding alcohol may help. Give fat-soluble vitamins (eg Multivite pellets®). Medium chain triglycerides (MCT oil®) may be given (they do not require lipase for absorption). Operations for unremitting pain include pancreatectomy, and end-to-end or side-to-side pancreaticojejunostomy.

Gastroenterology

Notes

12 Endocrinology

526

Relevant pages in other chapters:
Diabetic emergencies (p738); the diabetic patient undergoing surgery (p108); thyroid emergencies (p740); thyroid lumps (p150); Addison's disease, thyroid disease and surgery (p110); Addisonian crisis and hypo-pituitary coma (p742); hyperlipidaemia (p654).

The essence of endocrinology

- Define a clinical syndrome.
- Match it to a gland malfunction.
- Measure the output of the gland in the peripheral blood. Is release diurnal?
- Find a radiological technique to image the gland.
- Find a way of stimulating the gland, and define variations from the norm.
- Find a way of inhibiting the gland.

Measure intermediate metabolites to further define pathophysiology and guide treatment.

Diabetes mellitus: classification and diagnosis

Definition Diabetes mellitus (DM) is a syndrome caused by the lack or diminished effectiveness of endogenous insulin and, characterized by hyperglycaemia and derranged metabolism.

Type I (=insulin-dependent DM, IDDM) Usually juvenile onset and may be associated with other autoimmune disease. There is insulin deficiency. Concordance 30–50% in identical twins. Associated with HLA DR3 and DR4 and islet cell antibodies around the time of diagnosis. Patients *always* need some insulin: they are prone to ketoacidosis.

Type II (=non-insulin-dependent DM [NIDDM] or maturity onset DM) Older age group and often obese. ~100% concordance in identical twins. Due to impaired insulin secretion and insulin resistance.

Note: NIDDM may eventually require insulin. IDDM may develop at any age. Insulin is likely to be needed in those with ketonuria, glucose >25mmol/l, sudden onset, weight loss, dehydration, ketoacidosis.

Factors predisposing to DM *Drugs* (steroids and thiazides), *pancreatic disease* (chronic pancreatitis, surgery with >90% pancreas removed), haemo-chromatosis, cystic fibrosis), *endocrine disease* (Cushing's, acromegaly, phaeochromocytoma, thyrotoxicosis), *others* (acanthosis nigricans, congenital lipodystrophy with insulin receptor antibodies, and glycogen storage diseases).

Presentation of DM Patients may be asymptomatic. *Acute:* ketoacidosis (p738)—unwell, hyperventilation, ketones on breath); few weeks of weight loss, polyuria and polydypsia. *Subacute:* history as above but longer and in addition lethargy, infection (pruritis vulvae, boils). *Complications* may be the presenting feature: infections, neuropathy, retinopathy, arterial disease (eg MI or claudication).

Diagnosis 1 Fasting venous plasma glucose >7.8mmol/l on two occasions. 2 Glucose tolerance test (GTT): fasting glucose >7.8 and/or 2h glucose ≥11.1mmol/l. 3 Glycosuria: should prompt further investigation even if symptomless (sensitivity 32%, specificity 99%[1]).

Note: 1 If 2 fasting glucose levels >6 but <8mmol/l, give dietary advice (as for NIDDM, p482) and repeat glucose in 6 months. Do not label as having DM (because of insurance implications).

2 Screening for glycosuria is easy but is not a cost-effective way of screening entire populations of symptomless people.[2] ~1% of the population have low renal threshold for glucose; simultaneous blood and urine glucose measurements can be used for diagnosis.

3 Impaired glucose tolerance (IGT) (see p529) carries a higher risk of developing microvascular disease, and ~5% may develop frank DM.

1 J Yudkin 1990 *BMJ* i 1463   2 D Mant 1990 *BMJ* i 1053

Diabetes mellitus: classification and diagnosis
[*OTM* 9.51]

WHO diagnostic criteria for diabetes mellitus

Two fasting glucose estimations

<6mmol/l	DM excluded
>6mmol/l but <7.8mmol/l	Impaired glucose tolerance (IGT)
>7.8 mmol/l	Diabetes mellitus

Oral glucose tolerance test

Fast patient overnight and give 75g of glucose in 300ml water to drink. Venous plasma glucose measured before and 2h after drink.

DM diagnosed if fasting glucose >7.8mmol/l and/or 2h glucose ≥11.1mmol/l/

Impaired glucose tolerance diagnosed if fasting glucose ≥6 but <7.8mmol/l and/or 2h glucose >7.8mmol/l and <11.1mmol/l.

The newly diagnosed diabetic

▶ *Patient motivation and education are the keys to success*.

The aim of the treatment is normoglycaemia, but if life on the verge of hypoglycaemia is intolerable, this aim may be modified. Strict plasma glucose control does reduce renal, CNS and retinal damage, but the price is, on average, one hypoglycaemic coma a year, and a life-style compromised by frequent blood glucose tests.[1] Discuss this with your patient.

Initial treatment need not include hospital—even in IDDM. If there is ketonuria or patient is ill or dehydrated, admit. Children are liable to get ketosis rapidly, so prompt paediatric referral is a must (OHCS p262–5). If pregnant, share care with an interested obstetrician (OHCS p156). Stress the need for special pre-conception counselling (OHCS p94).

Education is crucial. Establish rapport and emphasize the need for dietary and drug compliance. Teach how to monitor blood and urine glucose. Consider exposing to a 'hypo', and show how to abort it with dextrose sweets (carry them always). Explain that when ill more (not less) insulin is needed. Introduce to a specialist nurse, chiropodist, and diabetic association. Emphasize need for regular follow-up (p532). Tell the patient to inform their *driving licence authority* (OHCS p468 for restrictions).

Diet See the *healthy diet* on p482 (saturated fats↓, sugar↓, starch-carbohydrate↑, moderate protein) adapted to meet tastes and needs (eg to prevent obesity). It may be important to ensure that some starchy carbohydrate (bread, potato, pasta) is taken at each meal. Dietitians can help. If any hint of renal failure (creatinine↑, microalbuminuria) restrict protein, and consider ACE inhibitor. ▶ Always carry glucose to counter drug-induced hypoglycaemia (including from ACE-inhibitors[2,3]).

Insulin Strength: 100 u/ml. Formulations of different durations: soluble insulin is short-acting (peak 2–4h, lasting ~8h). Longer-acting suspensions have onset of action of 1–2h, peak 4–12h and duration 16–35h. The hierarchy of increasing lengths of action is: isophane insulin, IZS (amorphous), IZS (mixed), IZS (crystalline), protamine zinc insulin. Tailor the insulin régime to the individual by checking control at different times of the day (see p531). NB: IZS = insulin zinc suspension.

Oral hypoglycaemics *Sulphonylureas* stimulate insulin secretion and increase insulin sensitivity. *Tolbutamide* is short-acting. Useful in elderly patients as hypoglycaemia is less likely. Dose 0.5–1g/12h. *Glibenclamide*'s action is intermediate. Beware prelunch hypoglycaemia. Dose 2.5–15mg/24h with breakfast. *Chlorpropamide* is long-acting and can be given once a day—avoid in renal impairment and in the elderly (risk of hypoglycaemia). Dose: 100–500mg/24h PO with breakfast (no alcohol: it may induce flushing). Other SE: headache, photosensitivity, Na+↓.

Biguanides increase insulin sensitivity but do not stimulate secretion. Also inhibit hepatic gluconeogenesis and often induce anorexia and diarrhoea—and may be best suited to obese subjects as an adjunct. They do not precipitate hypoglycaemia and therefore may be taken by HGV drivers. Avoid in hepatic and renal impairment. Watch for lactic acidosis (rare, see p739). Dose (*metformin*) 500mg/8h PO after food.

α-glucosidase inhibitors Consider *acarbose* as an alternative or an adjunct to oral hypoglycaemics. It inhibits α-glucosidase, so slowing the breakdown of starch to sugar (in practice its effect can be disappointing).[3]

1 Diabetes control & complications trial research group 1993 *NEJM* **329** 977 & 1995 *An Int Med* **122** 561 & E-BM 1995 **1** 9   **2** MJ Carella 1994 *Arch Int Med* **154** 649   **3** *Drug Ther Bul* 1994 **32** 51

Dose: 50mg chewed with the 1st mouthful of each meal (start with a once-daily dose: max 200mg/8h). SE: wind (can be terrible!), abdominal distension/pain, diarrhoea. CI: ● Pregnancy ● GI obstruction ● Hernias ● Crohn's/UC ● Past laparotomy ● Hepatic or severe renal failure.

Some commonly used insulin régimes

1 A single dose of very long acting insulin (Ultralente®) with three doses of short-acting insulin (15–30mins before each meal). This tends to be favoured by younger subjects—greater flexibility in life-style. Also may be of value in those with erratic control.

2 A mixture of short- and medium-acting insulin (eg soluble and Lente) at 7am and 6pm (ie ½h before meal). Give ⅔ of total insulin in the morning and ²/₃ as Lente. To improve control before breakfast, adjust evening Lente; before lunch—morning soluble; before supper—morning Lente; before bed—evening soluble.

3 Single dose of soluble and Lente may be sufficient in the elderly. In hospital, begin with at least 8U of soluble insulin before meals, monitoring the blood glucose, and transfer to one of the above when control is achieved.

4 'Pen' devices (eg NovoPen II®) are popular because of their ease of use (eg during parties). Short-acting insulin is given 15–30mins before each meal, with basal insulin requirements being provided by a conventional evening dose of long-acting insulin; mixed soluble (short) and isophane preparations (medium) are also available (eg PenMix® 10/90, 20/80, 30/70, 40/60, 50/50).

NB: With human insulins some patients report loss of hypoglycaemic awareness. Encourage regular home blood glucose monitoring.

Assessment of the established diabetic

Continuing assessment of the diabetic patient has 3 main aims: 1 To educate. 2 To find out what problems the patient is having with the diabetes (glycaemic control and morale). 3 To assess the long-term complications of DM.

Assess glycaemic control from:
1 Home glucose records.
2 History of hypoglycaemic attacks.
3 Glycated (glycosylated) Hb (=HbA$_1$c; use FBC bottle)—levels relate mean glucose level over previous 8 weeks (ie RBC half-life). If mean levels are persistently <10%, then proliferative retinopathy is rare.
4 Fructosamine (glycated plasma proteins)—levels relate to control over past 1–3 weeks; cheaper and quicker than HbA$_1$c measurement but may not be available.

Assessment of complications
Check injection sites for infection, lipoatrophy, or lipohypertrophy.

Vascular disease Look for evidence of cerebrovascular, cardiovascular and peripheral vascular disease. MI is 3–5 times more common in DM. Stroke is twice as common. Reduce other risk factors (see p432). Treat hypertension vigorously (thiazides raise blood glucose & lipids; enalapril aids insulin sensitivity and may protect against renal disease—see p396).

Eye disease Blindness is common (15–30% of IDDM) but preventable. Test visual acuity with Snellen chart or reading test-types (p33) with refraction corrected (pinhole or glasses). Refer to ophthalmologist if acuity decreased.
- Cataracts are more frequent in DM (both senile cataract and juvenile 'snowflake' cataracts). The osmotic changes induced in acute hyperglycaemia are reversible.
- Rubeosis iridis is new vessel formation in the iris. It occurs late and can lead to glaucoma.
- Fundoscopy with the pupils dilated is essential for assessing retinopathy (if no glaucoma). Use tropicamide 0.5% drops or phenylephrine 10% drops (not if on MAOI) to dilate; latter reversible with thymoxamine 0.5%. (See table on p533).

Neuropathy Cause is unknown though ischaemia and altered cellular Na$^+$ permeability have been suggested.[1] Somatic neuropathy: 1 Symmetric sensory polyneuropathy—distal numbness, tingling and pain, usually worse at night. 2 Mononeuritis multiplex—especially III and VI cranial nerves. 3 Amyotrophy—painful wasting of quadriceps; reversible. Autonomic neuropathy: postural hypotension, gastroparesis, nocturnal diarrhoea, urinary retention and impotence.

Kidneys (p396) Check U&E regularly. Some centres use 24h urine collection for creatinine clearance and for quantitating albuminuria.

Metabolic complications: see p738. Diabetic feet: see p534.

1 *Lancet* Ed 1989 i 1113

Assessment of the established diabetic

Diabetic retinopathy

Pathogenesis: Capillary occlusion→vascular leakage→microaneurysms. Arterial or venous occlusion→hypoxia+ischaemia→new vessel formation. High retinal blood flow caused by hyperglycaemia (and BP↑ and pregnancy) triggers these events, and causes capillary pericyte damage.[1]

Artery occlusion causes *cotton wool spots*, and there may be *blot haemorrhages* at the interface with perfused retina. *New vessels* form in the ischaemic area, proliferate, and are capped by fibrous tissue.

▶ Pre-symptomatic screening enables laser photocoagulation to be used. Find *where* lesions are. Darken the room; use 0.5% tropicamide mydriasis and a fine ophthalmoscope beam. The macula is the sacred place, being the only part of the retina with 6/6 vision. If there are signs here, refer at once. Background retinopathy (microaneurysm 'dots' + microhaemorrhage 'blots') are not urgent, but indicate that there is everything to be gained by achieving normoglycaemia, so lowering retinal flow.

533

---

1 EM Corner 1993 *BMJ* ii 1195

The diabetic foot

▶ Amputation is preventable: good care saves legs.

Feet of all patients with DM should be examined regularly. The feet are affected by a *combination* of peripheral neuropathy and peripheral vascular disease, though one or other may predominate.

Symptoms Numbness, tingling, and burning, often worse at night.

Signs Sensation↓ (esp. vibration) in 'stocking' distribution; absent ankle jerks; deformity (pes cavus, claw toes, loss of transverse arch, rocker-bottom sole). Neuropathy is patchy, so examine all areas. If the foot pulses cannot be felt, consider Doppler pressure measurement.

Any evidence of neuropathy or vascular disease puts the patient at high risk of foot ulceration. Educate (daily foot inspection, comfortable shoes—ie very soft leather, increased depth, cushioning insoles, weight-distributing cradles, extra cushioning—no barefoot walking, no corn-plasters). Regular chiropody. Treat fungal infection (p244).

Foot ulceration Usually painless, punched-out ulcer in an area of thick callous±superadded infection, pus, oedema, erythema, crepitus, odour.

Assess degree of: **1** Neuropathy (clinical). **2** ischaemia (clinical and Dopplers; consider angiography—even elderly patients may benefit from angioplasty). **3** Bony deformity eg Charcot joint (clinical, x-ray). **4** Infection (do swabs, blood culture, x-ray; probe ulcer to assess depth).

Management: Regular chiropody (ie at least weekly) initially to debride the lesion. Relieve high pressure areas with bed rest and special footwear. If there is cellulitis, admission is mandatory for IV antibiotics: start with benzylpenicillin 600mg/6h IV and flucloxacillin 500mg/6h IV ± metronidazole 500mg/8h IV, refined when microbiology results are known. Seek surgical opinion early.

Absolute indications for surgery
- Abscess or deep infection
- Spreading anaerobic infection
- Osteomyelitis
- Severe ischaemia—gangrene/rest pain
- Suppurative arthritis

The degree of peripheral vascular disease, patient's general health, and patient request will determine whether local excision and drainage, vascular reconstruction and/or amputation (and how much) is appropriate.

The diabetic patient with inter-current illness

The stress of illness tends to increase basal insulin requirement. If calorie intake↓ then increase long-acting insulin (eg by 20%) but reduce short-acting in proportion to meal size and check blood sugar frequently.

1 Insulin-treated; mild illness (eg gastroenteritis). Maintain calorie intake with oral fluids (lemonade etc). Continue normal insulin. Test blood glucose and urine ketones regularly (eg twice daily). Increase insulin if blood glucose consistently >10mmol/l.

2 Insulin-treated; moderate illness (eg pneumonia). Normal insulin and supplementary sliding scale of rapid acting insulin (p108), four times daily (before meals and bed-time snack).

3 Insulin-treated; severe illness (eg MI, severe trauma). IV soluble insulin by pump and IV dextrose (p108).

4 Diet and tablet treated; moderate/severe illness If on metformin, stop. If on sulphonylureas and illness likely to be self-limiting, keep on tablets, supplement with SC insulin and sliding scale or IV infusion (p108). Tail off insulin as patient recovers.

535

Hypoglycaemia

Definition Plasma glucose <2.5mmol/l. The threshold for symptoms varies markedly, especially in diabetic patients.

Symptoms *Autonomic*—Sweating; hunger; tremor.
Neuroglycopenic—Drowsiness; personality change; fits; rarely focal symptoms eg transient hemiplegia, loss of consciousness.

Two types:
1. Fasting hypoglycaemia (requires full investigation if documented).
Causes: By far the commonest cause is insulin or sulphonylurea treatment in the known diabetic. In the non-diabetic subject with fasting hypoglycaemia the following mnemonic is useful: EXPLAIN.
EX Exogenous drugs. Alcohol (often in alcoholics on a binge with no food), insulin, sulphonylureas (often paramedical or diabetic in the family).
P Pituitary insufficiency.
L Liver failure plus some rare inherited enzyme defects.
A Addison's disease.
I Islet cell tumours (insulinoma) and immune hypoglycaemia (eg anti-insulin receptor antibodies in Hodgkin's disease).
N Non-pancreatic neoplasms (especially retroperitoneal fibrosarcomas and haemangiopericytomas).
In addition *malaria* especially with quinine administration.

Diagnosis and investigations
1. Document the hypoglycaemia either by the patient taking finger-prick samples on filter-paper during attack or in hospital.
2. Exclude liver failure and malaria.
3. Admit for overnight fast. Take two separate samples for glucose, insulin, beta-hydroxybutyrate, and C-peptide.

Interpretation of results
1. Hypoglycaemia with high or normal insulin and no elevated ketones. *Causes:* insulinoma; sulphonylurea administration; insulin administration (no detectable C-peptide); insulin autoantibodies.
2. Insulin low or undetectable, no excess ketones. *Causes:* non-pancreatic neoplasm; anti-insulin receptor antibodies.
3. Insulin low or undetectable, ketones high. *Causes:* alcohol; pituitary/adrenal failure.

NB: if insulinoma suspected, confirm with a suppressive test: eg infuse IV insulin and measure C-peptide. Normally insulin suppresses C-peptide, but this suppression does not occur in patients with insulinomas. Localize the insulinoma using CT scanning. If none is visible sophisticated techniques such as intra-operative pancreatic vein sampling may be needed.

2. Post-prandial hypoglycaemia This occurs particularly after gastric surgery, and in those with mild type II diabetes.
Investigation: Prolonged OGTT (5h, p528).

Treatment See p738. Treat episodes with sugar orally if possible; otherwise with glucose 25–50g IV or glucagon 0.5–1mg SC (may be repeated after 20mins). If episodes frequent, advise many small meals high in carbohydrate. If post-prandial hypoglycaemia, meals should contain slowly absorbed carbohydrate (high fibre, complex carbohydrates). Insulinomas: surgical removal if possible; diazoxide and a thiazide diuretic.

Hypoglycaemia

Basic thyroid function tests

Abnormal hormone production is usually due to an abnormal thyroid. Primary abnormalities of thyroid-stimulating hormone (TSH) and thyrotrophin-releasing hormone (TRH) are very rare.

Basic physiology The hypothalamus secretes TRH, a tripeptide, which stimulates the production of TSH, a polypeptide, from the anterior pituitary. TSH increases the production and release of thyroxine (T_4 and triiodothyronine (T_3) from the thyroid. T_4 and T_3 reduce production of TSH which is the basis of the TRH test (p540). The thyroid produces mainly T_4, some of which is converted to T_3 in the blood or tissues. 85% of T_3 is produced this way; 15% is secreted from the thyroid. T_3 is five times as active as T_4. Most T_3 and T_4 in plasma is protein bound, mainly to thyroxine-binding globulin (TBG). It is the unbound portion which is active. T_3 and T_4 increase much cell metabolism. They are vital to normal growth and mental development. They also increase catecholamine effects.

Basic tests Write why you want the test. Different labs do different tests. In general, if you expect *hyperthyroidism* the most sensitive test is plasma T_3 (which is raised): if you expect *hypothyroidism* the most sensitive tests are *plasma T_4* (lowered) and TSH (raised).

1 Plasma T_4 (total T_4, ie total thyroxine)
Method: Collect any time, uncuffed.
Problems of interpretation: (for amiodarone see p544) False high (because bound, as well as free T_4 measured): pregnant; oestrogens; hereditary thyroid binding globulin excess.
False low: Salicylates; NSAID; phenytoin; corticosteroids; carbamazepine; thyroid binding globulin deficiency.

2 Plasma total T_3 *Method:* As for 1.
Problems of interpretation: (For amiodarone see p544.)
False high: Pregnancy; oestrogens.
False low: Severe infection; post-surgery; post-myocardial infarct; chronic liver disease; chronic renal failure; propranolol; phenytoin; salicylates; NSAID; carbamazepine.

3 Plasma basal thyroid-stimulating hormone (TSH)
Indications: Suspected hypothyroidism, risk of hypothyroidism.
Method: Collect blood at any time of day.
Problems of interpretation: Normal TSH is <5.7mU/l (some lab variation). >5.7mU/l with normal T_4 indicates *partial thyroid failure* caused by: Hashimoto's, drugs (lithium, antithyroids, excess iodine, eg expectorants), hyperthyroidism treatment, autoimmune disease (IDDM, pernicious anaemia, Addison's), iodine deficiency, dyshormonogenesis.

Immunometric assays (IMA) of TSH are now replacing radioimmunoassays and these are so sensitive that low TSHs may be quantified—so TRH tests are not now needed (p540) when borderline raised thyroid hormone levels suggest hyperthyroidism.[1] A TSH IMA level <~0.5U/l suggests hyperthyroidism. The normal range of TSH widens in the elderly (eg add 1mU/decade over 40yrs to the upper limit).[2]

If T_3 and T_4 low, but TSH not raised then diagnose *secondary* hypothyroidism (p544). Look for pituitary failure (p558).

1 *Drug Ther Bul* 1989 **27** 57–9   2 I Hay 1988 *Endocrinol Metab Clin North Am* **17** 473–509

Basic thyroid function tests
[*OTM* 10.33]

Special thyroid function tests

1 **Free T$_4$, Free T$_3$** may be useful when a false low or high T$_4$ or T$_3$ is suspected (p538). These tests are expensive. If unavailable consider: Free thyroxine index which is an estimate of free T$_4$ derived from measuring the unoccupied thyroxine binding sites on thyroxine binding globulin (TGB).

2 **The TRH test** *is indicated if:* Suspected hyperthyroidism (and other tests equivocal); T$_3$ marginally raised; suspected Graves' disease.
Method: Measure plasma TSH before, and 20mins and 60mins after, injection of 200µg TRH (protirelin).
Interpretation: If rise in TSH is >2.0mU/l after TRH, hyperthyroidism is excluded. If the rise is ≤2.0mU/l then: hyperthyroidism, euthyroid Graves' disease, multinodular goitre, solitary autonomous thyroid nodule, thyroxine replacement, or first 8 weeks of hyperthyroid treatment, will be the diagnosis.

Other tests of thyroid anatomy and pathology

1 **Thyroid scanning** *is indicated:* if there is an area of thyroid enlargement; if hyperthyroid without thyroid enlargement (is there diffuse uptake or a solitary nodule?); if hyperthyroid with one nodule (solitary nodule or multinodule?); to find the extent of retrosternal goitre; to detect ectopic thyroid tissue; to detect thyroid metastases (using whole-body scan); possible subacute thyroiditis.

Interpretation: The main question is: has the enlarged area increased (hot), or decreased (cold) or the same (neutral) uptake of pertechnetate as remaining thyroid? 20% of cold nodules are malignant. Few neutral and almost no hot nodules are malignant.

2 **Ultrasound** distinguishes cystic (usually benign) from solid (possibly malignant) 'cold' nodules.

3 **Thyroid autoantibodies** (antithyroid globulin; antithyroid microsomal) raised in Hashimoto's and some Graves' (if positive there is increased risk of later hypothyroidism).

Thyroid-stimulating immunoglobulins (against TSH receptor) may be raised in Graves' (this test is not widely available).

Serum thyroglobulin is useful in monitoring the treatment of carcinoma.

Special thyroid function tests
[*OTM* 10.33]

Hyperthyroidism (thyrotoxicosis)

Symptoms Weight↓, appetite↑, frequent stools, oligomenorrhoea, tremor, irrit-ability, frenetic activity, emotional lability, dislike of hot weather, sweating↑, itch. Infertility may be the presenting problem.

Signs Pulse↑ (even sleeping); warm peripheries; AF; fine tremor; lid lag (eyelid lags eye gaze as patient watches finger—hold your finger in front of and above patient's eyes and allow it to descend slowly and smoothly); thyroid enlargement, or nodules; thyroid bruit, myopathy. If BP↑ consider phaeochromo-cytoma. *Additional signs of Graves' disease*: Bulging eyes (exophthalmopathy, p543); pretibial myxoedema (oedematous swellings above lateral malleoli—the term *myxoedema* is confusing here).

Tests TSH ↓; free T_4 and free T_3 ↑ (p538–40). Consider ultrasound if goitre is present, or thyroid scan if subacute thyroiditis suspected; thyroid autoanti-bodies. If ophthalmopathy, test visual fields, acuity and eye movements. Can the lids close fully? If not, there is a risk of keratopathy.

Causes *Graves' disease:* Common, especially women aged 30–50yrs. Genetic influence. ♀/♂ ≈ 9. The disease probably results from antibodies against TSH-receptors. Signs as above, with diffuse thyroid enlargement. It may cause nor-mochromic normocytic anaemia, ESR↑, calcium ↑, abnormal LFTs. Associated with: IDDM, pernicious anaemia.

Toxic adenoma: ie nodule producing T_3 and T_4. On scanning, nodule is 'hot' (p540) and the rest of the gland is suppressed.

Subacute thyroiditis: Usually post-partum and self-limiting. Goitre (usually pain-ful). ESR↑. Probably viral cause. On scans, no radioiodine uptake.

Other causes: Toxic multinodular goitre; self-medication (detected by raised T_4, low T_3); follicular carcinoma of thyroid; choriocarcinoma; struma ovarii (ovarian teratoma containing thyroid tissue).

Treatment The options are drugs (carbimazole, propylthiouracil), partial thy-roidectomy, and radioactive iodine. *Carbimazole* may be given at 15mg/8h PO for 4 weeks, then gradually reducing according to TFTs every 1–2 months. Maintain on ~5mg/8h for 12–18 months then withdraw. >50% relapse. (Warn to seek advice if they get a rash or sore throat, because 0.1% on carbimazole get agranulocytosis: do FBC; headache, alopecia, pruritus, and jaundice are other side-effects.) Immediate symptomatic control is achievable with propranolol 40mg/6h PO.

As relapse is common with the above 'low-dose' carbimazole régime, many now receieve the *block-and-replace régime*[1]—eg carbimazole 40mg/24h PO until hypersecretion has stopped, then T_4 is added in (eg 50–150μg/24 h PO); this two-drug régime is continued for about 1 year. ~ 50% with Graves' disease will have a lasting remission. Monitor by measuring free T_4. The advantages of 'block-and-replace' is that hypersecretion is better controlled, and follow-up is less onerous for patients.[1] *Partial thyroidectomy* carries risk of damage to recur-rent laryngeal nerves and parathyroids. Furthermore, patients may be hypo- or hyperthyroid after surgery. *Radioiodine* can be repeated until euthyroid but the patient will almost always ultimately become hypothyroid. Decisions on treat-ment rest on local expertise and the patient's age. In general use carbimazole then surgery in the young; radioiodine in the elderly.

Ophthalmopathy: Control hyperthyroidism; steroids if ophthalmoplegia or gross oedema. *Consult eye surgeon early* (eg papilloedema, vision↓).

Atrial fibrillation: p290. Control hyperthyroidism.

Complications Heart failure, angina, osteoporosis, gynaecomastia.

In pregnancy and in infancy Get expert help. See *OHCS* p157.

---

1 AM McGregor 1996 *Oxford—Textbook of Medicine* OUP page 1612 (section 12.4)

Thyroid eye disease

This occurs in the presence of specific autoantibodies causing retro-orbital inflammation and lymphocyte infiltration.

History Decreased acuity, double vision, discomfort and/or protrusion.
▶ Decreasing acuity may mean optic nerve compression. Seek expert advice today. (Decompression may be needed.) Nerve damage need not go hand-in-hand with protrusion. If the eye cannot protrude for anatomical reasons, optic nerve compression will be all the more likely.

Signs Proptosis/exophthalmos (ie protruding eyes), conjunctival oedema, corneal ulceration, papilloedema, optic atrophy, lid lag, lid retraction, loss of colour vision, ophthalmoplegia (especially of upward gaze) from muscle tethering and fibrosis.

Investigations Rose Bengal eye-drops may show superior limbic keratitis. CT scan of orbit.

Management Hypromellose eye-drops 0.25% for lubrication as often as is needed. There is no upper limit to dose frequency or volume instilled. Lateral tarsorraphy; systemic steroids; 5% guanethidine eye-drops may reduce lid retraction. Surgical decompression uses space in the ethmoidal, sphenoidal, and maxilllary sinuses, via a medio-inferior approach. Orbital radiotherapy, cytotoxic drugs, and plasmapheresis have proved unreliable.[1]

NB: eye disease may pre-date other signs of Graves' disease, and does not always respond to treatment of thyroid status. Furthermore, it may develop for the first time following treatment of hyperthyroidism.

543

1 *Oxford Textbook of Medicine* OUP 1987 2 ed, page 23.17

Hypothyroidism (myxoedema)

▶ Thyroxine given to the elderly may precipitate angina.

Symptoms Insidious onset, weight gain, constipation, dislike of cold, menorrhagia, hoarse voice, lethargy, depression, dementia.

Signs Toad-like face, bradycardia, dry skin and hair, goitre, slowly relaxing reflexes, CCF, non-pitting oedema (eg eyelids, hands, feet).

Tests Diagnose by low T_4 (& T_3). TSH (high in thyroid failure, low in pituitary failure, p558). ECG, CXR. Fasting cholesterol and triglyceride may be raised. FBC may show a normochromic macrocytic anaemia. CK, AST and LDH may be raised due to abnormal muscle membranes.

Causes *Spontaneous primary atrophic hypothyroidism:* Common, autoimmune disease which is essentially Hashimoto's without the goitre and associated with IDDM, Addison's disease, or pernicious anaemia. ♀/♂=6.

After thyroidectomy or radioiodine treatment.

Drug-induced: Antithyroid drugs, amiodarone (below), lithium, iodine (eg in expectorants), para-aminosalicyclic acid. Treat by stopping drug or as below.

Iodine deficiency if dietary insufficiency.

Dyshormonogenesis: Autosomal recessive due usually to peroxidase deficiency. Diagnose by high radioimmune gland uptake displaced by potassium perchlorate.

Treatment If healthy and young: thyroxine 150µg/24h PO. Review in 3 months. Adjust dose according to clinical state and to keep TSH ≤5mU/l. Life-long therapy is needed. In the elderly start with 25µg/24h and slowly increase dose. NB: thyroxine's half-life is 7 days. If ischaemic heart disease, give propranolol 40mg/6h PO and thyroxine as for elderly.

Causes of goitrous hypothyroidism *Hashimoto's thyroiditis:* Autoimmune disease in which there is lymphocyte and plasma cell infiltration. Usually in women aged 60–70yrs. Often euthyroid. Autoantibody titres high. Treat as above if hypothyroid or to reduce goitre if TSH high.

Drugs: As above.

Effect of amiodarone[1] on the thyroid is complex and variable. It commonly causes ↑ free T_4, ↓ free T_3, but clinically euthyroid state. 2% of patients have clinically significant changes—which may be hyperthyroidism (eg weight loss) or hypothyroidism. Be guided more by clinical state than tests. Seek expert help. NB: long half-life (40–100 days), so problems persist after withdrawal.

Secondary hypothyroidism ie due to pituitary failure (p558) is very rare.

How to withdraw thyroxine If T_4 already started and diagnosis now in doubt, stop for 6wks, then measure T_4 and TSH to make diagnosis.

---

1 M Gammadge 1987 *Qu J Med* **238** 83

Hypothyroidism (myxoedema)

[*OTM* 10.46]

Hyperparathyroidism

This is an excess of parathyroid hormone (PTH, see p642).

Primary hyperparathyroidism Plasma PTH is inappropriately high with respect to plasma Ca^{2+} (ie calcium level is not low). *Presentation:* most commonly as $Ca^{2+}\uparrow$ on routine biochemistry. Any symptoms are usually due to: bone pain/fracture; renal stones; constipation; non-specific abdominal pain, duodenal ulcer, pancreatitis; depression. Hence mnemonic: 'bones, stones, abdominal groans and psychic moans'. Other presentations:

- Dehydration and confusion
- Thirst, nocturia, anorexia (∵$Ca^{2+}\uparrow$)
- Hypertension
- Stiff joints
- Walking problems
- Myopathy

Associations: Pancreatic islet cell tumour and pituitary adenoma (multiple endocrine neoplasia—MEN type I). Medullary carcinoma of thyroid and phaeochromocytoma (MEN type IIa); hypertension.

Causes: Single (90%) or multiple adenoma; carcinoma; hyperplasia.

Tests. Plasma $Ca^{2+}\uparrow$ (2 uncuffed fasting samples). Plasma phosphate low (unless renal failure). Biochemical evidence of increased bone turnover (alk phos ↑) and radiographic evidence of *osteitis fibrosa et cystica* (brown tumours) and subperiosteal resorption (especially on hand x-ray). Also do CXR, SXR ('pepper-pot skull'), pelvis x-ray. Plasma PTH (raised or normal) assay: this is technically difficult, consult lab first.

Treatment—if required—is surgery—neck exploration with removal of adenoma if present, or of most of gland if hyperplasia. Venous sampling or ultrasound is used to locate source of PTH if initial surgery is unsuccessful. Do plasma Ca^{2+} daily for ≥14 days' post-surgery (danger is $Ca^{2+}\downarrow$). Mild symptoms may not merit surgery—advise a high fluid intake to prevent stone-formation; review every 6 months.

Parathyroid-related protein (PTHrP) This is produced by some tumours, and causes some of the hypercalcaemia seen in malignancy. It increases osteoclast activity and the tubular resorption of urinary calcium.

Secondary hyperparathyroidism (raised PTH appropriate to low calcium). Causes: chronic renal failure; dietary deficiency of vitamin D.

Tertiary hyperparathyroidism is the continued secretion of large amounts of PTH after prolonged secondary hyperparathyroidism: the original cause of the secondary hyperparathyroidism (hypocalcaemia) has gone, but the parathyroids now act autonomously and cause hypercalcaemia. Treatment is as for primary hyperparathyroidism.

Hypoparathyroidism

Primary hypoparathyroidism PTH secretion↓, eg after neck surgery, causing $Ca^{2+}\downarrow$ with normal or raised phosphate and normal alk phos (p642). Associations: pernicious anaemia; Addison's; hypothyroidism; hypogonadism. Treatment is with alfacalcidol (p642). Lifelong follow-up.

Pseudohypoparathyroidism (failure of target cell response to PTH). Signs: Round face, short metacarpals and metatarsals. Symptoms and treatment as for 1° hypoparathyroidism. Plasma PTH & alk phos ↔ or ↑.

Pseudopseudohypoparathyroidism The morphological features of pseudohypoparathyroidism, but normal biochemistry.

Endocrinology

Hyperparathyroidism
Hypoparathyroidism
[*OTM* 10.59; 10.62]

Adrenal cortex and Cushing's syndrome

Physiology The adrenal cortex produces: *glucocorticoids* (eg cortisol) which affect metabolic functions (carbohydrate, lipid and protein metabolism), *mineralocorticoids* (eg aldosterone, p636), androgens, and *oestrogens*. Corticotrophin-releasing factor (CRF) from the hypothalamus stimulates ACTH (from pituitary) which stimulates cortisol production by the adrenal cortex. Cortisol is excreted as urinary free cortisol and as various 17 oxogenic steroids (17OGS). Metyrapone inhibits a step in cortisol synthesis resulting normally in loss of negative feedback, raised ACTH and raised steroids excreted as 17OGS.

Cushing's syndrome is due to chronic glucocorticoid excess. Clinical features are: wasting of tissues, myopathy, thin skin, purple abdominal striae, easy bruising, osteoporosis, water retention (plethoric *moon-face*, hypertension, oedema), obesity of trunk, head, neck (*buffalo-hump*), predisposition to infection, poor wound healing, hirsutism. Amenorrhoea. Hyperglycaemia (30%).

Tests: See p550. Firstly, confirm the hypercortisolaemia and consider the dexamethasone suppression test (p550). Secondly, establish the cause: CXR (carcinoma), AXR (calcified adrenal tumour). Consider IVU, adrenal venography, CT scan and radioisotope scans. If Cushing's disease suspected, investigate as for pituitary tumour (p560).

Causes and treatment:

1 *Corticosteroid (and ACTH) administration:* Treat by reducing dose.
2 *Cushing's disease* (adrenal hyperplasia due to excess ACTH from pituitary tumour. Commoner in women. Peak age: 30–50yrs). Treat in specialized centre. The treatment of choice is either surgery on the pituitary (transphenoidal or transfrontal approach) when it is sometimes possible to remove the adenoma selectively; or radiotherapy. Medical treatment is metyrapone.
3 *Alcohol excess.*
4 *Adrenal gland adenoma or carcinoma:* Treated by surgical removal of tumour, and metyrapone if residual glucocorticoid excess.
5 *Ectopic ACTH production:* (especially small cell bronchial carcinoma).

Prognosis: Untreated: 50% dead within 5yrs. Treated: much improved unless carcinoma.

Adrenal cortex and Cushing's syndrome
[*OTM* 10.69]

Assessment of suspected Cushing's

Screening tests (out-patient) See also p548.
- 24h urine collection for urine-free cortisol assay. Normal is <700nmol/24h. In Cushing's this is raised. More than one collection is needed.
- Overnight dexamethasone (DXM) suppression test: patient takes dexamethasone 2mg PO at 23.00h. Blood for cortisol taken at 09.00h day following. Normal <170nmol/l.

Inpatient tests
- Plasma cortisol measured at 22.00h and 09.00h. (Insert 'butterfly' earlier in day to avoid stress.) Cushing's suggested if level at 22.00h is about equal to that at 09.00h (it is normally lower).
- Several tests are run before results known to save time. Consult with lab about blood samples.

Day	Drugs given PO	Plasma cortisol	Plasma ACTH	24h urine Free cortisol	17 OGS	Other
0 (admit)	—	—	—	—	—	—
1	—	22.00h	09.00h 22.00h	+	+	OGTT (p528)
2	—	22.00h	22.00h	+	+	—
3	DXM 2mg/6h	—	—	—	—	—
4	DXM 2mg/6h	22.00h	22.00h	+	—	—
5	DXM 2mg/6h	22.00h	—	+	—	—
6	—	—	—	—	—	—
7	Metyrapone 750mg/4h with milk. 6 doses	—	after 4th dose	—	+	—
8	—	—	—	—	+	—

DXM = dexamethasone   +: do test;   —: do not do test

Interpreting the results
Cushing's disease: There is some, but not normal suppression of plasma cortisol with high dose DXM (ie 2mg/6h). Plasma ACTH is raised but <250ng/l. Metyrapone test: normal increment in urinary 17OGS on giving metyrapone is 35–135µmol/24h above basal. In Cushing's disease there is a larger increase.

Adrenal tumour: Usually no suppression of cortisol with high-dose DXM. Plasma ACTH is undetectable. There is usually no response to metyrapone test.

Ectopic ACTH secretion: No suppression of cortisol with high-dose DXM.

Plasma ACTH >250ng/l: No response to metyrapone test. Often there is hypokalaemic alkalosis.

False positives: Depressive illness; excessive alcohol.

Assessment of suspected Cushing's
[*OTM* 10.70]

Addison's disease

Disease of adrenal glands causing adrenocortical insufficiency.

Symptoms Insidious onset. Weakness, apathy, anorexia, weight loss, abdominal pain, infrequent periods.

Signs Hyperpigmentation (palmar creases, buccal mucosa), vitiligo, postural hypotension.

Tests: *General:* hyperkalaemia, hyponatremia, uraemia, mild acidosis, hypercalcaemia, eosinophilia.

Specific: (Contact lab first.) *Short ACTH stimulation test (screening):* Measure plasma cortisol before and 30mins after tetracosactrin 250µg IM. Addison's is excluded if: initial cortisol >140nmol/l, and second cortisol is both >500nmol/l and >200nmol/l above first. If not excluded proceed to:

Prolonged ACTH stimulation test: Measure plasma cortisol before and 6h after tetracosactrin depot (zinc phosphate complex) 1mg IM. Do this on three successive days. Addison's is diagnosed if plasma cortisol is <690nmol/l, 6h after 3rd injection. Steroid drugs may interfere with this assay: check with your local laboratory.

Measurement of ACTH & cortisol: Take blood for ACTH and cortisol at 09.00h and 22.00h. Diagnose Addison's if low cortisol with ACTH >300ng/l.

Adrenal antibodies, AXR and CXR (TB signs, eg calcification).

Causes 80% idiopathic in UK. Probably autoimmune (adrenal antibodies raised). Associated with Graves', Hashimoto's, IDDM, pernicious anaemia, hypoparathyroidism, ovarian failure.

Other causes: TB, metastases.

Treatment Treat cause. Replace steroids: *hydrocortisone* 20mg in morning, 10mg at bedtime PO. Adjust dose by measuring plasma cortisol at 0.5, 1, 2, 3h after morning dose. Aim for peak 700–850nmol/l. Evening dose is half morning dose. *Fludrocortisone* 0.05mg PO every second day to 0.15mg daily. Adjust on clinical grounds. If postural hypotension: increase. If hypertension, headache, oedema, hypokalaemia, alkalosis: decrease.

Addison's is often associated with other autoimmune disease—even at the time of diagnosis. Hyperthyroidism should be investigated, particularly if there is difficulty in maintaining adequate cortisol levels on treatment.

Warn against abruptly discontinuing treatment. Patients should have syringe at home for giving hydrocortisone IM in case vomiting prevents oral therapy. Emphasize that any doctor or dentist giving treatment must know that the patient is taking steroids. Counsel about the need to take more in intercurrent illness. Give *steroid card*. Advise about wearing a bracelet declaring that steroids are taken. For dental treatment double morning hydrocortisone. If severe URTI double morning and evening hydrocortisone while ill. If vomiting replace hydrocortisone by hydrocortisone sodium succinate 100mg IM; IV fluids if dehydrated.

Follow-up 6-monthly.

Prognosis Normal life-span if treated.

Addison's disease
[*OTM* 10.76]

Hyperaldosteronism

Primary hyperaldosteronism is excess production of aldosterone independent of the renin-angiotensin system. Consider this diagnosis when the following features are present: hypertension, hypokalaemia, alkalosis in someone not on diuretics. Sodium tends to be mildly raised.

Causes: 75% due to unilateral adrenocortical adenoma (*Conn's syndrome*). Also: bilateral adrenocortical hyperplasia; adrenal carcinoma.

Tests: Do 3 K^+ measurements (no diuretics, hypotensives, steroids, carbenoxolone, potassium, laxatives, metoclopramide for 4 weeks). If 1 or more measure is <3.7mmol/l, do plasma samples for: aldosterone, renin, angiotensin, U&E, on 3 occasions. Take blood in morning before patient sits up. *Renal venography, scans, CT or MRI to distinguish cause.*

Treatment: Conn's—surgery (spironolactone 300mg/24h PO for 4 weeks pre-op). *Hyperplasia:* dexamethasone 1mg/24h PO for 4 weeks. If BP remains high, stop dexamethasone, give spironolactone.

Secondary hyperaldosteronism is excess aldosterone as a result of high renin. It is caused by: renal artery stenosis, accelerated hypertension, diuretics, heart failure, hepatic failure.

Bartter's syndrome (primary hyperreninaemia) is recessively inherited and presents as failure to thrive, polyuria and polydipsia. BP is *normal* and there is no oedema. The chief feature is hypokalaemia with increased urinary K^+ excretion and hypochloraemic metabolic alkalosis. Urinary chloride is elevated. Plasma renin↑.

Treatments include K^+ replacement, indomethacin, amiloride, captopril.

554

Phaeochromocytoma

This is a rare, but treatable cause of hypertension. It is usually (90%) a benign tumour producing catecholamines, usually (90%) in the adrenal medulla, and these are usually (90%) unilateral. Inheritance may be autosomal dominant, and associated with neurofibromatosis and medullary thyroid carcinoma. Those not in the medulla may be in paraganglia (= phaeochrome bodies, ie collections of adrenaline-secreting chromaffin cells)—typically those by the aortic bifurcation (organ of Zuckerkandl).

Symptoms and signs: Hypertension (particularly episodic), cardiomyopathy, weight loss, hyperglycaemia. Periods of crisis, lasting about 15mins characterized by: fear, headache, palpitations, sweating, nausea, tremor, pallor, high BP, and paroxysmal thyroid swelling (rare).

Investigations: Look for glycosuria during an attack (seen in 30%). For screening test: 24h urine collection for 4-OH-3-methoxymandelate (HMMA, VMA) or total metadrenalines. For full investigation consult with specialized centre (consider MIBG scan (metiodobenzylguanidine) or pentolinium suppression test, CT of abdomen).

Treatment: Surgery. Carefully control BP from 14 days pre-op with phenoxybenzamine and propranolol. Consult with anaesthetist. Post-op, collect 24h urine as above and monitor BP (risk of hypotension).

Emergency treatment: p742. *Prognosis:* Normal life if surgery successful.

Hyperaldosteronism
Phaeochromocytoma
[*OTM* 10.76; 13.390]

555

Hirsutism, virilism, gynaecomastia, and impotence

Hirsutism is common (eg seen in 10% of women) and usually benign. It implies increased hair growth in women, in the male pattern. If menstruation is normal there is almost certainly no increased testosterone production. Treatment is by local removal of unwanted hair. If menstruation is abnormal the cause is usually *polycystic ovary syndrome* (Stein-Leventhal syndrome). This syndrome is: bilateral polycystic ovaries; secondary oligomenorrhoea; infertility; obesity; hirsutism. The cause is androgen hypersecretion. Diagnosis is made by ultrasound, a high LH:FSH ratio and less consistently a high testosterone and low oestradiol. Oligomenorrhoea and infertility are treated with clomiphene.

The management of hirsutism:[1]
- Do not abandon the patient once you have ascertained that there is no serious illness.
- Find out what her expectations are.
- Explain that she is not turning into a man.
- Has she tried depilation with wax or creams, or electrolysis (which is expensive, time-consuming, but *does* work)?
- Has she tried bleaching the area with 1:10 hydrogen peroxide?
- Would she be prepared to shave regularly?
- Oestrogens help by increasing serum sex-hormone-binding globulin—but always combine with a progesterone (= 'the Pill') to prevent excess risk of uterine neoplasia. Marvelon® contains desogestrel which is a fairly non-androgenic progestagen. Dianette® contains 2mg cyproterone acetate (an antiandrogen) with ethinyloestradiol 35μg, so will also give contraceptive cover.

Virilism is rare. It is characterized by: amenorrhoea, clitoromegaly; deep voice; temporal hair recession; hirsutism. This condition needs investigation for androgen-secreting adrenal and ovarian tumours.

Gynaecomastia is the presence of an abnormal amount of breast tissue in males (it may occur in normal puberty). It is due to an increase in the oestrogen/androgen ratio. It is seen in sydromes of androgen deficiency (eg Klinefelter's, Kallman's). It may result from liver disease or testicular tumours (oestrogens↑), or accompany hyperthyroidism. The commonest causes are drugs: oestrogens (especially stilboestrol); spironolactone; cimetidine; digoxin; testosterone; marihuana.

Impotence Failure in adult male to sustain adequate erection for vaginal penetration. It is common in old age. Psychological causes are common and are more likely if impotence occurs only in some situations, if there is a clear stress to account for the onset of the impotence, and if early morning erections occur (although these may persist at the onset of organic disease). Psychological causes may exacerbate organic causes. The major organic cause is *diabetes*. Other organic causes are:

Drugs: Antihypertensives (including diuretics), major tranquillizers, alcohol, oestrogens, antidepressants, cimetidine.

Diseases: Hyperthyroidism, hyperprolactinaemia, hypogonadism, vascular disease, MS, autonomic neuropathy, renal failure, cirrhosis, malignancy.

Hirsutism, virilism, gynaecomastia, and impotence
[*OTM* 10.84]

Tests: Plasma glucose; LFTs; nocturnal tumescence studies are not usually needed. If papaverine dose not induce an erection, the cause is probably vascular. Doppler may show reduced blood flow, but are rarely needed as vascular reconstruction is difficult.

Treatment depends on cause. Offer counselling even if an organic (but untreatable) cause ± mechanical (vacuum) aids, surgical implants or self injection of papaverine into the corpus cavernosum through a 26G needle. Experts usually start with 7.5mg, gradually increasing to 60mg, until a satisfactory erection occurs. Peyronie's disease may be induced by injecting more than once weekly. If the erection lasts >4h (priapism) patients must know what to do (contact a doctor for aspiration of the corpora ± metaraminol 1mg in 5ml 0.9% saline injected into the corpora).

557

Hypopituitarism

The six anterior pituitary hormones commonly measured are: adreno-corticotrophic hormone (ACTH), growth hormone (GH), follicle stimulating hormone (FSH), luteinizing hormone (LH), thyroid-stimulating hormone (TSH), prolactin (PRL). There are several opioids released, of uncertain clinical significance.

Causes of hypopituitarism Hypophysectomy, pituitary irradiation, pituitary adenoma (either non-functional or functional causing eg Cushing's or acromegaly together with hyposecretion of other hormones).

Symptoms Insidious onset, loss of libido, impotence, amenorrhoea, headache, depression, hypothyroidism.

Signs Atrophy of breast, small testes, reduction of all hair, thin flaky wrinkled skin (*monkey face*), hypotension, visual field defect.

Investigations Lateral SXR (pituitary tumour), CT scan, assessment of visual fields, basal T_4, PRL (5ml clotted blood, consult lab), testosterone (10ml heparin tube), U&E (dilutional hyponatraemia), FBC (normochromic, normocytic anaemia).

Triple stimulation test (can be done in out-patients).
Contraindications: Epilepsy, heart disease. Test done in morning (water only taken from 22.00h the night before). Have 50% glucose to hand and IV line open. Be alert throughout to hypoglycaemia. Discard first 3ml of blood samples (contain citrate) and flush cannula after. Consult lab.
- Insert IV cannula with 3-way tap (citrate, samples, injection).
- Attach 20ml syringe filled with citrate for flushing.
- Take samples for: T_4, TSH, oestradiol/ testosterone, FSH, LH, PRL, cortisol, glucose, GH.
- Inject insulin IV 0.15U/kg (0.3U/kg if acromegaly or Cushing's 0.05 U/kg if marked hypopituitarism).
- Flush and inject IV 200μg TRH and 50μg gonadotrophin-releasing hormone (GnRH). Use same syringe. Flush.
- Collect blood samples as follows: FSH, LH, TSH at 20mins and 60mins. GH, cortisol, glucose at 30, 60, 90, and 120mins.
- At the end give the patient a good breakfast.

Interpretation of triple stimulation test Glucose must fall below 2.2mmol/l. Normal values are: GH >20mU/l, peak cortisol >550mmol/l, TSH at 20mins 3.9–30mU/l, TSH at 60mins, 3.0–24mU/l. Values lower than these indicate some pituitary deficiency.

Other causes include: craniopharyngioma, sphenoid meningioma, Sheehan's syndrome (pituitary necrosis following post-partum haemorrhage), TB.

Treatment Hydrocortisone as for Addison's (p552). Thyroxine as for hypothyroidism (p544, TSH levels not helpful). Testosterone enanthate (men) 250mg IM every 3 weeks. Consider testosterone undecanoate 40mg/12h PO. Oestrogen (premenopausal women) combined contraceptive steroid.

Hypopituitarism
[*OTM* 10.16]

Pituitary tumours

▶ Symptoms are caused by local pressure, hormone secretion, or hypopituitarism (p558).

Pituitary tumours (almost always benign) account for 10% of intracranial tumours.

Histological classification There are 3 types.

1 *Chromophobe*—70%. Some secrete no hormone but cause hypopituitarism. Local pressure effect in 30%. Half produce prolactin (PRL); a few produce ACTH or growth hormone (GH).

2 *Acidophil*—15%. Secrete GH or PRL. Local pressure effect in 10%.

3 *Basophil*—15%. Secrete ACTH. Local pressure effect rare.

Classification by hormone secreted

PRL only	—35%	ACTH	—7%
GH only	—20%	LH/FSH/TSH	—1%
PRL and GH	—7%	No hormone	—30%

Clinical features of local pressure effect Headache; visual field defects (bilateral hemianopia, initially of superior quadrants); palsy of cranial nerves III, IV, VI; occasionally disturbance of temperature, sleep, feeding and erosion through floor of sella leading to CSF rhinorrhoea. Diabetes insipidus (p566) is a very rare result of pituitary tumour: it is more likely to result from hypothalamic disease.

Investigations Lateral SXR; MRI (or CT scan); accurate assessment of visual fields; triple stimulation test (p558); water deprivation test if diabetes insipidus suspected (p566).

Treatment Start hormone replacement as indicated by triple stimulation test before surgery.

Surgery: If intrasellar tumour, it is removed by transphenoidal approach. Suprasellar extension of tumour may need transfrontal approach.

Medical treatment: (ie bromocriptine) For prolactin-secreting and some GH-secreting tumours (p562).

Radiotherapy: Following surgery if complete removal of tumour has not been possible.

Peri-operative care Hydrocortisone sodium succinate 100mg IM, with pre-med, then every 4h for 72h. Then hydrocortisone 20mg PO mane. Retest pituitary function after few weeks (p558) to assess replacement needed.

Pituitary apoplexy Rapid expansion of a pituitary tumour due to infarction or haemorrhage may cause sudden local pressure effect. Suspect if sudden onset of headache in someone with known tumour, or if sudden headache and loss of consciousness (ie may present like subarachnoid haemorrhage). *Treatment* is urgent surgery under steroid cover.

Craniopharyngioma This is not, strictly speaking, a pituitary tumour as it originates from Rathke's pouch and is situated between pituitary and floor of third ventricle. 50% present with local pressure effects in childhood. *Investigations:* CT scan, lateral SXR (calcification). *Treatment:* surgery; irradiation; post-operative triple stimulation test (p558).

Infarction or haemorrhage may cause sudden local pressure effect.

Pituitary tumours
[*OTM* 10.16]

Hyperprolactinaemia

This is the most common biochemical disturbance of the pituitary. It tends to present early in women (amenorrhoea) but late in men.

Clinical (usually present to gynaecologist): menstrual disturbance (amenorrhoea; infertility; galactorrhoea) impotence in male, local pressure effects (p560).

Causes of raised basal plasma prolactin (>390mU/l)

Physiological: Pregnancy; breast-feeding; stress; sleep.

Drugs: Phenothiazines (including metoclopramide); haloperidol; α-methyldopa; oestrogens; TRH.

Disease: Pituitary disease; chronic renal failure; hypothyroid; sarcoid.

Investigations Basal plasma prolactin: Non-stressful venepuncture between 09.00–16.00h. Lateral skull x-ray plus tomography of pituitary fossa and CT scan.

Management[1] This depends in part on local surgical expertise. A reasonable approach is as follows:

Microprolactinomas (tumour <10mm diameter—defined on CT scan). If patient is post-menopausal or does not want to become pregnant consider no treatment but regular PRL levels. Otherwise, bromocriptine 2.5mg/24h (at 22.00h) PO increasing to 2.5mg/8h PO over 7 days, in order to decrease the chance of infertility.

The reasons for not pursuing surgery with microprolactinomas are that they rarely become macroprolactinomas, they rarely expand in pregnancy and they may recur after surgery.

Macroprolactinomas (>10mm diameter). If visual acuity failing then remove surgically (transphenoidal). Follow-up with serial PRL and CT scan. If these indicate further development consider bromocriptine and/or radiotherapy.

If visual acuity is not failing and pregnancy is not anticipated then a trial of bromocriptine is probably preferable to surgery. However, if pregnancy is anticipated then recommend surgery because otherwise the prolactinoma is likely to expand.

1 N Lawton 1987 *Qu J Med* **243** 557

Hyperprolactinaemia
[*OTM* 10.10; 10.88]

Acromegaly

This rare disease is due to hypersecretion of GH from a pituitary tumour. It usually presents between the ages of 30–50yrs.

Clinical features Insidious onset (look at old photos). Most features are due to growth of soft tissues. Coarse, oily skin, large tongue, prominent supraorbital ridge, prognathism, teeth spacing increased, increasing shoe size, thick spade-like hands, deepening voice, arthralgia, kyphosis, proximal muscle weakness, paraesthesiae (due to carpal tunnel syndrome—p464), progressive heart failure, goitre. Features of a pituitary tumour (p560). Sweating and headache are also common.

Complications DM, high BP and cardiomyopathy.

Investigations CT scan. X-ray of pituitary fossa. Oral glucose tolerance test (OGTT) with GH measurement. Method as described (p528) but insert IV cannula with three-way tap and citrate (see p558) 1h before test. Collect samples for GH, glucose, insulin at: 0, 30, 60, 90, 120, 150mins. Interpretation of OGTT: GH does not fall to <2mU/l in: acromegaly, anorexia nervosa, poorly controlled DM, hypothyroidism, Cushing's.

In-patient investigations
- Obtain old photos. New photo of full face, torso, hands on chest.
- T_4, basal serum PRL, testosterone (if male).
- ECG. Visual fields and acuity. Skin thickness.
- X-ray of pituitary fossa, hands, SXR, CXR, CT scan if not already done.
- On next day do OGTT with GH measurement (above).
- On next day do triple stimulation test (p558) if hypopituitarism suspected, or surgery likely.
- On next day do CT scan.

Treatment Surgery on younger patients or if pressure symptoms. Six weeks after surgery readmit for OGTT with GH measurement, T_4, PRL, testosterone, triple stimulation test. If GH fails to suppress below 2.0 mU/l irradiation or bromocriptine will be required. Steroids should be stopped before these tests and triple stimulation test done on day 4 only if no signs of steroid deficiency (postural hypotension, fever, nausea, anorexia). Follow-up every year [OGTT with GH measurement, T_4, PRL, ECG, visual fields, X-rays (see above), photos]. External irradiation: For older patients. Follow-up as for surgery. Bromocriptine as adjunct. Start with 2.5mg PO at night. Increase by 2.5mg every 3 days up to 5mg/6h. Follow-up 2 weeks after maximum dose as for surgery. SE: nausea, postural hypotension.

Acromegaly
[*OTM* 10.17]

Diabetes insipidus (DI)

This is due to impaired water resorption by the kidney because of too little ADH secretion from the posterior pituitary (cranial DI), or impaired response by the kidney to ADH (nephrogenic DI).

Clinical features Abrupt onset of polyuria (urine dilute), polydipsia (or water deficiency if not drinking, p636).

Causes of cranial DI Head injury, hypophysectomy, histiocytosis, metastases, pituitary tumour, sarcoid, vascular lesion, meningitis, inherited (autosomal dominant). Idiopathic DI (50%) is often self-limiting, and MRI may reveal an infundibuloneurohypophysitis (known to be lymphocytic, and possibly autoimmune[1]).

Causes of nephrogenic DI Low potassium, high calcium, drugs (lithium, demeclocycline), pyelonephritis, hydronephrosis.

Investigations U&E, Ca^{2+}. *Water deprivation test* (If first morning urine has osmolality >800mosmol/kg, DI is excluded.) Stop drugs (chlorpropamide, clofibrate, carbamazepine) before the test.
- Light breakfast, no tea, no coffee, no smoking.
- Weigh at 0, 4, 6, 7, 8h. Stop if >3% body weight lost.
- Supervise carefully to stop patient drinking.
- Empty bladder, then no drinks and only dry food for 8h. Collect urine hourly, measure volume. Measure osmolality at 1, 4, 7, 8h. Stop test if osmolality >800 mosmol/kg (DI is excluded).
- Venous sample for osmolality at: 0.5, 3.5, 6.5, 7.5h.
- If diuresis continues, desmopressin 20µg intranasally (or 1µg IM) at 8h.
- Water can be drunk after 8h. Measure urine osmolality at 8, 9, 10, 11, 12h.

Interpreting water deprivation test
Check plasma osmolality is >290mosmol/kg to ensure that the test has provided adequate stimulus for ADH release. The *normal* response is for urine osmolality to be >800mosmol/kg with a small rise only in osmolality after desmopressin. In *psychogenic polydipsia* the urine is also concentrated (>400mosmol/kg) but rather less than in the normal response. In *diabetes insipidus* the urine is abnormally dilute (<400mosmol/kg). However, in *cranial* DI the urine osmolality increases by more than 50% following desmopressin; whereas in *nephrogenic* DI it increases by less than 45% following desmopressin.

Treatment *Cranial DI:* Find cause, do triple stimulation test (p558). Give desmopressin 10–20µg/12–24h intranasally. *Nephrogenic:* Treat cause. If persists, bendrofluazide 5–10mg PO/24h.

Emergency treatment The diagnosis must be made. This can be done if: suitable cause, large urine output, urine osmolality low (around 150mosmol/kg), despite dehydration.
- Plasma U&E.
- IV fluids—5% dextrose, 2 litres in first hour.
- Continue fluids to rehydrate and then to keep up with urine output.
- Desmopressin 1µg IM (lasts 12–24h).

1 H Imura 1993 *NEJM* **329** 683

Diabetes insipidus (DI)

[*OTM* 9.51]

The pineal gland and circadian rhythms

There are 6 great rhythms in our lives harmonizing with day and night:
- Sleeping and waking.
- Cortisol secretion (highest at dawn).
- Intellectual performance (best at about noon).
- Body temperature (highest in the afternoon).
- Prolactin secretion (highest at night).
- Melatonin (*only* secreted at night).

Diurnal means recurring every day (Latin *diurnalis*, daily), and is chiefly used to describe the activities of heavenly bodies. Circadian designates physiological activity which recurs approximately every 24h, or the rhythm of such activity (Latin *circa*, about, plus *dies*, day).

Circadian happiness (synchrony) consists in keeping all these clocks in time with each other—and with the diurnal rhythms of the solar system. The biological clock is thought to be located in the suprachiasmatic nucleus of the hypothalamus, and depends on melatonin, the hormone of darkness[1], to inform it of time and its passing. Melatonin is secreted by the pineal gland—eg at 30µg/night; but if darkness fails to materialize, for example, while in intercontinental transmeridianal flight, or during a night on call, a most unhappy state of desynchrony exists—until melatonin is restored, either artificially, or by night.[2] Such jet lag is manifested by malaise, light headedness, insomnia, and reduced cognition—and its effects are minimized by avoiding insomnia during and immediately after transmeridianal flight. These solutions are not available to the person who is 'on call' continuously and for whom the notion of 'reduced cognition' is really a euphemism for *absent* cognition. It is unclear if such a person would benefit from exogenous melatonin administration: what *is* clear is that in these circumstances such administration is likely to cause immediate, irresistible sleep.

1 RV Short 1993 *BMJ* ii 952   **2** HS Yu 1993 *Melatonin Biosynthesis, Physiological Effects, and Clinical Applications*, Boca Raton, USA, CRC Press

Notes

570

Relevant pages in other chapters:
Haematology reference intervals (normal values)—p801
Blood transfusion & blood products—p102

Anaemia

Anaemia is a low haemoglobin due to a decreased red cell mass. If the low Hb is due to dilution from an increased plasma volume (as occurs in pregnancy), the 'anaemia' is called physiological. A low Hb (at sea level) is <13.5g/dl for men and <11.5g/dl for women. It may be due to reduced production or increased loss of RBC and has many causes. These will often be distinguishable by history, examination, and inspection of the blood film.

Symptoms These may be due to the underlying cause or to the anaemia itself and include fatigue, dyspnoea, palpitations, headache, tinnitus, anorexia and bowel disturbance.

Signs Pallor, retinal haemorrhages. In severe anaemia (Hb <8g/dl) there may be signs of a hyperdynamic circulation eg tachycardia, murmurs and cardiac enlargement. Later heart failure may occur and in this state rapid blood transfusion may be fatal.

Anaemia due to excessive red cell destruction is called haemolytic anaemia. Suspect haemolysis if there is a reticulocytosis, mild macrocytosis, bilirubin↑ and urobilinogen↑.

The next step in determining the cause of anaemia is to look at the MCV (normal 76–96 femptolitres, 10^{15}fl = 1 litre).

Low MCV (microcytic) The common cause is iron deficiency anaemia (IDA) with a microcytic, hypochromic blood film showing anisocytosis and poikilocytosis (p576). It may be confirmed by showing serum iron↓ and ferritin↓ (more representative of total body iron) with total iron binding capacity (TIBC)↑. Other causes include thalassaemia (suspect if the MCV is 'too low' for the level of anaemia and the red cell count is raised), and congenital sideroblastic anaemia (very rare). These are iron loading conditions and will show serum iron↑; ferritin↑; but TIBC↓.

Normal MCV (normocytic) The causes are anaemia of chronic disease (AoCD), bone marrow failure, haemolysis, hypothyroidism, renal failure and pregnancy. If there is a reduced white cell or platelet count suspect bone marrow failure and perform a bone marrow biopsy.

High MCV (macrocytic) Causes: B_{12} and folate deficiency, alcohol ingestion, liver disease, reticulocytosis (eg haemolysis), marrow infiltration, acquired sideroblastic anaemia and hypothyroidism. See p580.

Blood transfusion Avoid unless Hb dangerously low. The decision will depend on the severity and cause. If there is risk of haemorrhage (eg active peptic ulcer) transfuse up to 8g/dl. If there is severe anaemia and heart failure, transfusion is vital to restore Hb to safe level, eg 6–8g/dl, but must be done with great care. Give packed cells slowly with 40mg frusemide IV/PO with alternate units (do not mix with blood). Check for rising JVP and basal crepitations and consider CVP line. If CCF gets worse, and immediate transfusion is essential, try a 2–3 unit exchange transfusion, removing blood at same rate as transfused.

Anaemia
[*OTM* 19.62]

Iron-deficiency anaemia (IDA)

This is common, occurring in up to 14% of pre-menopausal women. The major regulator of body iron (total 5g) is GI absorption which averages 1mg/day. Anaemia may be due to dietary deficiency, malabsorption or blood loss. By far the commonest cause is blood loss, usually menorrhagia (eg if >80ml/cycle) or GI bleeding. In women with menorrhagia it is justifiable to start treatment without further tests, although follow up is vital. In men and post-menopausal women the commonest cause is GI bleeding (eg from peptic ulcer, gastric or colonic carcinoma, hiatus hernia or diverticulitis).

Dietary deficiency is rarely the major factor in the developed world but may occur in babies and those on unusual diets. Malabsorption may be a cause of IDA, eg in coeliac disease. Signs of chronic iron deficiency (tissue Fe↓): koilonychia (spoon-shaped nails), atrophic glossitis and post cricoid webs.

Investigations: Faecal occult blood, sigmoidoscopy, barium enema and microscopy for ova as appropriate. Hookworm is the major cause of GI blood loss outside the Western world.

Treatment: This is of underlying cause, then iron replacement eg ferrous sulphate 200mg/8h PO. SE: constipation; black stools; Hb should rise by 1g/dl/week (accompanied by a reticulocytosis). Continue until Hb is normal and for 3 months to replenish stores.

The anaemia of chronic disease

This is associated with a variety of diseases: eg infection, collagen vascular disease, malignancy and renal disease. There is mild normocytic anaemia (eg Hb>8g/dl), sometimes confused with iron-deficiency anaemia but TIBC↓ and serum ferritin normal or ↑. *Treatment* is of the underlying disease. Erythropoietin lack is a cause of the anaemia of renal failure—genetically engineered erythropoietin is effective in raising the haemoglobin level (dose example: 75–450 units/kg/week; SE: flu-like symptoms, hypertension, mild rise in the platelet count).

Sideroblastic anaemia

Hypochromic RBCs are seen on the blood film with sideroblasts in the marrow (erythroid precursors with iron deposited in mitochondria in a ring around the nucleus). It may be congenital (rare, X-linked) or acquired, and is usually idiopathic, but may follow alcohol or lead excess, myeloproliferative disease, malignancy, malabsorption, anti-TB drugs. *Treatment* is supportive; pyridoxine may help (10–100mg/24h PO). Iron loading (haemosiderosis, ie endocrine, liver and cardiac damage) can be a problem as GI iron absorption is increased.

Refractory anaemia

The usual reason for IDA to fail to respond to iron replacement is failure of compliance, often because of GI disturbance, and altering the dose of elemental iron may help. There may be continued blood loss, malabsorption or misdiagnosis (eg thalassaemia). Occasionally there is myelodysplasia with abnormal marrow maturation. This is refractory to most treatments and may be pre-leukaemic.[1]

1 T Hamblin 1987 *Br J Hosp Med* **38** 558

Haematology

The peripheral blood film

▶ Many haematological (and other) diagnoses can be made by careful examination of the peripheral blood film. It is also necessary for interpretation of the FBC indices.

Acanthocytes: RBCs show many spicules (in abetalipoproteinaemia).

Anisocytosis: Variation in size, eg in megaloblastic anaemia, thalassaemia as well as iron-deficiency anaemia (IDA).

Basophilic stippling of RBCs is seen in lead poisoning, thalassaemia and other dyserythropoetic anaemias.

Blasts: Nucleated precursor cells (eg in myelofibrosis or leukaemia). They are not normally seen in peripheral blood.

Burr cells: Irregularly shaped cells occurring in uraemia.

Dimorphic picture: A mixture of RBC sizes eg partially treated iron deficiency, mixed deficiency (Fe + B_{12}/folate), post-transfusion, sideroblastic anaemia.

Howell–Jolly bodies: Nuclear remnants seen in RBCs post-splenectomy; rarely leukaemia, megaloblastic anaemia, IDA.

Hypochromia: Less dense staining of RBCs seen in IDA, thalassaemia and sideroblastic anaemia (iron stores unusable).

Left shift: Immature white cells seen in circulating blood in any marrow outpouring, eg infection.

Leptocytes: See target cells.

Leucoerythroblastic anaemia: Immature cells (myelocytes and normoblasts) seen in film. Due to marrow infiltration (eg by malignancy), hypoxia or severe anaemia.

Leukaemoid reaction: A marked reactive leucocytosis. Usually granulocytic eg in severe infection, burns, acute haemolysis, metastatic cancer.

Myelocytes, promyelocytes, metamyelocytes, normoblasts: Immature cells seen in the blood in leucoerythroblastic anaemia.

Normoblasts: Immature red cells, with a nucleus. Seen in leucoerythroblastic anaemia, marrow infiltration, haemolysis, hypoxia.

Pappenheimer bodies: Granules of siderocytes, eg lead poisoning, carcinomatosis, post-splenectomy.

Poikilocytes: Variably shaped cells, eg seen in IDA.

Polychromasia: RBCs of different ages stain unevenly (the young are bluer). This is a response to bleeding, haematinics (eg ferrous sulphate, B_{12}), haemolysis or dyserythropoiesis.

Reticulocytes: (NR: 0.8–2% of RBCs) Young, larger RBCs signifying active erythropoiesis. Increased in haemolysis, haemorrhage, and if B_{12}, iron or folate is given to marrow which lack these.

Right shift: Hypersegmented polymorphs (>5 lobes to nucleus) seen in megaloblastic anaemia, uraemia and liver disease.

Rouleaux formation: Red cells stack on each other (the visual 'analogue' of a high ESR—see p618).

Schistocytes: Fragmented RBCs sliced by fibrin bands. Seen in intravascular haemolysis.

Spherocytes: Spherical cells; seen in haemolysis (or, rarely, in hereditary spherocytosis).

Target cells: (Also called Mexican hat cells) These are RBCs with central staining, a ring of pallor and an outer rim of staining seen in liver disease, thalassaemia or sickle-cell disease—and, in small numbers, in iron-deficiency anaemia.

The differential white cell count

The neutrophils 2–7.5 × 10⁹/l (40–75% of white blood cells: but absolute values are more meaningful than percentages).

Increased in: Bacterial infections, trauma, surgery, burns, haemorrhage, inflammation, infarction, polymyalgia, PAN, myeloproliferative disorders and drugs (eg steroids). Marked increase in leukaemias, disseminated malignancy, severe childhood infection.

Decreased in: Viral infections; brucellosis; typhoid; kala-azar; TB. Also drugs, eg carbimazole; sulphonamides. They may be used up during septicaemia or destroyed by hypersplenism or neutrophil antibodies (seen in SLE and rheumatoid arthritis). There is reduced manufacture in B_{12} or folate deficiency and in bone marrow failure (p610).

The lymphocytes 1.3–3.5 × 10⁹/l (20–45%).

Increased in: Viral infections—EBV, CMV, rubella; toxoplasmosis; whooping cough; brucellosis; chronic lymphatic leukaemia. Large numbers of abnormal ('atypical') lymphocytes are characteristically seen with EBV infection: these are T-cells reacting against EBV-infected B-cells. Other causes of 'atypical' lymphocytes: see p204.

Decreased in: Steroid therapy; SLE; uraemia; legionnaire's disease; AIDS; marrow infiltration; post chemotherapy or radiotherapy.

T-lymphocyte subset reference values: CD4 count: 537–1571/mm³ (low in HIV infection). CD8 count: 235–753/mm³; CD4/CD8 ratio: 1.2–3.8.

The eosinophils 0.04–0.44 × 10⁹/l (1–6%).

Increased in: Asthma and allergic disorders; parasitic infections (especially invasive helminths); PAN; skin disease especially pemphigus; urticaria; eczema; malignant disease (including eosinophilic leukaemia); irradiation; Löffler's syndrome (p704); during the convalescent phase of any infection.

The *hypereosinophilic syndrome* is seen when there is development of end organ damage (restrictive cardiomyopathy, neuropathy, hepatosplenomegaly) in association with a raised eosinophil count (>1.5 × 10⁹/l) for more than 6 weeks. Steroids or cytotoxic drugs may give temporary relief.

The *eosinophilia–myalgia syndrome* consists in myalgia, arthralgia, fever, rashes, swelling of the arms, and an intense eosinophilia. It is seen in the USA, and is thought to be due to contaminants in L-tryptophan-containing proprietary dietary supplements.

The monocytes 0.2–0.8 × 10⁹/l (2–10%).

Increased in: Acute and chronic infections (eg TB, brucellosis, protozoa); malignant disease (including M4 & M5 acute myeloid leukaemia—p602, and Hodgkin's disease); myelodysplasia.

The basophils 0–0.1 × 10⁹/l (0–1%).

Increased in: Viral infections; urticaria; myxoedema; post-splenectomy; CML; UC; malignancy; systemic mastocytosis (urticaria pigmentosa); haemolysis; polycythaemia rubra vera.

The differential white cell count

[*OTM* 19.156]

Macrocytic anaemia

Macrocytosis is a common finding in routine blood counts, often without anaemia. Only ~5% of these will be from B_{12} deficiency, so it is wise to approach a diagnosis in an organized manner.

Causes of macrocytosis MCV >110fl: Vitamin B_{12} or folate deficiency. MCV 100–110fl: Alcohol, liver disease, hypothyroidism, haemolysis, pregnancy, marrow infiltration, myelodysplastic (eg sideroblastic) states zidovudine (p212), azathioprine (dose related).

Consider the conditions above in the history and examination. The diagnosis may be obvious at this stage. Alcohol is the most likely cause of macrocytosis without anaemia.

Investigations The blood film may show hypersegmented polymorphs ($B_{12}\downarrow$) or target cells (liver disease). Other tests: ESR (malignancy), LFTs (including GGT), T_4, serum B_{12} (note: falsely low result if on antibiotics—microbiological assay), red cell folate (more reliable than serum folate—use an FBC bottle).

Bone marrow biopsy is indicated if the cause is not revealed by the above tests. It is likely to show one of these 4 states:
Megaloblastic—B_{12} or folate deficiency (or cytotoxic drugs). (A megaloblast is a cell in which cytoplasmic and nuclear maturation are out of phase—as nuclear maturation is slow.)
Normoblastic marrow—liver damage, myxoedema.
Increased erythropoiesis—eg bleeding or haemolysis.
Abnormal erythropoiesis—sideroblastic anaemia, leukaemia, aplastic anaemia.

If results indicate B_{12} deficiency consider Schilling test to help to identify the cause. This determines whether a low B_{12} is due to malabsorption (B_{12} is absorbed from the terminal ileum) or to lack of intrinsic factor—by comparing the proportion of an oral dose (1µg) of radioactive B_{12} excreted in urine with and without the concurrent administration of intrinsic factor. (The blood must be saturated by giving an IM dose of 1000µg of B_{12}.) If intrinsic factor enhances absorption, lack of it (eg pernicious anaemia) is likely to be the cause (if not, look for blind loops, diverticula and terminal ileal disease as the cause of the low B_{12}).

Causes of a low B_{12} Pernicious anaemia (p582); post-gastrectomy (no intrinsic factor); poor diet (eg vegans); more rarely disease of terminal ileum (where B_{12} is absorbed)—eg Crohn's disease, resection; blind loops; diverticula and worms (*Dyphyllobothrium*).

Causes of low folate Dietary deficiency (eg in alcoholics), increased need (pregnancy, haemolysis, dyserythropoiesis, malignancy, long-term haemodialysis), malabsorption especially coeliac disease, tropical sprue, drugs (phenytoin and trimethoprim).

Note In ill patients with megaloblastic anaemia (eg with CCF) it may be necessary to treat before the results of serum B_{12} and folate are to hand. Use large doses, eg hydroxocobalamin 1mg/24h IM, with folic acid 20mg/24h IM (▶ folate given alone may precipitate subacute combined degeneration of the spinal cord). Blood transfusions are very rarely needed, but see p572.

Macrocytic anaemia
[*OTM* 19.71; 19.93]

Pernicious anaemia

Essence This disease affects all cells of the body and is due to malabsorption of B_{12} resulting from lack of gastric intrinsic factor because of autoimmunity to parietal cells. Atrophic gastritis is usually present. In B_{12} deficiency synthesis of thymidine, and hence DNA, is impaired and in consequence red cell production is reduced. In addition the CNS, peripheral nerves, gut and tongue may be affected.

Common features
Tiredness and weakness (90%)
Dyspnoea (70%)
Paraesthesiae (38%)
Sore red tongue (25%)
Diarrhoea, dementia

Other features
Retinal haemorrhages
Lemon tinge to skin
Retrobulbar neuritis
Mild splenomegaly
Mild pyrexia (<38°C)

Subacute combined degeneration of the spinal cord: The posterior and lateral columns are often affected but not necessarily together. The onset may be insidious and associated with a peripheral neuritis. A typical picture is absent vibration sense, extensor plantars with absent knee jerks (p426). Neurological involvement may occur without anaemia.

Associations Thyroid disease, vitiligo, Addison's disease, myxoedema, carcinoma of stomach (so arrange endoscopy).

Diagnosis Hb low (3–11g/dl), MCV often >110fl and hypersegmented polymorphs (right shift). Megaloblasts are found in the marrow. These are abnormal red cell precursors with delicate nuclear chromatin, and in which nuclear maturation is slower than cytoplasmic maturation. There is often mild neutropenia and thrombocytopenia. Serum B_{12} is always low.

Antibodies to parietal cells are found in 90%; there may also be antibodies to intrinsic factor, either at the B_{12} binding site (50%) or at the ileal binding site (35%). A Schilling test may be appropriate occasionally.

Treatment Hydroxocobalamin (B_{12}) 5 doses of 1mg IM every 2–3 days to replenish stores, then a maintenance dose of 1mg IM every 3 months for life (child's dose: as for adult). Initial improvement is heralded by a marked reticulocytosis (although serum iron falls first, if looked for).

Practical hints
1 Beware of diagnosing pernicious anaemia in those under 40yrs old: look for GI malabsorption (small bowel biopsy, p522).
2 Watch for hypokalaemia, as treatment becomes established.
3 Pernicious anaemia with high output CCF may require exchange transfusion (p572) after blood for FBC, folate, B_{12} and marrow sampling.
4 As haemopoiesis accelerates on treatment, iron supplements should be given.
5 WCC and platelet count should normalize in 1 week. Hb rises ~1g/dl per week of treatment.

Pernicious anaemia
[*OTM* 19.97]

An approach to haemolytic anaemia

Haemolysis is said to occur when the mean RBC survival is less than 120 days. If the bone marrow does not compensate sufficiently a haemolytic anaemia will result.

Causes of haemolysis These are either genetic or acquired.

Genetic:
1 Membrane: hereditary spherocytosis or elliptocytosis.
2 Haemoglobin: sickling disorders (p588), thalassaemia.
3 Enzyme defects: G6PD and pyruvate kinase deficiency.

Acquired:
1 Immune: either iso-immune (haemolytic disease of newborn, blood transfusion reaction), autoimmune (warm or cold antibody mediated) or drug-induced.
2 Non-immune: trauma (cardiac haemolysis, microangiopathic anaemia, p586), infection (malaria, septicaemia), hypersplenism, membrane disorders (paroxysmal nocturnal haemoglobinuria, liver disease).

In searching for evidence of significant haemolysis (and, if present, its cause) try to answer these 4 questions:
- *Is there increased red cell breakdown?* Bilirubin↑ (unconjugated) urinary urobilinogen↑, haptoglobin↓ (binds free Hb avidly).
- *Is there increased red cell production?* Eg a reticulocytosis, polychromasia, macrocytosis, marrow hyperplasia.
- *Is the haemolysis mainly extra- or intravascular?* Extravascular haemolysis may lead to splenic hypertrophy. The features of intravascular haemolysis are methaemalbuminaemia, free plasma haemoglobin, haemoglobinuria and haemosiderinuria.
- *Why is there haemolysis?* See below, and p586.

History Ask about family history, race, jaundice, haematuria, drugs, previous anaemia.

Examination Look for jaundice, hepatosplenomegaly, leg ulcers (seen in sickle-cell disease).

Investigations FBC, reticulocytes, bilirubin, LDH, haptoglobin, urinary urobilinogen. Films may show polychromasia, macrocytosis, spherocytes, elliptocytes, fragmented cells or sickle cells.

Further investigations include the direct Coombs' test (DCT). This will identify red cells coated with antibody or complement and a positive result usually indicates an immune cause of the haemolysis. The RBC lifespan may be determined by chromium labelling and the major site of RBC breakdown may also be identified. Urinary haemosiderin (stains with Prussian Blue) indicates chronic intravascular haemolysis.

The cause of the haemolysis may now be obvious, but further tests may be needed. Membrane abnormalities are identified on the film, and Hb electrophoresis will detect Hb variants. Enzyme assays are reserved for when other causes have been excluded.

The Coombs' test will detect the immune acquired causes, and the non-immune group are usually identifiable by associated features.

An approach to haemolytic anaemia
[*OTM* 19.134]

Causes of haemolytic anaemia

Sickle-cell disease See p588.

Hereditary spherocytosis is inherited as an autosomal dominant (an autosonal dominant) and causes variable haemolysis with spherocytes on the film. Clinical features: mild anaemia (Hb 8–12g/dl), splenomegaly and risk of gallstones. RBCs show increased fragility. Diagnosis is by osmotic tests. Treatment: splenectomy.

Hereditary elliptocytosis Usually inherited as an autosomal dominant. The degree of haemolysis is variable. If needed, splenectomy may help.

Glucose-6-phosphate dehydrogenase deficiency is the commonest RBC enzyme defect. Inheritance is sex-linked with 100 million affected in Africa, the Mediterranean and the Middle and Far East. Neonatal jaundice may occur, but the majority are symptomless with normal Hb and blood film. They are susceptible to oxidative crises precipitated by drugs (eg primaquine or sulphonamides), exposure to the bean *Vicia fava* (favism) or illness. Typically there is rapid anaemia and jaundice with RBC Heinz bodies (denatured haemoglobin stained with methyl violet). Diagnose by enzyme assay. Treatment is avoidance of precipitants.

Pyruvate kinase deficiency Usually inherited as an autosomal recessive—homozygotes often present with neonatal jaundice, later chronic haemolysis with splenomegaly and jaundice. Diagnose by enzyme assay. Often well-tolerated. There is no specific therapy but splenectomy may help.

Drug-induced immune haemolysis Due to formation of new RBC membrane antigens (eg penicillin in prolonged, high dose), immune complex formation (many drugs, rare) or development of autoantibodies to the RBC (α-methyldopa, mefenamic acid, L-dopa).

Autoimmune haemolytic anaemia (AHA) Causes: warm or cold antibodies. They may be primary (idiopathic) or secondary, usually to lymphoma or generalized autoimmune disease, eg SLE. Warm AHA presents as chronic or acute anaemia. Treat by steroids (± splenectomy). Cold AHA: chronic anaemia made worse by cold, often with Raynaud's or acrocyanosis. Treatment: keep warm. Chlorambucil may help. Mycoplasma may produce acute cold haemolysis.

Paroxysmal cold haemoglobinuria is caused by Donnath–Landsteiner antibody (seen in mumps, measles, chickenpox, syphilis) sticking to RBCs in cold, which causes complement-mediated lysis on rewarming.

Cardiac haemolysis is due to cell trauma in a prosthetic aortic valve. It may indicate valve malfunction.

Microangiopathic haemolytic anaemia (MAHA) Suspect if marked fragmentation and many microspherocytes on the film. Includes haemolytic–uraemic syndrome and thrombotic thrombocytopenic purpura (see p398). Treat underlying disease; blood and fresh frozen plasma transfusions for support.

Paroxysmal nocturnal haemoglobinuria RBCs are sensitive to complement. The syndrome features pancytopenia, abdominal pain or thrombosis. Diagnosis: Ham's test (*in vitro* acid-induced lysis).

Factors exacerbating haemolysis Infection often leads to increased haemolysis. Parvoviruses (OHCS p214) cause cessation of marrow erythropoiesis (aplastic anaemia with no reticulocytes).

Causes of haemolytic anaemia

[*OTM* 19.134]

Sickle-cell anaemia

There is severe haemolysis from homozygous inheritance of a gene causing an amino acid substitution in haemoglobin (β_6 Glu→Val), so making HbS. It is common in black Africans and their world-wide descendants. The homozygote (SS) has sickle-cell *anaemia* and the heterozygote (AS) has sickle-cell *trait*, which causes no disability (and may protect from *falciparum* malaria) except in hypoxia—eg in unpressurized aircraft or anaesthesia, when veno-occlusive events may occur—so all blacks need a sickling test pre-op. Symptomatic sickling occurs in heterozygotes with genes coding other analogous amino acid substitutions (eg haemoglobin SC and SD diseases). Homozygotes (CC; DD) have asymptomatic, mild anaemia. NB: the normal adult AA genotype codes HbA (see p590).

Pathogenesis In the deoxygenated state the HbS molecules polymerize and cause RBCs to sickle. Sickle-cells are fragile, and haemolyse; they also block small vessels to cause infarction. Haemolysis is mild in those with high levels of HbF (it forms soluble hybrid polymers with HbS, so [Hb-s]↓—and this effect can be encouraged by agents such as hydroxyurea and erythropoietin, but their clinical effects may be slight[1]).

Tests Sickling tests detect HbS. Electrophoresis distinguishes SS, AS states, and other Hb variants. Film: all have target cells. Aim for diagnosis *at birth* (cord blood) so that pneumococcal prophylaxis (23-valent vaccine, p332, or penicillin V 125mg/12h PO) may be given.

The patient Typically has an Hb of 6–8g/dl; reticulocytes 10–20%: (haemolysis is variable) and jaundice *early on*, eg with painful swelling of hands and feet (hand and foot syndrome)—also splenomegaly (rare if >10yrs, as the spleen infarcts). Young sicklers alternate periods of good health with acute crises (below). *Later*, chronic ill-health supervenes from previous crises: ie renal failure; bone necrosis; osteomyelitis; leg ulcers; iron overload (and alloimmunization) from many transfusions.

Sickle-cell crises These may be from thrombosis—the aptly-named 'painful crises', haemolysis (rare), marrow aplasia or sequestration.

Thrombotic crises (precipitated by cold, dehydration, infection, ischaemia eg muscular exertion) are most common and can cause severe pain, often in the bones. They may mimic an acute abdomen or pneumonia. CNS signs: fits, focal signs. Priapism may occur (prolonged erections; if for >24h arrange prompt cavernosus-spongeosum shunting to prevent impotence (priapism also occurs in CML—p604).

Aplastic crises: Often due to parvoviruses; characterized by sudden lethargy and pallor and few reticulocytes. Urgent transfusion is needed.

Sequestration crises: These are particularly serious—spleen and liver enlarge rapidly due to RBC trapping. Anaemia is severe.

Management of sickle-cell crisis ► Seek expert help.

- Give *prompt*, generous analgesia, eg with opiates (p90).
- Crossmatch blood. FBC, reticulocytes, blood cultures, MSU, CXR.
- Rehydrate with IVI. ● Give O_2 by mask if PaO_2↓. ● Keep warm.
- Blind antibiotics (p182) if feverish, after infection screen.
- Measure PCV, reticulocytes, liver and spleen size twice daily.
- Give blood transfusion if PCV or reticulocytes fall sharply, or if there are CNS or lung complications—when the proportion of sickled cells should be reduced to <30%. If Hb <6g/dl it is safe to transfuse; if >9g/dl do a partial exchange transfusion.

NB ● Consider a histocompatible marrow transplant. ● Febrile children risk septicaemia: repeated admission is avoided by out-patient ceftriaxone (2 doses, 50mg/kg IV on day 0 & 1). Admit if Hb <5g/dl; WCC↓ or ↑ (>30 × 10⁹/l); T° >40°C; severe pain; dehydration; lung infiltration.[2]

Sickle-cell anaemia

Prevention Genetic counselling; prenatal tests (*OHCS* p210–12). Parental education can help prevent 90% of deaths from sequestration crises.[2]

Patient-controlled analgesia (PCA) An example with paediatric doses. First try warmth, hydration, and oral analgesia (ibuprofen 5mg/kg/6h PO, codeine phosphate 1mg/kg/4–8h PO up to 3mg/kg/day). If this fails, see on the ward and offer prompt morphine by IVI—eg 0.1mg/kg. Start PCA with morphine 1mg/kg in 50ml 5% dextrose, and try a rate of 1ml/h, allowing the patient to deliver extra boluses of 1ml when needed. Do respiration and sedation score every $\frac{1}{4}$h + pulse oximetry if chest/abdominal pain.[3]

589

1 1993 *Arch Dis Chi* **69** 176   2 J Williams *NEJM* 1993 **329** 472 & 501   3 R Grundy 1993 *Arch Dis Chi* **69** 256

Thalassaemia

Haemoglobin (Hb) consists of 2 different pairs of peptide chains (one called α, and the other β) with a haem molecule attached to each peptide. Hb is heterogeneous. Adults have 95% HbA ($\alpha_2\beta_2$) and a little HbA$_2$ ($\alpha_2\delta_2$). HbF ($\alpha_2\gamma_2$) predominates in fetal life but may persist in β thalassaemia.

The thalassaemias are a group of disorders resulting from reduced rate of production of one or more globin chains. This leads to precipitation of globin, and anaemia due to ineffective erythropoiesis and haemolysis. They are common in a band stretching from the Mediterranean to the Far East. The most important are α and β thalassaemia, in which there is reduced production of α chains and β chains respectively. It is known that there are 2 genes for β globin and 4 for α in each individual. It is possible to correlate clinical severity with genetic deficit.

β *Thalassaemia major:* This is the homozygous form and results in a severe anaemia presenting in the first year, often as failure to thrive. Death results in 1 year without transfusion. If adequate transfusion is given development is normal but symptoms of iron overload appear after 10 years as endocrine failure, liver disease and cardiac toxicity. Death usually occurs at 20–30 years due to cardiac siderosis. Long-term infusion of desferrioxamine may prevent the iron loading. If transfusion is inadequate there is chronic anaemia with reduced growth and skeletal deformity due to bone marrow hyperplasia eg bossing of skull. Also splenomegaly, bleeding and intermittent fever. The film shows very hypochromic, microcytic cells with target cells and nucleated RBCs. HbF↑↑, HbA$_2$ variable, HbA absent.

β *Thalassaemia minor:* This is the heterozygous state, recognized as MCV <75fl, HbA$_2$ >3.5% and mild anaemia (9–11g/dl) which is usually well-tolerated but may be worsened in pregnancy.

α *Thalassaemia:* 4 varieties are defined, depending on the number of defective genes. Absence of 1 or 2 genes is very common in black populations and produces a mild anaemia with a low MCV.

Hb H disease: HbH (β_4) is present at 5–30% throughout life with moderate haemolytic anaemia and splenomegaly. Prognosis: good.

Hb Barts: Infants are stillborn, or have neonatal jaundice (hydrops fetalis, OHCS p242). They lack all α genes. Hb Barts is γ_4, and physiologically useless. Obstetric problems are common.

Thalassaemia intermedia: This is a term used to describe cases of intermediate severity who are not transfusion dependent but have severe anaemia. It includes a wide range of Hb variants.

Diagnosis FBC, MCV, film, iron, HbA$_2$, HbF, Hb electrophoresis.

Treatment ● Transfusion to keep Hb >9g/dl. ● Iron chelating agent, eg desferrioxamine. ● Large doses of ascorbic acid also increase iron output. ● Perform splenectomy if hypersplenism exists. ● Give folate supplements. ● A histocompatible marrow transplant can offer the chance of a cure.[1]

Prevention Approaches are genetic counselling or antenatal diagnosis using fetal blood or DNA, then 'therapeutic' abortion.

1 DJ Weatherall 1993 *NEJM* **329** 877

Thalassaemia
[*OTM* 19.40]

Bleeding disorders

After trauma, 3 processes halt bleeding: constriction, gap-plugging by platelets, and the coagulation cascade. Disorders of haemostasis fall into these 3 groups. The pattern of bleeding is important—vascular and platelet disorders lead to prolonged bleeding from cuts, and to purpura and bleeding from mucous membranes. Coagulation disorders produce delayed bleeding after injury, into joints, muscle and the GI and GU tracts.

Vascular defects Causes may be congenital (Osler–Weber–Rendu syndrome, p706), acquired (eg senile purpura, steroid treatment, trauma, pressure, vasculitis, connective tissue disease, scurvy and painful bruising syndrome). Scurvy produces perifollicular haemorrhage. The painful bruising syndrome is seen in women who develop tingling under the skin followed by bruising over limbs/trunk, resolving without treatment.

Platelet disorders These are due to thrombocytopenia or disorders of platelet function. *Causes of thrombocytopenia:*
- Decreased production—marrow failure, megaloblastosis (p582).
- Decreased survival—immune thrombocytopenic purpura (ITP); viruses; DIC; drugs; SLE; lymphoma; thrombotic thrombocytopenic purpura; hypersplenism.
- Platelet aggregation—heparin causes this in 5% of patients.

ITP may be acute or chronic. Acute ITP is seen in children, often after infection, with rapid onset of self-limiting purpura. Treatment is rarely needed but steroids may be used. Chronic ITP runs a fluctuating course of bleeding, purpura and epistaxis, often in young women. If symptomatic or platelets $<20 \times 10^9/l$, consider steroids or splenectomy (cures ≤80%).

Abnormalities of platelet function occur with myeloproliferative disorders, drugs (eg aspirin), uraemia, Von Willebrand's disease.

Coagulation disorders Congenital: (eg haemophilia, Von Willebrand's—p710) or with: anticoagulation; liver disease; DIC; malabsorption (vitamin K↓).

Haemophilia A is due to deficiency of factor VIII, inherited as a sex-linked recessive in 1/10,000 ♂ births. *Presentation* depends on severity and is often early in life or after surgery/trauma—with bleeds into joints and muscle, leading to crippling arthropathy and haematomas with nerve palsies due to pressure. *Diagnose* by ↑KCCT and factor VIII assay. *Management:* Seek expert advice. Avoid NSAIDs and IM injections.
- Minor bleeding: pressure and elevation of the part. Desmopressin (0.4μg/kg/12–24h IVI in 50ml 0.9% saline over 20mins) raises factor VIII levels, and may be sufficient. • Major bleeding (eg haemarthrosis) requires factor VIII levels to be raised to 50% of normal. Life-threatening bleeding (eg causing airway obstruction) needs levels of 100%, eg with virally inactivated lyophilized factor VIII (cryoprecipitates are screened but are not guaranteed to be HIV free; they contain more vWF, see p710).

Haemophilia B (Christmas disease) is due to factor IX deficiency and behaves clinically like haemophilia A.

Liver disease produces a complicated bleeding disorder (synthesis of clotting factors↓, absorption of vitamin K↓, abnormalities of platelet function. *Malabsorption* leads to less uptake of vitamin K (needed for synthesis of factors II, VII, IX and X). Treatment is parenteral vitamin K (10mg) or fresh frozen plasma for acute haemorrhage.

Anticoagulants See p596. DIC See p598.

The intrinsic and extrinsic pathways of blood coagulation

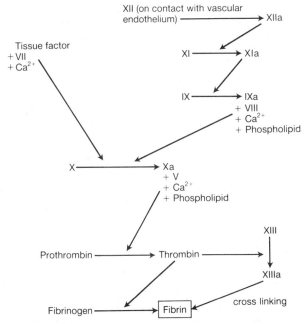

EXTRINSIC SYSTEM   INTRINSIC SYSTEM

The fibrinolytic system causes fibrin dissolution and acts via the generation of plasmin. The process starts by the release of tissue plasminogen activator (t-PA) from endothelial cells, a process stimulated by fibrin formation. t-PA converts inactive plasminogen to plasmin which can then cleave fibrin, as well as several other factors. t-PA and plasminogen both bind fibrin thus localizing fibrinolysis to the area of the clot.

Mechanism of fibrinolytic agents

1 *Alteplase* (≡rt-PA≡Actilyse®; from recombinant DNA) is a fibrinolytic enzyme imitating t-PA, as above. Plasma $t_{1/2} \approx 5$mins.

2 *Anistreplase* (≡anisoylated plasminogen streptokinase activator complex ≡APSAC®) is a complex of human plasminogen and streptokinase (so anaphylaxis is possible). Plasma $t_{1/2} \approx 90$mins.

3 *Streptokinase* is a streptococcal exotoxin and forms a complex in plasma with plasminogen to form an activator complex which forms plasmin from unbound plasminogen. Initially there is rapid plasmin formation which can cause uncontrolled fibrinolysis. However, plasminogen is rapidly consumed in the complex and then plasmin is only produced as more plasminogen is synthesized. The activator complex binds to fibrin and so produces some localization of fibrinolysis.

4 *Urokinase* is produced by the kidney and is found in urine. It also cleaves plasminogen but expense has limited its use.

An approach to bleeding

There are 3 sets of questions to be answered:

1 Is this case of bleeding very urgent—so that I must resuscitate and/or refer the patient immediately?
- Is the patient about to exsanguinate (shock, coma, p716–20)?
- Is there hypovolaemia (postural hypotension, oliguria)?
- Is there CNS bleeding (meningism, CNS and retinal signs)?

2 Why is the patient bleeding? You are trying to determine if the bleeding is a normal response to something else (eg trauma), or if the patient has a predisposition to bleed.
- Is there unexplained bleeding, bruising or purpura?
- Past or family history of bleeding—trauma, dentistry, surgery? Drugs (warfarin). Alcohol. Liver disease. Recent illness.
- Is the pattern of bleeding indicative of vascular, platelet or coagulation problems (p592)? Are venepuncture sites bleeding (DIC)? Look for associated conditions (eg with DIC, p598).
- Is a clotting screen abnormal? Check FBC, platelets, INR, KCCT and thrombin time. Consider bleeding time, fibrin degradation products (FDP) and factor VIII assay.

3 If a bleeding disorder, what is the mechanism?

The degree of initial investigation will depend on the severity of the bleeding. If there is shock the patient should be resuscitated with whole blood and fresh frozen plasma (FFP) in a ratio of 6 units to 2. If there is thrombocytopenia give platelet transfusion (aiming for >50 × 10⁹/l) except in ITP where steroids may be indicated: consult with experts. Is there overdose with anticoagulants (p748)? In haemophiliac bleeds, consult experts early (for coagulation factor replacement). *Never* give IM injections. Acutely, coagulation deficiencies may be treated by FFP and vitamin K (phytomenadione 2.5–20mg IV slowly).

Coagulation tests (Sodium citrate tube—false results if inaccurately filled)

1 *Thrombin time:* Thrombin is added to plasma to convert fibrinogen to fibrin. Normal range (NR): 10–15sec, prolonged in heparin treatment, DIC or afibrinogenaemia.

2 *Prothrombin time:* (PT). Thromboplastin is added to test the extrinsic system. PT is expressed as a ratio compared to control [International Normalized Ratio, INR, normal range (NR)=0.9–1.2]. Prolonged in coumarin (eg warfarin) treatment or liver disease (as above).

3 *Kaolin cephalin clotting time* (KCCT, or APTT, ≡PTT≡partial thromboplastin time) Kaolin activates the intrinsic system. Normal range 35–45sec, prolonged in heparin treatment or haemophilia.

Investigation of a bleeding disorder
Platelets: If low do FBC, film, bone marrow biopsy.
INR: If long, look for liver disease, or anticoagulant use.
KCCT: If long, consider factor VIII or IX deficiency, or heparin.
Bleeding time: If long, consider Von Willebrand's disease (p710), or platelet disorders. Aspirin prolongs the bleeding time.

Special tests are also available (eg factor assays: consult with a haematologist).

An approach to bleeding
[*OTM* 19.211]

Anticoagulants

▶ Consider bleeding as a cause of any symptom in anyone on anti-coagulants—and arrange for a prothrombin ratio (INR).

Indications[1]	Ideal INR	Duration of treatment
PAbove knee DVT; PE	2.0–3.0	3–6 months
PValve prostheses	3.0–4.5	Life (not needed for pig valves[2])
AMitral stenosis	2.0–4.0	Life (higher risk if AF or large L atrium; can stop post valvotomy)
EqMitral regurgitation	2.0–4.0	Only if large L atrium, heart failure or previous embolism
AAtrial fibrillation (▶ see p290)	1.4–2.8	Life (high risk if *intermittent* AF, CCF, T_3↑ or rheumatic AF; see p290)
AArterial embolism	3.0–4.5	? 6 months; life-long if a cardiac source persists
EqCABG (p278)	3.0–4.5	Usually only after endarterectomy
Recurrent DVT or PE	3.0–4.5	Life

Note: P=proven; A=accepted; Eq=equivalent; PE=*pulmonary embolism*.

Warfarin is the first choice (phenindione if warfarin sensitive). CI: peptic ulcer; bleeding disorders; severe hypertension; liver failure; infective endocarditis; cerebral aneurysms. Use with caution in the elderly and those with remote GI bleeds. For use in pregnancy, see OHCS p150. ▶ For interactions see p800.

Beginning anticoagulation Give heparin 5000iu IV over 5mins; then add 25,000iu to 50ml 0.9% saline (=500iu/ml) in a syringe pump. Give IVI at a rate of 2.8ml/h (=1400iu/h). Check APTT (activated partial thromboplastin time) at 6h. Institute the 5-day sliding scale below. (*Measure APTT every 10h—every 4h if APTT >7, and stop the IVI*).[3]

APTT	5–7	4–5	3–4	2½–3	1½–2½	1.2–1.4	<1.2
Change rate (iu/h) by:	−500	−300	−100	−50	0	+200	+400

Warfarin may be started at the same time (but takes 3 days to work). Give 10mg PO at 17.00h (if that morning's INR is <1.4). Measure the INR 16h later and if it is <1.8 give 10mg as the second dose at 17.00h. In the unlikely event of INR being >1.8 (warfarin sensitivity) give only 0.5mg PO. Then aim to keep the INR at 2–4 (see above) by giving a daily dose of warfarin PO as guided below.[3]

INR	<2	2	2.5	2.9	3.3	3.6	4.1
Third dose	10mg	5mg	4mg	3mg	2mg	0.5mg	0mg
Maintenance	>6mg	5mg	5.5mg	4.5mg	4mg	3.5mg	3mg*

*Miss a dose; give 1–2mg the next day (if INR >4.5, miss 2 doses).

Check INR daily for 5 days (stopping heparin when INR >2), then on alternate days until stable, tailing off to weekly or longer. Tell the lab if the patient is also on heparin (it alters the assay). Check platelet count if on heparin for more than 5 days. **Antidotes** Consult with a haematologist. Stop anticoagulation; if further steps are needed, protamine sulphate counteracts heparin: 1mg IVI neutralizes 100u heparin given within 15min. Max dose: 50mg (if exceeded, may itself have anticoagulant effect). Use phytomenadione (vitamin K) *slowly* IV for warfarin poisoning: 5mg for serious bleeding, 0.5–2mg for simple haematuria or epistaxis; consider 0.5mg if INR >7 without any bleeding. This takes some hours to act and may last for weeks. Give a concentrate of factors II, IX and X (+ VII if available). If unavailable, use fresh frozen plasma IVI.

1 Brit Heart Found guidelines (*Factfile* 7/87) 2 S Wessler 1984 *NEJM* **311** 645 3 Modified Fennerty formula, *Drug Ther Bul* 1992 **30** 77

Anticoagulants

Leukaemia and the house officer

Leukaemic patients often fall ill suddenly at inconvenient times. Our aim is to enable prompt, safe treatment to be started while awaiting advice.

In those with non-specific confusion/drowsiness, do blood cultures and exclude hypoglycaemia, as well as the tests on p446.

Neutropenic régime (for patients with a WCC ≤1.0 × 10^9/l)
► Close liaison with a microbiologist is essential.
● Full barrier nursing is often impossible, simple hand-washing is probably most important. Use a side room.
● Look for infection (mouth, perineum, axilla, IVI site). Do swabs.
● Avoid IM injections (the danger is an infected haematoma).
● Wash perineum after defecation. Swab moist skin with chlorhexidine. Avoid unnecessary rectal examinations. Give hydrogen peroxide mouth washes/2h and Candida prophylaxis (p500).
● TPR every 4h. High calorie diet; no salads (Pseudomonas risk).

After taking cultures (blood, urine, sputum, Hickman line) and CXR, treat any known infection. If T° >38°C for >4–6h or the patient is toxic start 'blind' broad spectrum antibiotics, eg piperacillin and netilmicin, with or without vancomycin (p180). Check local preferences.

If there are chest signs consider treatment for *Pneumocystis* (cotrimoxazole ie trimethoprim 20mg/kg and sulphamethoxazole 100mg/kg per day PO/IV in 4 divided doses) but remember TB.

Continue antibiotics until afebrile for 5 days and neutrophils recover (>0.5 × 10^9/l). If fever persists despite antibiotics consider CMV or fungal infection (eg Candida or Aspergillus, p244).

► Consider pneumococcal vaccine to prevent pneumococcal emergencies. There may be a rôle for genetically engineered recombinant human granulocyte-colony stimulating factor (rG-CSF, Filgrastim®) in preventing neutropenia, and allowing more intensive chemotherapy.[1]

Other dangers Prevent *hyperkalaemia* and *hyperuricaemia* following massive destruction of cells by giving a high fluid intake and allopurinol before cytotoxic therapy.

Inappropriate transfusion: If the WCC is >100 × 10^9/l WBC thrombi may form in brain, lung and heart (leucostasis). Transfusing here (before the WCC is reduced) increases blood viscosity, so increasing risk of leucostasis. If transfusion is essential, keep Hb <9g/dl.

Disseminated intravascular coagulation (DIC) This pathological activation of coagulation mechanisms may occur in any malignancy. (Other causes include infection, trauma or obstetric complications.)

Mechanism: The release of procoagulant agents leads to clotting factors and platelet consumption (consumption coagulopathy). Fibrin strands fill small vessels, slicing (haemolysing) passing RBCs.

Signs: Extensive bruising, old venepuncture sites start bleeding, renal failure, gangrene, bleeding anywhere (eg uterus, lungs, CNS).

Tests: Consult lab, as special bottles (kept in fridges) are used. Film: broken RBCs (schistocytes). Fibrin degradation products↑, clotting times↑, platelets↓. Blood may not clot in plain bottles.

Haematology

Leukaemia and the house officer
[*OTM* 19.15]

Treat the cause where possible and give supportive treatment eg: accurate fluid balance, monitor urine volume hourly by catheter. Give 2 units fresh frozen plasma IV at once while expert advice is sought; platelets and blood may be needed. The role of heparin is controversial: pre-chemotherapy low dose heparin may prevent DIC in acute promyelocytic leukaemia (a common cause of DIC).

1 *Drug Ther Bul* 1993 **31** 33

Acute lymphoblastic leukaemia (ALL)

This is a neoplastic proliferation of lymphoblasts.

Immunological classification

Common (c) ALL: (~75%) Defined by reaction with specific antilympho-blast antibody. Phenotypically pre-B (ie the cells carry the same surface antigen as the pre-B lymphocyte). Any age may be affected, commonly 2–4yr-olds. ♂/♀ = 1.

T-cell ALL: Any age but peak in adolescent males, eg presenting with a mediastinal mass and a high WCC.

B-cell ALL: Rare. Very poor prognosis. Surface immunoglobulins present on blast cells.

Null-cell ALL: Undifferentiated, lacking any markers.

An alternative classification is the French–American–British (FAB), types L1, L2 and L3.

Causes Unknown—but preconception paternal exposure to sawdust, benzene (eg in petrol) and radiation has been implicated.[1,2]

Clinical features These are due to marrow failure: anaemia, infection and bleeding. Also: bone pain, arthritis, splenomegaly, lymphadenopathy, thymic enlargement, CNS involvement—eg cranial nerve palsies.

Common infections: (Eg relating to pancytopenia following chemo-therapy): Zoster, CMV, measles, candidiasis, pneumocystosis, bacterial septicaemia. Consider use of immune serum for patients in contact with measles or zoster when on chemotherapy.

Diagnosis Characteristic cells in blood and bone marrow.

Treatment
● *Supportive care:* Blood and platelet transfusions, IV antibiotics at the first sign of infection.
● *Preventing infections:* The neutropenic régime (p598); barrier nurs-ing; prophylactic antibiotics (eg co-trimoxazole to prevent pneumo-cystosis.
● *Chemotherapy:* As in most leukaemias, patients are entered into national trials. A typical programme is in 3 steps:
  1 *Remission induction:* Vincristine, prednisolone, L-asparaginase and daunorubicin.
  2 *CNS prophylaxis:* Intrathecal methotrexate; cranial irradiation.
  3 *Maintenance chemotherapy:* Eg—mercaptopurine (daily), metho-trexate (weekly) and vincristine and prednisolone (monthly) for 2–3 years. Relapse is common in blood, CNS or testis. More details: *OHCS* p190.

Haematological remission means no evidence of leukaemia in the blood, a normal or recovering blood count and less than 5% blasts in a normal regenerating bone marrow.

Marrow transplant (p602) Consider in poor prognosis groups.

Quality of life Think of ingenious ways of making the patient's life better (especially while barrier nursed). Wigs for alopecia after chemotherapy. Avoid repeated venepuncture by using a Hickman CVP line—a central line with a Dacron® cuffed portion (to reduce the risks of infection) which lies subcutaneously.

Prognosis Cure rates for children are 70–90%; for adults only 35% (0–20% if >60yrs old, where there is a second peak in incidence)

Worse prognosis in:[3] Blacks, males, WCC >100 ×10⁹/l; Philadelphia translocation [t(9;22)(q34:q11)] (*OHCS* p190); CNS presentation.

1 M Gardner 1987 *BMJ* ii 822 2 P McKinney 1991 *BMJ* i 681 3 D Hoelzer 1993 *NEJM* **329** 1343

Acute lymphoblastic leukaemia (ALL)
[*OTM* 19.18]

Acute myeloid leukaemia (AML)

This is a neoplastic proliferation of blast cells derived from myeloid elements within the bone marrow. It is among the most rapidly progessive malignancies (eg death in ~2 months if untreated).

Morphological classification (FAB—ie French, American and British)
M1: Undifferentiated blast cells   M5: Monocytic
M2: Myeloblastic   M6: Erythroleukaemia
M3: Promyelocytic   M7: Megakaryoblastic leukaemia
M4: Myelomonocytic

Epidemiology 1 in 10,000. Increases with age. AML is getting more common, and is a long-term complication of chemotherapy eg for lymphoma.

Clinical features

Marrow failure	Leukaemic infiltration	Constitutional features
Anaemia	Bone pain; tender sternum	Malaise
Infection: often Gram −ve	CNS signs (cord compression, cranial nerve lesions)	Weakness
		Fever
Bleeding eg petechiae	Gums, testes, orbit (proptosis)	Other features:
	Hepatosplenomegaly	Polyarthritis
DIC	Lymphadenopathy	
	Skin and perianal involvement	

Diagnosis WCC is variable. Blast cells may be few in the peripheral blood, so diagnosis depends on bone marrow biopsy. Differentiation from ALL depends on finding granules in the abnormal cells—and the presence of Auer rods (thought to represent the amalgamation of granules).

Complications Leucostasis (white cell thrombi, common if WCC >200 × 10^9/l) may cause pulmonary and cerebral infarcts and death. To reduce the blast count quickly, a single dose of daunorubicin or doxorubicin may be given (60mg/m^2 IV) with allopurinol to counter the high uric acid level which this causes. Emergency cranial radiotherapy may also be given. Massive mediastinal lymphadenopathy may give rise to dyspnoea.

Treatment Supportive care: Blood and platelet transfusions. Barrier nursing, IV antibiotics. Pitfalls in diagnosis of infection: AML itself causes fever, common organisms present oddly, few antibodies are made, rare organisms—particularly fungi (especially Candida and Aspergillus).

Chemotherapy is very intensive, resulting in long periods of neutropenia and thrombocytopenia. The main drugs used include daunorubicin, cytosine arabinoside and thioguanine.

Bone marrow transplant (BMT) Allogeneic transplants from histocompatible siblings or from unrelated donors (accessed via international computer-held databases) or syngenetic transplants from a twin may be indicated in 1st or subsequent remissions. The idea is to destroy all leukaemic cells and the entire immune system by cyclophosphamide and total body irradiation, and then repopulate the marrow by transplantation from a matched donor infused IVI. Cyclosporin ± methotrexate may be used to reduce the effect of the new marrow attacking the patient's body (graft versus host disease). Complications: graft versus host disease; infections (CMV is common); relapse of leukaemia.

Prognosis Chemotherapy: Approximately 20% long-term survivors.
Bone marrow transplant: Perhaps 50% long-term survivors.

Acute myeloid leukaemia (AML)

[*OTM* 19.18]

Chronic lymphocytic leukaemia (CLL)

This is a monoclonal proliferation of well-differentiated lymphocytes. They are almost always (99%) B cells. The patient is usually over 40. Men are affected × 2 as often as women. CLL constitutes 25% of all leukaemias.

Staging (Correlates with survival) 0 Absolute lymphocytosis >15 × 10⁹/l. I Stage 0 + enlarged lymph nodes. II Stage I + enlarged liver or spleen. III Stage II + anaemia (Hb <11g/dl). IV Stage III + platelets <100 × 10⁹/l.

Symptoms (None in 25%.) Bleeding, weight↓ and anorexia.

Signs Enlarged, rubbery, non-tender nodes. Late hepatosplenomegaly.

Film Lymphocytosis may be marked. Often normochromic normocytic anaemia. Autoimmune haemolysis may contribute to this. Thrombocytopenia from marrow infiltration (rarely antiplatelet antibodies).

Complications 1 Autoimmune haemolysis 2 Infection—bacterial (mostly of the respiratory tract, as a result of hypogammaglobulinaemia)—or viral (altered cell mediated immunity). 3 Bone marrow failure.

Natural history Some patients remain in *status quo* for years, or even regress. Usually nodes slowly enlarge. Death is usually due to infection (zoster, pneumococcus, meningococcus, TB, Candida or aspergillosis).

Treatment Chemotherapy is not always needed, but may postpone marrow failure. Chlorambucil is used to ↓ lymphocyte count. Dose: eg 0.1–0.2mg/kg daily PO. Steroids are used, eg if there is autoimmune haemolysis.

Radiotherapy: Use for relief of lymphadenopathy or splenomegaly.

Supportive care: Transfusions, prophylactic antibiotics, occasionally, IV human immunoglobulin. Prognosis Often good: depends on stage.

Chronic myeloid leukaemia (CML)

CML is characterized by uncontrolled proliferation of myeloid cells. It accounts for 15% of leukaemias. It is a myeloproliferative disorder (p612) having features in common with these diseases—eg splenomegaly (often massive). It commonly presents with constitutional symptoms. It occurs most often in middle age. There is a slight male predominance.

Philadelphia chromosome (Ph¹) This is chromosome 22 which has lost about half of its long arm. In nearly all cases this is translocated to the long arm of chromosome 9 (9q+/22q-). The Philadelphia chromosome is present in granulocyte, RBC and platelet precursors in over 95% of those with CML. Those without Ph¹ have a particularly bad prognosis.

Symptoms Most are chronic and insidious, eg: weight↓, tiredness, gout, fever, sweats, haemorrhage or abdominal pain. 10% are detected by chance.

Signs Splenomegaly, variable hepatomegaly, anaemia, bruising.

Tests WBC↑↑ (often >100 × 10⁹/l), Hb↓ or↔, platelets variable. Leucocyte alk phos↓ (on stained film); plasma uric acid and alk phos↑; B₁₂↑.

Natural history Variable; typically 2 phases: chronic, lasting months or years of few, if any, symptoms, followed by blast transformation with features of acute leukaemia and usually rapid death.

Chronic lymphocytic leukaemia (CLL)
Chronic myeloid leukaemia (CML)

[*OTM* 119.32; 19.27]

Treatment Busulphan 2–4mg/24h PO is the mainstay of treatment in the chronic phase. α-interferon is under investigation; provisional results look promising. Monitor FBC to avoid pancytopenia. Stop treatment when WCC <20 × 10^9/l. Transformation may be signalled by lack of response to previously effective therapy. Treatment of transformed CML is poor. Some types have lymphoblastic characteristics (treat as for ALL). Treatment of myeloblastic transformation rarely achieves lasting remission. Another option is autologous bone marrow transplant starting with chemotherapy and whole-body radiotherapy followed by autografting of patient's previously stored haemopoietic stem cells. Allogeneic transplantation, from an HLA matched donor, should be considered if eg <55yrs, during chronic phase. ~50–60% of transplanted patients will be cured.

Hodgkin's lymphoma

Lymphomas are malignant proliferations of the lymphoid system. They are divided into Hodgkin's and non-Hodgkin's types by histology. In Hodgkin's lymphoma characteristic cells with mirror image nuclei are found (Reed–Sternberg cells).

Classification (In order of incidence)	Prognosis:
Nodular sclerosing	Good
Mixed cellularity	Good
Lymphocyte predominant	Good
Lymphocyte depleted	Poor

Clinical features The usual presentation is with enlarged, painless lymph nodes, usually in the neck or axillae. Rarely there may be alcohol-induced pain or features due to the mass effect of the nodes. 25% have constitutional symptoms such as fever, weight loss, night sweats, pruritus and loss of energy. If the fever alternates with long periods (15–28 days) of normal or low temperature, it is termed a Pel–Ebstein fever (other causes of this rare fever: TB, hypernephromas).

Signs: Lymphadenopathy (note position, consistency, mobility, size, tenderness). Look for weight loss, anaemia, hepatosplenomegaly.

Tests Lymph node biopsy for diagnosis. FBC, film, ESR, LFTs, uric acid, Ca^{2+}, CXR, bone marrow biopsy, abdominal CT or lymphangiography. Staging laparotomy involves splenectomy with liver and lymph node biopsy but may not influence final outcome and is rarely performed nowadays.

Staging (Influences treatment and prognosis.)
I Confined to single lymph node region.
II Involvement of 2 or more regions on same side of diaphragm.
III Involvement of nodes on both sides of diaphragm.
IV Spread beyond the lymph nodes.

Each stage is subdivided into A (no systemic symptoms) or B—presence of weight loss >10% in last 6 months, unexplained fever >38°C or night sweats. These indicate more extensive disease.

Treatment Radiotherapy for stages Ia and IIa. Chemotherapy for IIa → IVb. Chemotherapy is usually given with a 'MOPP'-type régime. (**M**ustine, **O**ncovin® (vincristine), **P**rocarbazine, **P**rednisolone. There are a number of variations on this. Recently, more intensive régimes have been used, particularly for advanced disease.

Complications of treatment: Radiation lung fibrosis and hypothyroidism. Chemotherapy—nausea, alopecia, infertility in men, infection and second malignancies, especially acute myeloid leukaemia and non-Hodgkin's lymphoma. Both may produce myelosuppression.

5-yr survival Overall 80%; depends on stage and grade.

Emergency presentations Infection; marrow failure; SVC obstruction (presents with JVP↑, a sensation of fullness in the head, dyspnoea, blackouts and facial oedema. ▶ Arrange same day radiotherapy).

Hodgkin's lymphoma
[*OTM* 19.170]

Non-Hodgkin's lymphoma

This group includes all lymphomas without the Reed–Sternberg cell and is a very diverse group. Most are B-cell proliferations. Pathological classification is complicated but the essential division is into low and high grade tumours. Low-grade lymphomas run a relatively indolent course but are incurable, whereas the high-grade types are more aggressive but long-term cure is achievable.

The low grade group includes lymphocytic (comparable to CLL but mainly in lymphoid tissue), immunocytic and centrocytic. In the high grade are centroblastic, immunoblastic and lymphoblastic.

Clinical features Rare before age 40yrs. Lymphadenopathy is common but extra-nodal spread occurs early so first presentation may be in skin, bone, gut, CNS or lung. Often symptomless. There may be pancytopenia due to marrow involvement and haemolysis. Infection is common. Systemic symptoms as in Hodgkin's.

Examine all over. Note nodes (if >10cm, staging is 'advanced'[1]). Do ENT exam if GI lymphoma (GI and ENT lymphoma often coexist).

Tests, diagnosis and staging As for Hodgkin's disease, but staging is less important as 70% have widespread disease at presentation. Always do node biopsy, CXR, abdomen and pelvis CT, FBC, U&E, LFT. Consider Ba meal, lymphangiogram, cytology of any pleural or peritoneal effusion, CSF cytology if CNS signs, or grade I or J.

Grading & % 5-yr survival (The numbers after each type denote per cent 5-yr survival. NB: survival is worse if elderly or symptomatic.)

Low	Intermediate	High grade or special*
A Small lymphocytic 59	D Follicular large 45	H Immunoblastic 32
B Follicular small cleaved 70	E Diffuse small cleaved 33	I Lymphoblastic*26
C Follicular mixed cell 50	F Diffuse mixed 38	J Small noncleaved* 23
[B and G are the commonest]	G Diffuse large cleaved 35	= Burkitt's lymphoma

Treatment Symptomless low grade tumours may not need treatment and occasionally show remission, but chlorambucil or cyclophosphamide may control symptoms. Splenectomy may help. Radiotherapy can be used for local bulky disease.

If high grade, but not advanced (mass <10cm, no B symptoms, stage <III, p606) optimum treatment is 6 weeks' doxorubicin (it is Adriamycin®-like), cyclophosphamide, vincristine (Oncovin®), bleomycin, and prednisolone (ACOB).[1] If lymphoblastic, treat as for ALL (600). In advanced disease, there is much more uncertainty.

Burkitt's lymphoma This is a lymphoblastic lymphoma occurring mainly in African children. It is associated with Epstein–Barr virus (EBV) infection and shows a 14q+/8q- chromosomal translocation. Jaw tumours are common, usually with GI involvement. Histology shows a 'starry sky' appearance (isolated histiocytes on background of abnormal lymphoblasts). Spectacular remission may result from a single dose of a cytotoxic drug—eg cyclophosphamide 30mg/kg IV.

Causes of lymphadenopathy See p50. Also angioimmunoblastic lymphadenopathy (skin rashes, fever—seen in the elderly; histology is characteristic; >50% progress to lymphoma).

1 S O'Reilly 1992 *BMJ* i 1682 and *Cancer* 1982 **49** 2112–35

Non-Hodgkin's lymphoma
[*OTM* 19.176]

Bone marrow, and bone marrow failure

The bone marrow is responsible for haemopoiesis and, in adults, is found in the vertebrae, sternum, ribs, skull and proximal long bones, although it may expand in anaemia eg thalassaemia. All blood cells are thought to arise from an early stem cell which produces a lymphoid stem cell and a mixed myeloid progenitor cell. These then undergo further differentiation before releasing formed elements into the blood. Bone marrow failure usually produces a pancytopenia, often with sparing of the lymphocyte count. Bone marrow biopsy will help find the cause.

Causes of pancytopenia Bone marrow failure, hypersplenism, SLE, megaloblastic anaemia, paroxysmal nocturnal haemoglobinuria.

Causes of bone marrow failure 1 Stem cell failure: eg aplastic anaemia. 2 Infiltration: from malignancy. 3 Fibrosis: eg myelofibrosis 4 Abnormal differentiation: eg myelodysplasia.

Aplastic anaemia This presents as pancytopenia with a hypoplastic marrow (ie the marrow stops producing cells). Causes are idiopathic (50%), radiation, drugs (eg cytotoxics, chloramphenicol and gold), viruses, congenital.

Incidence: 10–20 per million per year. Presents as bleeding, anaemia or infection. In approximately 50% of cases there is response to androgens ± immunosuppressive therapy.

Treatment of aplastic anaemia: Marrow transplantation is the best treatment if there is a histocompatible sibling donor. The marrow must be supported. In approximately 50% of cases there is response to androgens and antilymphocytoglobulin.

Bone marrow support The symptoms of bone marrow failure are due to the pancytopenia. Red cells survive for ~120 days, platelets for an average 8 days and neutrophils for 1–2 days so early problems are mainly from neutropenia and thrombocytopenia.

Erythrocytes: A one unit transfusion should raise the Hb by about 1g/dl. See p102. Transfusion may drop the platelet count so it may be necessary to give platelets before and after.

Platelets: Spontaneous bleeding is unlikely if platelets >20 × 10⁹/l but risk of traumatic bleeds is great if <40 × 10⁹/l. Platelets are stored at 22°C, should not be put in the fridge and require irradiation prior to transfusion. Indications for transfusion are counts <10 × 10⁹/l, excessive bleeds and DIC. Crossmatching is not needed but there is the possibility of rhesus sensitization and many patients become refractory after repeated transfusions. 4 units of fresh platelets, doubled if the platelets are >3 days old, should raise the count to >40 × 10⁹/l in an adult.

Neutrophils: Use a 'neutropenic régime' if the count <0.5 × 10⁹/l. See p598. The place of neutrophil transfusions is unclear. They may be indicated in proven, continued bacteraemia but are short-lived, expensive and may transmit infection or cause pneumonitis.

Bone marrow biopsy Ideally an aspirate and trephine should both be taken. Trephines may be taken from the posterior iliac crest, aspirates may be taken from the anterior iliac crest or sternum. Aspirates should be smeared promptly onto slides. Do at least 8. Thrombocytopenia is rarely a contraindication. Severe coagulation disorders may need to be corrected. Apply pressure afterwards (lie on that side for 1–2h if platelets are low).

Bone marrow, and bone marrow failure

The myeloproliferative disorders

These form a group of disorders characterized by proliferation of precursors for myeloid elements—RBC, WBC and platelets. While the cells proliferate they also retain the ability to differentiate. The 4 disorders share several features, including the fact that all may present with constitutional symptoms such as fever, weight loss, night sweats, itch and malaise.

Classification is by the cell type which is proliferating:

RBC	→ Polycythaemia rubra vera (PRV).
WBC	→ Chronic myeloid leukaemia (CML, p604).
Platelets	→ Essential thrombocythaemia.
Fibroblasts	→ Primary myelosclerosis (reactive and not part of the malignant clone).

Each may undergo transformation to acute leukaemia.

The blood count reflects 2 processes—the *proliferating cell line*, which may also involve other myeloid elements (eg in PRV, RBC↑, but there may be rises in WBC and platelets as well)—and *marrow infiltration* which may cause a decrease in normal cells.

Polycythaemia may be relative (plasma volume↓) or absolute. The way to distinguish these is by red cell mass estimation using radioactive chromium. Relative polycythaemia is often due to dehydration (eg alcohol or diuretics). Absolute polycythaemia may be primary (PRV) or secondary, eg from smoking, chronic lung disease, tumours (fibroids, hepatoma, hypernephroma), or altitude.

Polycythaemia rubra vera *Incidence:* 1.5/100,000/yr, peaks at 45–60yrs. *Signs:* ● PCV↑ ● ↑WBC, ↑platelets (or ↔) ● MCV↓ ● Spleen↑ (60%).
Presentation is determined by hyperviscosity (CNS signs, p620; angina; Raynaud's—p708; itch—typically after a hot bath) and ↑RBC turnover (gout). It may also present with bruising, or after a routine FBC.
Diagnosis: Red cell mass >125% of predicted (^{51}Cr studies). Marrow: cellularity↑; erythroid hyperplasia; megakaryocyte dysplasia. Leucocyte alk phos (LAP) usually↑ (LAP↓ in CML). B_{12}↑.
Treatment: Refer. Aim to keep PCV <50% by venesection. Iron-deficiency may result from repeated venesections. Drugs: hydroxyurea 10–20mg/kg/day PO, IV ^{32}P (eg if elderly—it is leukaemogenic) or busulphan 4–6mg/day PO.
Prognosis: Variable—many often remain well for many years but others will die from myelofibrosis, leukaemia, or thrombotic disorders from hyperviscosity and/or malfunctioning platelets. Do FBC every 3 months.

Essential thrombocythaemia is characterized by very high platelet counts, eg >800 × 10⁹/l. Platelet morphology and function is abnormal and presentation may be by bleeding or thrombosis.
Treatment of essential thrombocythaemia: Busulphan or hydroxyurea may be needed for symptoms or if the platelet count is >800 × 10⁹/l, otherwise the prognosis is good. Recently α-interferon has been used successfully.
Causes of a raised platelet count: Myeloproliferative or inflammatory disorders, bleeding, malignancy, post-splenectomy, Kawasaki disease.

The myeloproliferative disorders

Primary myelosclerosis There is intense marrow fibrosis with resultant haemopoiesis in the spleen and liver (myeloid metaplasia) causing massive splenomegaly. *Clinical features*: Variable—constitutional, splenomegaly, bone marrow failure.

Film: Leucoerythroblastic (p576); 'tear-drop' RBCs. Hb↓ (50%). Marrow tap: dry.

Treatment: supportive (p610), iron and folate supplements.

Other causes of marrow fibrosis: Any myeloproliferative disorder, lymphoma, secondary carcinoma, TB, leukaemia, irradiation.

Myeloma

Myeloma is a plasma cell neoplasm which produces diffuse bone marrow infiltration and focal osteolytic deposits. An M band (M for monoclonal, not IgM) is seen on serum and/or urine electrophoresis.

Incidence 5/100,000. Peak age: 70yrs. Sex ratio: equal.

Classification Based on the principal neoplastic cell product:

IgG   55%
IgA   25%
Light chain disease   20%.

60% of IgG and IgA myelomas also produce free Ig light chains which are filtered by the kidney and may be detectable as Bence Jones protein (these precipitate on heating and redissolve on boiling). They may cause renal damage or, rarely, amyloidosis.

Symptoms Bone pain is common, often postural, eg back, ribs, long bones and shoulder—not extremities. $Ca^{2+}\uparrow$ in 30% (p644). Pathological fractures may occur. Renal failure—due to light chain precipitation, hyperuricaemia and $Ca^{2+}\uparrow$. Anaemia. Infection. Neuropathy, blurred vision, bleeding (from hyperviscosity, p620). Amyloid.

Examination Bone tenderness, urine, fundoscopy (for hyperviscosity-induced haemorrhages and exudates), macroglossia (amyloid).

Investigations FBC, ESR$\uparrow$, bone marrow, serum/urine electrophoresis, Ig-myeloma class$\uparrow$ but others$\downarrow$; $Ca^{2+}\uparrow$, alk phos usually normal (unless healing fracture). Urea, creatinine and uric acid$\uparrow$ in 50%, CXR; skeletal x-rays show punched-out lesions (pepper-pot skull) and background osteoporosis. Bone scintigrams may be normal.

Diagnosis Abundant plasma cells in marrow. M band or urinary light chains (Bence Jones proteins, p650). Osteolytic bone lesions. NB: do not diagnose on basis of paraprotein alone.

Treatment *Supportive:* Bone pain, anaemia and renal failure are the main problems so give analgesia and transfusions as needed. Solitary lesions may be given *radiotherapy*, and may heal.

Chemotherapy: Intermittent melphalan is standard but chemotherapy is becoming more intensive. Monitor FBC and paraprotein. If no response or relapse, consider multiple chemotherapy. Common pattern: response followed by symptom-free plateau phase and then relapse. Death is commonly due to renal failure, infection or haemorrhage.

Prognosis 50% alive at 2yrs. Survival is worse if urea >10mmol/l or Hb <7.5g/dl.

Dangers

1 Hypercalcaemia: p644. Use IV saline 0.9% 4–6 litres/day with careful fluid balance. Consider steroids eg hydrocortisone 100mg/8h IV. Diphosphonates may be needed in refractory disease.

2 Hyperviscosity—causing mental impairment, disturbed vision, bleeding. May need plasmapheresis.

3 Acute renal failure may be precipitated by IVU. See p382.

Myeloma
[*OTM* 19.199]

Paraproteinaemia

Paraproteinaemia refers to the presence in the circulation of immuno-globulin produced by a single clone of plasma cells or their precursors. The paraprotein is recognized as a sharp M band (M for monoclonal, not IgM) on serum electrophoresis. There are 6 major categories:

1 *Multiple myeloma:* See p614.
2 *Waldenström's macroglobulinaemia:* This is a lymphoplasmacytoid malignancy producing an IgM paraprotein, lymphadenopathy and splenomegaly. CNS & ocular symptoms of hyperviscosity may occur (p620). Chlorambucil and plasmapheresis may help.
3 *Primary amyloidosis:* See below.
4 *Monoclonal gammopathy* (benign paraproteinaemia) is common (3% >70yrs) and may be misdiagnosed as myeloma (but paraprotein level is stable, immunosuppression and urine light chains are absent and there is little plasma cell marrow infiltrate).
5 *Paraproteinaemia in lymphoma or leukaemia:* Eg 5% of CLL.
6 *Heavy chain disease:* Production of free heavy chains. α chain disease is the most important, causing malabsorption from infiltration of small bowel wall. It may terminate in lymphoma.

Amyloidosis

This is a disorder characterized by extracellular deposits of an abnormal, degradation-resistant protein called amyloid. Various proteins, under a range of stimuli, may polymerize to form amyloid fibrils—which are detected by +ve staining with Congo Red and by showing apple-green birefringence in polarized light.

Classification

1 *Systemic:* ● Immunocyte dyscrasia (fibrils of immunoglobulin light chain fragments, known as 'AL' amyloid).
● Reactive amyloid ('AA' amyloid—a non-glycosylated protein).
● Hereditary amyloid (eg type 1 familial amyloid polyneuropathy).

2 *Localized:* ● Cutaneous. ● Cerebral. ● Cardiac. ● Endocrine.

AL amyloid (primary amyloidosis): This is associated with monoclonal proliferation of plasma cells—eg in myeloma. Clinical features include carpal tunnel syndrome, peripheral neuropathy, purpura, cardiomyo-pathy and macroglossia (a large tongue).

AA amyloid (secondary amyloidosis): This can occur in association with chronic infections (eg TB, bronchiectasis), inflammation (especially rheumatoid arthritis) and neoplasia. It tends to affect kidneys, liver and spleen and commonly presents as proteinuria, nephrotic syndrome and hepato-splenomegaly.

The diagnosis of amyloidosis is made after Congo Red staining of affected tissue. The rectum is a favourite site for biopsy.

Treatment rarely helps, but AA amyloid may improve with treatment of the underlying disease.

Paraproteinaemia
Amyloidosis

The erythrocyte sedimentation rate (ESR)

The ESR is a non-specific indicator of the presence of disease. It measures the rate of sedimentation of RBCs in anticoagulated blood over 1 hour. If certain proteins cover red cells these will stick to each other in columns (the same phenomenon as rouleaux on the blood film, p576)—and so they will fall faster. The ESR rises with age and anaemia. A simple, reliable way to allow for this is to calculate the upper limit of normal, using the Reference Westergren method, to be (for men) age in years÷2. For women the formula is (years+10)÷2. In those with a slightly abnormal ESR the best plan is probably to wait a month and repeat the test. The same advice does not hold true for patients with a markedly raised ESR (>100mm/h). In practice most will have signs pointing to the cause—usually malignancy (almost always disseminated), connective tissue diseases (eg giant cell arteritis), rheumatoid arthritis, renal disease, sarcoidosis or infection. However, there is a group of patients whose vague symptoms would have prompted nothing more than reassurance—were it not for a markedly raised ESR—and in whom there are no pointers to specific disease. The underlying disease (in one survey) turned out to include myeloma, giant cell arteritis, abdominal aneurysm, metastatic prostatic carcinoma, leukaemia and lymphoma. Therefore it would be wise (after history and examination) to consider these tests: FBC, plasma electrophoresis, U&E and creatinine, chest and abdominal x-rays and biopsy of bone marrow or temporal artery.

Some conditions *lower* the ESR, eg heart failure, polycythaemia, sickle-cell anaemia, and cryoglobulinaemia. Even a slightly raised ESR in these patients should prompt one to ask: 'What else is the matter?'.

If ready-prepared vacuum ESR tubes (eg Seditainer) are used, then lower values are recorded compared with the reference method, as follows:

Seditainer:	Reference:	Seditainer:	Reference:	Seditainer:	Reference:
10	11	30	40	50	75
15	18	35	47	55	87
20	25	40	56	60	100
25	32	45	65	65	118

For CRP (an alternative to the ESR), see p.650.

Hyperviscosity syndromes

These occur if the plasma viscosity rises to such a point that the microcirculation is impaired.

Causes: Myeloma (IgM increases viscosity more than the same amount of IgG), Waldenström's macroglobulinaemia, polycythaemia. High leucocyte counts in leukaemia may also produce the syndrome (leucostasis).

Presentation: Visual disturbance, retinal haemorrhages, headaches, coma, and GU or GI bleeding.

The visual symptoms ('slow-flow retinopathy') may be described as 'looking through a watery car windscreen'. Other causes of slow-flow retinopathy are carotid occlusive disease and Takayasu's disease (p710).

Treatment: Removal of as little as 1 litre of blood may relieve symptoms. Plasmapheresis may help.

The spleen and splenectomy

This was a mysterious organ for many years, but it is now known to play an important immunological role. Splenomegaly is not an uncommon problem and its causes can be divided into massive (into the RIF) and moderate.

Massive splenomegaly CML, myelofibrosis, malaria (hyperreactive malarial splenomegaly), schistosomiasis, leishmaniasis, 'tropical spleno-megaly' (idiopathic—Africa, SE Asia).

Moderate splenomegaly Infection (eg malaria, EBV, SBE, TB), portal hypertension, haematological (haemolytic anaemia, leukaemia, lymphoma), connective tissue disease (RA, SLE), storage diseases, idiopathic.

Splenomegaly can cause abdominal discomfort and may lead to *hyper-splenism* (ie pancytopenia as cells become trapped in the spleen's reticulo-endothelial system, with symptoms of anaemia, infection or bleeding). When faced with a mass in the left upper quadrant it is important to be able to recognize the spleen. Clinically the spleen moves with respiration, enlarges towards RIF, may have a notch and 'you can't get above it' (ie the top margin disappears under the ribs). Abdominal x-ray may help. When hunting the cause for enlargement look for lymphadenopathy and liver disease. Appropriate tests are FBC, ESR, LFTs and liver, marrow or lymph node biopsy.

619

The trend is towards doing fewer splenectomies than formerly but suitable indications remain trauma, haemolytic anaemias, ITP and occasionally for diagnosis. Post-operatively there may be a prompt, transient rise in the platelet count so patients should be mobilized early. All patients, especially children, remain at increased risk of infection, particularly severe pneumococcal septicaemia. Most children are advised to take prophylactic daily penicillin V until aged 20yrs (in adults, 5yrs is long enough unless there is co-existing haematological disease; broader spectrum agents such as amoxycillin may be more appropriate[1]). Give adults and unvaccinated children pre-op *Pneumovax II*® (p332) and Haemophilus vaccination (HiB, *OHCS* p208).[1]

1 PJ Flegg 1994 *BMJ* i 131

Thrombophilia[1,2]

Thrombophilia (inherited or acquired) is a primary coagulopathy resulting in a propensity to thrombosis. Note: thrombocytosis (platelets ↑) and polycythaemia also cause thrombosis. It is *not* rare, and it *is* treatable—and needs special precautions in *surgery*, *pregnancy*, and *enforced inactivity*. Be alert to it in non-haemorrhagic stroke eg (if <60yrs), thrombosis at <45yrs (or family history). Risk is increased by obesity, immobility, trauma (accidents or surgery), pregnancy, and malignancy.

Inherited —*Activated protein C (APC) resistance:* The commonest cause of inherited thrombophilia and, in many populations the commonest cause of thrombo-embolism. The molecular defect is a single point mutation in factor V (V Leiden). Thrombotic risk is increased in pregnancy and those on oestrogen-containing oral contraceptives (risk ↑ by 30–35-fold in FV:Q heterozygous carriage, and by several hundred-fold in homozygous carriage—and screening *might* be appropriate with the newly-available modified test for APC resistance[3]). There is ↑ risk of MI. *Antithrombin III deficiency:* This affects 1:2000. Heterozygotes' thrombotic risk is 4-fold greater than with protein C or S deficiency. Homozygosity is lethal. *Protein C and protein S deficiency:* These vitamin K dependent factors act together to neutralize factors V and VIII. Heterozygotes deficient for either protein risk thrombosis, and skin necrosis (especially if using oral anticoagulants). Homozygous deficiency for either protein causes neonatal purpura fulminans—fatal if untreated.

Acquired The commonest cause is *antiphospholipid syndrome* when patients have lupus articoagulant and-or anticardiolipin antibody in their blood predisposing to venous and arterial thrombosis, thrombocytopenia, and recurrent fetal loss in pregnant women. Most do not have SLE.

Tests *Consider special tests if recurrent or unusual thrombosis or:*

Venous thromboembolism <40yrs	FH venous thromboembolism
Arterial thrombosis <30yrs	Recurrent fetal loss
Skin necrosis (especially if warfarin)	Neonatal thrombosis

Liaise with a haematologist. Do FBC with platelets; prothrombin time; thrombin time; activated partial thromboplastin time; reptilase time and fibrinogen concentration. If thrombophilia is still suspected, investigations include: activated protein C resistance (ratio), lupus anticoagulant and anticardiolipin antibodies and functional assay for antithrombin III and proteins C and S (the phenotype is associated with heterozygosity or homozygosity for the factor V Leiden mutation in ≥95% of individuals).[3] Haematologists may recommend looking directly for the V Leiden mutation, eg if already on warfarin, and other results are confusing.[3]

Ideally investigate whilst well, not pregnant, and not anticoagulated.

Treatment Treat acute thromboses with heparin then warfarin. Those with antithrombin III deficiency may need unusually high doses of heparin (dose may be difficult to determine in those with lupus anticoagulant). Ensure full heparinization, then gradually introduce warfarin (no loading dose) in those with protein C and S deficiency. Treat 1st thromboses for ~3 months; it is not known if long-term warfarin prophylaxis is worthwhile; but it is wise to advise elastic stockings.[2,4] Recurrent thromboses may require long term anticoagulation—keep INR at 2–2.5 if warfarin used. Seek expert advice with relation to pregnancy. Avoid synthetic oestrogens (the Pill); HRT is probably all right.

Before medium-to-major (especially orthopaedic) surgery, liaise with a haematologist. Antithrombin III cryoprecipitate may be indicated, as well as more thorough prophylaxis against post-op thromboses.

1 *Drug Ther Bul* 1995 **33** 6 & 35   **2** Brit. Soc. Haematol Guidelies 1992 *J Clin Path* **46** 97–103   **3** AT Hattersley 1996 *Lancet* **348** 343   **4** B Dahlback 1996 *Lancet* **347** 134e

Hyperviscosity syndromes

14 | Biochemistry

Relevant pages in other chapters:
Biochemistry reference intervals (p802 & p803); IV fluids on the surgical wards (p100); acute renal failure (p384 & p386).

The laboratory and ward tests

▶ Laboratory staff like to have contact with you.

A laboratory decalogue
1. Interest someone from the laboratory in your patient's problem.
2. Fill in the request form fully.
3. Give clinical details, not your preferred diagnosis.
4. Ensure that the lab knows whom to contact.
5. Label specimens as well as the request form.
6. Follow the hospital labelling routine for crossmatching.
7. Find out when analysers run, especially batched assays.
8. Talk with the lab before requesting an unusual test.
9. Be thoughtful: at 4.30pm the routine results are being sorted.
10. Plot results graphically: abnormalities show sooner.

Artefacts and pitfalls in laboratory tests
- Do not take blood sample from arm with IV fluid.
- Repeat any unexpected result before acting on it.
- For clotting time do not use sample from heparinized IV catheter.
- Serum K+ is overestimated if sample old or haemolysed (this occurs if venepuncture is difficult).
- If using Vacutainers, fill *plain* tubes first—otherwise anticoagulant contamination from previous tubes can cause errors.[1]
- Calcium analysis is affected by albumin (p642) and tourniquets.
- INR may be underestimated if citrate bottle underfilled.
- Drugs may cause *analytic* errors (eg prednisolone cross-reacts with cortisol). Be suspicious if results unexpected.
- Food may affect result (eg bananas raise urinary HMAA—p554).

The use of dipsticks
Store dipsticks with container lid on in cool dry place, not refrigerated. If improperly stored, or past expiry date, do not use. For urine dipstick dip briefly in urine, run edge of strip along container and hold strip horizontal. Read at specified time—check instructions for the stick you are using.

Urine specific gravity
(SG) can be measured by dipstick. It is not a good measure of osmolality. Causes of low SG (<1.003) are: diabetes insipidus, renal failure. Causes of high SG (>1.025) are: DM, adrenal insufficiency, liver disease, renal failure, acute water loss. Hydrometers underestimate SG by 0.001 per 3°C above 16°C.

Sources of error in interpreting dipsticks results
Bilirubin: False +ve: phenothiazines. False −ve: urine not fresh, rifampicin.
Urobilinogen: False −ve: urine not fresh.
Ketones: L-dopa affects colour (can give false +ve).
Blood: False +ve: myoglobin, profuse bacterial growth. False −ve: ascorbic acid.
Urine glucose: Depends on test. Pads with glucose oxidase are not affected by other reducing sugars (unlike Clinitest®) but can give false +ve to peroxide, chlorine; and false −ve with ascorbic acid, salicylate, L-dopa.
Protein: Highly alkaline urine can give false +ve.
Blood glucose: Sticks use enzymatic method and are glucose specific. Major source of error is applying too little blood (large drop to cover pad is necessary), and poor timing. Reflectance meters increase precision but introduce new sources of error.

1 WA Bartlett 1993 *BMJ* ii 868

The laboratory and ward tests

The biochemistry report form

▶ Do not interpret biochemical results except in the light of clinical assessment (unless forced by examiners).

▶ If there is disparity trust clinical judgement and repeat biochemistry.

The reference intervals (normal range) are usually defined as the interval, symmetrical about the mean, containing 95% of results on the population studied. The more tests you run, the greater the probability of an 'abnormal' result of no clinical significance.

Artefacts (p642) Correct calcium for albumin (p642).

Anion gap reflects unmeasured anions (p630).

Major patterns of specific disease (↑ = raised, ↓ = lowered).
Dehydration: Urea ↑, albumin ↑ (useful to plot change in a patient's condition). Haematocrit (PCV) ↑, creatinine ↑.
Renal failure: Creatinine ↑, urea ↑, anion gap ↑, potassium ↑ (p632), bicarbonate ↓.
Thiazide & loop diuretics: Sodium ↓, bicarbonate ↑, potassium ↓, urea ↑.

Bone disease:	Ca^{2+}	PO_4^{3-}	Alk phos
Osteoporosis	↔	↔	↔
Osteomalacia	↓	↓	↑
Paget's	↔	↔	↑↑
Myeloma	↑	↑,↔	↔
Bone metastases	↑	↑,↔	↑
Primary hyperparathyroidism	↑	↓,↔	↔,↑
Hypoparathyroidism	↓	↑	↔
Renal failure (low GFR)	↓	↑	↔,↑

Hepatocellular disease: Bilirubin ↑, AST ↑, (alk phos, mildly ↑, albumin↓).
Cholestasis: Bilirubin ↑, GGT ↑↑, alk phos ↑↑, usually extrahepatic stasis if >350 IU/l (AST ↑).
Myocardial infarct: AST ↑ (also LDH ↑, CK ↑, p280).
Diabetes mellitus: Glucose ↑, bicarbonate ↓.
Addison's disease: Potassium ↑, sodium ↓.
Cushing's syndrome: May show potassium ↓, bicarbonate ↑, Na⁺ ↑.
Conn's syndrome: May present with potassium ↓, bicarbonate ↑ (and high blood pressure). Sodium ↑ or normal.
Diabetes insipidus: Sodium ↑ (both calcium ↑ and potassium ↓ may cause nephrogenic diabetes insipidus).
Inappropriate ADH secretion: Sodium ↓ with normal or low urea and creatinine.
Excess alcohol intake: Evidence of hepatocellular disease. Early evidence in GGT ↑, MCV↑, ethanol in blood before lunch.

Urgent treatment may be needed if: Potassium >6.5, or <2.5 mmol/l. Sodium >155, or <120 mmol/l. Calcium (corrected) >3.5, or <2.0 mmol/l.

The biochemistry report form

Intravenous fluid therapy

(See also p100 and 636.)

If fluids cannot be given orally they are normally given IV in a peripheral vein. An alternative is into a central venous line.

Three principles of fluid therapy
1 Give amount needed The amount and type of fluid lost is a guide. In practice the problem is usually whether to give saline or dextrose. Most body fluids (eg vomit) contain salt, but less than plasma.

Features indicating that water (5% dextrose is needed):
● Thirst.
● Low urine output with high (>4) ratio of urine:plasma osmolarity.
● Dry axilla and groin (ie decreased sweating).
● Raised plasma sodium.

Features indicating that saline is needed:
● Fluid lost (excess diuretic, vomit, burns, diarrhoea, sweat).
● Decrease tissue turgor (eg tongue furrowed).
● Postural hypotension.
● Tachycardia.
● Low JVP.
● Raised haematocrit.
● Low urine output, ratio of urine:plasma osmolarity around unity.

Features indicating need for another fluid:
● If acute blood loss give blood (p102).
● If patient shocked (p720) a plasma expander or colloid (eg plasma protein fraction, Dextran® or Haemaccel®) is normally given. NB: Dextran® interferes with platelet function and may prolong bleeding.

2 Allow for future abnormal loss Moderate, continuous sweating and raised temperature require extra fluid (roughly 10% extra per degree Celsius). Give 0.9% saline and 5% dextrose in equal parts.

3 Maintain normal daily requirements About 2500ml fluid containing roughly 100mmol sodium and 70mmol potassium per 24h are required. A good régime is 2 litres 5% dextrose and 1 litre normal saline every 30h with 20–30mmol potassium per litre of fluid. Post-operative patients may need more fluid and more saline depending on operative losses.

Patients with heart failure are at greater risk of pulmonary oedema if given too much fluid. They also tolerate less saline as Na^+ retention accompanies heart failure. If fluids IV must be given go cautiously.

A note on fluids 0.9% saline (normal saline): has about same sodium content as plasma (150mmol/l) and is isotonic with plasma. *5% dextrose* (dextrose is glucose) is a way of giving water. It contains no sodium and 278mmol/l glucose (50g/l). It provides little energy (220kcal/l).
Dextrose-saline contains 30mmol/l of sodium and 4% glucose (222mmol/l). It has roughly the concentration of saline required for normal fluid maintenance.

▶ Examine patient regularly for signs of heart failure (p296) which can result if excess fluid given. Excessive dextrose infusion may lead to water overload (p638).

▶ Daily weighing helps to monitor overall fluid balance.

Intravenous fluid therapy

Acid–base balance

First decide what acid–base status patient is likely to have based on clinical picture. The arterial blood gas measure will give you pH and $PaCO_2$. Find the position defined by these on the diagram below. The patient's acid–base status is then determined. The $PaCO_2$ reflects ventilation.

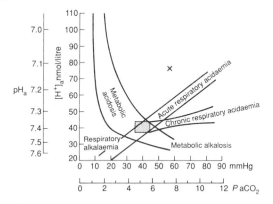

The shaded area represents normality. This method is very powerful. The result represented by point X, for example, indicates that the acidaemia is in part respiratory and in part metabolic. Seek a cause for each.

Metabolic acidaemia causes a decrease in total bicarbonate concentration. To help diagnosis work out *anion gap* (AG):

$$AG = [K^+] + [Na^+] - [Cl^-] - [HCO_3^-] \text{ (plasma concentrations)}$$

The normal range is 8–16mmol/l. It is a measure of the difference between unestimated anions (eg phosphate, ketones, lactate) and cations. If >16mmol/l consider renal failure or hyperlactataemia (eg from shock).

Causes of metabolic acidaemia and increased anion gap: High plasma lactic acid (exercise, shock, hypoxia, liver failure, biguanides, trauma). Uraemia. Acid ingestion (eg aspirin, ethylene glycol). Acid from metabolism, eg ketoacidosis (DM, alcohol, starvation).

Causes of metabolic acidaemia and normal anion gap: failure to excrete acid (renal tubular disease, Addison's). Loss of base (diarrhoea, proximal tubular acidosis, carbonic anhydrase inhibitor, ureterosigmoidostomy).

Metabolic alkalaemia causes an increase in total bicarbonate.
Causes: Vomiting. K$^+$ depletion. Ingestion of base. Burns.

Respiratory acidaemia ($PaCO_2\uparrow$, pH$\downarrow$). *Causes:* any respiratory, neuromuscular or physical cause of respiratory failure (p352).

Respiratory alkalaemia ($PaCO_2\downarrow$, pH$\uparrow$) is a result of hyperventilation.
CNS causes: Stroke, subarachnoid haemorrhage, or meningitis.
Other causes: Fever, hyperthyroidism, pregnancy, anxiety.

Acid—base balance

The biochemistry of kidney function[1]

The kidney controls the elimination of many substances. It also makes erythropoietin, renin and 1,25-dihydroxycholecalciferol. Sodium from urine is exchanged with potassium and hydrogen ions by a pump in the distal tubule. Glucose spills over into urine when plasma concentration is above renal threshold ($\approx$10mmol/l, but varies from person to person, and is lower in pregnancy).

Creatinine clearance is a measure of glomerular filtration rate (GFR) which is the volume of fluid filtered by the glomeruli per minute. About 99% of this fluid is reabsorbed. Creatinine once filtered is only slightly reabsorbed. Thus:

$$[\text{Creatinine}]_{\text{plasma}} \times \text{creatinine clearance} = [\text{creatinine}]_{\text{urine}} \times \text{urine flow rate}.$$

To measure creatinine clearance (normal value is >100ml/min). Collect urine over 24h. Empty bladder just before start. Take sample for plasma creatinine once during 24h. Use formula above. Take care with units. Major sources of error are calculation (eg units) and failure to collect all urine. If urine collection unreliable use formula:[2]

$$\text{creatinine clearance (ml/min)} = \frac{(140 - \text{age in years}) \times (\text{wt in kg})}{72 \times \text{serum creatinine in mg/dl}}$$

For women multiply above by 0.85.
Unreliable if: unstable renal function; very obese; oedematous.[3]

A better estimate of GFR is by ^{51}Cr-EDTA clearance.

Abnormal kidney function
There are three major biochemical pictures.
● *Low GFR (classic acute renal failure)*
Plasma biochemistry: The following are raised: urea, creatinine, potassium, hydrogen ions, urate, phosphate, anion gap.
The following are lowered: calcium, bicarbonate.
Other findings: Oliguria.
Diagnosis: Low GFR (creatinine clearance).
Causes: Early acute oliguric renal failure (p384), long-standing chronic renal failure (p388).
● *Tubular dysfunction* (damage to tubules)
Plasma biochemistry: The following are lowered: potassium, phosphate, urate, bicarbonate. There is acidosis. Urea and creatinine are normal.
Other findings: (highly variable): Polyuria with glucose, amino acids, proteins (lysozyme, β_2-microglobulin), and phosphate in urine.
Diagnosis: Test renal concentrating ability (p566).
Cause: Recovery from acute renal failure. Also: hypercalcaemia, hyperuricaemia, myeloma, pyelonephritis, hypokalaemia, Wilson's disease, galactosaemia, metal poisoning.
● *Chronic renal failure:* As GFR reduced; creatinine, urea, phosphate and urate all increase. Bicarbonate (and Hb) decrease. Eventually potassium increases and pH decreases. There may also be osteomalacia.

NB: Assessment of renal failure may need to be combined with other investigations to reach diagnosis, eg urine microscopy (p370); radiology (p372), or renal biopsy (in glomerulonephritis), or ultrasound.

The biochemistry of kidney function

Creatinine clearance: worked example

Suppose:
urine creatinine concentration = u mmol/l;
plasma creatinine concentration = p μmol/l;
24h urine volume = v ml.
NB: There are 1440mins per 24h (used below to convert urine rate from volume per 24h into volume per minute). $p/1000$ is used to convert micromoles to millimoles.

Creatinine clearance = $u \times v/1440 \div p/1000$ ml/min.
$$u \times v/p \times 0.7$$
Thus, if: $u = 5$ mmol/l; $p = 120$ μmol/l; $v = 2500$ ml;
creatinine clearance = $5 \times (2500/120) \times 0.7$
$$= 73\text{ml/min}$$

1 R Gabriel *Postgraduate Nephrology* 3 ed, Butterworths **2** D Cockcroft 1976 *Nephron* **16** 31 **3** M Reidenberg 1985 *NEJM* **313** 816

Urate and the kidney

Causes of hyperuricaemia High levels of urate in the blood (hyper-uricaemia) may result from increased turnover or reduced excretion of urate. Either may be drug-induced.

Examples of *drugs* causing hyperuricaemia: cytotoxics; diuretics; ethambutol.

Examples of *increased turnover:* lymphoma; leukaemia; psoriasis; haemolysis; muscle necrosis (p394).

Examples of *reduced excretion:* primary gout (p666); chronic renal failure; lead nephropathy; hyperparathyroidism.

In addition, hyperuricaemia may be associated with hypertension and hyperlipidaemia. It may be raised in disorders of purine synthesis such as the Lesch–Nyhan syndrome (*OHCS* p752).

Hyperuricaemia and renal failure Severe renal failure from any cause may be associated with hyperuricaemia, and very rarely this may give rise to gout. Sometimes the relationship of cause and effect is reversed so that it is the hyperuricaemia which causes the renal failure. This can occur following cytotoxic treatment (eg in leukaemia) and in cases of muscle necrosis.

How urate causes renal failure In some instances ureteric obstruction from urate crystals occurs. This responds to retrograde ureteric catheterization and lavage. More commonly, urate precipitates in the renal tubules. This may occur at plasma levels ≥ 1.19 mmol/l.

Prevention of renal failure Before starting chemotherapy, ensure good hydration. Alkalinize the urine. Allopurinol (a xanthine oxidase inhibitor) prevents sharp rises in urate following chemotherapy. The dose is: 200–800mg/24h. NB: it increases the toxicity of azathioprine. There is a remote risk of inducing xanthine nephropathy.

634 Treatment of hyperuricaemic acute renal failure Prompt rehydration and alkalinization of the urine after excluding bilateral ureteric obstruction. Once oliguria is established, haemodialysis is required and should be used in preference to peritoneal dialysis.

Gout See p666.

Urate and the kidney

Electrolyte physiology

Most sodium is extracellular: it is pumped out of the cell.
Molarity is the number of moles per litre of solution.
Molality is the number of moles per kg of solvent.
Λ *mole* is the molecular weight expressed in grams.

To estimate plasma osmolality: $2[Na^+] + 10$. If the measured osmolality is greater than this (eg an osmolar gap of >10mmol/l), consider: DM, renal failure, high blood ethanol, methanol or ethylene glycol.

Fluid compartments For 70kg man: *total fluid*=42 litres (60% body weight). *Intracellular fluid*=28 litres (67% body fluid), extracellular fluid=14 litres (33% body fluid). *Intravascular component*=3 litres plasma (5 litres of blood).

Distribution between intra- and extravascular compartments is determined osmotically by proteins. Electrolytes determine the distribution between intra- and extracellular compartments.

Fluid balance over 24h is roughly:

Input (ml water)	*Output (ml water)*
drink: 1500	urine: 1500
in food: 800	insensible loss: 800
metabolism of food: 200	stool: 200
total: 2500	total: 2500

Control of sodium *Renin* is produced in the kidney in response to decreased renal blood flow (due usually to decrease intravascular volume) and converts angiotensinogen to angiotensin I which is converted in the lungs by angiotensin converting enzyme to angiotensin II. This both stimulates the adrenal cortex to produce aldosterone, and causes peripheral vasoconstriction. Aldosterone activates the pump in the distal renal tubule leading to reabsorption of sodium and water from the urine, in exchange for potassium and hydrogen ions.
High GFR (p632) results in high sodium loss.
High renal tubular blood flow and haemodilution decrease sodium reabsorption in the proximal tubule.

Control of water Controlled mainly by sodium concentration. An increased plasma osmolality causes thirst, and the release of antidiuretic hormone (ADH) from the posterior pituitary which increases the passive water reabsorption from renal collecting ducts.

Clinical features
Low body sodium (leads to cellular overhydration): confusion, fits.

Water excess: Hypertension, cardiac failure, oedema, anorexia, nausea, muscle weakness, haemodilution (PCV <40% in men, <35% in women).

Increased body sodium (cellular dehydration): Thirst, confusion, coma.

Water deficiency: Hypotension, low pulse volume, decreased skin turgor, dry skin, peripheral vasoconstriction, tachycardia, raised PCV (>55% in men, >50% in women), raised plasma protein, uraemia.

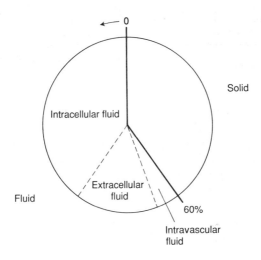

Sodium

Hyponatraemia[1]

▶ Do not base treatment on plasma sodium concentration alone.

Clinical features of low sodium, and water excess (p636).

Diagnosis See tree opposite (p639). The key question is: is the patient dehydrated? History and urine analysis are your guides.

Causes of hyponatraemia The most common are thiazide diuretic excess and water excess. For other causes see tree opposite.

Management Treat specific cause. Assess renal function: if poor, dialysis may be needed.

If not dehydrated and renal function good, then if Na >125 mmol/l, treatment rarely needed. If Na <125mmol/l, restrict water to 0.5–1 litre/day if tolerated. Consider frusemide 40–80mg/24h IV slowly/PO for few days only. SIADH (below) is occasionally treated by producing nephrogenic diabetes insipidus (eg with demeclocycline).

If dehydrated and kidney function good, 0.9% saline can be given. In *emergency* (seizures, coma) consider rapid IVI of 0.9% saline or hypertonic 5% saline at 70mmol Na$^+$/h. Aim for a gradual increase in plasma sodium to about 125mmol/l. Watch out for heart failure (and central pontine myelinosis). Seek expert help.

Syndrome of inappropriate ADH secretion (SIADH)
This is an important cause of hyponatraemia. The diagnosis is made by finding a concentrated urine (sodium >20mmol/l) in presence of hyponatraemia (<125mmol/l) or low plasma osmolality (<260mmol/kg), and absence of hypovolaemia.

Causes: (1) *Malignancy* (lung—small cell; pancreas; duodenum; thymus; prostate; adrenal; ureter; nasopharynx; lymphoma; leukaemia). (2) CNS *disorders* (meningoencephalitis; abscess; stroke; subarachnoid, subdural haemorrhage; head injury; vasculitis—eg SLE; Guillain–Barré). (3) *Chest disease* (TB; pneumonia; abscess; aspergillosis). (4) *Metabolic* disease (porphyria, trauma), (5) *Drugs* (opiates, chlorpropamide, psychotropics, cytotoxics).

Hypernatraemia

Clinical features of increased body sodium with or without water deficiency (p636).

Causes Due usually to water loss in excess of sodium loss.
- Fluid loss without water replacement (eg unconscious patient with fluid lost from diarrhoea, vomit, respiration, sweat, burns).
- Incorrect IV fluid replacement.
- Diabetes insipidus (p566). Suspect if large urine volume. This may follow head injury.
- Osmotic diuresis. (NB diabetic coma; p738).
- Primary aldosteronism. Suspect if hypertension, hypokalaemia, alkalosis (raised bicarbonate).

Management: Give water orally if possible. Otherwise dextrose 5% IV slowly (about 5 1itres/24h) guided by urine output and plasma sodium. Some authorities recommend giving normal saline since this causes less marked fluid shifts and is hypotonic in a hypernatraemic patient.

---

1 M Jamieson 1985 *BMJ* **290** 1723

Sodium

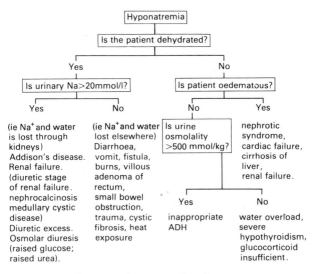

NB: in cirrhosis, hyponatremia may precede oedema

Potassium

General points Most potassium is intracellular. The concentrations of potassium and hydrogen ions in extracellular fluid tend to vary together. This is because these ions compete with each other in the exchange with sodium which occurs across most cell membranes (sodium is pumped out of the cell) and in the distal tubule of the kidney (sodium is absorbed from the urine). Thus if hydrogen ion concentration is high, potassium ions will be excreted less into the urine. Similarly K$^+$ will compete with H$^+$ for exchange across cell membranes and extracellular K$^+$ will accumulate. The resulting hyperkalaemia may not reflect total body potassium, conversely for low hydrogen ion concentration.

Hyperkalaemia

A plasma potassium >6.5mmol/l needs urgent treatment (p386) but first ensure that this is not an artefact (p624).

Signs and symptoms Cardiac arrhythmias. Sudden death.

ECG: Tall tented T-waves; widened QRS complex; small P-wave.

Causes Usually due to metabolic acidosis. Renal failure, diabetic keto-acidosis, excess potassium therapy, potassium sparing diuretics (spironolactone, amiloride, triamterene); artefact (potassium leaks out of red blood cells even without cell lysis), Addison's disease, sodium depletion, anorexia, severe tissue damage, anoxia, massive blood transfusion.

Treatment Treat underlying cause. In emergency see p386.

Hypokalaemia

If <2.5mmol/l needs urgent treatment. NB: hypokalaemia exacerbates digoxin toxicity.

Signs and symptoms Muscle weakness, hypotonia, cardiac arrhythmias, cramps and tetany.

ECG: Small or inverted T-waves; prominent U-wave (after T-wave); prolonged P–R interval; depressed ST segment.

Causes Often due to diuretics. If bicarbonate raised then loss probably longstanding with low intracellular potassium. Replacement will take days. High plasma bicarbonate is a more sensitive indication of potassium depletion than plasma potassium in long-term diuretic therapy. Vomiting, diarrhoea, purgative and liquorice abuse, intestinal fistula, villous adenoma of rectum, carbenoxolone, glucose given with insulin, renal tubule failure (p632), pyloric stenosis, alkalosis, low intake, hyperaldosteronism, Cushing's syndrome, Conn's syndrome, steroid or ACTH therapy are other causes. In hypokalaemic periodic paralysis intermittent weakness lasting up to 72h appears to be caused by K$^+$ shifting from the extracellular to the intracellular fluid.

Suspect Conn's syndrome if hypertensive, hypokalaemic alkalosis in someone not taking diuretics (p554).

Treatment If mild (>2.5mmol/l, no symptoms) give oral potassium supplement (at least 80mmol/24h). If patient taking thiazide diuretic, potassium >3.0mmol/l rarely needs treating.

If severe (<2.5mmol/l, dangerous symptoms) give IV potassium cautiously, not more than 20mmol/h, not more than 40mmol/l. Do not give potassium if oliguric.

Potassium
[*OTM* 18.30]

Calcium physiology and hypocalcaemia

General points Take blood specimens uncuffed (remove tourniquet after needle in vein, but before taking blood sample) and fasting. About 40% of plasma calcium is bound to albumin. Usually it is total plasma calcium which is measured although it is the unbound portion which is important. Therefore *adjust calcium level for albumin as follows:* Add 0.1mmol/l to calcium concentration for every 4g/l that albumin is below 40g/l, and a similar substraction for raised albumin. However, many factors affect binding (eg other proteins in myeloma, cirrhosis, individual variation) so be cautious in your interpretation.[1]

The control of calcium metabolism
- *Parathyroid hormone (PTH):* A rise in PTH causes a rise in calcium and a fall in phosphate plasma levels. This is due to PTH increasing calcium phosphate reabsorption from bone and calcium reabsorption from kidney tubule, increasing the action of vitamin D, and decreasing phosphate reabsorption from the kidney tubule. PTH secretion is itself controlled by ionized plasma calcium levels.
- *Vitamin D:* 25-Hydroxycholecalciferol (25-HCC) is made in the liver and converted to the much more active 1,25-di-HCC in the kidney. A rise in vitamin D causes an increase in calcium and a decrease in phosphate by increasing calcium absorption in the gut and phosphate excretion in urine. Disordered regulation of 1,25-di-HCC underlies familial normocalcaemic hypercalcuria which is a major cause of calcium oxalate renal stone formation (p376).
- *Calcitonin:* Made in C cells of the thyroid, this causes a decrease in plasma calcium and phosphate. Its importance in unclear. It is a marker for medullary carcinoma of the thyroid.
- Thyroxine may increase plasma calcium although this is rare.

642 Hypocalcaemia

NB: Apparent hypocalcaemia may be artefact of hypoalbuminaemia (see above).

Clinical effects Tetany, depression, perioral paraesthesiae, carpo-pedal spasm (wrist flexion and fingers drawn together) especially if brachial artery occluded with blood pressure cuff (*Trousseau's sign*), neuro-muscular excitability eg tapping over parotid (facial nerve) causes facial muscles to twitch (*Chvostek's sign*). ECG: prolonged Q–T interval.

Causes It may be a consequence of thyroid or parathyroid surgery. *If phosphate raised* then either chronic renal failure (p388) or hypopara-thyroidism or pseudohypoparathyroidism (p546). If phosphate normal or low then either osteomalacia (high alkaline phosphatase), overhydration or pancreatitis.

Treatment If *symptoms mild give* calcium 5mmol/6h PO. Do daily plasma calcium levels. ▶ In chronic renal failure see p384. If necessary add alfacalcidol 0.5–5µg/24h PO. If *symptoms severe*, give 10ml (2.32mmol) calcium gluconate 10% IV over 3mins. Repeat as necessary.

1 J Kanis 1985 *BMJ* i 728

Calcium physiology and hypocalcaemia
[*OTM* 10.64; 10.68]

Hypercalcaemia

Clinical effects ('Bones, stones, groans and psychic moans') Abdominal pain; nausea; vomiting; constipation; polyuria, depression; anorexia; weight loss; polydipsia; tiredness; weakness; BP↓; clouding of consciousness; pyrexia; sudden cardiac arrest; renal calculi; renal failure; corneal calcification. ECG: Q–T interval↓.

Causes & diagnosis Most commonly malignancy (bone metastases, PTHrP↑, p546) and hyperparathyroidism. Pointers to malignancy are: low plasma albumin, lowish chloride, hypokalaemia, alkalosis, raised phosphate and raised alkaline phosphatase. Other investigations (eg isotope bone scan, CXR, FBC) may also be of diagnostic value.

Treatment Treat the underlying cause.
If Ca >3.5mmol/l, or BP↓, severe abdominal pain, vomiting, pyrexia, clouding of consciousness, then aim to reduce calcium to abolish these as follows:
- Rehydrate with IVI 0.9% saline eg 4–6 litres in 24h as needed.
- Measure plasma Na^+, K^+, Mg^{2+}, urea.
- Correct hypokalaemia and hypomagnesaemia by IVI (mild metabolic acidosis does not need treatment).
- Frusemide 125mg IV slowly (<4mg/min) up to every 3h. Monitor CVP, U&E.
- Salmon calcitonin 8U/kg/8h IM. This inhibits osteoclasts.
- Plicamycin (≤25μg/kg/24h for 3 days) or bisphosphonates may be used if a tumour is the cause: single-dose pamidronate (30mg IVI over 4h in 0.9% saline, p764) is more effective than etidronate.[1] It inhibits bone resorption and osteoclasts. In multiple myeloma, sarcoidosis and vitamin D overdose, hydrocortisone (300mg/24h IVI) may be effective.

Magnesium

644   Magnesium is distributed 65% in bone and 35% in cells. Its level tends to follow those of calcium and potassium.

Magnesium deficiency Clinical features: Paraesthesiae, fits, tetany, arrythmias. It can exacerbate digitalis toxicity. Causes: Severe diarrhoea; diabetic ketoacidosis; alcohol abuse; total parenteral nutrition (monitor levels weekly); accompanying hypocalcaemia; accompanying hypokalaemia (especially with diuretics). Treatment: Replace with magnesium salts.

(Magnesium excess usually caused by renal failure, but rarely requires treatment in its own right.)

Zinc

Zinc deficiency This may occur in parenteral nutrition or inadequate diet. Rarely it is due to a genetic defect.
Clinical features: Red, crusted skin lesions especially round nostrils and corners of mouth.
Diagnosis: Therapeutic trial of zinc (plasma levels are unreliable as they may be low, eg in infection or trauma, without deficiency).

1 S Ralston 1989 Lancet ii 1180

Hypercalcaemia
Magnesium
Zinc
[*OTM* 10.67]

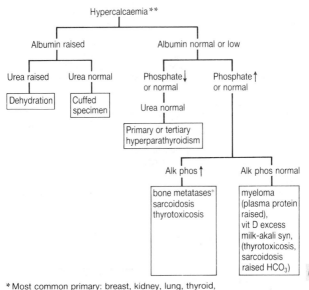

* Most common primary: breast, kidney, lung, thyroid,
prostate, ovary, colon.

NB: the best discriminating features between bone metastases and
hyperparathyroidism (the commonest two causes of hypercalaemia) are
low albumin, low chloride, alkalosis (all suggesting metastases).
Raised plasma PTH strongly supports hyperparathyroidism.

645

**This biochemical flow diagram must be taken in conjunction with
clinical picture and is a guide only.

Metabolic bone disease: 1. Osteoporosis

Osteoporosis refers to bone appearance, implying reduction in amount of bone mineral per unit volume of anatomical bone (exceeding natural decline with age). If trabecular bone is mostly affected, crush fractures of vertebrae are common (causing the littleness of little old ladies); if cortical bone is mostly affected, fracture of a long bone is more likely, eg of the femoral neck—a major cause of death and orthopaedic expense.

Prevalence: 5%. $♀/♂ ≈ 4$. *Risk of osteoporotic fracture* ↑ if:

Slender or anorectic	Early menopause	Osteoporosis in family
Smoker or alcoholic	Cushing's disease	Primary biliary cirrhosis
Prolonged rest	Malabsorption	Rheumatoid arthritis
Hyperparathyroidism	Thyrotoxicosis	Hypogonadism
>5mg/day prednisolone	Myeloma	Past low-trauma fracture
Vertebral deformity	Amenorrhoea	Mastocytosis (OHCS p602)

Diagnosis: x-ray (easier with hindsight afforded by bone fracture). Serum Ca^{2+}, PO_4^{3-}, and alk phos normal. Bone densitometry may be used, but its rôle is unclear: screening is probably not effective, and it is less effective than taking into account of risk factors in the prediction of hip fracture. Biopsy specimens may be unrepresentative.

Prevention: Good mineralization in youth (primary prevention, ie *good diet, sunlight, exercise*). If postmenopausal ↑risk (above) consider bone densitometry, and hormone replacement therapy (HRT) to ↓bone loss. Unopposed oestrogens risk endometrial cancer—so give combined HRT eg *conjugated oestrogens* 0.625mg/24h PO continuously; start 1st day of cycle (or at any time if cycles have ceased), with *norgestrel* 0.15mg daily from days 17–28 (eg Prempak C 0.625® maroon and brown tablets, respectively). NB not *all* women will be candidates for HRT, eg if past DVT or PE, or breast cancer; or ↑risk of breast cancer—see OHCS p18. In malabsorption, give *vitamin D and calcium* supplements.

In vertebral osteoporosis consider ~3yrs of osteoclast inhibition with *etidronate*, eg Didronel PMO®, each pack lasts 90 days, with 14 days of etidronate (400mg/day, 2h ac; SE angio-oedema, constipation, headache, paraesthesia, $Ca^{2+}↓$, $PO_4^{3-}↑$, WCC↓) + 76 tablets of 1250mg effervescent calcium carbonate. This ↑ bone density.[1,2] *Alendronate* ~10mg/24h ½h ac is an alternative here, and is known to reduce risk of hip and wrist fracture too (SE: rashes, constipation, oesophagitis—so patients *must* wash tablets down with plenty of water, and remain sitting up for 30mins.)[3]

2. Paget's disease of bone

There is increased bone turnover associated with abnormal hypernucleated osteoclasts causing remodelling, bone enlargement, deformity and weakness. Rare in the under-40s. Incidence rises with age. Commoner in temperate climes, and Anglo-Saxons; it may be asymptomatic or cause pain and enlargement of skull, femur, clavicle—and bowed (*sabre*) tibia—also pathological fractures, nerve deafness (bone overgrowth) and high-output CCF. *X-rays:* Patchy changes in bone density (eg sclerotic '*white*' vertebrae) with remodelling. *Blood biochemistry:* Ca^{2+} and PO_4^{3-} normal but Ca^{2+} goes up with bed rest; alk phos markedly raised.

Complications: Bone sarcoma (1% of those affected for >10yrs).

Biochemistry

Treatment: If analgesia fails, *disodium etidronate* (reduces bone turn-over) may be tried (5mg/kg/24h PO ac for 6 months; some specialists give higher doses for lesser periods eg 20mg/kg/ daily PO ac for 4 weeks[1]). SE: focal osteomalacia, fractures.[2] Consider also *salcatonin* 50–100u 3 times weekly IM/SC to help pain and lower alk phos (SE: nausea, faints, flushes).

1 N Watts 1990 *NEJM* **323** 73   **2** *Drug Ther Bul* 1992 **30** 45   **3** D Black 1996 *Lancet* **348** 1535

Metabolic bone disease:
3. Osteomalacia

In osteomalacia there is a normal amount of bony tissue but a reduced quantity of mineral content. Thus there is an excess of uncalcified osteoid and cartilage. Rickets is the result if this process occurs during the period of bone growth; osteomalacia is the result if it occurs after fusion of the epiphyses.

Types *Renal osteomalacia:* Renal failure leads to 1,25-dihydroxycholecalciferol (1,25-DHCC—p642) deficiency.

Vitamin D deficiency: Due to gastrointestinal disease; poor diet; or lack of sunlight.

Drug-induced: Anticonvulsants may induce liver enzymes which leads to a breakdown of 25-hydroxycholecalciferol (25-HCC).

Vitamin D resistance: A number of mainly inherited conditions in which the osteomalacia responds to high doses of vitamin D.

Altered vitamin D metabolism in the liver: Cirrhosis (p506).

Investigations *Plasma:* Mildly low calcium; low phosphate; raised alkaline phosphatase; 25-HCC low except in resistant cases; 1,25-DHCC low in renal failure; PTH (p642) high.

Biopsy: Shows incomplete mineralization.

X-ray: Cupped, ragged metaphyseal surfaces are seen in rickets. In osteomalacia there is a loss of cortical bone; in addition, apparent partial fractures without displacement may be seen especially on the lateral border of the scapula, inferior femoral neck and medial femoral shaft (Looser's zones).

Clinical features *Rickets:* Knock-kneed; bow-legged. Features of hypocalcaemia (p642). Children with rickets are ill.

Osteomalacia: Bone pain; fractures (neck of femur); proximal myopathy (waddling gait).

Treatment Calcium with vitamin D (400 units) tablets: 1–2 tablets/day.
  If due to malabsorption give parenteral calciferol 7.5mg monthly.
  If vitamin-D-resistant give calciferol 10,000 units/24h PO.
  If due to renal disease give alfacalcidol 1μg/24h PO and adjust dose according to plasma calcium.
  Monitor plasma calcium.

▶ Vitamin D therapy (especially alfacalcidol) can cause dangerous hypercalcaemia.

Hypophosphataemic (vitamin D resistant) rickets is X-linked and dominantly inherited, and is due to renal (proximal tubular) unresponsiveness to vitamin D. As well as rickets, there is a low plasma phosphate, high alkaline phosphatase, and phosphaturia. Treatment is with high dose vitamin D compounds and phosphate.

Metabolic bone disease
3. Osteomalacia
[*OTM* 17.14]

Protein

Electrophoresis distinguishes a number of bands (see figure).

Albumin is synthesized in the liver and has a $t_{1/2}$ of ~20 days. Albumin binds *bilirubin*, *free fatty acids*, *calcium* and a number of *drugs*.
Low albumin results in oedema. Causes: Liver disease, nephrotic syndrome, burns, protein-losing enteropathy, malabsorption, malnutrition, late pregnancy, artefact (eg from arm with IVI), posture (5g/l higher if upright), genetic variations, malignancy.
High albumin—Causes: dehydration; artefact (eg haemostasis).

α_1 Zone Absence or obvious reduction suggests α_1-antitrypsin deficiency, an autosomal recessive condition associated with emphysema (unopposed phagocyte proteases; exacerbated by smoking and infection) and cirrhosis (hepatocytes unable to secrete the protein—p506).

α_2 Zone is mainly α_2 macroglobulin and haptoglobin. May be increased in nephrotic syndrome with an associated decrease in other bands.

β Zone is *low* in active nephritis, glomerulonephritis and SLE.

γ Zone is *diffusely raised* in: chronic infections, liver cirrhosis (with a low albumin and α-globulin), sarcoidosis, SLE, RA, Crohn's, TB, bronchiectasis, PBC, hepatitis and parasitaemia. It is *low* in: nephrotic syndrome, malabsorption, malnutrition, immune deficiency (severe illness, DM, renal failure, malignancy or congenital).

Paraproteinaemia See p616.

Acute phase response The body responds to a variety of insults with, amongst other things, the synthesis, by the liver, of a number of proteins (normally present in serum in small quantities)—eg α_1-antitrypsin, fibrinogen, complement, haptoglobin and C-reactive protein. An increased density of the α_1- and α_2-fractions, often with a reduced albumin level, is thus seen with conditions such as infection, malignancy (esp. α_2-fraction), trauma, surgery and inflammatory disease.

The C-reactive protein (CRP) is normally present in serum at less than 10mg/l. It rises within 8h of an inflammatory stimulus (eg post-surgery or MI) and is waning by ~3 days. Monitoring the CRP can help answer questions like: is Crohn's currently active; is there a bacterial infection on top of leukaemia; is the infection responding to antibiotics? Leukaemia, SLE, viruses, osteoarthritis, ulcerative colitis, steroids, oestrogens and pregnancy have little effect on CRP levels.

IN URINE (If protein >0.15g/24h then pathological. See p370.)

Albuminuria Usually caused by renal disease.

Microalbuminuria (defined as excretion of 30–150µg/min) is not detected by dipsticks, but can be measured using radioimmunassay. It is an early indicator of diabetic renal disease (p396, p532).

Bence Jones protein consists of light chains excreted in excess by some patients with myeloma (p614). They are not detected by dipsticks and may occur with normal serum electrophoresis.

Also see haemoglobinuria (p586) and myoglobinuria (p394).

C-Reactive protein

Persistent elevation	Normal slight elevation
Bacterial infection	Viral infection
Crohn's disease	Ulcerative colitis
Connective tissue diseases	SLE
(except SLE)	Steroids/oestrogens
Neoplasia	
Trauma	
Necrosis (eg MI)	

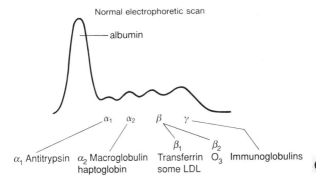

Normal electrophoretic scan

albumin

α_1 α_2 β γ

β_1 β_2

α_1 Antitrypsin α_2 Macroglobulin Transferrin O_3 Immunoglobulins
haptoglobin some LDL

Plasma enzymes

▶ Reference intervals vary from laboratory to laboratory.

Raised plasma levels of specific enzymes can be very useful indications of specific disease. But remember that, for example, 'cardiac enzymes' may be raised for reasons other than cardiac pathology. The major causes of raised enzymes are given below.

● Reference intervals are given on p802.

Acid phosphatase Prostatic carcinoma. NB: the prostatic enzyme can be measured separately from total acid phosphatase. Not sufficiently specific for screening, but valuable in following response to treatment.

Alkaline phosphatase Liver disease (suggesting cholestasis). Bone disease (isoenzyme can be distinguished, and reflects the laying down of osteoid—ie osteoblast activity) especially Paget's, growing children, osteomalacia, metastases, hyperparathyroidism and renal failure (p626). The placenta secretes its own isoenzyme which is raised in pregnancy.

Alanine-amino transferase (ALT; SGPT). Liver disease (suggests hepatocyte damage). Also raised in shock.

Aldolase Skeletal muscle. Also heart and liver disease.

α-Amylase Acute pancreatitis. Also severe uraemia, diabetic ketoacidosis.

Aspartate-amino transferase (AST; SGOT). Liver disease (suggesting hepatocyte damage). Following myocardial infarct (p280). Also skeletal muscle damage.

Creatine kinase Following myocardial infarct (p280). Skeletal muscle damage (even following IM injection and defibrillation). Cardiac isoenzyme can be distinguished (known as CKMB: normally <5% total).

Gamma-glutamyl transpeptidase (GGT) Liver disease and probably more specific for alcohol-induced damage.

Lactate dehydrogenase (LDH) Following myocardial infarct (p280). Liver disease (suggests hepatocyte damage). May also be raised in kidney and blood diseases.

Tumour markers

Tumour markers are rarely sufficiently specific to be of diagnostic value. Their main value is in following the course of an illness and the effectiveness of treatment.

Alpha-fetoprotein (α-FP). Raised in primary hepatoma (hepatocellular carcinoma, p510). Is also usually raised in choriocarcinoma (see below) including testicular teratoma.

Human chorionic gonadotrophin raised in hydatidiform moles and choriocarcinoma (tumours arising from the chorion, OHCS p26).

Acid phosphatase See above.

Carcino-embryonic antigen (CEA) is useful in following the progression of gastrointestinal neoplasms but is not of diagnostic value.

Plasma enzymes
Tumour markers

653

Hyperlipidaemia

This is a major risk factor for coronary disease in middle age but take into account: smoking, BP↑, DM, lipids, past and family history).

▶ *Be cautious of trigger-happy screening*[1]—unless the person wants to modify his life-style (diets may be a major imposition)—and you and your patient are good at handling uncertainty, and you are prepared to prescribe a great deal. Half the UK population have a cholesterol level which is thought to contribute markedly to CHD. A real danger is to turn healthy people into patients chronically anxious about a hyperlipidaemia. *For some, it may be healthier not to know.* NB: good evidence that it is worth treating hyperlipidaemia even if there is no existing heart disease comes from a trial ($N = 32,216$) indicating that treating 1000 such patients with pravastatin (20–40mg/24h) would result (in 5 yrs) in ~27 fewer MIs and deaths than in a placebo arm—but the cost is very high, so strategies focusing on screening those at highest risk may be the most sensible.

Identify primary hyperlipidaemia (opposite) as risk of CHD↑↑. Lipids travel in blood packaged with proteins as lipoproteins. There are 4 classes: chylomicrons (mainly triglyceride); low density lipoprotein (LDL, mainly cholesterol, the lipid correlating most strongly with CHD); very low density lipoprotein (VLDL, mainly triglyceride); high density lipoprotein (HDL, mainly phospholipid, correlating inversely with CHD—it's good).

Do plasma lipids in those <70 if: ● Family history of hyperlipidaemia.
● Family or personal history of CHD before 60yrs—or risk ↑, eg DM, BP↑.
● Xanthomata or xanthelasmata. ● Corneal arcus before 50yrs old.
Do not measure from 24h to 3 months after stress (eg stroke; MI). Ideal cholesterol range might be 3.5–5.2mmol/l (triglyceride 0.5–1.9mmol/l). Use uncuffed samples, eg after a 14h fast, but while on usual diet.

Xanthomata These yellowish lipid deposits may be: eruptive (itchy nodules in crops in hypertriglyceridaemia); tuberous (yellow plaques on elbows and knees); planar—also called palmar (orange-coloured streaks in palmar creases), virtually diagnostic of remnant hyperlipidaemia; or deposits in tendons, eyelids (xanthelasmata), or cornea (arcus).

Primary hyperlipidaemias Opposite. Secondary hyperlipidaemias are caused by DM; alcohol abuse; T₄↓; renal failure; nephrosis; cholestasis.
Management If cholesterol >5.2 mmol/l and aged <60–70yrs:
● Ask questions to uncover secondary hyperlipidaemias (seen in <20%).
● Suggest HRT if appropriate.
● Aim for a body mass index of 20–25 (p502); use a diet with <10% of calories from saturated fats and plenty of gel-forming fibre (p482).
● Encourage exercise (HDL↑). If after 3 months cholesterol remains >6.5mmol/l or triglycerides >2.3mmol/l, consider a <7% saturated fats diet + drugs—the advice of the British Hyperlipidaemia Society—but *not* of all meta-analysists[2]). Fibrates are often used (*bezafibrate* 200mg/8h PO—action: triglycerides/LDL↓; HDL↑). *Nicotinic acid* 100mg–2g/8h PO also has this dual action and is much cheaper (but SE of flushing limits its use). Alternatives: resins (eg *colestipol* 5–30g/day PO in liquid, and *not* with meals) bind bile acids, preventing their reabsorption, so the liver uses more LDL for more bile acid synthesis. Resins may upset absorption of fat-soluble vitamins. β-hydroxy-β-methylglutaryl-coenzyme A reductase inhibitors (eg *simvastatin*, 10–40mg PO at night—inhibits cholesterol synthesis by up to 40%) produces the best-documented fall in all-cause mortality in those with CHD;[3] do CK (risk of myositis) + LFT (pre-treatment, and when maintenance dose is reached).
● Drugs rarely help premopausal ♀. Avoid drugs if pregnant. Avoid lipid-raising drugs: diuretics, unselective β-blockers, norgestrel.

1 M Dunnigan 1993 *BMJ* i 1355   2 J Shepherd 1995 *NEJM* 333 1301   3 '4S' study 1994 *Lancet* 344 1388

Classification of primary hyperlipidaemias

Type	Plasma cholesterol (mmol/l)	Plasma triglyceride (mmol/l)	Relative prevalence	Inheritance	Xanthomas	Risk of atherosclerosis
Familial hypercholesterolaemia	7.5–15 (homozygotes up to 30)	0.55–3	++	autosomal dominant	tendon	+++
Familial combined hyperlipidaemia	6–10	2–7	+++	? dominant	—	++
Familial hypertriglyceridaemia	7–15	6–60	+	autosomal dominant	eruptive	+?
Lipoprotein lipase deficiency; apo-C11 deficiency	7–10	20–100	v. rare	autosomal recessive	eruptive	—
Remnant Hyperlipoproteinaemia (a familial condition with marked combined hyperlipidaemia)	10–15	5–12	+	apo-E_3 deficiency with other factors	tuberous planar	++
Common hypercholesterolaemia	>7.8	0.55–1.9	++++	polygenic	—	+

655

Porphyrias

These are a group of (mainly) inherited diseases in which there are errors in the pathway of haem biosynthesis resulting in the excretion or accumulation of porphyrias.

There are two usual presentations of these rare diseases.

The first clinical presentation—acute intermittent With the following clinical features: colicky abdominal pain with vomiting or constipation (mimicking the *acute abdomen*, especially as there may be mild fever and a leucocytosis); proteinuria, peripheral neuritis (especially motor); paralysis; psychosis; hyponatraemia; hypokalaemia. These features are intermittent, and may be precipitated by a wide range of drugs (see below). If this is suspected then send a urine sample to the laboratory to look for porphobilinogens which are raised during attacks and often (50%) between them (the urine may go deep red on standing).

The suspected diagnosis is: *acute intermittent porphyria*, a dominantly inherited disease.

Drugs to avoid in acute intermittent porphyria are legion (because they may precipitate above symptoms and quadriplegia, see OTM); they include: *alcohol*; *several anaesthetic agents* (barbiturates, halothane); *antibiotics* (chloramphenicol, sulphonamides, tetracyclines); *pain killers* (pentazocine); *oral hypoglycaemics* (tolbutamide, chlorpropamide); *contraceptive pill* (oestrogens, progesterones).

Treatment of acute intermittent porphyria
- Remove precipitating factors.
- IV fluids to correct electrolyte imbalance.
- High carbohydrate intake (eg Hycal®) by NG tube if necessary.
- Nausea controlled with promazine 50mg PO/IM.
- Sedation if necessary with chlorpromazine 50–100mg PO/IM.
- Pain control with: aspirin, dihydrocodeine or morphine.
- Convulsions controlled with diazepam.
- Treat tachycardia and hypertension with propranolol.
- Physiotherapy if muscle weakness.

The second clinical presentation With photosensitive skin lesions and any of the above symptoms and signs. Specimens of faeces, urine, and blood should be sent to the laboratory for measurement of porphyrins. There are several congenital causes (*porphyria variegata*, prevalent in the Afrikaner population of South Africa; *hereditary coproporphyria*, and the *erythropoietic porphyrias*). There is also an acquired cause *porphyria cutanea tarda symptomatica* resulting from severe liver disease (especially alcoholic cirrhosis).

Porphyrias
[*OTM* 9.136]

Biochemistry

Notes

Notes

15 | Rheumatology

Relevant pages in other chapters: Charcot's joints (p476).
Eponymous syndromes: Behçet's disease (p694); Sjögren's syndrome
(p708); Wegener's granulomatosis (p712).

Points of note in the rheumatological history

Age, occupation, ethnic origin.

Presenting symptoms

Joints
- Pain
- Morning stiffness
- Swelling
- Loss of function
- Joints affected over course of disease

General
- Nodules or lumps
- Red eyes, Dry eyes or mouth
- Fever, rashes, ulcers
- Raynaud's phenomenon

Past history

- Trauma
- Infection
- Diarrhoea, Inflammatory bowel disease
- Venereal disease
- Gout
- Previous operations

Treatments

- Current and previous
- Adverse reactions to medications

Social and family history

Domestic situation.
What can the patient do or not do?
Family history of rheumatoid arthritis, osteoarthritis, gout, back disease.

Arthritis

Clinical features Pain, stiffness (especially early morning), loss of function, signs of inflammation at one or more joints.

Differential diagnosis
► In acute monoarthritis, always exclude septic arthritis first.

Monoarthritis:	Polyarthritis:
Septic arthritis	Viral illness
TB arthritis	Rheumatoid arthritis
Trauma	Osteoarthritis
Gout	Spondyloarthritides
Pseudogout	SLE
Spondyloarthritides	Henoch–Schönlein purpura
Osteoarthritis	TB
Rheumatoid (rarely)	Drug allergies
Haemarthrosis	Acute rheumatic fever
Local malignant deposit	Gonorrhoea

Assess Extent of joint involvement (include spine), symmetry, disruption of joint anatomy, limitation of movement (by pain or contracture), effusions and periarticular involvement. Associated features: dysuria or genital ulcers, skin or eye involvement, lungs, kidneys, heart, GI (eg mouth ulcers, bloody diarrhoea) and CNS.

Investigation Radiology: Affected joints looking for erosions, calcification, widening or loss of joint space, changes in underlying bone, sacroiliac joints if a spondyloarthritis is possible; CXR in RA, SLE, and TB.
Joint aspiration for culture, microscopy for blood, pus and crystals.
Blood: Culture in septic arthritis, FBC and ESR. Uric acid, urea and creatinine if systemic disease. Rheumatoid factor, antinuclear antibody, and other autoantibodies. In trauma, arthroscopy may be of value.

Septic arthritis

Pyogenic infection of a joint is usually spread haematogenously, but may develop from adjacent osteomyelitis. It may completely destroy the joint within 24 hours of onset, so if treatment is not prompt, the patient is at risk of a bony ankylosis.

Causes: Commonest are *Staph aureus* (80%), *Haemophilus influenzae*, and gonococci. Others: TB, *Brucella*, *Salmonella typhi*. Patients with damaged joints (eg RA) or on steroids are at greater risk.

Clinical features: Usually a monoarthritis in which the joint is so swollen, red and painful that no movement is tolerated.

Investigation: ► Whenever you see a hot, swollen joint, aspirate it (OHCS p656–8). Pus usually means infection or pseudogout (p664—the ordinary light microscope may show phagocytosed crystals in WBCs). Obtain urgent Gram stain. Culture blood and aspiration fluid. x-ray will show no changes. FBC. ESR. C-reactive protein (falls more promptly than ESR in response to treatment).

Treatment: Analgesia; support the joint; regular aspiration as needed. 'Blind' antibiotics: flucloxacillin 500mg/6h IV plus *either* benzylpenicillin 600mg/6h (>5yrs) or (if <5yrs, for *Haemophilus*) ampicillin 250mg/6h. Treat IV for 10 days then PO for 2 weeks.

Viral arthritis Rubella, mumps, hepatitis B, enterovirus, AIDS.

Rheumatology

Arthritis
Septic arthritis
[*OTM* 16.1]

Rheumatoid arthritis (RA)

Typically a persistent, symmetrical, deforming, peripheral arthropathy. Peak onset in 4th decade. 3:1 female predominance. It affects 3% of the population and is HLA DR4 linked.

Presentation Typically with swollen, painful and stiff hands and feet, especially in morning. This gradually gets worse and larger joints become involved. Less common presentations are:
1 Relapsing and remitting monoarthritis of different large joints (palindromic). 2 Persistent monoarthritis (often of knee). 3 (Commoner in men.) Systemic illness (pericarditis, pleurisy, weight loss) with minimal joint problems at first. 4 Vague limb girdle aches. 5 Sudden onset widespread arthritis.

Signs At first, sausage-shaped fingers and MCP joint swelling. Later, ulnar deviation and volar subluxation (partial dislocation) at MCP joints, Boutonnière and swan-neck deformities of fingers (OHCS p670), or Z-deformity of thumbs. The wrist subluxes and the radial head becomes prominent (piano-key). Extensor tendons in the hand may rupture and adjacent muscles waste. Similar changes occur in the feet. Larger joints may be involved. Atlanto-axial joint subluxation may threaten the cord.

Extra-articular Anaemia. Nodules. Lymphadenopathy. Vasculitis. Carpal tunnel syndrome. Multifocal neuropathies. Splenomegaly (5%; but only 1% have Felty's syndrome: leucopenia, lymphadenopathy, weight loss—see p698). Eyes: episcleritis, scleritis, keratoconjunctivitis sicca. Pleurisy. Pericarditis. Pulmonary fibrosis. Osteoporosis. Amyloidosis.

Tests X-ray of RA joints show soft tissue thickening, juxta-articular osteoporosis, loss of joint space, then bony erosions, and subluxations. Complete carpal destruction may be seen. Rheumatoid factor +ve in 75% (cf Sjögren's 100%, SLE 30%, MCTD 30%, PSS 30%, p670), ANA +ve in 30%.

Management
- Regular exercise. ● Physiotherapy. ● Occupational therapy.
- Household aids and personal aids (orthoses), eg wrist splints.
- Intralesional steroids (for joint injection technique, see OHCS p656–8).
- Surgery—to improve function, not for cosmetic effect.
- Oral drugs: if no contraindication (asthma, active peptic ulcer) start an NSAID eg ibuprofen 400–800mg/8h pc (cheapest and least likely to cause GI bleeds) or naproxen 250–500mg/12h pc (useful as twice daily dose, and is more anti-inflammatory than ibuprofen, and may cause less GI bleeding than other NSAIDs). Diclofenac 25–50mg/8h pc is similar to naproxen and is available combined with misoprostol (Arthrotec®)—an ulcer-protecting agent (p490), which should be considered if history of dyspepsia and NSAIDs vital. One cannot predict which NSAID a patient will respond to. If 1–2 week trials of 3 NSAIDs do not control pain, or if synovitis for >6 months, try disease-modifying drugs (DMDs). DMD side-effects are many and serious. ▶ Regular monitoring is essential. If DMDs fail, consider cytotoxics (eg azathioprine, methotrexate).
- DMDs and side-effects: *Gold:* Marrow suppression, proteinuria, LFTs↑. *Penicillamine:* Marrow suppression, proteinuria, taste↓, oral ulcers, myasthenia, Goodpasture's. *Sulphasalazine:* Marrow suppression, sperm count↓, hepatitis, oral ulcers. *Chloroquine:* Permanent retinopathy, tinnitus, headache. Rash with any of them. See OHCS p672 for dosages.

Rheumatoid arthritis (RA)

[*OTM* 16.1]

Osteoarthritis (OA)

OA, the commonest joint condition, is symptomatic three times more often in women, and the mean age of onset is 50yrs. It is usually primary, but may develop secondary to any joint disease.

Clinical features Pain on movement, worse at end of day; background pain at rest; stiffness; joint instability. Most commonly affected are DIP joints, 1st metacarpophalangeal, 1st metatarsophalangeal, cervical and lumbar spine, next the hip, then knee. Joint tenderness, derangement, and bony swelling (eg Heberden's nodes), poor range of movement, effusions.

Radiology Loss of joint space, subchondral sclerosis and cysts, marginal osteophytes. Other investigations are normal.

Treatment Simple analgesics (eg paracetamol). If the pain is very bad, give NSAIDs (see p664). Consider joint replacement. Reduce weight.

Crystal arthropathy

Gout In the acute stage there is severe pain, redness and swelling in the affected joint—often the metatarsophalangeal joint of the big toe (podagra). Attacks are due to hyperuricaemia (p634) and the deposition of sodium monourate crystals in joints, and may be precipitated by trauma, surgery, starvation, infection and diuretics. After repeated attacks, urate deposits (tophi) are found in avascular areas, eg pinna, tendons, joints—and chronic tophaceous gout is said to be present. 'Secondary' causes of gout: polycythaemia, leukaemia, cytotoxics, renal failure.

Diagnosis depends on finding urate crystals in tissues and synovial fluid (serum uric acid not always↑). Microscopy of synovial fluid: negatively birefringent crystals; neutrophils (with ingested crystals). x-rays may show only soft-tissue swelling in the early stages. Later well-defined 'punched out' lesions are seen in juxta-articular bone. There is no sclerotic reaction, and joint spaces are preserved until late.

Treatment of acute gout is with an NSAID such as naproxen 750mg as early as possible, followed by 250mg/8h PO pc. If contraindicated (eg peptic ulcer), give colchicine 1mg PO initially, then 0.5mg/2h PO until the pain goes or D&V occurs, or 10mg has been given.

Prevention of attacks: Avoid purine-rich foods (offal, oily fish), obesity and alcohol excess. No aspirin (it increases serum urate). Consider reducing serum urate with long-term allopurinol, but not until 3 weeks after an attack. Start with 100mg/24h PO pc, increasing as necessary up to 600mg/24h. Monitor serum urate levels. SE: rash, fever, leucopenia. If troublesome, substitute a uricosuric, eg probenecid 0.5g/12h PO.

Pseudogout (Calcium pyrophosphate arthropathy). Risk factors: ● Old age ● Dehydration ● Intercurrent illness ● Hyperparathyroidism ● Myxoedema; DM ● PO₄³⁻↓; Mg²⁺↓ ● Any arthritis (RA, OA) ● Haemochromatosis ● Acromegaly

Clinical features: 1 Acute pseudogout: less severe and longer-lasting than gout; affects different joints (mainly the knee).
2 Chronic calcinosis: Destructive changes like OA, but more severe; affects different joints (knees, also wrists, shoulders, hips).

Tests: Calcium deposition on x-ray (eg triangular ligament in wrist). Joint crystals are weakly positively birefringent in plane polarized light.

Treatment of pseudogout: This is symptomatic.

Osteoarthritis (OA)
Crystal arthropathy
[*OTM* 16.1]

667

Spondyloarthritides

These conditions are sometimes called seronegative arthritides, but exclude seronegative RA. They have certain common features:
- Involvement of spine (spondylo-) and sacroiliac joints (SIJs).
- Asymmetrical oligo- or monoarthritis of large joints.
- Inflammation then calcification of tendinous insertions into bone (enthesopathy), eg plantar fasciitis, Achilles tendinitis, costochondritis.
- Extra-articular manifestations: uveitis, aortic regurgitation, upper zone pulmonary fibrosis, amyloidosis.
- HLA B27 association (88–96% of those with ankylosing spondylitis).

The spondyloarthritides show considerable overlap with one another. They are treated with physiotherapy, occupational therapy, advice about posture, NSAIDs, and genetic counselling.

Ankylosing spondylitis (AS) Suspect this in a young man presenting with morning stiffness in his back. *Prevalence:* 1 in 2000. More severe in men; but not as rare in women as was thought. *Clinical features:* There is progressive spinal fusion with loss of movement and ultimately the patient is left with a fixed, kyphotic spine and hyperextended neck (question mark posture), spino-cranial ankylosis and restricted respiratory excursion.

The typical patient is a young man who presents with morning stiffness and progressive loss of spinal movement. The main lesion is spinal ankylosis with sacroiliac joint involvement. Other features: reduced thoracic excursion, chest pain, hip and knee involvement, periostitis of the calcaneum or ischial tuberosities, plantar fasciitis, iritis, amyloidosis, apical lung fibrosis.

Tests: Radiology may show: a 'bamboo spine', squaring of the vertebra, erosions of the apophyseal joints, obliteration of the sacroiliac joints (sacroiliitis also occurs in Reiter's disease, Crohn's disease and chronic polyarthritis). Other tests: FBC (normochromic anaemia); ESR↑.

Treatment: Backache requires exercise, not rest. If able to co-operate with an intensive exercise régime to maintain posture and mobility, he may keep happy and employed, despite chronic progressive illness. NSAIDs may relieve pain and stiffness. Rarely, spinal osteotomy is useful.

Enteropathic arthropathies Inflammatory bowel disease and GI bypass surgery are associated with a spondyloarthritis and a large joint mono- or oligo-arthropathy. Peripheral arthritis also occurs. Whipple's disease (p712) may be one of this group.

Psoriatic arthritis See *OHCS* p586.

Chronic arthritis in children may take several forms:
- Juvenile ankylosing spondylitis (mainly boys over 9 years);
- Juvenile rheumatoid arthritis (mainly post-pubertal girls);
- Juvenile psoriatic arthritis;
- Juvenile chronic arthritis or Still's disease (commonest; no adult equivalent) may be systemic illness, polyarticular, or pauciarticular (associated with ANA and chronic uveitis). See *OHCS* p758.

Other causes of arthritis Behçet's (p694), leukaemia (p600), pulmonary osteoarthropathy (p342), Lyme disease (p233), endocarditis (p312), acne, acromegaly (p564), Wilson's disease (p712), familial Mediterranean fever (p186), sarcoid (p360), sickle-disease (p588), haemochromatosis (p507).

Reiter's syndrome *Presentation:* A triad of *urethritis*, *conjunctivitis* and *seronegative arthritis*. A typical patient is a young man with recent NSU—which may be asymptomatic; it may also follow dysentery (and epidemics have occurred eg in war). Often large joint mono- or oligo-arthritis; it may be chronic or relapsing.

Other features: Iritis, keratoderma blenorrhagica (brown, aseptic abscesses on soles and palms); mouth ulcers; circinate balanitis (painless serpiginous rash); enthesopathy (plantar fasciitis, Achilles tendonitis) and aortic incompetence (rare).

Tests: First of 2-glass urine test shows debris in urethritis. Neutrophils, and macrophages containing neutrophils (Peking cells) in synovial fluid. x-rays: periostitis at ligamentous insertions; rheumatoid-like changes if chronic.

Management: Bed rest, splintage of affected joints and NSAIDs. Recovery may take months.

Connective tissue diseases—1

Essence The connective tissue diseases (CTDs or collagen vascular diseases) form an unholy alliance, like Yorkist pretenders to a Lancastrian throne. And they often seem as sinister to the student. Unholy, because the rightful heirs to the title (like the true diseases of collagen, eg Ehlers–Danlos syndrome, OHCS p746) are excluded. Alliance, because clinically they have much in common. They overlap with each other, affect many organ systems, are associated with systemic fever and malaise, tend to run a chronic course, often respond to steroids, and are often associated with anaemia of chronic disease and a raised ESR. But their pathologies vary.

Included in this group are: systemic lupus erythematosus (p672), RA (p664), progressive systemic sclerosis, polymyositis, dermatomyositis, mixed connective tissue disease, Sjögren's syndrome (p708), relapsing polychondritis, polyarteritis nodosa (p674), polymyalgia rheumatica (p674), giant cell arteritis (p675), Wegener's granulomatosis (p710), Behcet's syndrome (p694), Takayasu's arteritis (p708).

Systemic lupus erythematosus The group 'prototype' (p672).

Progressive systemic sclerosis (PSS) There is progressive fibrosis and tightening of the skin (scleroderma). This may be the only abnormality. The skin may be globally affected; morphoea (localized skin sclerosis) only rarely, if ever, develops into PSS. The CREST syndrome (**C**alcinosis of subcutaneous tissues, **R**aynaud's phenomenon, disordered o**E**sophageal motility, **S**clerodactyly, and **T**elangiectasia) is associated with a more benign course. Generalized progressive systemic sclerosis affects particularly the kidney (causing severe hypertension), lungs (fibrosis), and GI tract. There is sometimes a polyarthritis and myopathy. ANA stains nuclei in a characteristic nucleolar pattern and CREST has anti-centromere ANA. No treatment works; give symptomatic support. Prognosis varies from weeks (accelerated-phase hypertension) to years.

Scleroderma diagnostic criteria: Proximal skin scleroderma, or any 2 of: sclerodactyly, digital pitting scars, pulp loss, bibasilar lung fibrosis.

Mixed connective tissue disease (MCTD) is a mixture of features of SLE, PSS, and polymyositis. Renal or CNS involvement is rare. MCTD is characterized by the presence of anti-RNP (ribonuclear protein) antibody without other types of ANA, and a speckled pattern of staining. Prognosis is probably the same as SLE.

Relapsing polychondritis attacks the pinna, nasal septum and larynx, the last causing stridor. It is associated with aortic valve disease, arthritis and vasculitis. Treat with steroids.

Polymyositis and dermatomyositis

Insidious, symmetrical, proximal muscle weakness results from muscle inflammation. Dysphagia, dysphonia, or respiratory weakness may develop. 25% have a purple (heliotrope) rash on cheeks, eyelids and light-exposed areas. Raynaud's, lung involvement, polyarthritis, retinitis (like cotton wool patches) and myocardial involvement also occur. There is an associated malignancy in >10% of patients >40yrs, but this figure is controversial. Diagnosis is by muscle enzyme (CK) levels, electromyography (EMG—shows fibrillation potentials) and muscle biopsy. Rest and prednisolone help (start with 60mg/24h PO). A more aggressive form with prominent vasculitis occurs in children.

Connective tissue diseases—1
Polymyositis and dermatomyositis
[*OTM* Section 16]

Systemic lupus erythematosus

SLE is a non-organ-specific, autoimmune disease characterized by antinuclear antibodies (ANA) and vasculitis. Other features (eg cutaneous and some CNS signs) are not vasculitic. It is nine times commoner in women, especially in pregnancy, and almost ten times commoner in West Indian blacks. Onset 15–25yrs. It is a disease of relapses and remissions.

Clinical features
- *Musculoskeletal* (95%): Arthralgia, myalgia, myositis, proximal myopathy. Non-erosive polyarthritis with periarticular and tendon involvement. Deforming arthropathy rarely may occur due to capsular laxity (Jaccoud's arthropathy). Aseptic bone necrosis.
- *Skin* (81%): Photosensitive malar butterfly rash (hyperkeratosis, follicular plugging over nasal bridge, spreading to cheeks), scarring alopecia, livedo reticularis (net-like blush), Raynaud's, purpura, oral ulceration, urticaria, conjunctivitis. Discoid lupus (eg on ears, cheeks, scalp, forehead, chest) is a 3-stage rash: erythema→pigmented hyperkeratotic oedematous papules→atrophic depressed lesions.
- *CNS* (≤18%): Depression, psychosis, fits, hemi- or paraplegia, cranial nerve lesions, cerebellar ataxia, chorea, meningitis.
- *Renal* (<50%): Proteinuria, oedema, hypertension, renal failure.
- *Pulmonary* (48%): Pleurisy (±pleural effusion), pneumonia, obliterative bronchiolitis, fibrosing alveolitis, pulmonary oedema (rare).
- *CVS* (38%): Pericarditis (±effusion), Libman–Sacks endocarditis.
- *Blood:* Normochromic, normocytic anaemia (75%), haemolysis (rare, 15% are Coombs' +ve), leucopenia, lymphopenia, thrombocytopenia.
- *Other:* ● Fever (77%) ● Splenomegaly ● Leucopenia ● Lymphopenia.

Immunology 80% are ANA +ve. High titre of anti-double-stranded DNA is almost exclusive to SLE. 40% are rheumatoid factor +ve. 11% have a false +ve syphilis serology. Antibodies to Ro (SS-A), La (SS-B), and U1 ribonuclear protein may help define overlap syndromes (see Sjögren's syndrome, p708).[1] Low levels of complement (plasma C3, C4) are better markers of disease activity than ESR. The acute-phase response is normal.

Treatment Refer to expert. Advise a sun-block cream (OHCS p599).
- NSAIDs to relieve symptoms. Use *sun screens* to protect the skin.
- *Hydroxychloroquine* if joint or skin symptoms are not controlled by NSAID, eg 6mg/kg/24h PO for 4 weeks, then 200mg/24h. (SE: irreversible retinopathy—check vision at least annually).
- *Prednisolone* for exacerbations (0.75–1mg/kg/24h PO for 6–10 weeks, then reduce) combined with hydroxychloroquine increased to 6mg/kg/24h.
- *Low dose steroids* may be of value in chronic disease.
- *Cyclophosphamide* helps renal function more than steroids.[1] Dose example: 0.5–3mg/kg/day PO; intermittent pulses of 20mg/kg/month IV—give fewer SE. Its rôle in CNS disease is unclear. *Azathioprine* 1–2.5mg/kg/day PO is used as a steroid-sparing drug (SE: lymphoma).
- *Cyclosporin* or *methotrexate* may be recommended by experts.

Drug-induced lupus occurs with isoniazid, hydralazine (if >50mg/24h in slow acetylators), procainamide, chlorpromazine, anticonvulsants. Renal and CNS involvement is rare, but pulmonary disease and rashes are common. It remits when drug is stopped. Remember that sulphonamides and the contraceptive steroids may exacerbate idiopathic SLE.

Antiphospholipid syndrome SLE may be seen with arterial or venous thromboses, stroke, migraine, miscarriages, myelitis, myocardial infarct, multi-infarct dementia—and ↑*cardiolipin* (a phospholipid) antibody. Aspirin and anticoagulation (to give INR≈3) may be needed.[2]

Connective tissue diseases—2
Systemic lupus erythematosus
[*OTM* 16.20; 18.48]

1 PJ Venables 1993 *BMJ* **ii** 663   **2** MA Khamashta 1993 *BMJ* **ii** 883

Vasculitis

Essence Vasculitis occurs in non-organ specific autoimmune diseases (eg RA, SLE), but may be less important than the glomerulonephritis or synovitis. It is the principal feature in certain other connective tissue diseases (CTDs) which may or may not be autoimmune (eg PAN, Wegener's). There is a third group of CTDs in which the importance of vasculitis is unknown (eg progressive systemic sclerosis). Vasculitis also occurs in conditions that are not usually included in the CTDs (eg drug reactions).

Polyarteritis nodosa (PAN) Commoner in males by 4:1. Some cases are associated with HBsAg. It is a necrotizing vasculitis which causes aneurysms of medium-sized arteries.

Clinical features:
● General: Fever, malaise, weight↓, arthralgia.
● Renal: (75%) Main cause of death. Hypertension, haematuria, proteinuria, renal failure, intra-renal aneurysms.
● Cardiac: (80%) Second biggest cause of death. Coronary arteritis and consequent infarction. ↑BP and failure. Pericarditis. In Kawasaki's disease (childhood variant, OHCS p750) coronary aneurysms occur.
● Pulmonary: Pulmonary infiltrates and asthma occur in vasculitis (some say that lung involvement is incompatible with PAN, calling it then Churg–Strauss syndrome)
● CNS: (70%) Mononeuritis multiplex, sensorimotor polyneuropathy, fits, hemiplegia, psychoses.
● GI: (70%) Abdominal pain (any viscus may infarct), malabsorption because of chronic ischaemia.
● Skin: Urticaria, purpura, infarcts, livedo (p672), nodules.
● Blood: WCC↑, eosinophilia (in 30%), anaemia, ESR and CRP raised.

Diagnosis is most often made from clinical features in combination with renal or mesenteric angiography. ANCA may be +ve (p690).

Treatment: Treat hypertension meticulously. Refer to experts. Use high dose prednisolone, and then cyclophosphamide.

Other vasculitides Wegener's granulomatosis (p710). Behçet's disease (p694). Takayasu's arteritis (p708).

674

Polymyalgia rheumatica (PMR)

Common in old ladies, *presenting* as aching and morning stiffness in the proximal limb muscles for at least a month—± a mild polyarthritis, depression, fatigue, fever, anorexia. It overlaps with giant cell arteritis. Both may cause CNS signs, claudication, angina, pulmonary infarcts and hypopituitarism.[1] *Tests* ESR usually >40mm/h; CK usually normal. *Treat* with steroids: start with prednisolone 15mg/24h PO and reduce to 7.5–10mg/24h over 2–3 months. Thereafter, reduce dose slowly. Some people require daily steroids for 2–3 years. Monitor with ESR.

1 M Radhamanohar 1992 *Care of the Elderly* **4** 162

Connective tissue diseases—3: Vasculitis
Polymyalgia rheumatica (PMR)
Giant cell (cranial) arteritis (GCA)

Giant cell (cranial) arteritis (GCA)

This vasculitis (cranial arteritis) is associated with polymyalgia in 25%. Common in the elderly, it is rare if <55yrs. *Symptoms:* headache, scalp and temporal artery tenderness (eg on combing hair), jaw claudication, amaurosis fugax, or sudden blindness in 1 eye. *Tests:* ESR↑, platelets↑, alk phos↑, Hb↓. ▶ If you suspect the diagnosis, take blood for an ESR, start prednisolone 60mg/24h PO immediately, and consider a temporal artery biopsy in the next few days. Skip lesions occur, so do not be diverted by a negative biopsy. Use 80–120mg if visual symptoms (phone an ophthalmologist). Reduce prednisolone over 4–6 weeks but increase dose if symptoms recur. Typical course: 2yrs. Monitor need for treatment with ESR.

Back pain

▶ This is usually self-limiting; but be alert to sinister causes.

History Key points: 1 Onset: sudden (related to trauma?) or gradual? 2 Are there motor or sensory symptoms? 3 Are bladder or bowel affected? Pain worse on movement and relieved by rest is often mechanical. If it is worse after rest, an inflammatory disease is likely.

Examination 1 Movements of the spine: mark skin points 10cm above and 5cm below L5 in midline with patient erect. Ask patient to bend forward as far as possible. Measure distance between points. If <20cm, movement is restricted; 2 Neurological deficits: perianal sensation; UMN and LMN signs in legs; 3 Signs of generalized disease may suggest malignancy.

Dorsal root irritation causes lancinating pain in the relevant dermatome, made worse by coughing and bending forward. *Lasègue's sign* is positive if straight leg raising on supine patient is painful and restricted to <45°. It suggests lumbar disc prolapse.

Neurosurgical emergencies ● *Acute cauda equina compression:* Alternating or bilateral root pain in legs, saddle anaesthesia (ie bilaterally around anus) and disturbance of bladder or bowel function.

● *Acute cord compression:* Bilateral pain, LMN signs at level of compression, UMN & sensory signs below, sphincter disturbance. Causes (same for both): bony metastasis (look for missing pedicle on x-ray), myeloma, cord or paraspinal tumour, TB (p198), abscess.

▶ Urgent treatment is needed to prevent irreversible loss: laminectomy for disc protrusions; decompression for abesses; radiotherapy for tumours.

Features suggestive of sinister pathology
● Progressive constant pain; local bony tenderness; painful restriction of lumbar spine in all directions. These are suggestive of bony pathology eg tumour, infection.
● Raised ESR—eg in tumours, myeloma, infection, TB.

Investigations Anatomical basis elucidated with *plain spinal* x-ray, *radiculography* (reveals encroachment on roots), *myelography* (reveals encroachment on or expansion of cord). MRI ideally replaces CT scans and myelography in diagnosing cord compression, myelopathy, intraspinal neoplasms, cysts, haemorrhages and abscesses.[1] FBC, ESR, U&E, acid phosphatase and *technetium* scan 'hot spot' may support diagnosis of neoplastic or inflammatory lesion.

Causes Age determines the most likely causes.
● 15–30yrs: Prolapsed disc, trauma, fractures, ankylosing spondylitis (p668), spondylolisthesis (L5 shifts forward on S1), pregnancy.
● 30–50yrs: Degenerative joint disease, prolapsed disc, malignancy (lung, breast, prostate, thyroid, kidney).
● >50yrs: Degenerative, osteoporosis, Paget's, malignancy, myeloma. Lumbar artery atheroma (which may itself cause disc degeneration[2]).

Rarer causes: Spinal stenosis (bony encroachment), cauda equina tumours, spinal infection (usually staphylococcal, also *Proteus*, *E coli*, *S typhi*, TB). Often no systemic signs of infection.

Treatment Cause usually remains uncertain. Treat empirically: exercise; avoid precipitants (eg improper lifting). Consider: bed rest, analgesia, traction, plaster jackets, manipulation, and occasionally laminectomy. Specific causes need specific treatment.

---

1 P Armstrong 1991 *BMJ* i 105   **2** L Kauppilla 1993 *BMJ* i 1267

Rheumatology

Notes

Notes

16 Miscellaneous topics

681

Alcoholism[1,2]

An alcoholic is one whose repeated drinking leads to harm in his work or social life. ▶ Denial is a leading feature of alcoholism, so be sure to question relatives. Screening tests: MCV↑; gamma-GT↑. (NB: alcohol can do you good in low doses, eg <21 units per week in men, see p482.[2])

Probing questions [Useful mnemonic: **control**] Can you always **co**ntrol your drinking? Has alcohol ever led you to **n**eglect your family or your work? What **t**ime do you start drinking? Do you drink before this? Do friends comment on how much you drink or ask you to **r**educe intake? Do you ever drink in the mornings to **o**vercome a hangover? Go through an average day's alcohol, **l**eaving nothing out.

Alcohol's effects and: ● *The liver* (It is normal in 50% of alcoholics.)
Fatty liver: Acute and reversible; but may progress to cirrhosis if drinking continues (also seen in obesity, diabetes mellitus, and with amiodarone).
Hepatitis: (Fever, jaundice and vomiting)—80% will progress to cirrhosis (hepatic failure in 10%). Biopsy: Mallory bodies ± neutrophil infiltrate. *Cirrhosis* (p506): 5yr survival is 48% if drinking continues (if not, 77%).
● *The CNS* Poor memory and cognition: ▶ may respond to multiple high-potency vitamins IM; cortical atrophy; retrobulbar neuropathy; fits; falls; neuropathy; Korsakoff's and Wernicke's encephalopathy (p702 & 712).
● *The gut* Obesity, diarrhoea; gastric erosions; peptic ulcers; varices (p496); pancreatitis (acute and chronic).
● *The blood* MCV↑; anaemia from: marrow depression, GI bleeds, alcoholism-associated folate deficiency; haemolysis; sideroblastic anaemia.
● *The heart* Arrhythmias; BP↑; cardiomyopathy.

Withdrawal signs Pulse↑; BP↓; tremor; fits; hallucinations (*delirium tremens*)—may be visual or tactile, eg of animals crawling all over one.

Alcohol contraindications Driving, hepatitis; cirrhosis; peptic ulcer; drugs (eg antihistamines); carcinoid; pregnancy (fetal alcohol syndrome—IQ↓, short palpebral fissure, absent filtrum and small eyes).

Managing alcohol withdrawal Admit; do BP + TPR/4h. Beware BP↓. For the first three days give generous diazepam, eg 10mg/6h PO or PR if vomiting, or IVI in seizures, p736), then reduce diazepam (eg 10mg/8h PO for days 4–6, then 5mg/12h for 2 more days). NB: the once-preferred chlormethiazole readily causes addiction and respiratory depression; also phenothiazines are problematic. ● Vitamins may be needed (p712).

Prevention Alcohol-free 'beers'; alcohol-free pubs; low-risk drinking (≤20u/week if ♂; ≤15u/week if ♀—there are no absolutes—risk is a continuum. NB: higher limits are controversial.) 1u is 9g ethanol, ie 1 measure of spirits, 1 glass of wine, or ½ pint of beer.

'Treatment' of the established alcoholic may be rewarding, particularly if he really wants to change. If so, group therapy or self-help (eg 'Alcoholics Anonymous') may be useful. Encourage the will to change. Think about graceful ways of declining a drink, eg "I'm seeing what it's like to go without for a bit." Suggest the patient does not buy him or herself a drink when it is his turn. Suggest: "Don't lift your glass to your lips until after the slowest drinker in your group takes a drink. Sip, don't gulp." Give follow-up and encouragement. Reducing pleasure that alcohol brings (and withdrawal craving) with naltrexone 50mg/24h PO can halve relapse rates.[4] SE; vomiting, drowsiness, dizziness, cramps, arthralgia; CI: hepatitis, liver failure. It is costly. Confer with experts on its use.

1 *Lancet* 1995i 16–43   **2** R Doll 1994 *BMJ* ii 911   **3** J Gaziano 1995 *BMJ* ii 3   **4** J Volpicelli 1995 *Lancet* i 456

Alcoholism
[*OTM* 12.245]

The eye in systemic disease

Systemic disease often manifests itself in the eye and, in some cases, eye examination will first suggest the diagnosis.

Vascular disease The retina may be variably affected by BP↑. Grade I 'copper wiring' of arteries. Grade II arteriovenous nipping. Grade III 'flame' and 'blot' haemorrhages, 'cotton wool' exudates. Grade IV papilloedema. The assumption is that this grading correlates with BP—but this was not confirmed in a careful trial of grade I and II retinopathy—which, confusingly, may occur in normotensive (non-diabetic) subjects.[1] In some patients hypertension presents with an intraocular bleed.

Emboli passing through the retina produce *amaurosis fugax*—seeing 'a curtain passing across the eyes'. They may arise from atheromatous plaques (listen to carotids) or from heart (from valves or infarct site). Treat with aspirin (p436). Retinal haemorrhages may be seen in leukaemia; comma-shaped conjunctival haemorrhages and retinal new vessel formation may occur in sickle-cell disease; optic atrophy in pernicious anaemia.

Metabolic disease Diabetic eyes: see p532. Hyperthyroidism and exophthalmos: see p543. In myxoedema, eyelid and periorbital oedema is quite common. Superficial corneal opacities, small cortical lens opacities, and optic neuritis are rare. Lens opacities may also occur in hypoparathyroidism, whereas conjunctival and corneal calcification may occur in hyperparathyroidism. In gout monosodium urate deposited in the conjunctiva may give sore eyes, if deposited in the sclera it may cause anterior uveitis.

Granulomatous disorders (TB, sarcoid, leprosy, brucellosis, and toxoplasmosis) can all produce inflammation in the eye; TB and syphilis producing iritis, the others posterior uveitis. TB, congenital syphilis, sarcoid, CMV and toxoplasmosis may all produce choroidoretinitis. In sarcoid there may be nerve palsies.

Collagen diseases These also cause inflammation. Conjunctivitis is found in SLE and Reiter's syndrome; episcleritis in polyarteritis nodosa and SLE; scleritis in rheumatoid arthritis; and uveitis in ankylosing spondylitis and Reiter's. In dermatomyositis there is orbital oedema with retinopathy (haemorrhages and exudates).

Keratoconjunctivitis sicca (Sjögren's syndrome, p708) There is reduced tear formation (Schirmer filter paper test), producing a gritty feeling in the eyes. Decreased salivation also gives a dry mouth (xerostomia). It occurs in association with collagen diseases. Treatment is with artificial tears (tears naturale, or hypromellose drops).

AIDS Those with HIV infection may develop CMV retinitis, characterized by cotton-wool retinal spots. This may be asymptomatic but can cause sudden, untreatable, visual loss. Candidiasis of the aqueous and vitreous is hard to treat. Kaposi's sarcoma may affect the lids or conjunctiva.

1 S Dimmitt 1989 *Lancet* i 1103

Differential diagnosis of 'red-eye'

	Conjunctiva	Iris	Pupil	Cornea	Anterior chamber	Intraocular pressure	Appearance
Acute glaucoma	Both ciliary and conjunctival vessels injected. Entire eye is red	Injected	Dilated, fixed, oval	Steamy, hazy	Very shallow	Very high	
Iritis	Redness most marked around cornea. Colour does not blanch on pressure	Injected	Small, fixed	Normal	Turgid	Normal	
Conjunctivitis	Conjunctival vessels injected, greatest toward fornices. Blanch on pressure. Mobile over sclera	Normal	Normal	Normal	Normal	Normal	
Subconjunctival hemorrhage	Bright red sclera with white rim around limbus	Normal	Normal	Normal	Normal	Normal	

After RD Judge, GD Zuidema, FT Fitzgerald 1989 *Clinical Diagnosis* 5 ed, Little Brown, Boston/Toronto.

685

Skin manifestations of systemic diseases

Erythema nodosum Painful, erythematous nodular lesions on anterior shins (± thighs/forearms). Associated with: sarcoid; drugs (sulphonamides, oral contraceptive, dapsone); bacterial infections (streptococcus, mycobacterium—TB, leprosy). Less common associations: Crohn's; UC; BCG vaccination; leptospirosis; various viral and fungal infections.

Erythema multiforme 'Target' lesions (symmetrical distribution; circular, often with central blister and symmetrical, including palms + soles). Usually on limbs. Often with mouth, genital and eye ulcers and fever (severe = Stevens–Johnson syndrome). Associated with: drugs (barbiturates; sulphonamides); infections (herpes; mycoplasma; orf); collagen disorders. 50% idiopathic. Give steroids in severe cases.

Erythema marginatum Pink coalescent rings on trunk which come and go. Associated with rheumatic fever (or rarely other causes eg drug reaction).

Erythema chronicum migrans This starts as a small papule, developing into a slowly enlarging red ring (≤50mm across) with a raised border. It fades from the centre. It lasts from 2 days to 3 months, and may be multiple. The cause is Lyme disease (p233).

Vitiligo Presents as bilateral, symmetrical white patches sometimes with a hyperpigmented border. Sunlight may cause itch. Associations: organ-specific autoimmune disorders (eg Graves' & Hashimoto's disease, diabetes mellitus, alopecia areata, Addison's disease, hypoparathyroidism, premature ovarian failure). Treat by camouflage cosmetics.

Pyoderma gangrenosum Nodule or pustule ulcerates with tender red/blue necrotic edge. Any site (commonly calf, abdomen, face). Associations: UC; Crohn's; RA; paraproteinaemia; Wegener's granulomatosis; myeloma. Treat with high dose oral steroids.

Specific diseases and their manifestations

Crohn's disease Perianal/vulval ulcers. Erythema nodosum. Pyoderma gangrenosum.

Diabetes mellitus Recurrent infections; ulcers; *necrobiosis lipoidica* (shiny area on shins with yellowish skin and telangiectasia); *granuloma annulare* (p 578); fat necrosis at site of insulin injection.

Gluten-sensitive enteropathy (coeliac disease) *Dermatitis herpetiformis* (itchy blisters eg in groups on knees, elbows and scalp). It is postulated that gluten induces circulating immune complexes, which react with elements in the dermis. The itch (which can drive patients to suicide) responds to dapsone 100-200mg/24h PO within 48h—and this may be used as a diagnostic test. The maintenance dose may be as little as 50mg/week. In 30% this will need to be continued, in spite of having a gluten-free diet. SE (dose related): haemolysis, hepatitis, agranulocytosis.

Malabsorption Dry pigmented skin, easily bruised, hair loss, leuconychia.

Miscellaneous topics

Skin manifestations of systemic diseases
Specific diseases and their manifestations

[*OTM* Section 20]

Hyperthryoidism *Pretibial myxoedema* (red oedematous swellings above lateral malleoli which progress to thickened oedema of legs and feet), thyroid acropachy (clubbing + subperiosteal new bone in phalanges).

Neoplasia *Acanthosis nigricans:* pigmented rough thickening of skin in axillae or groin with warty lesions, associated especially with carcinoma of the stomach; *dermatomyositis* (p670); *skin metastases*; *acquired icthyosis:* dry scaly skin associated with lymphoma; *thrombophlebitis migrans:* successive crops of tender nodules affecting blood vessels throughout the body, associated with carcinoma of the pancreas (especially body and tail tumours).

Liver disease Reddened palms; spider naevi; gynaecomastia; decrease in secondary sexual hair; jaundice; bruising; scratch marks.

Skin diagnoses not to be missed

Malignant melanomas These rapidly metastasizing, increasingly common tumours occur at any age (after puberty), affecting any part of the body. Excess sunlight is a major risk factor. Some will arise in pre-existing moles. Suspect this malignancy if a pigmented lesion shows: ● Rapid enlargement. ● Bleeding. ● Increasing variegated pigmentation—especially blue-black or grey. ● Ulceration. ● A notched or indistinct border. ● Persistent itching. ● 'Satellite' lesions around the principal lesion.

Smooth, regular lesions with a clear demarcation with normal skin are unlikely to be melanomas. For the distinction between melanomas and 'fried egg-like' dysplastic naevi, see OHCS p584. *Treatment:* prompt excision with wide margins is required.

Squamous cell carcinoma This usually presents as an ulcerated lesion, with hard raised edges. They may begin in solar keratoses (below), or be found on the lips of smokers or in long-standing gravitational leg ulcers. Metastases are rare, but local destruction may be extensive. Treatment: total excision. Note: the condition may be confused with a keratoacanthoma—a fast-growing, benign, self-limiting papule plugged with keratin.

Basal cell carcinoma (Rodent ulcer) This epithelioma has malignant potential—in an indolent fashion. If left untreated, extensive local destruction may occur. Caucasians who have enjoyed decades in the sun are particularly at risk. The lesion is an ulcerated nodule with raised, pearly margins—usually on the face above a line joining the chin to the ear lobe. Early removal eg by curettage is usually curative. Cryotherapy is an alternative. Radiotherapy is valuable in the elderly, but should be avoided in the young because of scarring and telangiectasia.

Senile or actinic keratoses These occur on sun-exposed areas, appearing as crumbly yellow-white crusts. Malignant change may occur. Treatment: cautery; cryotherapy; 5% 5-fluorouracil cream which is used twice daily (with this sequence of events: erythema → vesiculation → erosion → ulceration → necrosis → healing epithelization). Healthy skin is unharmed. Treatment is usually for 4 weeks, but may be prolonged. There is no significant systemic absorption if the area treated is <500cm². Avoid in pregnancy. The hands should be washed after applying the cream.

Secondary carcinoma The most common metastases to skin are from breast, kidney and lung. The lesion is usually a firm nodule, most often on the scalp. See acanthosis nigricans (p687)

Mycosis fungoides is a lymphoma (cutaneous T-cell lymphoma, as in Sézary syndrome) which is usually confined to skin. It causes itchy, red, round plaques.

Leukoplakia This appears as white patches (which may fissure) on oral or genital mucosa. Frank carcinomatous change may occur.

Leprosy Suspect in any anaesthetic hypopigmented lesion.

Syphilis Any genital ulcer is syphilis until proved otherwise. Secondary syphilis: papular rash—including palms (p218).

Others Kaposi's sarcoma (p702); Bowen's disease and Queyrat's erythroplasia (p696); Paget's disease of the breast (p706).

Skin diagnoses not to be missed
[*OTM* Section 20]

Plasma autoantibodies: disease associations

Antinuclear antibody	*+ve in*
Systemic lupus erythematosus	99%
Rheumatoid arthritis	32%
Juvenile rheumatoid arthritis	76%
Chronic active hepatitis	75%
Sjögren's syndrome	68%
Progressive systemic sclerosis	64%
'Normal' controls	0–2%
(↑ with age: 20% at 70yrs)	

Smooth muscle antibody	*+ve in*
Chronic active hepatitis	40–90%
Primary biliary cirrhosis	30–70%
Idiopathic cirrhosis	25–30%
Viral infections	≤80% (low titres)
'Normal' controls	3–12%

Gastric parietal cell antibody	
Pernicious anaemia (adults)	90–100%
Atrophic gastritis: females	60%
males	15–20%
Autoimmune thyroid disease	33%
'Normal' controls	2–16%

Antibody to mitochondria	
Primary biliary cirrhosis	60–94%
Chronic active hepatitis	25–60%
Idiopathic cirrhosis	25–30%
'Normal' controls	0.8%

Antibody to reticulin	
Coeliac disease	37%
Crohn's disease	24%
Dermatitis hepetiformis	17–22%
'Normal' controls	0–5%

Thyroid antibodies	*Microsomal*	*Thyroglobulin*
Hashimoto's thyroiditis	70–91%	75–95%
Graves' disease	50–80%	33–75%
Myxoedema	40–65%	50–81%
Thyrotoxicosis	37–54%	40–75%
Juvenile lymphocytic thyroiditis	91%	72%
Pernicious anaemia	55%	?
'Normal' controls	10–13% (50% in older women)	6–10%

Figures from the Department of Biochemistry, Crawley Hospital, UK (1986)

Rheumatoid factor	*+ve in*
Rheumatoid arthritis	70–80%
Systemic lupus erythematosus	up to 40%
Sjögren's syndrome	up to 100%
Felty's syndrome	up to 100%
Progressive systemic sclerosis	30%
Still's disease	rarely +ve
Infective endocarditis	up to 50%
'Normal' controls	5–10%

It is also associated with viral hepatitis, infectious mononucleosis, TB and leprosy.

Anti-neutrophil cytoplasmic antibody (ANCA)

Wegener's disease >95% +ve.

(2 types: cANCA against elastase I and pANCA against myeloperoxidase. The latter is associated with non-granulomatous small vessel vasculitis. There is a cross-reaction between ANA and pANCA.)

Plasma autoantibodies: disease associations

Miscellaneous topics

Notes

Notes

17 | A dictionary of eponymous syndromes

▶ *Common things happen commonly, but many people have rare diseases.*
Note: for paediatric and orthopaedic syndromes, see *OHCS* p742–58.

Alice in Wonderland syndrome Disturbance of one's view of oneself (intrapsychic time passes too fast—seen in epilepsy, migraine, or on falling asleep).

Arnold–Chiari malformation The cerebellar tonsils and medulla are malformed and herniate through the foramen magnum. This may cause progressive hydrocephalus with mental retardation, optic atrophy and ocular palsies, and spastic paresis of the limbs. It may also cause syringomyelia (p476) or focal cerebellar and brainstem signs such as ataxia, dysphagia, oscillopsia, and nystagmus. There may be an association with bony abnormalities of the base of the skull (basilar impression). MRI is better than CT in aiding diagnosis.

Baker's cyst Posterior herniation of the capsule of the knee joint leads to escape of synovial fluid into one of the posterior bursae, stiffness and knee swelling (transilluminable in the popliteal space). *Tests:* Arthrogram. *Differential:* DVT (may coexist). Aspiration is possible, but recurrence is common.

Barrett's oesophagus In chronic reflux oesophagitis squamous mucosa undergoes metaplastic change and the squamocolumnar junction (*ora serrata*) appears to migrate caudally. The length affected may be a few cm only or all the oesophagus. It carries a 40-fold ↑ risk of adenocarcinoma. Loss of the tumour-suppressor gene p53 may be important. *Presentation:* ● Retrosternal pain, radiating to neck, worsened by hot or cold foods. ● Dysphagia ● Vomiting ● Haematemesis ● Melaena. Once diagnosed, surveillance programmes vary depending on age and general health. If pre-malignant changes are found, some advocate oesophageal resection especially in younger, fit patients; others favour laser ablation with frequent endoscopic ultrasound monitoring. *Management:* Regular endoscopy and intensive anti-reflux measures, including long-term proton pump inhibitors ± epithelial laser ablation or photodynamic therapy (PDT) aiming for resolution of metaplasia. PDT involves light-induced activation or an orally administered photosensitizer such as 5-aminolaevulinic acid which causes the accumulation of protoporphyrin IX in GI mucosal cells. Local laser light at 630nm then causes necrosis—which is confirmed by finding squamous re-epithelialization. This can prevent the need for surgery.

Bazin's disease Skin TB (local areas of fat necrosis with ulceration and an indurated rash, characteristically on adolescent girls' legs = *erythema induratum*).

Behçet's disease ♂/♀ ≈2; HLA-B5/51 associated progressive systemic disease of unknown cause affecting many organs: *Joints:* Arthritis. *Eyes:* Pain, blurred vision, floaters, iritis ± hypopyon, retinal vein occlusion *via* retinal vasculitis and retinal infiltrates. *Mouth, scrotum, labia:* Painful ulcers (heal by scarring). *CNS:* Meningoencephalitis, ICP↑, brainstem signs, dementia, myelopathy, encephalopathy, cerebral vein thrombosis. *Treatment:* Prednisolone (start with 20–60mg/24h PO) or colchicine; Steroid cream for ulcers. In non-responsive uveitis try chlorambucil 0.2mg/kg/day PO or cyclosporin. [*Halushi Behçet, 1919.*]

Berger's disease (IgA nephropathy) The most common glomerulonephritis (p380) in the West—causing episodic haematuria with viral infections. IgA is deposited in glomeruli; microscopy shows mesangial proliferation. Heavy proteinuria and hypertension indicate a poor prognosis. 15–20% go on to end-stage renal failure. Secondary IgA nephropathy may be associated with ankylosing spondylitis, coeliac disease, and HIV. *Treatment:* Alternate-day prednisolone *may* help, as may phenytoin (IgA levels ↓), as may tonsillectomy.

Bickerstaff's brainstem encephalitis Acute, progressive cranial nerve dysfunction, ataxia, coma and apnoea may lead to a (reversible) brain death picture (not true brain death: no structural brain damage has been proven).

Bornholm disease (Devil's grip) This is caused by coxsackie B virus. It usually presents in young children or adults with sudden severe pain in the chest and abdomen, on one side. There may be no pain in other muscles. Rhinitis is usually present. *Differential diagnosis:* MI, acute abdomen. *Treatment:* Analgesics.

Dictionary of eponymous syndromes

Bowen's disease (Intraepidermal carcinoma-*in-situ*) This starts as a slow growing, non-invasive scaly red plaque, but later it may change into a squamous cell epithelioma. Confirm by biopsy. *Treatment:* excision, cautery cryosurgery, radiotherapy—or topical 5-fluorouracil. Note: penile Bowen's disease is called Queyrat's erythroplasia.

Brown-Séquard syndrome This results from a lesion in one (lateral) half of the spinal cord (eg hemisection). There is ipsilateral paralysis and joint position sense loss below the lesion. There is ipsilateral analgesia and thermoanaesthesia at the level of the lesion, but there is contralateral loss of these modalities a few segments below the lesion. The reflexes are brisk and the tone is increased ipsilaterally. There are also sphincter disturbances. Causes: trauma, infection, neoplasia, MS and degenerative diseases.

Budd–Chiari syndrome This is caused by hepatic vein obstruction (usually by thrombosis or tumour). It may present with sudden epigastric pain and shock, or more insidiously with signs of portal hypertension, jaundice, and cirrhosis. It may follow the use of oral contraceptives. *Treatment:* surgery may be indicated.

Buerger's disease (Thromboangiitis obliterans; endarteritis obliterans) This is inflammation of arteries, veins and nerves with thrombosis in the middle sized arteries found *only* in male cigarette smokers. It may lead to gangrene.

Caplan's syndrome This is rheumatoid arthritis coupled with necrotic lung granulomata in coal miners. The cause cough, dyspnoea, and haemoptysis. CXR: bilateral nodules (0.5–5cm). *Treatment:* steroids are used (carefully exclude TB first).

Charcot–Marie–Tooth syndrome (Peroneal muscular atrophy) This presents at puberty or in early adult life and begins with foot drop and weak legs. The peroneal muscles are the first to atrophy. The disease spreads to the hands and then the arms. Sensation is usually diminished, as are the reflexes. The cause is unknown. It seldom becomes totally incapacitating.

Curtis–Fitz-Hugh syndrome Chlamydial perihepatitis in sexually active women. It may simulate biliary pain and can cause chronic peritonitis with ascites.

Devic's syndrome (Neuromyelitis optica) This condition is increasingly being considered as a variant of multiple sclerosis. There is massive demyelination of the optic nerves and chiasm and also in the spinal cord. The prognosis is variable, and complete remission may occur.

Dressler's syndrome (Post-myocardial infarction syndrome) This develops 2–10 weeks after an MI or heart surgery. It is thought that myocardial necrosis stimulates the formation of autoantibodies against heart muscle. *Clinical features:* fever and chest pain, pleural and pericardial rub (from serositis). Cardiac tamponade may complicate the picture so avoid anticoagulants. *Treatment:* steroids and anti-inflammatory agents are used.

Dubin–Johnson syndrome A familial metabolic disorder which leads to failure of excretion of conjugated bilirubin. There is intermittent jaundice with pain in the right hypochondrium. It does not cause hepatomegaly. *Investigations:* normal alkaline phosphatase. There is bile in the urine. Liver biopsy shows diagnostic pigment granules.

Dupuytren's contracture The palmar fascia contracts so that the fingers cannot extend. The little finger of the right hand is the most commonly affected. There is a nodular thickening of the connective tissue over the 4th and 5th fingers. *Prevalence:* ~5%, increasing with age, and a positive family history. Associations: heavy manual labour, trauma, diabetes mellitus, phenobarbitone treatment, alcoholic liver disease, Peyronie's disease, AIDS. It may be thought of as being a marker of disturbed metabolism of free radicals derived from O_2—like this: ischaemia (the primary event) increases xanthine oxidase activity, which reduces oxygen to yield superoxide free radicals who induce fibroblast proliferation, and, therefore, type III collagen and palmar fibrosis (M Bower 1990 *BMJ* i 164). This explains the observation that allopurinol (it binds xanthine oxidase) is associated with the resolution of Dupuytren's contracture (G Murrell 1987 *Speculations in Sci & Tec* **10** 107–12). [*Baron Guillaume Dupuytren, 1831.*]

696

Dictionary of eponymous syndromes

Eisenmenger's syndrome A congenital heart defect which is at first associated with a left to right shunt may lead to pulmonary hypertension and shunt reversal. If so, cyanosis develops (± heart failure and respiratory infections), and Eisenmenger's syndrome is said to be present. It is irreversible, and carries a poor prognosis.

Ekbom's syndrome (Restless legs syndrome) The patient (who is usually in bed) is seized by an irresistible desire to move his legs in a repetitive way accompanied by an unpleasant sensation deep in the legs. The pathogenesis is unknown, but a defect in the basal ganglia has been suggested (C Clough 1987 *BMJ* i 262). *Treatment:* many drugs have been tried; only carbamazepine (p450) and clonazepam (1–4mg PO nocte) have shown any benefit.

Fabry's disease An X-linked disorder of glycolipid metabolism due to deficiency of galactosidase-A. Ceramide trihexoside is deposited in the skin (angiokeratoma corporis diffusum), kidneys, and vasculature. Most die in the 5th decade. Cardiac manifestations include infarction, angina, short PR interval, rhythm disturbances, LVH, CCF, mitral valve lesions, congestive and hypertrophic obstruction cardiomyopathies.

Fanconi anaemia Aplastic anaemia associated with congenital absence of the radii. The disease may terminate in acute myeloid or monomyelocytic leukaemia.

Fanconi syndrome The cardinal features are generalized aminoaciduria, glycosuria, hyperphosphaturia and rickets (or osteomalacia). The defect is in renal tubular reabsorption. Causes are multiple eg heavy metal poisoning, vitamin deficiency (B$_{12}$, C, D), myeloma, Wilson's disease, cystinosis. Other features: bone pain, fractures, hypouricaemia, kaliuresis (excess urine K$^+$), plasma K$^+$↓. *Treatment:* vitamin D (p648). Check plasma Ca^{2+} often. [*Guido Fanconi*, 1936.]

Felty's syndrome This is rheumatoid arthritis, splenomegaly and leucopenia. Usually in long-standing disease. Recurrent infections are common. Hypersplenism also produces anaemia and thrombocytopenia. Lymphadenopathy, pigmentation and persistent skin ulcers occur. Rheumatoid factor often in high titre. Splenectomy may improve the neutropenia. [*Augustus Felty, 1934*].

Foster Kennedy syndrome This is optic atrophy of one eye with papilloedema of the other, caused by a tumour on the inferior surface of the frontal lobe—on the side of the optic atrophy.

Friedreich's ataxia An hereditary condition (autosomal recessive or rarely sex-linked) of idiopathic cord degeneration. The spinocerebellar tracts degenerate causing cerebellar ataxia and dysarthria, nystagmus and dysdiadochokinesis. Loss of the corticospinal tracts occurs and causes weakness and extensor plantars, but there is also peripheral nerve damage, so the tendon reflexes are paradoxically depressed (differential diagnosis p426). There is dorsal column degeneration resulting in loss of postural sense. Pes cavus is characteristic and there is often a scoliosis. Cardiomyopathy may cause cardiac failure. Patients usually die at about 50 years of age. [*Nikolaus Friedreich, 1863*].

698

Froin's syndrome CSF protein↑ and xanthochromia in the presence of a normal cell count—seen in CSF below a block in spinal cord compression.

Gardner's syndrome A dominantly-inherited disorder in which there are multiple premalignant colonic polyps, bone tumours (benign exostoses), and soft-tissue tumours such as epidermal cysts, dermoid tumours, fibromas and neurofibromas. Fundoscopy reveals black spots (congenital hypertrophy of retinal pigment epthelium) and is valuable in detecting carriers of the gene (which is on the long arm of chromosome 5) before symptoms develop. Median age at onset: 20yrs. Symptoms: bloody diarrhoea. Careful follow-up is needed. Subtotal colectomy with fulguration of rectal polyps aims to prevent malignancy (References: P Chapman 1989 *BMJ* ii 353).

Gélineau's syndrome The patient, usually a young man, succumbs to irresistible attacks of inappropriate sleep. Associations: encephalitis; MS; HLA DR2. *Treatment:* methylphenidate (*OHCS* p206) 10mg PO after breakfast and lunch. It is an amphetamine, and may cause dependence and psychosis.

Dictionary of eponymous syndromes

Gerstmann's syndrome Finger agnosia, left/right disorientation, dysgraphia and acalculia. Although these features do not occur together more often than predicted by chance, when they do, this may point to a lesion in the dominant parietal lobe.

Gilbert's syndrome This is an inherited metabolic disorder leading to increased unconjugated hyperbilirubinaemia. Population incidence is estimated at 1–2%. The onset is shortly after birth, but it may be unnoticed for many years. Liver biopsy is normal, but should rarely be required clinically. Jaundice occurs during intercurrent illness. A rise in bilirubin on fasting or after IV nicotinic acid can confirm the diagnosis. Prognosis is excellent. [*Nicolas Gilbert, 1901.*]

Gilles de la Tourette syndrome *Presentation:* Motor tics (p428) with irrepressible, explosive, occasionally obscene verbal ejaculations. It usually begins in childhood and remains for life (with remissions). There may be a witty, innovatory, phantasmogoric picture, 'with mimicry, antics, playfulness, extravagence, impudence, audacity, dramatizations, ... surreal associations, uninhibited affects, speed, 'go', vivid imagery and memory, and hunger for stimuli'.[1] *Diagnosis:* Motor or vocal tics for >1yr (may not be concurrent). *Association:* Obsessive-compulsive disorder. *Prevalence:* 1:2000. A dominant gene with variable expression may be responsible. *Treatment:* Haloperidol (eg 3mg/4h PO) or clonidine (50μg/12h PO) may help the severely affected. Tics may be brought under control through the medium of dance. (1 O Sacks 1992 *BMJ* ii 1515)

Goodpasture's syndrome Proliferative glomerulonephritis with lung symptoms (haemoptysis) caused by antibasement membrane antibodies which bind both the kidney's basement membrane and the alveolar membrane. *Tests:* CXR: infiltrates, often in the lower zones. Kidney biopsy: crescentic nephritis. Many die in the first 6 months. *Treatment:* vigorous immunosuppressive treatment and plasmapheresis may be of value (References: *NEJM* 1992 **326** 373 & 1380). [*Ernest Goodpasture, 1919.*]

Guillain–Barré syndrome (Acute post-infective polyneuritis) A variety of infections, often mild (eg CMV, EBV, campylobacter enteritis) are followed after an interval of days or weeks by a mainly motor polyneuropathy. Back pain may be prominent. The paralysis is often progressive and ascending but it may come on rapidly and affect all 4 limbs simultaneously. Unlike other peripheral neuropathies, proximal muscles may be more affected. Similarly trunk, respiratory, and cranial nerves (especially VII) may be affected. Sensory symptoms are common but signs are usually difficult to detect. It is the progressive respiratory involvement that is the most common cause of serious problems. *Tests:* These patients should have their vital capacity measured 4–6 hourly and they should be ventilated sooner rather than later, eg if FVC <1.5 litres; PaO_2 <10kPa; $PaCO_2$ >6kPa. CSF shows normal cell count but protein may be raised as much as 10g/l. *Pathology:* Inflammation, oedema and demyelination of peripheral nerves and roots. *Treatment:* Supportive; consider plasma exchange. *Prognosis:* Most recover in 3–6 months, but have a permanent deficit. Complete paralysis is compatible with complete recovery. [*Georges Guillain & Jean Barré, 1916.*]

Henoch–Schönlein purpura This presents with a purpuric rash (ie purple spots which do not disappear on pressure—signifying intradermal bleeding) often over the buttocks and extensor surfaces. There may be a nephritis (in one-third of patients), joint involvement, or abdominal pain, which may be severe enough for the patient to present as an 'acute abdomen'. The fault lies in the vasculature; the platelets are normal. It often follows an acute respiratory tract infection, and it usually follows a benign course over weeks or months. Complications include massive GI haemorrhage, ileus, and renal failure (rare). [*Johann Schönlein, 1832, and Eduard Henoch, 1865*].

Horner's syndrome There is pupillary constriction (miosis), a sunken eye (enophthalmos), ptosis and ipsilateral loss of sweating (anhydrosis) due to interruption of the sympathetic supply to the face. This may occur in the brainstem (*demyelination; vascular disease*), the cervical cord (*syringomyelia*), the thoracic outlet (*Pancoast's tumour*, p706), or on the sympathetic nerves' journey on the internal carotid artery into the cranium (*surgery, carotid artery aneurysm*), and thence to the orbit. [*Johann Horner, 1869.*]

Dictionary of eponymous syndromes

Huntington's chorea is an autosomal dominant condition (gene on chromosome 4) with full penetrance. Onset is usually in middle age. Thus the child of an affected parent lives under a Damocles' sword, having a 50% chance of becoming affected. Onset is insidious, and the course is relentlessly progressive with: chorea; personality change (eg more irritable) preceding dementia and death. Epilepsy is common. *Pathology:* marked reduction in GABA-nergic and cholinergic neurones in the corpus striatum. *Treatment:* no treatment prevents progression. Counselling and support of the patient and the family are of the most value. For the treatment of chorea, see p40. [*George Huntington, 1872.*]

Jakob–Creutzfeldt disease (JCD, CJD) is a human spongiform encephalopathy (scrapie is a form of the disease found in sheep, and bovine spongiform encephalopathy, BSE, is a related encephalopathy found in cows). The cause is a 'prion'. This is a protein (PrPSc) which, it is believed, is an altered form of a normal protein (PrPc) having the unusual property of causing the normal protein to become transformed to the abnormal prion protein (thus accounting for the infectious nature of the disease). The abnormal protein accumulates leading to neuronal damage and tiny cavities. In the genetically determined form of the disease (Gerstmann–Sträussler syndrome), it is thought that the 'normal' protein is abnormally unstable and readily transforms to the abnormal protein. Spread between species is possible but often only after a lengthened incubation period, eg between cow and sheep, and, as it seems, between cow and man—probably from eating infected offal (scrapie-infected protein was fed to cows in the UK and elsewhere) in the late 1980s before offal was 'banned' (in the UK) from the food chain. Note that not all abattoirs will have been equally scrupulous in removing offal from carcasses. In the apparently BSE-related form of JCD, age of onset is younger (median 27yrs). Other sources of infection: corneal transplantation, hormones extracted from human pituitaries. After an incubation period which may be 20yrs, features develop eg non-specific anxiety/depression, seizures, focal CNS signs, myoclonus and visual problems. Tests: liaise with a neurologist. EEG—periodic spike wave. A diagnostic gel electrophoresis CSF test is available. [*Hans Creutzfeld, 1920, and Alfons Jakob, 1921.*]

Kaposi's sarcoma (KS) This indolent sarcoma is either derived from capillary endothelial cells or from fibrous tissue associated with a new serologically identifiable human γ-herpesvirus (KSHV=HHV-8). It presents as purple papules or plaques on the skin and mucosa (any organ). It metastasizes to lymph nodes. There are 2 types: 1 Endemic (Central Africa), featuring peripheral, slow-growing lesions in older men with rare visceral involvement and having a good response to chemotherapy. 2 KS in HIV +ve patients (particularly in homosexual men), where it is diagnostic of AIDS (p212). Pulmonary Kaposi's sarcoma may present as breathlessness. Lymphatic obstruction predisposes to cellulitis. Other groups affected: Jews and transplant patients. *Diagnosis* is by biopsy. *Treatments:* (eg if extensive disease or for cosmetic reasons) local radiotherapy, interferon α (10–18mU-day), and chemotherapy with doxorubicin, bleomycin, vinblastine (intralesional) and dacarbazine—have all been used with success in open trials. [*Moricz Kaposi, 1887*]

702

Korsakoff's syndrome This is a profound inability to acquire new memories, and it may follow Wernicke's encephalopathy. It is due to thiamine deficiency (eg in alcoholics). The patient may have to relive his grief each time he hears of the news of the death of a family member. He will confabulate to fill in the gaps in his memory. [*Sergei Korsakoff, 1887.*]

Leriche's syndrome Absent femoral pulses, intermittent claudication of the buttock muscles, pale cold legs, and impotence. It is due to aortic occlusive disease (eg a saddle embolism at its bifurcation). [*René Leriche, 1940.*]

Löffler's eosinophilic endocarditis A restrictive cardiomyopathy, in which there is an eosinophilia (eg ≥126 × 10⁹/l). These infiltrate many organs. It may be an early stage of tropical endomyocardial fibrosis. It also overlaps with the idiopathic hypereosinophilic syndrome (HES), but is probably distinct from eosinophilic leukaemia. It presents with increasingly severe heart failure (75%), and mitral regurgitation (49%). **Treatment:** Digoxin and diuretics often only help if combined with suppression of the eosinophilia (eg with prednisolone or hydroxyurea).

Dictionary of eponymous syndromes

Löffler's syndrome (Simple pulmonary eosinophilia) This is an allergic infiltration of the lungs by eosinophils. Allergens include: *Ascaris lumbricoides, Tricinella spiralis, Fasciola hepatica, Strongyloides stercoralis, Ankylostoma duodenale, Toxocara canis*, sulphonamides, hydralazine, nitrofurantoin, chlorpropamide. Often there are no symptoms, and the diagnosis is suggested by an incidental CXR (diffuse fan-shaped shadows), or there may be cough, fever, and an eosinophilia (eg 20%). *Treatment:* eradicate allergens. If idiopathic, steroids are tried.

Lown–Ganong–Levine syndrome Acceleration of the conduction of the cardiac impulse with a normal QRS complex. Atrial impulses may bypass the AV node using a fast-conducting accessory pathway, but rejoin the bundle of His—so producing a short P–R interval and atrial arrhythmias (*but* no delta wave).

McArdle's glycogen storage disease (type V) The cause is myophosphorylase deficiency (as shown on muscle biopsy). Inheritance: autosomal recessive. Stiffness follows exercise. Venous blood from exercised muscle shows low levels of lactate and pyruvate. There may be myoglobinuria. *Treatment:* avoid extreme exercise. Oral glucose and fructose may help.

Mallory–Weiss syndrome This is haematemesis from a tear in the oesophagus, brought on by prolonged vomiting (any cause).

Marchiafava–Bignami syndrome Wine induces degeneration of the corpus callosum, leading to fits, ataxia, tremor, excitement and apathy.

Marchiafava–Micheli syndrome (Paroxysmal nocturnal haemoglobinuria) An intracorpuscular defect causes haemolysis (hence haemoglobinaemia and haemoglobinuria) during the night, in young adults.

Marfan's syndrome An autosomal dominant connective tissue disease, producing arachnodactyly (long, spidery fingers), high-arched palate, armspan > height, lens dislocation and aortic incompetence. Aortic incompetence may occur, particularly during pregnancy. Overall life expectancy is (about) halved, from the cardiovascular risks. Echocardiogram screening may be helpful. Similar morphological appearances occur in homocystinuria. [*Bernard Marfan, 1896.*]

Meigs' syndrome The association of a pleural effusion with a benign ovarian fibroma or thecoma. The effusion is usually right-sided but may be bilateral or left-sided. It is a transudate. Ascites may also be present. The mechanism of the effusion is unknown. [*Joseph Meigs, 1937.*]

Ménétrier's disease There is hyposecretion of gastric acid, gross mucosal hypertrophy into folds up to 4cm high, epigastric pain, GI bleeding, excessive protein loss from the stomach, diminished gastric mobility, oedema, weight loss, diarrhoea, and vomiting. It may be due to a trophic effect of a hormone, or it may be a response to regurgitation of the small bowel contents through the incompetent pylorus. Total gastrectomy may be indicated. Gastric carcinoma may follow. Less severe forms of the disease are known as Schindler's disease. (References: J Simpson 1985 *BMJ* ii 1298)

Meyer–Bet syndrome (Paroxysmal myoglobinuria) A necrotizing disease of muscle which occurs more commonly in men. It is precipitated by exercise in 50% of cases. There are cramps, muscle weakness, chills, pallor, abdominal pain, fever, and shock. First the urine turns pink, and then as more myoglobin is excreted it becomes deep red or brown. Oliguria or anuria may follow. *Diagnosis:* muscle biopsy, serum myoglobin↑.

Mikulicz's syndrome A variant of Sjögren's syndrome in which there is epithelial hyperplasia of the salivary glands, blocking of ducts and symmetrical enlargement. The patient complains of a dry mouth. [*Johann von Mikulicz-Radecki, 1892.*]

Milroy's syndrome (Lymphoedema praecox) An inherited malfunction of the lymphatics causing asymmetric swelling of young girls' legs. [*William Milroy, 1892.*]

704

Marfan's	vs	Homocystinuria
• autosomal dominant		• autosomal recessive
• upwards lens dislocation		• downwards dislocation
• aortic incompetence		• heart rarely affected
• normal mentality		• mental retardation
• scoliosis, flat feet, herniae		• recurrent thromboses
NB: Life expectancy is (about)		• osteoporosis
halved from cardiovascular risks		• +ve urine cyanide-nitroprusside test
		• response to pyridoxine

Münchausen's syndrome The patient gains many hospital admissions through deception, feigning surgical illness, hoping for a laparotomy (laparotimophilia migrans), or bleeding alarmingly (haemorrhagica histrionica) or presenting with curious fits (neurologica diabolica) or false heart attacks (cardiopathia fantastica). (References: R Asher 1972 *Talking Sense*, Pitman.)

Myalgic encephalomyelitis (*Synonyms:* Royal Free disease, epidemic neuromyasthenia, postviral fatigue syndrome, Iceland disease) There is unaccountable fatigue and debility, which may last for many months. Routine tests are all normal (although there may be raised titres to coxsackie B virus) and many patients are sent to psychiatrists. One theory is that there is an abnormal immunological response to any of a number of different organisms. Persisting enteroviral infection has been found in some patients (G Yousef 1988 *Lancet* i 146). There is some evidence of muscle membrane dysfunction (A Hughson 1988 *BMJ* i 1067). Case clustering occurs. *Treatment:* rest (which may shorten the illness) and support (Myalgic Encephalomyelitis Association, PO box 8, Stanford-le-Hope, Essex). See *OHCS* p471.

Nelson's syndrome If bilateral adrenalectomy is performed, feedback inhibition of ACTH is removed. Browning of the skin due to this excess ACTH production is known as Nelson's syndrome.

Ogilvie's syndrome There is functional ('pseudo') GI obstruction caused by malignant retroperitoneal infiltration. (Other causes of pseudo-obstruction include lumbar spine fracture, U&E imbalance.) A water-soluble contrast enema or colonoscopy allows decompression, and excludes mechanical obstruction. Management is conservative. Correct U&E. Caecal exteriorization is rarely needed (References: H Dudley *BMJ* 1987 i 1157).

Ortner's syndrome This is recurrent laryngeal nerve palsy caused by a large left atrium associated with mitral stenosis.

Osler–Weber–Rendu syndrome This is hereditary haemorrhagic telangiectasia. There are punctiform lesions on mucous membranes. It presents with epistaxis or GI bleeding. The natural history is variable.

Paget's disease of the breast This is an intra-epidermal, intraductal carcinoma. Any red, scaly lesion around the nipple should bring to mind thoughts of Paget's disease, and a biopsy is indicated. ► Never diagnose eczema of the nipple without doing such a biopsy first. *Treatment:* mastectomy is usually required. [*Sir James Paget, 1874.*]

Pancoast's syndrome Apical lung carcinoma coupled with ipsilateral Horner's syndrome, caused by invasion of the cervical sympathetic plexus. Also shoulder and arm pain (brachial plexus invasion C8–T2) and hoarse voice/bovine cough (unilateral recurrent laryngeal nerve palsy and vocal cord paralysis). [*Henry Pancoast, 1932.*]

Peutz–Jeghers' syndrome Intestinal polyposis and black freckles on lips, oral mucosa, and other areas. There may be massive GI haemorrhage. Gastric and duodenal polyps may become malignant. Autosomal dominant.

Peyronie's disease There is local fibrosis in the shaft of the penis leading to angulation which makes coitus most inconvenient. Associations: Dupuytren's contracture, premature atherosclerosis. Surgery and prostheses aid penetration. External penile splints ('supercondoms') also help. [*François de la Peyronie, 1743.*]

Pott's syndrome This is tuberculosis of the spine. Rare in the West, this tends to affect young adults giving pain, and stiffness of all back movements. ESR is raised. Abscess formation and spinal cord compression may occur. Disc spaces may be affected in isolation or with vertebral involvement either side (usually anterior margins early). x-rays show narrowing of disc spaces and local vertebral osteoporosis early, with bone destruction leading to wedging of vertebrae later. Whereas in the thoracic spine paraspinal abscesses may be seen on x-ray and kyphosis on examination, with lower thoracic or lumbar involvement abscess formation may be by the psoas muscle in the flank, or in the iliac fossa. The bacillus reaches the spine via the blood. T10–L1 are the most commonly affected vertebrae. *Treatment:* is with antitubercular therapy (p200). [*Sir Percival Pott, 1779.*]

Dictionary of eponymous syndromes

Prinzmetal (variant) angina This is angina occurring at rest, due to coronary artery spasm. ECG shows ST *elevation*. *Treatment*: as for ordinary angina (p278); nifedipine is said to be particularly useful.

Raynaud's syndrome This is intermittent vasospasm of the arterioles of the distal limbs after exposure to cold or emotional stimuli. There is usually colour change (pale→blue→red) and pain in the affected part. It may be idiopathic (Raynaud's disease) or secondary to an underlying cause (Raynaud's phenomenon) eg scleroderma, SLE, RA, thoracic outlet obstruction, trauma, atherosclerosis or cryoglobulins. *Treatment*: keep the limb warm; discourage smoking. Nifedipine 10–20mg/8h PO is helpful for some. Electrically heated mittens are available (Raynaud's Association, 40 Blagden Crescent, Alsagar, Cheshire, ST2 2BG). Sympathectomy may help severe disease. Relapse is common. [*Maurice Raynaud, 1862.*]

Refsum's syndrome This is the association of polyneuritis, nerve deafness, night blindness due to retinitis pigmentosa, cerebellar ataxia, with a high CSF protein. There is a disorder of lipid metabolism leading to excessive phytanic acid in the liver, kidney, fat, and nerves. Plasmalogens are lowered (a kind of phospholipid). Autosomal recessive. Plasma phytanic acid and pipecolic acid are raised. *Treatment*: eat a chlorophyll-free diet; plasmapheresis in severe disease.

Rotor syndrome There is a defect in the excretion of conjugated bilirubin, which produces cholestatic jaundice. Inheritance: autosomal dominant with impaired penetrance.

Sjögren's syndrome This is the association of a connective tissue disease (rheumatoid arthritis in 50%) with keratoconjunctivitis sicca (diminished lacrimation causing dry itchy eyes) or xerostomia (diminished salivation causing dry mouth). Lymphocytes and plasma cells infiltrate secretory glands (also lungs and liver—causing fibrosis). The connective tissue disease is usually severe and rheumatoid factor is always present. Anti-Ro (SSA) and anti-La (SSB) antibodies are variably present. Gland biopsy shows sialadenitis. Involvement of other secretory glands is common causing dyspareunia, dry skin, dysphagia, otitis media, and pulmonary infection. Other features: peripheral neuropathy, renal involvement, hepatosplenomegaly. There is an association with other connective tissue disorders, renal tubular acidosis, adverse drug reactions and lymphoma. *Tests*: conjunctival dryness may be quantified by putting a strip of filter paper under the lower eyelid and measuring the distance along the paper that tears are absorbed (Schirmer's test). *Treatment*: use hypromellose drops (artificial tears) and consider occluding the punctum which drains the tears. [*Henrik Sjögren, 1933.*]

Stevens–Johnson syndrome This is usually a response to a drug (eg sulphonamide, penicillin, some sedatives), viral or other infection (eg orf, *Herpes simplex*), neoplasia or other systemic disease. The patient is systemically very ill with fever, arthralgia, myalgia, and possibly pneumonitis and uveitis. Vesicles develop in the mucosa of the mouth, GU tract, and/or the conjunctivae. The skin develops the typical target lesions of erythema multiforme, often on the palms. They may blister in the centre. The signs may also include polyarthritis and diarrhoea. *Treatment*: calamine lotion for the skin, steroid drops for the eyes, and systemic steroids. Prognosis is good, but the illness may be severe for 10 days (with resolution over 30 days). [*Albert Stevens & Frank Johnson, 1922.*]

Sturge–Weber syndrome The association of a facial port wine stain (haemangioma) with contralateral focal fits. There may also be glaucoma, exophthalmos, strabismus, optic atrophy, and spasticity with a low IQ. The fits are caused by a corresponding capillary haemangioma in the brain. Skull x-ray shows cortical calcification, but angiography is usually normal.

Takayasu's arteritis (The aortic arch syndrome) An idiopathic arteritis affects the first few cm of the innominate, carotid, and subclavian arteries with the adjacent portion of the aorta. These and the renal arteries become partially occluded. The typical patient is a woman 20–40yrs old. *Clinical features*: eyes—amblyopia, blindness, cataracts, atrophy of the iris, optic nerve and retina. CNS—hemiplegia, headache, vertigo, syncope, convulsions. CVS—pulselessness; systolic murmurs above and below the clavicle are common; aortic regurgitation; aortic aneurysms. If the renal arteries are involved, hypertension may develop.

Dictionary of eponymous syndromes

Diagnosis: ESR >40mm/h (if the disease is active); aortography. *Treatment:* prednisolone (starting dose example: 40mg/24h PO); reconstructive surgery and endarterectomies may be tried. *Prognosis:* 10yr survival: ~90%. [*Mikito Takayasu, 1908*].

Tietze's syndrome An idiopathic costochondritis. There is pain in the costal cartilage, often localized. It is enhanced by motion, coughing, or sneezing. The second rib is most commonly affected. Tenderness is a prominent feature. *Treatment:* simple analgesia, eg aspirin. Its importance is that it is a benign cause of what at first seems to be alarming (eg cardiac) chest pain. In protracted illness local steroid injections may be needed. [*Alexander Tietze, 1921.*]

Todd's palsy Focal CNS signs (eg hemiplegia) following a fit. The patient seems to have had a stroke, but recovers within 24 hours. [*Robert Todd, 1856.*]

Vincent's angina A pharyngeal infection with an ulcerative gingivitis caused by *Borrelia vincentii* (a spirochaete) together with fusiform bacilli: Gram −ve, non-sporing bacteria variously classified as *Bacteroides, Fusobacterium,* or *Fusiforms.* The condition responds to penicillin V 250mg/6h PO combined with metronidazole 400mg/8h PO. [*Jean-Hyacinthe Vincent, 1898*].

Von Hippel–Lindau syndrome The association of haemorrhagic cystic lesions of the lateral lobes of the cerebellum with aneurysms and tortuosities of the retinal vessels, leading to subretinal haemorrhages. The disease presents in young adults with headache, dizziness, unilateral ataxia, and blindness. It is associated with renal cysts, renal carcinoma, pancreatic cysts, cystic tumours of the epididymis and phaeochromocytomas.

Von Willebrand's disease This is an autosomal dominant deficiency of factor VIII which behaves like a platelet abnormality. Because of the diminished levels of factor VIII related antigen (VIIIR:Ag), there is diminished platelet adherence to connective tissue. There is therefore post-operative, traumatic and mucosal bleeding (nose bleeds; menorrhagia) but haemarthroses and muscle haematomas are rare. *Tests:* There is a prolonged bleeding time, but normal platelet count and whole blood clotting time. There are low levels of factor VIII clotting activity (VIII:C) and VIIIR:Ag and deficient platelet aggregation with ristocetin. *Treatment:* factor VIII concentrates.

Waterhouse–Friedrichsen's syndrome The association of haemorrhage into the adrenal cortex with a fulminant necrotizing meningococcaemia, which causes a purpuric rash, fever, meningitis, coma, and DIC. There is shock as normal vascular tone requires cortisol to set the level and activity of alpha and beta adrenergic receptors—and aldosterone is needed to maintain extracellular fluid volume.

Sepsis is not the only cause of adrenal haemorrhage: it may follow coagulopathic states, diseases of pregnancy, shock and other stresses.
Treatment: benzylpenicillin 1.2g/4h IV, hydrocortisone 200mg/4h IV, and specific support, for complications such as renal failure. (References: C Homcy 1989 *NEJM* 321 1595)

Weber's syndrome Ipsilateral third nerve palsy with contralateral hemiplegia, due to midbrain infarction after occlusion of the paramedian branches of the basilar artery (which supplies the cerebral peduncles).

Wegener's granulomatosis This is a potentially fatal granulomatous vasculitis. *Diagnostic criteria:* ● Necrotizing granulomas in respiratory tract ● Generalized necrotizing arteritis ● Glomerulonephritis. Any organ *may* be involved—eg nasal ulceration, rhinitis, sinus involvement, otitis media, multiple cranial nerve lesions, lung symptoms and variable shadows on CXR (eg multiple nodules), hypertension and glomerulitis. The patient is also systemically ill. *Eye signs* (seen in 50%): proptosis ± ptosis (orbital granuloma), conjunctivitis, corneal ulcers, episcleritis, scleritis, uveitis, retinitis. *Tests:* ANCA may help diagnostically and in disease monitoring. *Treatment:* high dose steroids may be backed up by cyclophosphamide which has revolutionized prognosis in these patients. Dose example: 3–4 pulses of 500mg cyslophosphamide IV separated by 7–10 days with mesna to reduce cyclophosphamide-induced chemical cystitis.

Wernicke's encephalopathy A thiamine deficiency state (often in alcoholics) with a triad of nystagmus, ophthalmoplegia (external recti commonly) and ataxia. Other eye signs such as ptosis, abnormal pupillary reactions, and altered consciousness may occur. Consider this diagnosis in all those with any of the above signs: it may present with headache, anorexia, vomiting, and confusion. *Tests:* red cell transketolase↓ or plasma pyruvate↑. *Treatment:* urgent thiamine to prevent irreversible Korsakoff's syndrome (p702). Dose examples: thiamine 200–300mg/24h PO; maintenance 25mg/24h PO. If the oral route is quite impossible, consider Pabrinex®, 2–3 pairs of high-potency ampoules/8h IV over 10 mins for ≤2 days, then daily for 5–7 days. An IM (gluteal) preparation is available (dose example: 1 pair/12h IM for up to 7 days). (Have resuscitation facilities to hand as anaphylaxis can occur.) [*Karl Wernicke, 1875.*]

Whipple's disease A cause of GI malabsorption which usually occurs in men >50yrs old. Other features: arthralgia, pigmentation, weight↓, lymphadenopathy. Jejunal biopsy shows intact villi, but the cells of the lamina propria are replaced by macrophages which contain PAS +ve glycoprotein granules. Similar cells are found in nodes, spleen, and liver. *Cause: Tropheryma whippelii. Treatment:* prolonged tetracycline may bring about a remission. [*George Whipple, 1907.*]

Wilson's disease (Hepatolenticular degeneration) This recessively inherited disorder of copper metabolism leads to copper deposition in liver and brain, causing cirrhosis and basal ganglia destruction. *Prevalence:* 29/million. *Presentation:* The patient may be a child or young adult with either tremor, chorea, dysarthria, dysphagia, drooling, weakness, fits, and mental deterioration—or just cirrhosis. Examination may reveal muscle wasting. Diagnosis is by clinical evidence and demonstration of copper in tissues (eg liver, or cornea—seen by expert using slit lamp). Kayser–Fleischer rings are brown pigmentation in periphery of cornea due to copper deposit. They are sometimes seen. Serum caeruloplasmin is usually reduced together with *total* serum copper. *Tests:* Contact your laboratory. High plasma copper (eg >21μmol/l) and low caeruloplasmin (eg <25mg/dl). Urine: copper >100μg/24h. *Treatment* with penicillamine (to aid copper ion elimination) is effective if given early. 500mg/6–8h PO ac for first year then 250mg/6–8h. Monitor FBC and platelets, and test urine for blood and protein frequently. [*Samuel Wilson, 1912*].

Zollinger–Ellison syndrome This is the association of peptic ulcer with a gastrin-secreting pancreatic adenoma (or simple islet cell hyperplasia). Gastrin excites excessive acid production which can produce many ulcers in the duodenum and stomach. Rarely, adenomas are located in the stomach or duodenum. 50–60% are malignant, 10% are multiple and 30% of cases are associated with multiple endocrine adenomatosis (type 1). Incidence: 0.1% of patients with duodenal ulcer disease.

Suspect in those with multiple peptic ulcers resistant to drugs, particularly if there is associated diarrhoea and steatorrhoea or a family history of peptic ulcers (or of islet cell, pituitary or parathyroid adenomas). *Tests:* raised fasting serum gastrin level (>1000pg/ml) with a raised basal gastric acid output of >15mmol/h (the latter test is of less importance).

Treatment: cimetidine (p490) helps to heal the peptic ulcers, although high doses may be needed (eg up to 6g/day to reduce basal gastric acid secretion to <10mmol/h before the next dose is due). Possibly more effective is omeprazole. This is a proton pump inhibitor which binds irreversibly with parietal cell potassium hydrogen ATPase. SE: headaches, rash, diarrhoea. Dose: 20mg/24h–60mg/12h PO (start with 60mg/day and adjust according to response). The hazards of long-term use are unknown.

Neither gastrectomy nor cimetidine alters the prognosis and death is more likely to be due to malignant disease than the effects of hypersecretion of acid. It is therefore worthwhile trying to excise tumours after location by ultrasound, CT scans or angiography—provided there is not already disseminated malignancy (usually in the liver).

Prognosis: 5yr survival with metastases: ~20%. (References: M Langman 1991 *BMJ* ii 482)

Dictionary of eponymous syndromes

18 Emergencies

Relevant pages in other sections: Cardiovascular emergencies: Myocardial infarction (p282); arrhythmias (p286–294); aortic dissection (p118); malignant hypertension (p300).

Respiratory emergencies: Respiratory failure (p352)—from myasthenia gravis and Guillain–Barré (p472 & p700); pneumothorax (p356); superior vena cava obstruction (p764).

Neurological emergencies: Cord compression (p676); subdural and extradural haemorrhage (p440); subarachnoid haemorrhage (p438).

Renal and biochemical emergencies: Acute renal failure (p384–6); hyperkalaemia (p386); hypokalaemia (p640); hypercalcaemia (p644); hypocalcaemia (p642); hypernatraemia (p638); hyponatraemia (p638).

Infectious diseases emergencies: Meningitis (p442–4; for the treatment of TB meningitis, see *OHCS* p260); tetanus (p226); Lassa fever (p188); Waterhouse–Friedrichsen's syndrome (p710).

Gastrointestinal emergencies: GI bleeding (p494); bleeding varices (p496); fulminant hepatic failure and hepatic coma (p504).

Surgical emergencies: The acute abdomen (p112); bowel obstruction (p130); ruptured abdominal aortic aneurysm (p118); femoral artery embolism (p128); acute pancreatitis (p116); torsion of the testis (p132); paediatric surgery (p132); rupture of the spleen (p112).

Oncology emergencies: (p764).

Others: Stridor: (p58)—after thyroidectomy (p94); giant cell arteritis (p675); hypothermia (p76); DIC (p598).

An approach to the very ill patient

▶ *Don't go so fast: we're in a hurry!* (Talleyrand to his coachman.)

What makes the topics listed opposite *emergencies* is that prompt action saves lives. Even within emergencies there are gradations, so that for most, there is time to assess your patient in some detail. But other emergencies demand the promptest action, such that when summoned from sleep, there is no time even to dress: for example, cardiac arrest, acute stridor, shock and tension pneumothorax. So when one is the doctor who is to be called first should such emergencies arise, not only have you to choose your night-wear carefully, but you need to know *by instinct* what to do when you arrive, bleary eyed, at the bedside. This means that no opportunity is to be lost in watching experts deal with emergencies. Our aim here is to allow the reader to prepare his mind so that he can make the most of such observations.

To the inexperienced, the following treatments may look easy to execute; but this is rarely so as emergencies do not present themselves in the tidy compartments listed opposite. The problem is that a life is slipping through your fingers, and you must find out what the matter is to save the life—but the patient is too ill to survive a detailed history and examination. The trick is to buy time by supporting vital functions, as follows.

● Is the patient *breathing*, and is a *pulse* present? ● If not, remove any respiratory obstruction and start cardiorespiratory resuscitation. ● If there is a pulse and some respiratory effort, decide if these are effective: *is the tongue blue?* ● If so, give O_2 and assisted ventilation as needed). ● Now ask: is the systolic BP recordable? ● If not (or <80mmHg), and if myocardial infarction (MI) is unlikely to be the cause (eg on surgical wards) put up an IV line and infuse a plasma-like substance (eg Haemaccel®) fast, while you look for the cause as described on p720. But if an MI is likely (eg after a collapse in the street) avoid fast IV fluids: the danger is precipitating pulmonary oedema. ● If the patient is very breathless always consider the diagnosis of a tension pneumothorax. If this cannot be excluded quickly, put a relieving needle (p753) into the side of the chest in which the breath sounds are quieter, and the percussion note more resonant.

Next decide if wheeze is present. If so, treat as asthma, unless there is chest pain, a big heart (a displaced apex beat, p26), oedema, and an 'ischaemic' ECG, when the cause is more likely to be heart failure (confirm by CXR). If in doubt, treat both.

In coma, the way to buy time is to place the patient semi-prone (the coma position) and to treat those things which are treatable and most likely to be rapidly fatal. So give 50ml of 50% dextrose IV to counter hypoglycaemia (unless you know the blood glucose is >3mmol/l) and, if the pupils are small, give 1.2mg naloxone IV to reverse opiate narcosis. If the patient is toxic and could have septicaemia or meningitis, treat accordingly (p182 & p442–4).

Once ventilation and circulation are adequate you have bought time to start assessing your patient in the traditional way.

715

▶▶ Coma

Ask an experienced nurse to help. Nurse the patient semi-prone.

Quick examination *Airway:* Clear (ventilate if cyanosed). *Pulse and BP* (is shock causing the coma—p720)? Assess: *pupils, breath* (alcohol, ketones, hepatic fetor), *coma level* (p718).

Set up an IV line. Keep open with 0.2ml of heparin 1000U/ml). Blood for glucose (ward test & lab), urgent FBC, U&E. Give 50ml of 50% **glucose** IV (unless certain that blood glucose is normal). If pupils are constricted, give **naloxone** 0.4–1.2mg IV stat, to counter opiate poisoning.

Quick history from family or ambulance men: Onset of coma? How found? If injured suspect cervical spinal injury. Do not bend spine (OHCS p726). Is he a known diabetic, asthmatic, opiate or alcohol abuser? Any travel? Is there head injury? Any suicide note? Any known diseases?

Examination ● Any shock, anaemia, jaundice, cyanosis, myxoedema?
● Pupil size/reaction. Fundi: papilloedema, hypertensive retinopathy.
● Are there venepuncture marks and small pupils (narcotic abuse)?
● Any signs of meningism (p420)? ● Is there blood or CSF in ears or nose?
● Feel the skull for fracture 'step' deformities. Is there bruising?
● Examine for CNS asymmetry (tone, spontaneous movements, reflexes).
● Examine abdomen (ascites, peritonism, aneurysm). PR: any melaena?
● Low-reading rectal thermometer (hypothermia). ● Any wheeze?
● Test the brainstem by the gag reflex, pupil responses, corneal reflex, and, if spine trauma excluded, 'doll's eye' movements (if the eyes stay looking at the same point as you swivel the head, the brainstem works).
● Dysconjugate gaze: VI palsy—medial deviation; III palsy—lateral.
● If conjugate deviation suspect a frontal lesion on the side of the gaze.
● Pontine lesions: all eye movement may vanish; pupils may be pin-point.
● The oculovestibular test powerfully stimulates the midbrain, and may be used even in cervical spine trauma. If the ear drum is intact, and there is no wax, instill 20ml of ice-cold water. Eliciting conjugate eye movement toward the stimulated ear shows that the midbrain works.

Further management Note the *absence* of signs, as well as their presence—eg the *absence* of pin-point pupils in a known heroin addict, or diabetic patient whose breath does *not* smell of acetone. Monitor pupils, BP and TPR/15mins. Cardiac monitor. Do CXR, ECG, blood gases (is O_2 indicated, p352?). Spare no effort in getting a detailed history.

Taking stock The diagnosis may be clear, eg hyperglycaemia, alcohol excess (do blood alcohol levels), drug poisoning, hypoxia, uraemia, pneumonia, hypertensive or hepatic encephalopathy (p504). Are there focal CNS signs?—*supratentorial* (focal motor signs) suggesting stroke, tumour or abscess; *infratentorial* (pupil responses abnormal) suggesting brainstem stroke or ICP†. If pupils symmetrical and there are no focal signs the cause is probably metabolic. In all undiagnosed comas consider:
● CT scan: if skull fracture with fits or any focal neurological signs, persistent coma or confusion, deteriorating consciousness, compound depressed fractures, skull base fractures, penetrating injuries.
● LP (unless focal signs) for meningitis or subarachnoid haemorrhage.
● Blood cultures. ● Blood for drug analysis and blood alcohol level.
● Is shock the cause of the coma (p720)?
● Tropical coma: malaria (thick films), typhoid, rabies, trypanosomiasis.
● *Herpes simplex* encephalitis: if coma of gradual onset, with no obvious cause and particularly if there is fever, fits or temporal lobe signs, consider giving acyclovir 10mg/kg/8h by IVI over 1h (p955 & p203).

Coma
[*OTM* 31.46]

The Glasgow coma scale (GCS)[1]

This gives a reliable, objective way of recording the conscious state of a person. It can be used by medical and nursing staff for initial and continuing assessment. It has value in predicting ultimate outcome. 3 types of response are assessed:

Best motor response This has 6 grades:

6 Carrying out request ('obeying command'): The patient does simple things you ask (beware of accepting a grasp reflex in this category).

5 Localizing response to pain: Put pressure on the patient's finger-nail bed with a pencil then try supraorbital and sternal pressure: purposeful movements towards changing painful stimuli is a 'localizing' response.

4 Withdraws to pain: Pulls limb away from painful stimulus.

3 Flexor response to pain: Pressure on the nail bed causes abnormal flexion of limbs—decorticate posture.

2 Extensor posturing to pain: The stimulus causes limb extension (adduction, internal rotation of shoulder, pronation of forearm)—decerebrate posture.

1 No response to pain.
NB: record the best response of any limb.

Best verbal response This has 5 grades.
5 Oriented: The patient knows who he is, where he is and why, the year, season, and month.

4 Confused conversation: The patient responds to questions in a conversational manner but there is some disorientation and confusion.

3 Inappropriate speech: Random or exclamatory articulated speech, but no conversational exchange.

2 Incomprehensible speech: Moaning but no words.

1 None.
NB: record level of best speech.

Eye opening This has 4 grades.
4 Spontaneous eye opening.

3 Eye opening in response to speech: Any speech, or shout, not necessarily request to open eyes.

2 Eye opening to response to pain: Pain to limbs as above.

1 No eye opening.

An overall score is made by summing the score in the 3 areas assessed. Eg: no response to pain+no verbalization+no eye opening=3. Severe injury, GCS ≤8; moderate injury, GCS 9–12; minor injury GCS 13–15.

Note: an abbreviated coma scale, AVPU, is sometimes used in the initial assessment ('primary survey') of the critically ill
- A = **a**lert
- V = responds to **v**ocal stimuli
- P = responds to **p**ain
- U = **u**nresponsive

Some centres score GCS out of 14, not 15, omitting 'withdrawal to pain'.

1 G Teasdale 1974 *Lancet* **ii** 81

The Glasgow coma scale (GCS)
[*OTM* 21.186]

Shock

Essence If the blood pressure falls so that vital organs are not perfused adequately, the patient is shocked. There are signs of end-organ under-perfusion (pallor, cold peripheries, faints, restlessness, capillary refill >2sec, oliguria) as well as tachycardia (unless on β-blocker, or in spinal shock—*OHCS* p728) and hypotension (eg systolic BP <90mmHg). In the young and fit, the systolic BP may remain normal, although the *pulse pressure* will narrow, with up to 30% blood volume depletion—making diagnosis difficult. The above end-organ signs are more reliable markers than BP and central venous pressure.[1]

Quickly examine BP; carotid pulse; pallor; cyanosis; jaundice; heart and chest sounds; aortic aneurysm and melaena (do PR).

Causes
Cardiogenic: See p722.

Anaphylaxis: Hypersensitivity, eg to penicillin or bee sting. Signs: wheeze, cyanosis, soft tissue swelling (eg eyelids, lips). Specific remedy: 1ml of 1:1000 adrenaline IM (which may need repeating every 10mins, until improvement), chlorpheniramine 10mg IV, hydrocortisone 100mg IV, ± IVI (0.9% saline). Remove the cause.

Endocrine failure: Addison's disease; hypothyroidism (p740).

Endotoxins (bacterial lipopolysaccharides): Gram −ve (or +ve) septicaemic shock from vasodilatation may be sudden and devastating, with coma but no signs of infection (fever, WCC↑). Patient will be hypotensive, but warm peripheries. Specific remedy (if no clue to source, p182): IV cefuroxime 1.5g/8h (after blood culture)—or gentamicin 1.0–1.5mg/kg/8h slowly IV (do levels; reduce in renal failure, see p796–7) *plus* an antipseudomonal penicillin (eg ticarcillin 5g/8h IVI). Give a colloid such as Haemaccel® by IVI. Consider dopamine in a 'renal' dose (2–4µg/kg/min IVI, p722). Human monoclonal IgM antibody (HA-1A, Centoxin® 100mg IVI) can neutralize many endotoxins—but is expensive (£2200/dose), and might only have a marginal effect on survival,[2] and needs further evaluation. This is an example of immunotherapy.

Bleeding: Eg trauma, ruptured aortic aneurysm, ectopic pregnancy. Remedy: IVI until crossmatched blood arrives. Use group ORh −ve blood in severe bleeding (p102). Treat the underlying cause.

Fluid loss: Vomiting (eg GI obstruction), diarrhoea (eg cholera), burns, pools of sequestered (unavailable) fluids ('third spacing', eg in pancreatitis). Remedy: IVI 0.9% saline until BP rises (p100).

The use of hypotensive agents: Anaesthetics, antihypertensives.

Note Except for the first 3 causes given above, the immediate need is for fluid to be given IV quickly (see guidelines, p721).

---

1 G Ramsay 1988 *BMJ* i 1422   **2** P Richard, S Warren 1992 *NEJM* **326** 1151 & 1153

▶▶ Resuscitation in hypovolaemic shock

- Use the largest veins with which you are familiar, i.e. antecubital fossa—otherwise do cut down (p752) or lastly central venous cannulation (p754). Rate of flow is inversely proportional to length of cannula—and cvp lines are long.
- Use 2 ×large-bore cannulae (14G or 16G): it takes 20mins to infuse 1 litre through 18G Venflon®, but only 3mins through 14G!
- Raise drip stand and squeeze bag or use pressure device to increase infusion rate.
- Estimate and anticipate the blood loss from fractures:

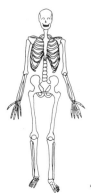

ribs	= 150 ml
pelvis	= 2000 ml
femur (shaft)	= 1500 ml
tibia	= 650 ml

DOUBLE LOSSES IF COMPOUND FRACTURE

- Crystalloid (eg Hartmann's solution) has intravascular $t_{1/2}$ of only 30–60mins and must be given in 3 times the volume to that lost; colloid (eg Haemaccel®) lasts several hours and is replaced 1:1 with blood lost. Consider blood if haemorrhage and over 15% volume lost (adult blood volume = 70ml/kg; child = 80–90ml/kg)—crossmatched blood is ideal, but time-consuming; group compatible is a compromise, only takes 10mins to prepare and is preferable to ORh−ve.
- Fluid should be given rapidly and slowed only when pulse falls, BP rises and urine flows (>30ml/h adult and >1ml/kg/h child). Catheterize the bladder. Consider cvp line to monitor progress.
- In children give initial bolus of 20ml/kg Hartmann's repeated twice depending on response; give further fluid in trauma as blood in 10ml/kg boluses. If IV access difficult use intraosseous infusion just below and medial to tibial tuberosity (use specific needles); any fluid and all common drugs can be given via this route. See OHCS p310 for further details.
- Do not forget to give oxygen (at 12 litres/min via tight-fitting face mask with reservoir in all trauma, unless strong contraindication); to reduce peripheral blood loss by pressure and elevation; and to splint fractures (traction very important in reducing blood loss in fractured femur).
- Monitor pulse, BP and ECG continuously. Remain at bedside.
- ▶ Ensure that someone takes time to talk to the relatives and explain that the patient is seriously ill, and may die.

721

▶▶ Cardiogenic shock

This has a 90% mortality. ▶ Ask a senior physician's help both in formulating an *exact* diagnosis and in guiding treatment.

Cardiogenic shock is shock (p720) caused primarily by the failure of the heart to maintain the circulation. It may occur suddenly, or after progressively worsening heart failure.

Causes *Rapidly reversible:* Arrhythmias, cardiac tamponade, tension pneumothorax. Others: Myocardial infarction; myocarditis; myocardial depression (drugs, hypoxia, acidosis, sepsis); valve destruction (endocarditis), pulmonary embolus, aortic dissection.

Management
- In Coronary Care Unit if available.
- Lie patient in most comfortable position.
- Give diamorphine 5mg IV for pain and anxiety (with prochlorperazine 12.5mg IM).
- Monitor: ECG; urine output by catheter; blood gases; U&E; CVP. Do a 12-lead ECG every hour until the diagnosis is made.
- Consider monitoring: pulmonary wedge pressure (Swan–Ganz catheter); arterial pressure; cardiac output.
- Correct arrhythmias (see p286–94) and U&E imbalance.
- Give O_2 (guided by blood gases). Correct acid–base balance.
- Give positive inotropic drugs: eg dobutamine 2.5–10μg/kg/min IVI, adjusted to keep systolic BP >80mmHg.
- Promote renal vasodilatation with low dose dopamine: 2–4μg/kg/min IVI.
- If pulmonary wedge pressure <15mmHg give HPPF (human plasma protein fraction) 100ml every 15mins IV. Aim at pressure 15–20mmHg.
- Consider vasodilators, eg sodium nitroprusside if filling pressure is adequate.
- If the cause is myocardial infarction, consider streptokinase, eg 1.5 million units in 0.9% saline by IVI over 30mins (over 50mins if body weight <60kg), starting as soon as possible after the diagnosis is made.[1] Start heparin (eg 40,000u/day IVI). CI: active bleeding; stroke; pre- or post-operatively.
- Consider surgery for: acute VSD, mitral or aortic incompetence.
- Consider intra-aortic balloon pump if you expect the underlying condition to improve, or you need time awaiting surgery.

Cardiac tamponade *Essence:* Pericardial fluid collects rapidly, increasing intrapericardial pressure and reducing cardiac filling.

Signs: Muffled heart sounds + hypotension + distended neck veins = Beck's triad; JVP↑ on inspiration (Kussmaul's sign); pulsus paradoxus (pulse fades on inspiration). Echocardiography is diagnostic. CXR: globular heart; left heart border convex or straight; right cardiophrenic angle <90°. ECG: voltage↓; changing axis.

Causes: Neoplastic or tuberculous effusions, trauma, rupture of the heart or great vessels after surgery or placing of a pacemaker, bleeding (eg over-anticoagulated).

Management: ▶▶ Enlist expert help for prompt pericardiocentesis (drainage of the effusion). Relief is swift. If there is reaccumulation, pericardiectomy may be needed.

1 H White 1987 *NEJM* **317** 850

Needle pericardiocentesis

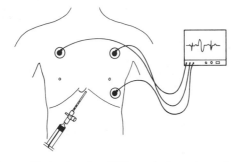

- Equipment: 20ml syringe, long 18G cannula, 3-way tap, ECG monitor, skin cleanser.
- If time allows use aseptic technique. Ensure have IV access and full resuscitation equipment before starting.
- Introduce needle at 45° to skin just below and to left of xiphisternum, aiming for tip of left scapula. Aspirate continuously and watch ECG. Frequent ventricular ectopics or an injury pattern (ST segment↓) on ECG imply myocardium breached—withdraw slightly.
- Evacuate pericardial contents through syringe and 3-way tap. Removal of only a small amount of fluid (eg 20ml) can produce marked clinical improvement.
- Cannula may be left *in situ* temporarily to allow repeated aspiration if fluid reaccumulates.

Complications: laceration of ventricle or coronary artery (± subsequent haemopericardium); aspiration of ventricular blood; arrhythmias (ventricular fibrillation); pneumothorax; puncture of aorta, oesophagus (± mediastinitis) or peritoneum (± peritonitis).

723

▶▶ Cardiorespiratory arrest

Causes MI, PE, trauma, electrocution, drugs (adrenaline, digoxin), hypotension, hypoxia, hypercapnia, hypothermia, U&E imbalance.

Action Confirm diagnosis (unconscious, apnoeic, absent carotid pulse). Note what time it is. Begin cardiopulmonary resuscitation (CPR):

- Thump the chest. This may terminate ventricular arrhythmias. Recheck carotid pulses. If still absent, proceed as follows.
- Enlist help urgently (eg via special 'arrest' phone number).
- Clear the airway. Extend the neck and grasp the nose. Give 2 breaths of mouth-to-mouth to inflate lungs.
- Now give 15 chest compressions centred over the lower $\frac{1}{3}$ of the sternum. Use the heel of your hand, and the weight of your trunk transmitted via unbent elbows. Rate: 80/min.
- Only stop compressions to give mouth-to-mouth (ratio 15:2).
- Once help arrives, carry on chest compressions except during intubation and defibrillation. During these times count aloud each second as it passes. Never count to >10 without restarting compressions. If intubation fails, bag & mask ventilate and restart compressions. Allow anaesthetist another go after 15 compressions.
- Defibrillate as soon as possible (200–360 joules). This takes precedence over all else because it is rapid, and has a good chance of terminating VF—the commonest fatal arrhythmia, and the most amenable to correction. Use the 'QUICK READ' function with defibrillator paddles.
- Ask someone to take over respirations: 100% O_2 via *Ambu-bag* + *reservoir*.
- Ask someone to put up a CVP line or IV line (not in a leg). If only a peripheral (arm) line is used, elevate and massage the arm after each drug, to aid its journey to the heart.
- Place ECG leads—more convenient than reading through paddles. What is the rhythm? See opposite for actions determined by the ECG. (Note: electromechanical dissociation means the rhythm is normal, but that there is no output.) Wait 1min after each drug before going to the next stage.
- Do not stop CPR (eg getting distracted, gazing at monitors).
- Ask one person to draw up emergency drugs: 10ml 1:10,000 adrenaline; atropine 3mg IV—3 × 1mg IMS (International Medical Systems) syringes; lignocaine 100mg IV; 50ml 8.4% sodium bicarbonate only after >10mins of resuscitation.

Other points One person should be in charge. He or she feels for pulses (eg after each defibrillation), uses defibrillator and diagnoses rhythms—but does nothing else other than direct others. It is vital to have one person to take an overview, and make all the decisions.

Action • If there is any question of hypoxaemia causing the arrest (eg in asthma), the top priority is to give 100% O_2 via an ET tube.
- Use all personnel who arrive (up to 5). Do not just shout 'Adrenaline!': point to the person who is to do each task.
- Ask a nurse to find the patient's notes and the patient's usual house physician. Do not try to resuscitate the terminally ill, or the elderly who are known to be against resuscitation.
- Unless either a very young child, hypothermic or poisoning may be involved, give up after 20 mins or so, if no response.
- If the patient recovers, place semi-prone in the recovery position. Monitor vital signs.
- Phone the next of kin. The nurses will be grateful for this.

Further reading: P Baskett 1992 *Br J Anaesthesia* **69** 182–93

Cardiorespiratory arrest protocols

Ventricular fibrillation:
- Precordial thump
- Defibrillate 200J; check pulse; 15 chest compressions while recharge; see monitor
- Defibrillate 200J; check pulse; 15 chest compressions while charge to 360J; see monitor
- Defibrillate 360J; check pulse; see monitor
- Intubate the trachea
- Get IV access

♥
- Adrenaline 1mg IV
- 10 CPR sequences of 5:1 compression:ventilation
- Defibrillate 360J
- Defibrillate 360J
- Defibrillate 360J
- Go back to ♥ and complete 3 loops
- After 3 loops consider:
  - Bicarbonate
  - Anti-arrhythmics (lignocaine;bretylium; procainamide)
  - Change paddles to antero-posterior and give 360J
  - Change defibrillator

Asystole:
- Precordial thump
- Check leads
- Check gain:
- VF *cannot* be excluded:
  200J
  ↓
  200J
  ↓
  360J
- If VF excluded: intubate and give 100% O_2
- IV access

♥♥
- Adrenaline 1mg IV
- 10 CPR sequences of 5:1 compression: ventilation
- Atropine 3mg IV (once only)
- Electrical activity? YES = pace NO = return to ♥♥ and complete 3 loops
- After 3 loops consider adrenaline 5mg IV

Electro-mechanical dissociation:
- Exclude: tamponade, tension pneumothorax, hypovol-aemia, massive PE, drug overdose, hypothermia
- Intubate; give 100% O_2
- Get IV access

♥♥♥
- Adrenaline 1mg IV
- 10 CPR sequences 5:1 compression:ventilation
- Return to ♥♥♥ and continue loop
- Consider:
  - Pressor agents
  - Calcium (if Ca²⁺↓; K⁺↑; overdose of calcium antagonists)
  - Bicarbonate
  - Adrenaline 5mg IV
- Doses:
  Adrenaline 1mg (10ml 1:10,000)
  Bicarbonate 1mmol/kg (1ml 8.4% = 1mmol; 50ml is standard dose)
  Bretylium tosylate 5mg/kg; (if repeat: 10mg/kg)
  Lignocaine 100mg IV
  Procainamide 100mg IV

▶ If IV access fails, double or treble the dose and give down the endotracheal tube (adrenaline or atropine)

▶ The interval between shocks (3) and (4) should not be >2 minutes. *Do not* interrupt CPR for more than 10sec, except to defibrillate.

Adrenaline is the drug of first choice: it increases myocardial and cerebral blood flow during cardiac compressions—so give it every 2–3 minutes during prolonged resuscitation.

Adequate ventilation is preferred to bicarbonate in relieving acidosis: without ventilation bicarbonate may *worsen* intracellular acidosis (extracellular CO_2 rapidly moves back into the cell in response to the high extracellular HCO_3^{3-}), and in excess can itself produce intractable dysrhythmias. Use only in prolonged resuscitation.

Intracardiac injections are not recommended. They rarely reach the intended ventricle and may cause long interruptions in chest compressions (as well as having numerous complications—pneumothorax; myocardial/coronary artery laceration; dysrhythmias with intramyocardial adrenaline).

Pacing is only appropriate when there is some evidence of electrical activity (eg P-waves). Use transvenous ventricular pacing wire (or transthoracic pacing, but less effective). While preparing to pace in complete heart block use fist pacing ± atropine (0.6mg/5mins to keep pulse above 60 bpm, max. 2mg) ± isoprenaline infusion (start at 2µg/min).

Post-resuscitation care: Involve a senior doctor in the decisions. Transfer to intensive care for monitoring (perform 12-lead ECG; arterial blood gases; U&E; CXR; urine output) and treatment (give 100% O_2 and consider elective ventilation; lignocaine infusion (p292) after VF or VT).

Principal sources: T Evans 1990 ABC of Resuscitation 2 ed, Br Med Assoc; European and UK Resuscitation Council Guidelines, 1992 Resuscitation **24** 1992 103

▶▶ Severe pulmonary oedema

History (if possible) Note: drugs; precipitating factors (symptoms of myocardial infarct, infection); pre-existing lung disease; hypertension; previous heart disease.

Examination Signs of pulmonary oedema: pink frothy sputum; pulsus alternans, fine crackles on auscultation of chest; triple/gallop rhythm. (NB: wheezes can be a feature of heart failure as well as of asthma.)

Baseline tests and markers of progress: BP; heart rate; cyanosis; respiratory rate; JVP; peripheral oedema; enlarged liver.

Investigations CXR (signs of pulmonary oedema); ECG (look for signs of myocardial infarct); blood gases; U&E; 'cardiac' enzymes.

Management If severe begin treatment before investigations:
● Sit the patient up.
● Give O_2 by face mask: 100% if no pre-existing lung disease.
● Insert an IV cannula.
● Drugs: *Frusemide* 40–80mg IV (slowly). *Diamorphine* 5mg IV slowly, unless liver failure or chronic obstructive airways disease.
  Give GTN spray 2 puffs SL or 2 × 0.3mg tablets SL.
  If there is fast AF, give *digoxin* 0.75–1.25mg PO (or 0.5mg IV, slowly, if very urgent). Treat other arrhythmias (p286–94).
● Complete investigations, examination, and history.
● Consider isosorbide dinitrate 2–7mg/h IVI if systolic BP >110mmHg or dobutamine 2.5–10μg/kg/min if BP <100mmHg.
NB: if the BP is >180mmHg treat, for hypertensive LVF (p302).

Continuing management: Frequent check of BP and pulse rate, heart sounds; restrict fluids; measure urine output; further IV frusemide may be given; U&E and ECG daily initially.

If improving Change to oral frusemide; sequential CXR; daily weight; 6-hourly BP and pulse.

If the patient's condition is worsening
If blood pressure falls, the danger is *cardiogenic shock* (p722).
▶▶ Consider venesection to remove 500ml blood quickly. Consider ventilation. Obtain further advice.

It is important to examine the patient repeatedly, and perform serial ECGs. Are there any new murmurs? For example, the patient who has severe pulmonary oedema following a myocardial infarction may go on to develop a VSD or mitral regurgitation.

Differential diagnosis Bronchospasm; pneumonia. It is often difficult to make a definite diagnosis and all 3 conditions may coexist, especially in the elderly. Do not hesitate to treat all 3 simultaneously (eg with salbutamol nebulizer, frusemide IV, diamorphine, ampicillin—p176).

Severe pulmonary oedema
[*OTM* 13.334; 14.5]

▶▶ Massive pulmonary embolism

▶ Always suspect pulmonary embolism (PE) in sudden collapse 1–2 weeks after surgery.

Mechanism Venous thrombi, usually from DVT, pass into the pulmonary circulation and block blood flow to lungs. The source is often occult.

Prevention Following surgery, mobilize early; consider heparin prophylaxis 5000U/8h SC. Avoid contraceptive pill if at risk eg major or orthopaedic surgery.

Clinical features Classically, it presents 10 days post-op, with collapse and sudden breathlessness while straining at stool—but PE may occur after any period of immobility, or with no predisposing factors. Other features: pleuritic pain, dyspnoea, shock, tachypnoea, cyanosis, haemoptysis, gallop rhythm, JVP↑, pleural rub, right ventricular rub, loud P_2. Note: multiple small emboli may present less dramatically with pleuritic pain, haemoptysis, and gradually increasing breathlessness.

Signs of shock—tachycardia; thready pulse; low BP.

Look for a source of emboli—especially DVT (is a leg swollen?)—but DVT signs may occur only *after* the embolism.

Tests ECG (commonly normal or sinus tachycardia); may be deep S-waves in I, Q-waves in III, inverted T-waves in III ('SɪQɪɪɪTɪɪɪ'), strain pattern V1–3 (p266), right axis deviation, RBBB, AF.

CXR—often normal; decreased vascular markings.
Blood gases (hyperventilation and impaired gas exchange, p328).
Lung ventilation–perfusion (V/Q) scan. Look for ventilation–perfusion mismatch. Pulmonary angiography.

Diagnosis Usually clinical. If good story and signs, make the diagnosis. Start IVI heparin (see below) prior to confirmatory radiology (ventilation–perfusion scans, p788, or pulmonary angiography).

Management If cardiorespiratory arrest or critically ill consider immediate surgery. Otherwise:
- Give up to 100% O_2.
- Morphine 10mg IV if the patient is in pain or very distressed.
- IV cannula: give heparin 10,000U IV over 5min, then 1000–2000U/h IVI (15–25U/kg/h, or 250U/kg/12h SC) as guided by APPT (p 596).
- Consider streptokinase (p282).
- Begin warfarin: eg 10mg/24h PO (p596). Continue for 3 months.
- Try to prevent further emboli with antiembolic stockings.

Prognosis Good if the patient survives the acute episode.

If no obvious cause is found, is a predisposing disease present (eg polycythaemia, malignancy, SLE, familial clotting disorders)?

Massive pulmonary embolism
[*OTM* 13.355]

▶▶ Acute severe asthma[1]

▶ An atmosphere of calm helps cure the patient.

History Normal and recent treatment; previous acute episodes.

Signs of severe asthma Wheeze prevents finishing sentences in 1 breath; >25 breaths/min; pulse ≥110; peak flow <50% of predicted; cyanosis; pulsus paradoxus >10mmHg. *Life-threatening signs:* ● Peak flow <33% of predicted ● Silent chest ● Coma ● BP↓ ● Pulse↓ ● PaO_2 <8kPa ● $PaCO_2$↑ ● pH↓.

Differential diagnosis Respiratory obstruction from inhaled foreign body, swollen epiglottis, mediastinal (p326) and neck (p148) lumps. These usually cause inspiratory stridor, rather than expiratory wheeze.

Tests CXR; PEFR (but may be too ill); blood gases; pulse oximetry.

Immediate management Start treating before investigations. Warn ITU.
● Sit up (only the comatose are found lying flat: most will be sitting up, supporting their thorax on outstretched arms—to aid respiration).
● Give oxygen in high dose, eg 60%.
● Salbutamol: 5mg nebulized at 6–8 litres/min in O_2 (air if COAD and $PaCO_2$↑). Maximum dose: 10mg if side-effects allow—tachycardia, arrhythmias, tremor, K↓—or 500µg/4h SC.
● Hydrocortisone 200mg IV and prednisolone 30mg/24h PO.

If life-threatening features are present:
● Add ipratropium 0.5mg to the nebulized salbutamol.
● Aminophylline IV—see opposite (dose example, if not already on a theophylline: 250mg over 20mins). Keep finger on pulse during injection. Alternative drug: salbutamol 250µg IV over 10mins.

Subsequent management if improving ● 40–60% O_2. ● Prednisolone 30–60mg/24h PO. ● Nebulized salbutamol every 4h. Monitor peak flow.

If patient not improving after ~20mins Continue 60% O_2 ± NIPPV (p354).
● Nebulized salbutamol up to every 15–30mins. ● Ipratropium 0.5mg every 6h.
● Aminophylline infusion (p731, opposite)—do levels if IVI lasts >24h. Alternative: salbutamol IVI, eg 3–20µg/min (start with 5µg/min).
● Hydrocortisone 200mg/6h IV.
● IVI: dextrose-saline 4 litres/24h to maintain urine output.
● Consider antibiotics (eg erythromycin 500mg/8h PO) and physiotherapy.
● Take to ITU (accompanied by a doctor prepared to intubate the trachea) if there is drowsiness, confusion, coma, respiratory arrest. Mechanical ventilation may be needed if PaO_2 <8.0kPa; $PaCO_2$ >6.5kPa; pH <7.3; systolic BP <90mmHg). $PaCO_2$ in normal range is ominous.
● Consider ventilation if patient exhausted or had cardiac arrest.

Once patient improving ● Stop hydrocortisone 6h after first PO dose.
● Wean down and stop aminophylline over 12–24h. Reduce nebulized salbutamol. Switch to inhaler. PEFR every 4h: is there any deterioration on reduced treatment? Beware early morning dips in PEFR.

Before discharge Check inhaler technique, ensure peak flow is >75% of predicted and diurnal variation <25%. Give written instructions on how to tail off oral steroids. Give a peak flow meter, and agree a written self-management plan. Follow up: GP in 1 week + consultant within 4 weeks.[1]

1 BTS Guidelines *BMJ* 1993 i 779 & *Thorax* 1993 **48** suppl. 1–24

IV aminophylline: dosage and therapeutic monitoring

Guidelines for continuous aminophylline IVI (mg/kg/h)—after a loading dose* of 5mg/kg IV over 30mins. (For dose of theophylline, multiply by 0.85.)

Age	mg/kg/h	Type of adult	mg/kg/h
Neonates with apnoea (see OHCS p232)	0.15	Adults who smoke	0.70
		Healthy, non-smoking adults	0.47
<6 months	0.47	Adults with heart failure	0.23
6–11 months	0.82	Adults with liver failure	0.23
1–9 years	0.94	Adults taking cimetidine,	
Children >9yrs	0.70	erythromycin, propranolol, ciprofloxacin, and contraceptive steroids	~0.23**

The above is one of the more conservative of the available régimes.

* If theophyllines have been taken in the last 24h, avoid the loading dose—but if the plasma theophylline concentration is known, and is subtherapeutic, an additional loading dose may be given. Expect an extra dose of 1mg/kg to increase the plasma concentration by 1.7–3.5µg/ml.

** These drugs increase the half-life of aminophylline. Drugs which shorten the half-life: phenytoin, carbamazepine, barbiturates, rifampicin. Adjust the dose according to plasma concentrations.

▶ Aim for a plasma concentration of 10–20µg/ml (55–110µmol/l).[1,2] Serious toxicity (hypotension, arrhythmias, cardiac arrest) can occur at concentrations ≥25µg/ml. Measure plasma K^+: theophyllines may cause K^+↓.

731

1 J Aronson 1992 BMJ ii 1355   2 D Reynolds 1993 BMJ i 457

Head injury

Emergency care Clear airway; ventilate if necessary; protect cervical spine. ▶ If the pupils are unequal, diagnose rising intracranial pressure (ICP), eg from extradural haemorrhage and summon urgent neurosurgical help. Burr holes (on the side of the dilated pupil) may be needed. Other signs of ↑ICP: deepening coma (or a lucid interval, then relapse), a rising BP and a slowing pulse (Cushing reflex), Cheyne–Stokes breathing, apnoea, fixed dilated pupils (ipsilateral at first) and extensor posture. Try 20% mannitol 5ml/kg IV over 15mins. Enlist an anaesthetist's help with transfer to a neurosurgical centre for urgent CT±craniotomy. Aim to offer the neurosurgeon a well-perfused, well-oxygenated brain.

Initial management Write careful notes. Record times accurately.
● Stop blood loss. Take BP; treat shock with Haemaccel® (p720).
● History: When? Where? How? Had a fit? Lucid interval? Alcohol?
● Search for other injuries. Do blood alcohol, gases, U&E and FBC.
● If comatose, assess level on Glasgow coma scale (p718).
● If alert, record the most difficult thing he can do.
● Assess anterograde amnesia (loss from the time of injury) and retrograde amnesia—for pre-injury events (its extent correlates with the severity of the injury, and it never occurs without anterograde amnesia).
● Examine the CNS. Chart pulse, BP, T°, respirations + pupils/¼h.
● If there is an open fracture or CSF rhinorrhoea, give antibiotic and tetanus prophylaxis (p226). Refer to neurosurgeons.
● Skull x-ray (posterior/anterior, Townes', and lateral views): look for sinus fractures and intracranial air. If the pineal is shifted or fractures cross blood vessels (eg middle meningeal artery—the vessel of extra-dural haemorrhage), do a CT scan.
● Nurse semi-prone if no spinal injury. Meticulous attention to airway, eyes, bladder.
● If not severely injured: sit up when headache allows (if severe, suspect a complication). Treat pain with aspirin. *Criteria for admission:* difficult to assess (child; post-ictal; alcohol intoxication); abnormal neurology (including mental state) at time of examination; severe headache or vomiting; fracture. Loss of consciousness does **not** require admission if well and attendant responsible adult.
● Be alert to complications: early—extradural or subdural haemorrhage (p440), cranial nerve injury, fits. Later—subdural haemorrhage, fits, diabetes insipidus, parkinsonism, dementia.

Comatose trauma patients smelling of alcohol Do skull x-ray (+ CT if fracture or focal signs). Alcohol is an unlikely cause of coma if plasma alcohol <44mmol/l. If unavailable, estimate alcohol from the osmolar gap (p636). (If blood alcohol≈40mmol/l, osmolar gap≈40mmol/l.)

Improvement following impairment of brain function (concussion) Increasing pulse volume→deeper respirations→return of brainstem reflexes→eyes open→vomiting→consciousness→restlessness→complaint of headache→amnesia→post-concussional syndrome (headache, dizziness, inability to concentrate, headaches, poor memory).

Indicators of a bad prognosis Increasing age, decerebrate rigidity, extensor spasms, prolonged coma, hypertension, PaO_2↓, T° >39°C. 60% of those with loss of consciousness of >1 month will survive 3–25yrs, but may need daily nursing care.

Risk of intracranial haematoma (in adults). Fully conscious, no skull fracture = <1:1000; confused, no skull fracture = 1:100; fully conscious, skull fracture = 1:30; confused, skull fracture = 1:4.

▶▶ Rising intracranial pressure (ICP↑)

Presentation Headache; drowsiness; vomiting; seizures. Trauma.

Signs Listlessness; irritability; drowsiness; falling pulse; rising BP; coma; irregular breathing; papilloedema is an unreliable sign.

Causes Meningoencephalitis; head injury; haemorrhage (subdural, extradural, subarachnoid; intracerebral); cerebral oedema. There are 3 types of cerebral oedema:
- Vasogenic—increased capillary permeability (tumour, trauma, ischaemia, infection).
- Cytotoxic—cell death, from hypoxia.
- Interstitial (eg obstructive hydrocephalus).

What happens The tentorium separates the cerebral hemispheres from the brainstem and the cerebellum. The upper brainstem passes through the tentorial hiatus. As pressure above the tentorium rises, the nearest part of the cerebral hemispheres, the uncus (temporal lobe), is squeezed through the tentorial hiatus, alongside the brainstem. This affects the brainstem and the nearby third cranial nerve, so producing these progressive CNS signs:

Consciousness: Stupor→deepening coma.
Ventilation: Slow deep breaths→Cheyne–Stokes breathing→apnoea.
Pupils: Normal→ipsilateral dilatation→bilateral dilatation.
Posture: Decorticate→decerebrate.

A sixth nerve palsy may be an early false localizing sign. Its long intracranial course makes it vulnerable to stretching as the brainstem is displaced downwards.

A similar chain of events may occur after lumbar puncture in the presence of raised ICP. The cerebellar tonsils move downwards so that they impact within the foramen magnum, so compressing the medulla ('coning').

Neurosurgical decompression To be effective this must occur before the pupils become fixed and dilated. This is usually achieved by drilling burr holes in the skull on the side of the dilated pupil.

The management of cerebral oedema Ensure principally there is adequate oxygenation. Aim to increase serum osmolarity with mannitol 20%, 1g/kg (5ml/kg) IV over 15mins (this lasts ~4h); frusemide may also be used, but discuss first with neurosurgeon. Dexamethasone is valuable for vasogenic cerebral oedema (see above)—give 4mg/8h IV.

Avoid fluid overload. Facilitate venous drainage by keeping the head in the midline, elevated at ~40°. Mechanical hyperventilation to maintain a $PaCO_2$ around 3.5kPa, which causes cerebral vasoconstriction, can be helpful. In rare instances (eg Reye's syndrome, *OHCS* p756) monitoring ICP with an extradural sensor and the use of cerebral protective drugs (eg thiopentone 2mg/kg IV after a small test dose) to lower cerebral blood flow and decrease cerebral metabolism may be lifesaving.[1]

1 R Ede 1988 *BMJ* **i** 517

Rising intracranial pressure (ICP↑)

[*OTM* 21.76; 13.369]

►► Status epilepticus

This term is applied to prolonged seizures or multiple seizures occurring without recovery of consciousness between them. Mortality and risk of permanent brain damage increase with the length of the attack. Aim to terminate the seizure as soon as possible, and certainly within 20mins.

It usually occurs in known epileptics. If the patient presents for the first time with status epilepticus, the chance of a structural brain lesion is high (>50%). ► Always ask yourself: *Is the patient pregnant* (pelvic mass)? If so, eclampsia is the likely diagnosis: call a senior obstetrician—immediate delivery may be needed (*OHCS* p96). Alert theatre.

Diagnosis of tonic–clonic status is usually clear. Non-convulsive status (eg absence status or continuous partial seizures with preservation of consciousness) may be more difficult: an EEG can be very helpful here.

Management

- Remove false teeth. Insert a Guedel or nasopharyngeal (if trismus) airway. Give O_2 if needed.
- 50ml of 50% glucose as a bolus (unless certain that blood glucose is ↔).
- Have a laryngoscope, endotracheal tube, O_2, and suction to hand.
- Diazepam (as Diazemuls®—less risk of thrombophlebitis) 10mg IV over 2mins (beware respiratory depression). Wait 5mins. (Use time to prepare other drugs.) The rectal route is an alternative if IV access difficult.
- If seizures continue, set up a 0.9% saline IVI—if possible, avoiding veins crossing joints: splints are likely to fail. Monitor ECG.
- Give diazepam IVI (5mg/min) until seizures stop or 20mg has been given—or significant respiratory depression occurs.
- At the same time as diazepam, start phenytoin IVI, up to 50mg/min to a total of 1000mg. Use a different IV line to diazepam (they do not mix). Beware BP↓ and do not use if bradycardia or heart block.
- If seizures continue, get expert help. Give a further diazepam IVI: dilute 100mg in 500ml of 5% dextrose, infusing at 40ml/h. It is most unusual for seizures to remain unresponsive. If they do, allow the idea to pass through your mind that they could be pseudoseizures (p706).
- If the seizure continues for 60mins after it began, paralyse and ventilate under anaesthesia. Monitor EEG continuously.

Note: chlormethiazole is a very helpful and reliable alternative to phenytoin for use when diazepam alone has failed. Give 40–100ml of the ready prepared 0.8% chlormethiazole infusion over 5–10mins IV, and then by IVI according to response (eg 10–15 drops/min). A significant advantage of this method is that the solution is ready prepared. It needs to be kept in a fridge. Care must be taken with prolonged administration as the solution contains no electrolytes. Monitor vital signs and state of consciousness continuously.

Tests Pulse oximetry, cardiac monitor, glucose, blood gases, U&E, Ca^{2+}, FBC, platelets, ECG. Consider anticonvulsant levels, toxicology screen, lumbar puncture, culture blood and urine, EEG, CT, carbon monoxide level.

When conscious, start oral therapy (p450). Ask yourself what the cause was (p448—eg hypoglycaemia, pregnancy, alcohol, drugs, CNS lesion or infection, hypertensive encephalopathy, inadequate anticonvulsant dose).

Status epilepticus
[*OTM* 21.64]

▶▶ Diabetic emergencies

Hyperglycaemic ketoacidotic coma This typically presents with a 2–3-day history of gradual deterioration, perhaps precipitated by infection, with dehydration, acidosis and coma. The patient hyperventilates. The breath smells ketotic. The dehydration is more life-threatening than the hyperglycaemia—so its correction takes precedence.

Management of ketoacidosis: ● Set up a 0.9% saline IVI. Give ≥1.5 litres/h for 2h, and then 500ml/h over the next 4h. Then reduce to 500ml/2h.
● Blood samples: glucose, U&E, bicarbonate, osmolality, blood gases, FBC, blood culture. Urine tests: ketones, MSU.
● Pass an NGT—to prevent gastric dilatation and aspiration.
● Detailed history (from relatives) and examination. Do CXR.
● Give IV soluble insulin by pump, 50u in 50ml of 0.9% saline infused at 6ml/h (10u/h if infection is present). When the blood glucose is <14mmol/l, halve infusion rate. If there is no pump, give 6u/h IM.
● Measure plasma K⁺ every hour. The danger is hypokalaemia as glucose enters cells taking K⁺ with it. Aim for 4–5mmol/l. If K⁺ <3mmol/l, give 40mmol K⁺ over 1h IVI and recheck K⁺. If K⁺ 3–4mmol/l, give 30mmol K⁺ over 1h; if 4–5mmol/l, give 20mmol. Monitor urine output. Withhold K⁺ if oliguric (rare) or if initial plasma K⁺↑ (it will fall fast, as glucose enters cells).
● Use dextrose saline for IVI when blood glucose is <12mmol/l.
● If unconsciousness is prolonged, give heparin 5000u/6h SC.
● Monitor vital signs and blood glucose every hour.
● Treat any manifest infection after culturing urine and blood.
● Be alert to shock, cerebral oedema (use mannitol), DVT, DIC (p598).

Note: If the acidosis is severe (pH <7.1), some physicians give IV bicarbonate (eg 100mmol over 1h); others never give it because of its effect on the oxyhaemoglobin dissociation curve.

Hyperglycaemic hyperosmolar non-ketotic coma This presents with a long history (eg 1 week), marked dehydration and a glucose >35mmol/l. Acidosis is absent as there has been no switch to ketone metabolism—the patient is often old, and presenting for the first time. The osmolality is >340mmol/kg. Focal CNS signs may occur. The risk of DVT is high, so give heparin (eg 5000u/12h IVI). Long-term insulin may not be needed.

Treat as for ketoacidosis—but use 0.45% saline IVI (not 0.9%) if plasma Na⁺ >150mmol/l. Infuse insulin at 3u/h. Anticoagulate.

Hyperlactataemia is a rare but serious complication of DM (eg after septicaemia or biguanide use). Blood lactate: >5mmol/l. Seek expert help. Give O₂. Treat any sepsis vigorously.

Hypoglycaemia This presents with altered behaviour (eg aggression), sweating, pulse↑, fits and coma of *rapid* onset. See p716. Give 50–100ml 50% dextrose IV fast. This harms veins, so follow by 0.9% saline flush. Expect prompt recovery. If not give dexamethasone 4mg/4h IV to combat cerebral oedema after prolonged hypoglycaemia. Dextrose IVI may be needed for severe hypoglycaemia. If IV access fails try glucagon (1–2mg IM). Once conscious, give sugary drinks.

Pitfalls in diabetic ketoacidosis

1 **An elevated WCC** (even a leukaemoid reaction) may be seen in the absence of infection.
2 **Infection:** often patients are apyrexial. Perform an MSU, blood cultures and CXR. Start broad spectrum antibiotics early if infection is suspected.
3 **Creatinine:** some assays for creatinine cross-react with ketone bodies, so plasma creatinine may not reflect true renal function.
4 **Hyponatraemia** may occur due to the osmotic effect of glucose. If <120mmol search for other causes eg hypertriglyceridaemia. **Hypernatraemia** (>150mmol/l) may be treated with up to 1 litre of 0.45% saline. Continue with 0.9% saline thereafter.
5 **Ketonuria** does not equate with ketoacidosis. Normal individuals may have ketonuria after an overnight fast. Not all ketones are due to diabetes—consider alcohol if glucose normal. Test plasma with Ketostix/Acetest to demonstrate ketonaemia.
6 **Acidosis** but without gross elevation of glucose may occur, but consider overdose (eg aspirin) and lactic acidosis (in elderly diabetic patients).
7 **Serum amylase** is often raised (up to ×10) and non-specific abdominal pain is common even in the absence of pancreatitis.

To calculate approximate plasma osmolarity: $2[Na^+] + [urea] + [glucose]$ mmol/l

▶▶ Thyroid emergencies

Myxoedema coma
Clinical: Looks hypothyroid; >65 years old; hypothermia; hyporeflexia; bradycardia; comatose; fits.

History: Prior surgery or radioiodine for hyperthyroidism.

Precipitants: Infection; myocardial infarct; stroke, trauma.

Examination: Goitre; cyanosis; heart failure; precipitants.

Treatment: Preferably in intensive care.
- Take venous blood for: T_3, T_4, TSH, FBC, U&E, cultures.
- Take arterial blood for PaO_2.
- Give high flow O_2 if cyanosed.
- Give T_3 (triiodothyronine) 5–20µg/12h IV slowly (may be given every 4h if necessary). Be cautious: this may worsen ischaemic heart disese.
- Give hydrocortisone 100mg/8h IV, especially important if pituitary hypothyroidism is suspected (ie no goitre, or previous radioiodine, or thyroid surgery).
- IVI 0.9% saline. Be sure to avoid precipitating LVF.
- If infection suspected give antibiotics, eg cefuroxime 750mg–1.5g/8h IVI.
- Treat *heart failure* as appropriate (p298).
- Treat *hypothermia* with warm blankets in warm room. Beware complications (hypoglycaemia, pancreatitis, arrhythmias). See p76.

Continuing therapy: T_3 5–20µg/4–12h IV until sustained improvement (2–3 days) then thyroxine 50µg/24h PO. Continue hydrocortisone.
- IV fluids as appropriate (hyponatraemia is dilutional).

Hyperthyroid crisis (thyrotoxic storm)

Clinical: Severe hyperthyroidism: fever, agitation, confusion, coma, tachycardia, AF, D&V. It may mimic the 'acute abdomen'.

Precipitants: Recent thyroid surgery or radioiodine; infection; trauma.

Diagnosis: Confirm with technetium uptake if possible, but do not wait for this if urgent treatment is needed.

Treatment: Enlist expert help.
- IVI 0.9% saline, 500ml/4h. NG tube if vomiting.
- Take blood for: T_3, T_4, cultures (if infection suspected).
- Sedate if necessary (chlorpromazine 50–100mg PO/IM).
- Give propranolol 40mg/8h PO (maximum IV dose: 1mg over 1min, repeated up to ten times at ⩾2min intervals).
- Carbimazole 15–25mg/6h PO (or down NGT, if necessary). After 1h give Lugol's solution 0.5ml/8h PO or radiographic contrast agent to block thyroid.
- Dexamethasone 4mg/6h PO.
- Treat suspected infection with eg cefuroxime 750g/8h IVI.
- Adjust IV fluids as necessary. Keep cool with tepid sponging and paracetamol.

Continuing treatment: After 5 days reduce carbimazole to 15mg/8h PO. After 10 days stop propranolol and iodine. Adjust carbimazole (p542).

Thyroid emergencies
[*OTM* 10.39]

▶▶ Addisonian crisis

Clinical: Shock (tachycardia; peripheral vasoconstriction; postural hypotension; oliguria; weak; confused; comatose) usually in someone known to have Addison's disease or in someone on long-term steroids who has forgotten to take their tablets.

Precipitating factors: Infection, trauma, surgery.

Management: If suspected, treat before biochemical results.
- Take blood for cortisol (10ml heparin) and ACTH if possible (10ml heparin, to go straight to laboratory).
- IVI: use a plasma expander first, for resuscitation, then 0.9% saline in 5% dextrose.
- Hydrocortisone sodium succinate 100mg IV stat.
- Blood, urine, sputum for culture.
- Give antibiotic (eg cefuroxime 750mg/8h IVI).
- Monitor blood glucose: the danger is hypoglycaemia.

Continuing treatment • Glucose IV may be needed if hypoglycaemic.
- Continue IV fluids, more slowly. Be guided by clinical state.
- Continue hydrocortisone sodium succinate 100mg IM every 6h.
- Change to oral steroids after 72h if patient's condition good. (NB: tetracosactrin test impossible while on hydrocortisone).
- Fludrocortisone is needed only if hydrocortisone dose <50mg per day and the condition is due to adrenal disease.
- Search for the cause, once the crisis is over.

Hypopituitary coma

Usually develops gradually in a person with known hypopituitarism. Rarely, the onset is rapid due to infarction of a pituitary tumour (pituitary apoplexy)—as symptoms include headache and meningism, subarachnoid haemorrhage is often misdiagnosed.

Presentation: Headache; ophthalmoplegia; consciousness↓; hypotension; hypothermia; hypoglycaemia; signs of hypopituitarism (p558).

Tests: T_4; cortisol; TSH; ACTH; glucose. Pituitary fossa x-ray.

Treatment: • Hydrocortisone sodium succinate 100mg IV/6h.
- Only after hydrocortisone begun: T_3 10μg/12h PO.
- Prompt surgery is needed if the cause is pituitary apoplexy.

▶▶ Phaeochromocytoma emergencies

Stress, abdominal palpation, general anaesthetic, or contrast media used in radiography may produce dangerous *hypertensive crises* (pallor, pulsating headache, hypertension, feels 'about to die').

Treatment • Phentolamine 2–5mg IV. Repeat to maintain safe BP.
• Labetolol is an alternative agent (p302). • When BP controlled, give phenoxybenzamine 10mg/24h PO (increase by 10mg/day as needed, up to 0.5–1mg/kg/12h PO). Propranolol (eg 20mg/8h PO) is also given at this stage. Prepare for surgery.

Emergencies

Acute poisoning—general measures

Emergency measures Clear airway. Consider ventilation (p354). Treat shock (p720). If unconscious, nurse semi-prone.

Assess patient Diagnosis mainly from history. Patient may not tell truth about what has been taken. Tablet identification helped by MIMS Colour Index, and by description in BNF and MIMS. Identification helps plan specific treatment.

Assess level of consciousness See p718. Do full physical examination. (NB: pinpoint pupils suggest opiates taken.)

Treat the patient not drug level except for paracetamol and salicylate poisoning.

Pointers to further pathology Papilloedema; recent head injury; signs of injury (consider head injury); lateralizing neurological signs; behaviour change, nausea, headache before overdose; no improvement after 12h. Do urine and plasma drug screen, look for papilloedema, do skull x-ray.

If you are not familiar with the poison Get more information. The *Data Sheet Compendium* is useful. If still in doubt how to act phone Poisons Information Service (p746).

Take blood as appropriate (p746). Always check paracetamol and salicylate levels.

Empty stomach if appropriate (p746). Deep-freeze 20ml of aspirate, urine and blood in case of criminal investigations.

Consider antidote (p748) or oral activated charcoal (p750).

Continuing care Measure temperature, pulse, BP regularly. If unconscious nurse semi-prone, turn regularly, keep eyelids closed. Urinary catheter needed if bladder distended, renal failure suspected or forced diuresis undertaken.

Psychiatric assessment Be sympathetic despite the hour! Interview relatives and friends if possible. Aim to establish:
● *Intentions at time:* Was the act planned? What precautions against being found? Did the patient seek help afterwards? Does the patient think the method was dangerous? Was there a final act (eg suicide note)?
● *Present intentions.* ● *What problems* led to the act: do they still exist? ● *Was the act* aimed at someone? ● Is there a *psychiatric disorder* (depression, alcoholism, personality disorder, schizophrenia, dementia)? ● What are his *resources* (friends, family, work, personality)?

The assessment of suicide risk: The following factors increase the chance of future suicide: original intention was to die; present intention is to die; presence of psychiatric disorder; poor resources, previous suicide attempts; socially isolated; unemployed; male; >50yrs old. See OHCS p328 & p338.

Referral to psychiatrist depends partly on local resources. Ask advice if presence of psychiatric disorder or high suicide risk.

Acute poisoning—general measures
[*OTM* 6.2; 25.19]

Acute poisoning—specific points

Poisons Information Services Belfast: 0232 240503; Cardiff: 0222 709901; Dublin: 0001 379964; Edinburgh: 031 229 2477; Leeds: 0532 430715; London: 071 635 9191; Newcastle 091 232 5131.

Taking blood 10ml in lithium-heparin tube to laboratory for storage to keep until patient discharged unless there is a strong possibility of the patient being HbSAg or HIV +ve.

Emergency blood drug level (in lithium-heparin tube) for all unconscious patients (paracetamol level) and for any patient where the following may have been taken: paracetamol (eg in Distalgesic®, co-proxamol), iron, lithium (use plain tube), paraquat. Consider if salicylate has been taken (see below).

When to empty the stomach Within 4h of poisoning unless only small amounts of a relatively safe drug (eg a benzodiazepine) has been taken. It is worth doing up to 6h after ingestion of *opiates* and *anticholinergics* and up to 12h after *salicylates*, *tricyclic antidepressants*, and *aminophylline*.
► *Do not empty stomach* if petroleum products or corrosives have been ingested (*exception:* paraquat) or if the patient is unconscious (unless you protect the airway).

How to empty stomach

1 *Induce vomiting:* Do not do if patient likely soon to lose consciousness. Give 15–30ml syrup of ipecacuanha followed by 200ml water. Repeat after 20mins if there has not been any vomiting. (Note: this is not a very efficient way of reducing the absorption of poisons, and the use of activated charcoal, p750, is to be preferred.[2])

2 *Gastric emptying and lavage:* If the patient is unconscious, or has no gag reflex, protect the airway with a cuffed endotracheal tube.

- Have suction apparatus to hand and working.
- Position the patient in left lateral position.
- Raise the foot of the bed by 20cm.
- Pass a lubricated tube (14mm external diameter) via the mouth, asking the patient to swallow.
- Confirm position in stomach—blow air down, and auscultate over the stomach.
- Siphon the gastric contents.
- Perform gastric lavage using 300ml tepid water at a time. Massage the left hypochondrium.
- Repeat until no tablets in siphoned fluid.
- When pulling out tube, occlude end to prevent aspiration of fluid in tube.

Note: if the patient is known to be unconscious because of alcohol intoxication and the use of only a small number of benzodiazepine tablets (the commonest agents), it is probably wisest to avoid gastric lavage, with its attendant risks of aspiration, and to allow the patient to sleep off the effects of the poison.

1 A Proudfoot 1982 *Diagnosis and Management of Acute Poisoning*, Blackwell
2 J Vale 1986 *BMJ* ii 1321

Acute poisoning—specific points
[*OTM* 6.2]

Acute poisoning—antidotes

Benzodiazepines Flumazenil (for respiratory arrest) 200µg over 15sec; then 100µg at 60sec intervals if needed. Usual dose range: 300–600µg IV over 3–6mins.

Beta-blockers Severe bradycardia or hypotension. Try atropine 0.3mg IV. Give glucagon 5–10mg IV bolus if atropine fails.

Botulinum toxin (from tinned foods). It causes diplopia, blurred vision, photophobia, ataxia, pseudobulbar palsy, and sudden cardiorespiratory failure. GI symptoms may be absent. Nurse on ITU. Intubate early. Seek expert help. Give botulinum antitoxin IM (eg 50,000u of types A and B with 5000u of type E), as early as possible. Also give it to those who have ingested the toxin but who have not yet developed symptoms.

Carbon monoxide In spite of hypoxaemia the skin is pink (or pale), not cyanosed, because of the formation of carboxyhaemoglobin (COHb). This displaces O_2 from Hb binding sites. ►► Remove the source. Give 100% O_2 (±IPPV). Metabolic acidosis will usually respond to correction of hypoxia. If severe, anticipate cerebral oedema. Give mannitol IVI (p734), and dexamethasone 4mg/6h IV. Confirm diagnosis with a heparinised blood sample (COHb↑) quickly as levels may soon return to normal. Monitor ECG. Hyperbaric O_2 may help—even many hours post-exposure.[1] Consider this if there is a history of coma, neuropsychiatric symptoms, CVS complications, pregnancy or COHb is >40%. Ring the Royal Navy to locate the nearest compression chamber (0705 822351 or 0705 818888).

Co-proxamol Contains dextropropoxyphene (opiate)+paracetamol (p750).

Cyanide Dicobalt edetate 300mg IV over 1min followed by 50ml 50% dextrose IV. Repeat up to 2 times.

Digoxin This may be inactivated by digoxin-specific antibody fragments. Give 60 ×total digoxin load (antibody fragments are 60 ×heavier than digoxin molecules). The load (mg) is the plasma digoxin concentration (ng/ml) × body weight (kg) × 0.0056. If the load or level is unknown give 20 vials (800mg)—adult or child. Consult data sheet. Dilute in water for injections (4ml/40mg vial) and 0.9% saline (to make a convenient volume), and give by IVI over 30mins, via a filter with 0.22µm pores.

Heavy metals Dimercaprol eg 2.5–3mg/kg IM up to 6 doses/24h for 48h—then twice daily for 10 days.

Iron Desferrioxamine 2g IM and desferrioxamine IV continuous infusion up to 15mg/kg/h, not exceeding 80mg/kg/24h.

Opiates (Many analgesics contain opiates). Give naloxone 0.4–1.2mg IV repeated every 2mins until breathing is adequate (it has a short half-life, so it may need to be given often—up to 10mg). Note: naloxone may precipitate features of opiate withdrawal. Symptoms of diarrhoea and cramps will normally respond to diphenoxylate and atropine (Lomotil®—eg 2 tablets/6h PO). Sedate with thioridazine 25–50mg PO as needed. High dose opiate misusers may need methadone (eg 10–30mg/12h PO) to combat withdrawal.[2] Register opiate addiction (OHCS p362), and refer for help.

Acute poisoning—antidotes

Oral anticoagulants If not normally on anticoagulants and not bleeding, give phytomenadione (vit. K) 10mg/24h IM. If bleeding, give fresh frozen plasma (FFP) IV. If normally on warfarin and bleeding, give vit. K, 5–10mg IV. But if it is vital that he remains anticoagulated, use FFP, and enlist expert help. Warfarin can normally be restarted within 2–3 days. Major bleeding should be treated with vit. K 10–20mg IV: life-threatening bleeding with vit. K, 50mg IV, followed by infusion of vitamin-K-dependent clotting factors. Cholestyramine 4g/6h PO aids elimination.[3]

Organophosphate insecticides These inactive cholinesterase—the resulting increase in acetylcholine causes D&V, colic, sweating, salivation, wheeze, twitches, fits, pupil constriction (giving headache) and respiratory depression. *Treatment:* Decontaminate (remove clothes; wash skin). Remember to wear gloves, mask, apron & goggles. For symptoms give atropine IV 2mg every 30mins until full atropinization (dry mouth, pulse >70). Up to 3 days' treatment may be needed. Also give pralidoxime 30mg/kg slowly IV (dilute with >10ml Water for Injections). Repeat as needed every 30min.

1 *Drug Ther Bul* 1988 **26** 77 **2** DoH 1984 *Guidelines of Good Clinical Practice in the Treatment of Drug Misuse*, London **3** S Renowden 1985 *BMJ* **ii** 513

Salicylate poisoning

Empty the stomach (p746) if >15 tablets (=4.5g) have been taken within 12h. Do an emergency salicylate level, particularly if symptoms are present (tinnitus, deafness, nausea, vomiting, tremor, sweating, hyperventilation, confusion).

If the level is <4.3mmol/l (600mg/l) simply increase oral fluids. If ≥4.3mmol/l, use alkaline diuresis (under expert guidance in ITU, as there is significant morbidity) or activated charcoal, which is a much safer means of enhancing the elimination of salicylates.[1] However, charcoal is unpalatable—and patients may be vomiting after salicylate poisoning, so give it by NGT. A suitable régime is 50g activated charcoal on admission, or after gastric lavage, followed by 25g/4h, until recovery or until plasma drug concentrations have fallen to safe levels.[2]

An alternative is urinary alkalinization (in ITU, eg with boluses of 50–100ml sodium bicarbonate 1.2% IV; do catheter urine pH every 15mins—aim for ~8). If none of these measures is practicable, try haemodialysis or charcoal haemoperfusion, particularly if plasma salicylate is >7.2mmol/l (1000mg/l). Seek expert help.

Paracetamol poisoning

The main danger is liver damage. Do an emergency blood level, but not before 4h have elapsed since ingestion. Empty the stomach if >7.5g has been taken. If the plasma paracetamol is above the line on the graph opposite, or if the patient presents 8–15h after ingestion of >7.5g paracetamol, take action to prevent liver damage, as follows.

Give N-acetylcysteine by IVI, 150mg/kg in 200ml of 5% dextrose over 15mins. Then 50mg/kg in 500ml of 5% dextrose over 4h. Then 100mg/kg/16h in 1 litre of 5% dextrose. SE: shock, vomiting, bronchospasm (in up to 10%). An alternative is methionine 2.5g PO/4h for 12h (total: 10g), but absorption is unreliable if there is vomiting.

Treatment starting >15h from ingestion may be beneficial; seek expert help.

▶ Beware of hypoglycaemia. Monitor blood glucose hourly.

Criteria for transfer to a specialist unit:[3]
- Encephalopathy and rising intracranial pressure. Signs of cerebral oedema: BP >160/90 (sustained) or brief rises (systolic >200mmHg), bradycardia, decerebrate posture, extensor spasms, poor pupil responses. ICP monitoring can help. See p734.
- INR >2.0 at <48h—or >3.5 at <72h (so measure INR every 12h). Peak elevation: 72–96h. LFTs are *not* good markers of hepatocyte death. NB: if INR is *normal* at 48h, the patient may go home.[4]
- Renal impairment (creatinine >200μmol/l). Monitor urine flow. Daily U&E & serum creatinine (use haemodialysis if >400μmol/l).
- Blood pH <7.3 (lactic acidosis—from tissue hypoxia).
- Hypotension (systolic BP <80mmHg).

1 R Hillman 1985 *BMJ* ii 1472   2 *Lancet* Ed 1987 ii 1013   3 *Drug Ther Bul* 1988   4 *Drug Ther Bul* 1988 26 97   5 Reproduced, with permission, from J Vale & T Meredith 1985 *Concise Guide to the Management of Poisoning* 3 ed, Churchill Livingstone

Salicylate poisoning
Paracetamol poisoning

Graph for use in deciding who should receive N-acetylcysteine[5]
▶ 1 tablet of paracetamol = 500mg

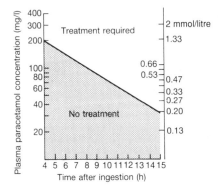

Emergency procedures: 1

There is no substitute for learning by experience. Often it is wiser to wait for someone to come and carry out an urgent procedure than to try for the first time by oneself—but some procedures must occasionally be performed at once. They are:

Cut down onto a vein ● Indication: shock, eg exsanguination, when veins are collapsed, and IV access fails.
● Aim: to expose a vein and cannulate it under direct vision.
● Equipment: scalpel, artery forceps, cannula, dressing pack.
● Procedure: use the long saphenous vein as it courses anterior to the medial malleolus. Make a 3cm transverse skin incision. Free the vein using fine artery forceps. Insert and secure the cannula. It does not matter if you cannot use stitches: haemostasis can be achieved by pressure and elevation of the leg.
Note: experts may choose a CVP line rather than do a 'cut down'. However, the insertion of CVPs is hazardous, and should not be attempted without prior supervision. But a 'cut down' should certainly be attempted if expert help is not immediately available. In these circumstances, they have saved many lives.

Relieving a tension pneumothorax
● Aim: to release air from the pleural space. (In a tension pneumothorax air is drawn in, intrapleurally, with each breath, but cannot escape due to a valve-like effect of the tiny flap in the parietal pleura. The increasing pressure progressively embarrasses the heart and the contralateral lung.)
● Equipment: IV cannula (eg Venflon®), three-way tap, syringe.
● Insert the cannula through an intercostal space anywhere on the affected side. The three-way tap is connected between the syringe and the cannula. By aspirating on the syringe and expelling the intrapleural air through the three-way tap, the tension is relieved, and the lung is able to re-expand.
● Proceed to formal chest drainage (p754).

Cardioversion/defibrillation ● Indications: ventricular fibrillation or tachycardia, fast AF supraventricular tachycardias if other treatments (p290–4) have failed.
● Aim: to completely depolarize the heart using a direct current.
● Procedure: do not wait for a crisis before familiarizing yourself with the defibrillator.
  —Set the energy level (eg 200 joules for ventricular fibrillation or ventricular tachycardia; 100J for atrial fibrillation; atrial flutter 50J).
  —Place conduction pads (Littman™ Defib Pads) on chest, one over apex and one below right clavicle—less chance of skin arc than jelly.
  —Make sure that no one else is touching the patient.
  —Press the button(s) on the electrode to give the shock.
  —Watch ECG. Repeat the shock at a higher energy if necessary.
● Note: for AF and SVT it is necessary to synchronize the shock on the R wave of the ECG (by pressing the 'SYNC' button on the machine). This ensures that the shock does not initiate a ventricular arrhythmia. *If the SYNC mode is engaged in* **VF** *the defibrillator will not discharge!*
● It is only necessary to anaesthetize the patient if conscious.
● After giving the shock, monitor ECG rhythm. Consider anticoagulation, as the risk of emboli is increased.
▶ For children use 2J/kg in VF/VT; if >10kg use *adult* paddles.

Cricothyroidotomy

Essence An emergency procedure to overcome upper airway obstruction above the level of the larynx.

Indications Upper airway obstruction when endotracheal intubation not possible, eg irretrievable foreign body; facial oedema (burns, angio-oedema); maxillofacial trauma; infection (epiglotittis).

Procedure Lie the patient supine with neck extended (eg pillow under shoulders). Run your index finger down the neck anteriorly in the mid-line to find the notch in the upper border of the thyroid cartilage: just below this, between the thyroid and cricoid cartilages, is a depression—the cricothyroid membrane.

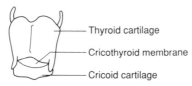

— Thyroid cartilage

— Cricothyroid membrane

— Cricoid cartilage

a. *Needle cricothyroidotomy:* Pierce the membrane with large bore cannula (14G) attached to syringe: withdrawal of air confirms position. Slide cannula over needle at 45° to skin in sagittal plane. Use Y-connector or improvise connection to O_2 supply and give 15 litre/mins: use thumb on Y-connector to allow O_2 in 1sec and CO_2 out 4sec ('transtracheal jet insufflation'). Preferred method in children <12yrs. Will only sustain for 30–45mins before CO_2 builds up.

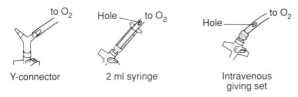

Y-connector   2 ml syringe   Intravenous
                                giving set

b. *Mini-Trach II®:* This contains a guarded blade, introducer, 4mm uncuffed tube (slide over introducer) with ISO connection and binding tape. Patient can breathe spontaneously or be ventilated via bag (high resistance). Will sustain for 30–45mins.

753

c. *Surgical cricothyrotomy:* Smallest tube for prolonged ventilation is 6mm. Introduce high volume low pressure cuff tracheostomy tube through horizontal incision in membrane.

Complications Local haemorrhage; posterior perforation of trachea ± oesophagus; laryngeal stenosis if membrane over-incised in childhood; tube blockage; subcutaneous tunnelling.

▶ Note: needle and Mini/Trach® are temporary measures pending formal tracheostomy.

Emergency procedures: 2

Inserting a chest drain (p356, p358)

- Preparation: trolley with dressing pack, iodine, needles, 10ml syringe, lignocaine 1% (10ml), scalpel (no 15), suture, chest drain (eg 28F, smaller ones block or kink), drainage bottle, connection tubes, sterile H_2O, elastoplast). Incontinence pad under patient. Swab extensively.
- Choose insertion site:[1] 4th–6th intercostal space, mid-axillary line; if you need to drain fluid also, use the 7th space posteriorly.
- Infiltrate down to pleura with 10–20ml of 1% lignocaine. Wait 3mins.
- Make 2cm incision above 6th rib, so avoiding neurovascular bundle under 5th rib. Bluntly dissect with forceps down to pleura. Puncture pleura with scissor/forceps then sweep a finger inside chest to clear adherent lung and exclude (eg in blunt abdominal trauma) stomach in the chest!
- Insert 2 horizontal sutures over the hole. Leave ends free to make a seal once the drain is finally removed.
- Before inserting the drain withdraw metal trochar 2cm; introduce the drain *atraumatically*. There is often no need for a trochar at all.
- Advance the tip upwards to the apex. Stop when you meet resistance. Then withdraw the introducer completely and attach the drain via the tubing to the bottle. Ensure the longer tube within the bottle is under-water and bubbling with respiration. If patient is to be moved to another hospital substitute Heimlich flutter valve or drainage bag with flap valve for underwater drain. Never clamp a chest drain.[2]
- Fix the drain with a second suture tied around the tube like a 'Roman gaiter'. Bandage the drain to prevent it slipping.
- Request CXR to check the position of the drain. Give analgesia (PO/IM).

Inserting a subclavian venous cannula

- Preparation: set trolley (dressing pack, iodine, needles, scalpel (no. 15), CVP pack, suture, Opsite® dressing); put incontinence pad under patient who should be lying flat with the bed tilted 10° head down; swab chest and neck; surround area with green towels.
- Choose insertion point: 1cm below the right clavicle, $^2/_3$ of the way along its length from the sternoclavicular joint. Infiltrate with ligno-caine. Wait 3mins. Puncture skin with scalpel.
- Insert the CVP pack's large needle into the hole. Advance in the direction of the suprasternal notch, under the clavicle, exerting mild continuous suction on the syringe. Keep the needle parallel with the floor to minimize the risk of pneumothorax.
- You may require several passes at slightly different angles to enter the vein. Stop when you can easily aspirate blood.

- What to do now depends on the type of CVP pack. If there is a sheath over the large needle, advance it, remove the needle and secure the sheath. If not, keep the needle very still and remove syringe. There will be a trickle back of blood. Insert the guide wire (soft end first) into the needle. It should advance easily. Remove the needle. There will be a sheath which comes mounted on a tapering dilator. Advance this over the wire into the vein. Holding the sheath, remove wire and dilator. Secure the sheath.
- CXR: check end of the line is in the SVC; exclude pneumothorax.

1 Br Thoracic Society 1993 *BMJ* ii 114   **2** PS Wong 1993 *BMJ* ii 443

Emergency procedures: 2

Gunshot injury

Serious injury from guns is not confined to the battlefield and reflects an increasingly violent society. The injury inflicted by a bullet is dependent on the energy expended: in turn, this depends on the inherent energy of the bullet (kinetic energy $= \frac{1}{2}mv^2$) and its retardation within tissues. All bullets will lacerate tissue in their track, but high velocity missiles (ie >speed of sound) also make 'temporary cavities' (milliseconds only) 30–40 × the bullet size, causing damage distant from the track and sucking in debris/bacteria on collapsing.

Remember
- Small surface wounds may hide serious internal injury.
- All wounds need exploration under GA to debride wound track and repair damaged structures: never be tempted to sew up the wound without exploration! It is good surgical practice anyway to leave debrided wounds open for several days ('delayed primary suture').
- These wounds are contaminated: give antibiotic + tetanus prophylaxis.
- Victims are often young adults who may not show obvious signs of shock (systolic BP↓) until >25% blood volume is lost (p720).

Chest injuries Exclude tension pneumothorax (absent breath sounds; hyperresonance; deviated trachea) and treat with 14G cannula in second intercostal space, mid-clavicular line. ► **Do not** delay for CXR. Leave open until formal chest drain inserted (p754). If there is a sucking chest wound (open pneumothorax) air will exchange across this defect in preference to the long trachea and impair gas exchange: seal immediately with waterproof dressing (eg plastic bag) on three sides only, thus allowing seal to function as flutter valve. Massive haemothorax requires chest drain after intravenous resuscitation started: drainage of >1500ml stat or >200ml/h should prompt thoracotomy.

Cardiac tamponade See p722. Always consider mediastinal involvement with penetrating injury medial to nipple or scapula, or if orientation of entrance + exit wounds suggestive. Needle pericardiocentesis may be lifesaving (p723). Leave cannula taped *in situ* to allow repeated aspiration. Exploratory thoracotomy is mandatory.

Electromechanical dissociation (EMD, p725) If EMD is present with chest trauma, check neck veins: if distended check the trachea; if deviated think tension pneumothorax, if not think cardiac tamponade. If the neck veins are flat think hypovolaemia (**but** be aware may be hypotension and tension pneumothorax/tamponade: a high index of suspicion is needed).

Abdominal injuries These are associated with a high incidence of visceral damage. A conservative approach is inappropriate and all should be explored at laparotomy. Give antibiotics eg benzylpenicillin (vs *C welchii*) + piperacillin (vs coliforms) + metronidazole (vs anaerobes).

Limb injuries May threaten limb viability. Life-threatening haemorrhage can usually be controlled by direct pressure and elevation; a tourniquet is rarely needed, will augment ischaemic damage and if incorrectly used increases venous blood loss. Similarly, haemostats often cause more local damage and should be avoided. Check distal pulses (but presence does not exclude arterial damage), temperature, motor and sensory function: evidence of vascular insufficiency after a few hours may indicate compartment syndrome (measure pressure directly: >30mmHg abnormal).

Resuscitation ► Secure airway. Give ≥35% O_2. ► Put up 2 large IV lines (p721). ► Crossmatch ≥6u blood. Continuously monitor ECG, BP, respirations.

Blast injury

Blast injury may be encountered in domestic (eg gas explosion) or industrial (eg mining) accidents or as the result of a terrorist bomb. Terrorism world-wide is responsible for 10,000 injured or killed in the last 20 years and 95% of casualties are from bombs. Death may occur without any obvious external injury, often due to air emboli, the correct aetiology first being recognized by Pierre Jars in 1758 as a 'dilatation d'air' (ie blast wave). Explosions cause injury in six ways:

1 Blast wave A transient (milliseconds) wave of overpressure expands rapidly away from the point of explosion, its intensity inversely proportional to the distance cubed. It produces (a) cellular disruption at air-tissue interface ('spalling')—ie perforated ear-drum at 100kPa, 'blast lung' at 175kPa; (b) shearing forces along tissue planes—submucosal/subserosal haemorrhage; (c) re-expansion of compressed trapped gas—bowel perforation, fatal air embolism (coronary artery or cerebral).

Blast lung is often delayed (up to 48h). It is **rare** in survivors—only 0.6%. Suspect it if there is a perforated drum, but this is **not** a prerequisite (as position of drum in relation to blast wave is critical). Intra-alveolar haemorrhage causes ARDS (p350).

2 Blast wind Air displaced by the explosion will totally disrupt a body in the immediate vicinity. Others may suffer avulsive amputations. Bodies can be carried by the wind with deceleration injuries on landing. Glass, wood, stones and other objects are also carried and act as secondary missiles.

3 Missiles Penetration or laceration from missiles are by far the commonest injuries. Missiles arise from the bomb (casing or preformed fragment—nails, nuts and bolts), or are secondary (as above, glass and wood particularly).

4 Flash burns These are usually superficial and occur on exposed skin (hands/face) in those close to explosion.

5 Crush Injuries result from falling masonry.

6 Psychological Acute fear and panic is the aim of the terrorist. Chronically, intrusive thoughts, anxiety and poor concentration may form the basis of a post-traumatic stress disorder (PTSD, *OHCS* p347).

Treatment Approach the same as any major trauma with priority to airway and cervical spine control, breathing and circulation with haemorrhage control. Rest and observe any suspected of exposure to significant blast, but without other injury. Sudden death or renal failure may follow release of a limb after prolonged crush (hyperkalaemia and myoglobinuria): ensure continuous ECG and adequate hydration. Facial burns may compromise airway, which should be secured by intubation or surgical airway (p753). Psychological support will be required.

757

Principal source: SG Mellor Blast injury. In: I Taylor, CD Johnson (eds) *Recent Advances in Surgery 14*, Churchill Livingstone, London 1991 53–68

Burns

Most hospitals do not have their own burns unit. Resuscitate and organise expeditious transfer for all major burns. Always assess site, size, and depth of the burn; smaller burns in the young, the old, or at specific sites, may still require specialist help.

Assessment The size of the burn must be estimated to calculate fluid requirements. Ignore erythema. Consider transfer to a burns unit if >10% burn in children or elderly, or >20% in others. Use a Lund & Browder chart or the 'rule of nines':

Arm (all over) 9%	Front 18%	Head (all over) 9%	Palm 1%
Leg (all over) 18%	Back 18%	Genitals 1%	

(NB: in children, head = 14%; leg = 14%)

Estimating burn thickness: Partial thickness is painful, red, and blistered; full thickness is painless and white/grey.

Refer full thickness burns >5% of body area or full/partial thickness burns involving face, hands, eyes, genitalia to a burns unit.

Resuscitation ● Airway: beware of upper airway obstruction developing if inhaled hot gases (only superheated steam will cause thermal damage distal to trachea)—suspect if singed nasal hairs or hoarse voice. Consider early intubation or surgical airway (p753).
● Breathing: give 100% O_2 if you suspect carbon monoxide poisoning (may have cherry-red skin, but often not) as $t_{1/2}$ COHb falls from 250mins to 40mins (consider hyperbaric oxygen if pregnant; CNS signs; >40% COHb). SaO_2 measured by pulse oximeter is unreliable (falls eg 3% with up to 40% COHb). Do escharotomy bilaterally in anterior axillary line if thoracic burns impair chest excursion (OHCS p689).
● Circulation: >10% burns in a child and >15% burns in adults require IV fluids. Put up 2 large-bore (14G or 16G) IV lines. Do not worry if you have to put these through burned skin. Secure them well: they are literally lifelines. It does not matter whether you give crystalloid or colloid: the volume is what is important. Use a formula, eg:
● Muir and Barclay formula (popular in UK):

[weight (kg) × %burn]/2 = ml colloid (eg Haemaccel®) per unit time

Time periods = 4h, 4h, 4h, 6h, 6h, 12h.

● Parkland formula (popular in USA)

4 × weight (kg) × %burn = ml Hartmann's solution in 24h, half given in first 8h.

▶ NB: you must replace fluid from the time of burn, not from the time first seen in hospital. You must also give 1.5–2.0ml/kg/h 5% dextrose if Muir and Barclay formula is used. Formulae are only guides: adjust IVI rate according to clinical response and urine output (aim for >30ml/h in adults; >1ml/kg/h in children). Catheterize the bladder.

Treatment ▶ Do *not* apply cold water to extensive burns: this may intensify shock. Do not burst blisters. Covering extensive partial thickness burns with sterile linen prior to transfer to a burns unit will deflect air currents and relieve pain. Use morphine in 1–2mg aliquots IV. Suitable dressings include Vaseline gauze, or silver sulphadiazine under absorbent gauze; change every 1–2 days. Hands may be covered in silver sulphadiazine inside a plastic bag. Ensure tetanus immunity. Give 50ml whole blood for every % of full-thickness burn, half in the second 4h of IV therapy and half after 24h.

Burns

Lund and Browder charts

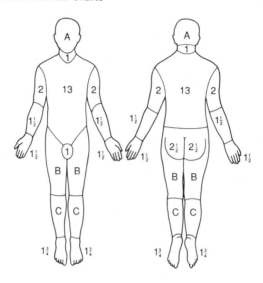

Relative percentage of body surface area affected by growth

Area	Age 0	1	5	10	15	Adult
A=1/2 of head	$9\frac{1}{2}$	$8\frac{1}{2}$	$6\frac{1}{2}$	$5\frac{1}{2}$	$4\frac{1}{2}$	$3\frac{1}{2}$
B=1/2 of one thigh	$2\frac{3}{4}$	$3\frac{1}{4}$	$4\frac{1}{2}$	$4\frac{1}{2}$	$4\frac{1}{2}$	$4\frac{3}{4}$
C=1/2 of one leg	$2\frac{1}{2}$	$2\frac{1}{2}$	$2\frac{3}{4}$	3	$3\frac{1}{4}$	$3\frac{1}{2}$

The major disaster

Planning For a hospital to be prepared to cope with multiple casualties there must be planning. Each hospital will produce a detailed *Major Accident Plan*, but additionally the tasks of key personnel can be distributed on individual *Action Cards*.

At the scene The designated hospital provides a mobile medical team (eg surgeon + anaesthetist + 2–4 nurses); doctors from the local voluntary BASICS scheme (British Association for Immediate Medical Care—usually GPs) specialize in rescue and they should be requested to the scene.

Safety: Is paramount, for yourself as well as the victims. Be visible (luminous monogrammed jacket) and wear protective clothing where appropriate (safety helmet; waterproofs; boots; respirator in chemical environment).

Triage: Is the process of sorting casualties into priorities for treatment. It is dynamic and must be reapplied at every stage of the evacuation chain. Use triage labels with interchangeable colour-coded flaps—IMMEDIATE (RED: eg obstructed airway; hypovolaemic shock); URGENT (YELLOW: eg multiple long bone fractures; significant burns); DELAYED (GREEN: walking wounded).

Communications: Are essential. The police are in overall charge of the scene. Each emergency service will dispatch a control vehicle and will have a designated incident officer for liaison. Support medical staff from hospital report to the ambulance incident officer. The medical incident officer is usually the first doctor on the scene: his job is to assess then communicate to the designated hospital the number and severity of casualties, to organize resupply of equipment and to replace fatigued staff. He must resist temptation to treat casualties as this compromises his rôle.

Equipment: Must be portable (in small cases/backpacks) and include: intubation and cricothyrotomy equipment; intravenous fluids (colloid); bandages and dressings; chest drain (plus flutter valve); amputation kit (when required ideally two doctors should concur); drugs—*analgesic:* morphine; *anaesthetic:* ketamine 2mg/kg IV over >60sec (0.5mg/kg is a powerful analgesic without depression respiration), *specific antidote* if a chemical threat; cardiac resuscitation drugs; to cover common medical emergencies: eg GTN spray, salbutamol inhaler); limb splints (inflatable are compact, but are not robust and give no traction); defibrillator/monitor; ± pulse oximeter.

Evacuation: Remember that with immediate treatment on scene, the priority for evacuation may be reduced (eg a tension pneumothorax—RED— relieved can wait for evacuation—becomes YELLOW), but those who may suffer by delay at the scene must go first (eg unconscious closed head injury; MI).

At the hospital a 'major incident' is declared. The *designated* hospital receives the bulk of the casualties; the *support* hospital(s) will cope with the overflow and may provide mobile teams so that staff are not depleted from the designated hospital. A control room is established and the medical coordinator ensures staff have been summoned, nominates a triage officer and supervises the best use of in-patient beds and ITU/theatre resources.

Principal source: S Miles 1990 Major Accidents *BMJ* **301** 923

The major disaster

Relevant pages in other chapters:
The leukaemias and lymphomas (p598–603); dying at home (*OHCS* p442); facing death (p7). For specific cancers, see the chapters relevant to the specific organ—such as gastroenterology (p480), chest medicine (p322) etc.

Oncological emergencies

▶ A patient with cancer who becomes acutely ill may be made more comfortable using simple measures. Pain control is important (p90 & OHCS p778), but treat specific problems with specific treatment.

Hypercalcaemia Often seen in carcinoma of breast, bronchus, prostate; also myeloma. It is due to bone metastases or ectopic PTH secretion.

Signs: Malaise; nausea/vomiting; polydipsia; polyuria; dehydration; weight↓; constipation; fits; psychosis; confusion; drowsy; coma.

Management: 1 Treat malignancy, stop thiazide diuretics, and consider loop diuretic. Mobilize patient if possible. 2 Hydrate: 3–4 litres of 0.9% saline IV over 24h. If mild, give phosphate tablets 500mg/12h PO. If no improvement in 24h, 3 Give biphosphonate, eg disodium pamidronate (below)—up to 90mg/course. *Infuse slowly*, eg 30mg in 300ml 0.9% saline over 3h. Expect response in 3–5 days.

Superior vena caval obstruction Typically with lung carcinoma (80%) or lymphoma (17%). If tests do not clinch diagnosis of underlying cancer, treat on basis of most likely cause (eg elderly smoker likely to have lung cancer; but lymphoma more likely in fit young men).

Clinical features: Distended thoracic/neck veins ± dyspnoea, oedema, feeling of fullness in head, headache; JVP↑; plethoric face; tachypnoea.

Investigations: Sputum cytology (eg might detect cells from small cell carcinoma); CXR; CT thorax. NB: bronchoscopy might be hazardous.

Management: Dexamethasone 4mg/6h PO. Chemotherapy (small cell carcinoma and lymphoma). Radiotherapy (squamous cell carcinoma).

Spinal cord compression Thoracic 70%; lumbosacral 20%; cervical 10%.

Clinical features: Back pain; weakness; upper motor neurone and sensory signs; (cord level can often be found). Urine retention (treat urgently).

Tests: Plain x-ray of spine; myelogram; CT scan; MRI.

Management: Discuss with neurosurgeon and radiotherapist. Dexamethasone (4–8mg/6h PO). Radiotherapy. Surgery indicated if: no previous histology; progression of condition while on radiotherapy; previous radiotherapy to maximum dose; spine mechanically unstable.

Raised intracranial pressure (p734) In setting of cancer consider dexamethasone; radiotherapy; occasionally surgery to remove solitary metastasis. Surgery also indicated for some primary CNS tumours.

Tumour lysis syndrome Lysis of rapidly growing tumour (especially lymphomas, leukaemias, and myeloma) due to effective treatment can cause rise in uric acid, potassium and phosphate leading to renal failure. Prevention is with adequate hydration before starting therapy and allopurinol 300mg /24h PO started the day before therapy. Treat hyperkalaemia; haemodialysis if renal failure.

Inappropriate ADH secretion p638. **Febrile/neutropenic régime** p598.

Treatment of hypercalcaemia with disodium pamidronate

Calcium (mmol/l)	Pamidronate (mg)
up to 3	15–30
3–3.5	30–60
3.5–4	60–90
>4	90

Symptom control in severe cancer[1,2]

Pain ▶ Do not be miserly with analgesia: aim to eliminate the pain.

Types of pain: There are many types of pain and several causes (see table). Patients will often be suffering with more than one pain. Listen carefully to the patient's story; try to identify each pain.

Management 1. Pain is affected by mood, morale, and meaning. Explain its origin to both the patient and relatives. 2. Modify the pathological process where possible, for example with: radiotherapy; hormones; chemotherapy; surgery. 3. Plan rehabilitation goals. 4. Give analgesics, by mouth if possible. Do not use the patient's pain to prompt the drug round—aim to prevent pain with regular prophylactic doses (eg 4-hourly). Work up the ladder until pain is relieved (see table). Monitor response carefully. Laxatives (eg co-danthrusate 10ml PO at night) and anti-emetics (eg haloperidol 1.5mg) are often needed with analgesic.

Giving oral morphine: Start with aqueous morphine 10mg/4h PO (or 30mg rectal suppository). Double the dose at night (the aim is to promote 8 hours' sleep). Most patients need no more than 30mg/4h PO. A few need much more. It is usually possible to change to slow release morphine (MST) given every 12h, when daily morphine needs are known. If the disease process modifies morphine's metabolism to its more active 6-glucuronide form it may become ineffective (morphine-resistant pain, also confusingly termed 'paradoxical pain')—and methadone, which has a different pathway, may be effective (p342).[3] NSAIDs and ketamine are alternatives.[4]

Nausea and vomiting Give oral relief if possible. If severe vomiting prevents this, give drug rectally or subcutaneously by syringe driver.

Drug treatment

Cause:	Anti-emetic:
Drugs, radiotherapy, chemotherapy	Ondansetron 4–8mg/8–12h PO/IV
Uraemia, hypercalcaemia	Haloperidol 1–20mg/24h PO
Intracranial pressure ↑	Cyclizine 50–100mg/24h PO
Intestinal obstruction	Cyclizine 50–100mg/24h PO
Oesophageal reflux	Metoclopramide 10mg/8h PO (≤0.5mg/kg/day)
Delayed gastric emptying	Metoclopramide 10mg/8h PO (≤0.5mg/kg/day)
Gastric irritation	Stop drugs (eg NSAIDs)

Pruritus (itching) Do not underestimate the unpleasantness of itching. *Causes:* drug reaction; cholestatic jaundice; renal failure; malignant disease; diabetes (with candidosis); psychological reaction; coexistent skin disease. *Management:* Discourage scratching (cut nails); no soap (use aqueous cream and emollient, eg 50% liquid paraffin and 50% soft white paraffin); dry skin by gentle patting with towel. Occasionally an antihistamine is needed in addition.

Breathlessness Consider supplementary oxygen or morphine. Is there a malignant pericardial effusion? If so, only start treatments such as pericardiocentesis, pericardiectomy, pleuropericardial windows, external beam radiotherapy, percutaneous balloon pericardiotomy, and pericardial instillation of immunomodulators or sclerosing bleomycin if there is a good prospect for improving life's quality.

Hiccup 4 ways to stop the attack: 1 Stimulate pharyngeal nerve (cold key down back; drink from wrong side of cup; 2 teaspoons of granulated sugar). 2 Reduce gastric distension (peppermint water; metoclopramide 10mg/6h PO). 3 Elevate $PaCO_2$ (rebreathe paper bag; hold breath). 4 Haloperidol (above).

Symptom control in severe cancer

Common pains in cancer

Bone	Caused by cancer
Nerve compression	Caused by cancer
Soft tissue	Caused by cancer
Visceral	Caused by cancer
Myofascial	Related to cancer or disability
Constipation	Caused by cancer treatment
Muscle spasm	Related to cancer or disability
Low back pain	Caused by concurrent disorder
Chronic post-op	Caused by cancer treatment
Capsulitis of shoulder	Related to cancer or disability

The analgesic ladder

rung one	*non-opioid*	aspirin; paracetamol; NSAID
rung two	*weak opioid*	codeine; dihydrocodeine; dextropropoxyphene
rung three	*strong opioid*	morphine; diamorphine; buprenorphine (sublingual)

NB: If one drug fails to relieve pain move up ladder; do not try other drugs at the same level.

767

1 R Twycross 1990 *Symptom Control in Terminal Cancer*, Sobell Publications, Oxford
2 R Twycross, S Lack 1990 *Therapeutics in Terminal Care*, Churchill Livingstone, Edinburgh
3 D Bowsher 1993 *BMJ* i 473   4 R Twycross 1993 *BMJ* i 793

Cancer treatments

Invasive cancer affects 30% of the population and 20% die from cancer. Management often involves close liaison between the patient, surgeons, physicians, oncologists, haematologists and paediatricians.

Surgery Usually, a biopsy is offered to establish diagnosis. Surgery may be the most common, and sometimes the only, treatment—as is the case for some tumours in the GI tract, soft tissue sarcomas and gynaecological tumours and for advanced cancers of the head and neck.

Radiotherapy Uses ionizing radiation to kill tumour cells. Radiation is by naturally occurring isotopes, or by artificially produced x-rays. The quantity of radiation is the adsorbed dose; its SI unit is the gray (Gy). Sometimes germ-cell tumours (seminoma, teratoma) and lymphomas appear to be particularly sensitive to irradiation which is therefore a major part of management particularly for local disease. Many head-and-neck tumours, gynaecological cancers, and localized prostate and bladder cancers are curable with radiotherapy. Radiotherapy may also be used in reducing symptoms, for example the pain from bone metastases.

Unwanted effects of radiation: Generalized effects: lethargy; loss of appetite. Skin: erythema; dry desquamation with itching; moist desquamation. Patient should keep treated area dry and clean and avoid soap. Radiotherapy markings should not be removed during treatment. Mouth and GI tract: prior dental assessment if treatment to oral cavity; antiseptic mouth washes and avoidance of smoking and spicy food. Aspirin mouth wash for pain, and treat oral thrush (p500). Treat dysphagia with Mucaine® mixture 15mins before meals. Pelvic area: Low residue diet helpful if change in bowel habit. Treat diarrhoea with codeine phosphate. Culture MSU if urinary frequency or dysuria, and give Mist potassium citrate.

Chemotherapy The choice of agent, its administration and dosage require expert guidance. Over twenty substances are in common use. Some major classes are: alkylating agents (eg cyclophosphamide, clorambucil, busulphan); antimetabolites (eg methotrexate); vinca alkaloids (eg vincristine, vinblastine); antitumour antibiotics (eg actinomycin D).

Chemotherapy is an important treatment in: germ-cell tumours and some leukaemias and lymphomas; ovarian cancer (following surgery, although only 20% will be cured); and small cell lung cancer (although most patients will still die within 18 months). Chemotherapy has a part to play in the treatment of: some breast cancer; advanced myeloma; sarcomas and some childhood cancers (such as Wilms' tumour).

Common side-effects: Nausea and vomiting; marrow suppression; alopecia. Each substance has its own spectrum of unwanted effects. *Extravasation of chemotherapeutic agent* (ie agent infused into subcutaneous tissue). Maintain a high index of suspicion. Signs and symptoms: pain; burning; swelling; infusion slows. Management: Stop infusion. Withdraw 5ml of blood. Remove needle. Elevate limb and apply cold pack over area for about 15mins. Do not use same site for infusion until all signs have disappeared. If ulceration seek advice from plastic surgeon.

Cancer treatments

Communication ▶ *Include the patient in the decision-making process.* It is known that younger, female patients particularly want this—and giving of information and the sharing of decisions is known to reduce treatment morbidity. So although this is exhausting (sometimes the same ground needs to be covered many times) it is definitely worth spending time in this area. One widely-liked aid in this regard is for the specialist to video-tape her consultation, to give to the patient for repeated viewing. A huge amount is forgotten or fails to register the first time.

Relevant pages in other chapters:
The chest x-ray (p326); IVU (p372); ultrasound in obstetrics (*OHCS* p188).

Radiology and imaging

CT (computerized tomography) has revolutionized the practice of neurology and neurosurgery. For other areas of the body, the decision whether to use conventional radiology, nuclear medicine, ultrasound or CT is not always clear-cut and may depend on local availability. Generally, do plain films before contrast, and do ultrasound where possible (cheap and non-invasive). The more expensive CT and MRI (magnetic resonance imaging) are to be held in reserve for when the preliminary investigations are indecisive or inadequate. However, because a hospital bed is such an expensive resource, if an expensive test can make a quick diagnosis, money may be saved. An example of using expensive tests in cost-containment is using early CT in the investigation of abdominal masses, rather than leaving CT to be the last in a long line of often inconclusive tests.[1] This philosophy is all the more telling in some emergencies—eg severe head injury with Glasgow coma scale <8 (p718), which requires CT as the first test.

Every physician should be able to read the plain chest and plain abdominal film, and recognize common medical disorders presenting on hand and skull radiographs. There is no substitute for examining the x-rays of your patient yourself to gain experience in the appearances of normal and abnormal films.

There are 4 basic steps in interpreting any radiograph:

1. Check the film is technically correct Is the film named, dated, right/left orientated, and marked as to whether AP (anteroposterior), PA (posteroanterior), erect, or supine. Penetration is important (vertebral bodies just visible through heart on a CXR) when commenting on diffuse shadowing. Check for rotation (eg asymmetry of clavicles on a CXR) as this may affect the appearances of normal structures (eg the hila).

2. Describe the abnormalities seen This may be a change in the appearance of normally visualized structures, or an area of increased opacity or translucency. There are 4 main radiographic densities: bone, air, fat, water (ie soft tissue, which is mostly water). A border is only seen at an interface of 2 densities, eg heart (water) and lung (air); this 'silhouette' is lost if air in the lung is replaced by consolidation (water). This is the silhouette sign and can be used to localize pathology (eg right middle lobe pneumonia or collapse causing loss of distinction of the right heart border).

3. Translate these into gross pathology (eg pleural fluid, consolidation, collapse).

4. Suggest a differential diagnosis.

The order of scrutiny is unimportant, but it is advisable to have a structured approach so as not to miss any major abnormality.

▶ Remember . . . *Treat the patient not the film.*

1 J Husband 1985 *BMJ* i 527

Radiology and imaging

The chest film

(▶ See also p326)

Check that the film is technically correct, only then:
- Comment on any obvious abnormality.
- Systematically consider: heart, mediastinum, lung fields, the diaphragm, soft tissues (remember both breasts) and bones.
- Re-study the 'review areas': the apices, the hila, behind the heart, and the costo-phrenic angles (for small effusions).
- Consider if you need a lateral film.

The diaphragm Check expansion (normally 6 ± 1 anterior ribs or 9 ± 1 posterior ribs). Overinflation may suggest COAD or asthma. Right hemidiaphragm is higher than the left by up to 3cm in 95% of cases. The lateral costo-phrenic angles should be sharp and acute—ill-defined in hyperinflation or with effusion. On a lateral view, the right hemidiaphragm is the one that passes through the heart shadow to the anterior chest wall while the left ends at the posterior heart border.

The root of the neck and trachea The trachea is central, but slight deviation to the right may be seen inferiorly. A paratracheal line is commonly seen—loss of this suggests paratracheal lymphadenopathy.

Mediastinum and heart. Look for the landmarks. The cardio-thoracic ratio is the ratio of the heart width to the chest width. It should be <50%—(essential to have a PA film as the heart is magnified on AP or supine films). Observe patterns of cardiac enlargement (see fig.). Mediastinal widening may be due to aortic dissection, lymph nodes, thymus, thyroid or tumour.

The hila Composed of pulmonary arteries and veins with nodes and airways. Left usually higher than right (~1cm) but of equal density. Look for change in density—tumour, lymph nodes or just rotated film?

The lung fields Increased translucency may be:
1. Pneumothorax—absent vascular markings and lung edge visible.
2. Bullous change—eg in emphysema.
3. Pulmonary embolus—localized oligaemic lung fields.
4. Pulmonary hypertension—p366.
5. Hyperinflation in COAD.

Abnormal opacities may be classifed as:
1. *Consolidation:* Diffuse margins, or opacity+air bronchogram within it (silhouette sign) but little volume change (unlike collapse).

2. *Collapse:* Characteristic patterns are seen (see p327), and loss of volume causes shift of the normal landmarks (hila, fissures, etc.).

3. *'Coin' lesions:* (ie a solitary pulmonary nodule) have an enormous differential diagnosis. Treat as tumour until proved otherwise. See p343.

4. *'Ring' shadows:* Either airways (pulmonary oedema, bronchiectasis), or cavitating lesions, eg pulmonary infarct (triangular with a pleural base), abscess (bacterial, fungal, amoebic, hydatid) or tumour.

5. *Linear opacities:* Septal lines (Kerley B lines—interlobular lymphatics seen with fluid, tumour, dusts, etc.). Atelectasis.

6. Diffuse lung shadowing: See p326.

Some structures seen on a normal PA chest radiograph

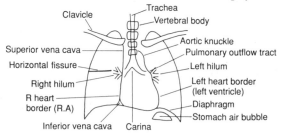

- Clavicle
- Trachea
- Vertebral body
- Aortic knuckle
- Superior vena cava
- Pulmonary outflow tract
- Horizontal fissure
- Left hilum
- Right hilum
- Left heart border (left ventricle)
- R heart border (R.A)
- Diaphragm
- Stomach air bubble
- Inferior vena cava
- Carina

Some structures seen on a normal left lateral CXR

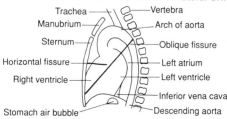

- Trachea
- Vertebra
- Manubrium
- Arch of aorta
- Sternum
- Oblique fissure
- Horizontal fissure
- Left atrium
- Right ventricle
- Left ventricle
- Inferior vena cava
- Stomach air bubble
- Descending aorta

Patterns of cardiac enlargement in standard PA and lateral chest radiographs

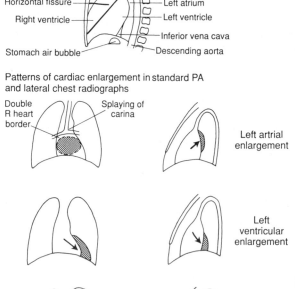

Double R heart border
Splaying of carina

Left artrial enlargement

Left ventricular enlargement

Right ventricular enlargement

775

The plain abdominal film

- Check the name, the date and orientation (right/left).
- Is it erect or supine?—the gas pattern is best seen on a supine film, but many surgeons prefer an erect film to demonstrate fluid levels. Free intraperitoneal air is best seen on an erect CXR.
- If any contrast evident? Always ask for the control film. Calcification is otherwise easy to miss.
- Observe the gas pattern and position Small bowel is recognized by its central position and valvulae conniventes, which reach from one wall to the other. Larger bowel is more peripheral and the haustrae go only part of the way across. Abnormal position may point to pathology eg gas is central in ascites; displaced to the left lower quadrant by splenomegaly.
- Look for extraluminal gas For example, in the liver or biliary system (after passing a stone, after ERCP (p498), or with gas-forming infection), the GU system (entero-vesical fistula), in the peritoneum, the colonic wall (pneumatosis coli, infective colitis) or in a subphrenic abscess.
- Look for calcification in the abdomen or pelvis Calcification in arteries (atherosclerosis, the egg-shell calcification in an aneurysm), lymph nodes, phleboliths (smooth, round), renal calcification or ureteric stones (usually jagged; look along the line of the ureter), adrenal (TB), pancreatic (chronic pancreatitis), liver or spleen, gallstones (only 10% opaque), bladder (stone, tumour, TB, schistosomiasis), uterus, dermoid cyst, which may contain teeth.
- Bones of spine and pelvis Look for metastases (osteolytic or osteoblastic), Paget's disease, Looser's zones (osteomalacia), collapse, osteoarthritis; the 'rugger-jersey' spine of osteomalacia.
- Soft tissues The psoas lines are obliterated in retroperitoneal inflammation, haemorrhage or peritonitis. Note kidney size and shape (normally 2–3 vertebral bodies length and parallel to the psoas line).
- Meteorism Localized peritoneal inflammation can cause a localized ileus. This may be seen on the plain film as a 'sentinel loop' of intraluminal gas and can provide a clue to the site of pathology.

Cholecystitis	Pancreatitis
Appendicitis	Diverticulitis

(NB: in young patients, even localized inflammation can produce a generalized ileus).

The plain abdominal film

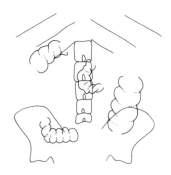

Plain abdo film (structures visible)

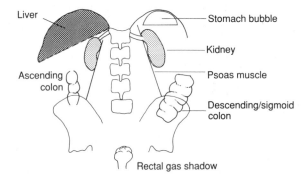

Plain skull views (SXR)

The most common reason for requesting a skull x-ray is trauma.

▶ If the Glasgow coma scale score is ≤8/15 after acute head injury, the patient needs an urgent CT scan: *do not delay this for skull x-rays*.

The most important things to look for are:

- Presence of a linear skull fracture (increases likelihood of intracranial haematoma from 1:1000 to 1:30 if alert, and from 1:100 to 1:4 if consciousness impaired).
- Presence of a depressed skull fracture (needs elevation if depressed > width of vault).
- Presence of a calcified pineal, shifted >3mm from the midline—but not all the population has a conveniently calcified gland.
- The status of the cranio-cervical junction.

Skull radiographs may be useful in other clinical situations:

- **Bone density and thickness** Diffuse increase in vault density is seen in Paget's disease and fluorosis (excessive fluoride ingestion, eg from insecticides, causing mottled tooth enamel, and osteofluorosis, which is a combination of osteosclerosis and osteomalacia); diffuse increase in thickness is seen in acromegaly, thalassaemia and meningiomas (especially parasaggital or sphenoidal ridge) and Paget's disease; localized increase in bone density is seen in Paget's disease, osteomyelitis, malignancy (eg leukaemia and histiocytosis); lucent areas are seen in trauma, malignancy (myeloma), Paget's disease, hyperparathyroidism.

- **Intracranial calcification** May be normal (pineal) or may reflect a tumour (7% of gliomas, 10% of meningiomas, >70% of craniopharyngiomas are calcified), vascular malformation (sinuous tram-line occipital calcification seen in Sturge–Weber syndrome; ring calcification of cerebral aneurysms) or old infection (TB, toxoplasmosis).

- **The sella turcica** Normally, widest AP diameter of the pituitary fossa is 11–16mm and depth 8–12mm in adults. Enlargement occurs with pituitary adenoma; erosion of the posterior clinoids with raised intracranial pressure; erosion of the lamina dura of the dorsum sellae with tumour or aneurysm.

- **Air sinusus** Thickened mucosa is seen in chronic sinusitis; enlarged sinuses in acromegaly; fluid level after trauma (eg in maxillary sinus with infraorbital fracture; sphenoid sinus with basal skull fracture—you will miss this if you do not turn lateral SXR on its side, ie the position in which the film was taken).

The hand x-ray

Cast your eye over the entire film looking for:

- Deformities Ulnar deviation and subluxation (rheumatoid arthritis), syndactyly, polydactyly, short metacarpals (p546).
- Joints Erosions and bone destruction of the arthritides, effusions (seen as joint space widening and pericapsular soft tissue). Check the joints involved and symmetry (DIP, PIP, wrists).
- Loss of bone density—diffuse: osteoporosis; localized: cysts, (osteo-arthritis—Heberden's nodes at DIP joints, Bouchard's nodes at PIP joints; simple cyst; sarcoid), avascular necrosis, erosions (below).
- Erosions eg of terminal phalangeal tufts seen in sarcoid, hyper-parathyroidism, and scleroderma (the latter allowing the nail to curve—'pseudoclubbing') and subperiosteal erosions along the radial border of the middle phalanges (an early sign of hyperparathyroidism). Coarse trabeculations are seen in chronic haemolytic anaemia, Paget's disease, and the lipidoses (eg Gaucher's syndrome, OHCS p748).
- Soft tissues—generalized increase in thickness seen in acromegaly (spade-like hands), localized thickness, eg gouty tophi, arterial, peri-capsular or soft-tissue calcification (eg CREST, p670).

The spine

In major trauma the first x-ray to be performed after resuscitation is a cross-table lateral of the cervical spine. All seven cervical vertebrae must be seen, along with C7–T1 junction: do not accept an incomplete x-ray—try traction on the arms, or a swimmer's view. Occasionally subluxations do not show up without flexion and extension views (perform only on the advice of a senior colleague).

▶ If a cervical spine injury is suspected you must immobilize the neck until it is excluded.

▶ A cross-table portable in the best hands will miss up to 15% of injuries.

When examining the film, follow 4 simple steps:

● *Alignment:* Of anterior and posterior vertebral body, posterior spinal canal and spinous processes. A step of <25% of a vertebral body implies unifacet dislocation; if >50% it is bifacetal. 40% of those <7yrs old have anterior displacement of C2 on C3 (pseudosubluxation); still present in 20% up to 16yrs. 15% show this with C3 on C4. In this physiological subluxation, the posterior spinal line is maintained. Angulation between vertebrae >10% is abnormal.

● *Bone contour:* Trace around each vertebra individually. Look for avulsion fractures of the body or spinous process. A wedge fracture is present if the anterior height differs from posterior by >3mm. Check odontoid (open mouth view but can be seen on lateral view); don't mistake vertical cleft between incisors for a peg fracture!—epiphyses in children can be mistaken for fractures; the distance between the odontoid and anterior arch of C1 should be <3mm (may be increased in children). Type 1 fractures are of the odontoid peg, type 2 of its base, and type 3 fractures extend into the body of C2.

● *Cartilages:* Check that intervertebral disc space margins are parallel.

● *Soft tissues:* If the space between the lower anterior border of C3 and the pharyngeal shadow is >5mm suspect retropharyngeal swelling (haemorrhage; abscess)—this is often useful indirect evidence of a C2 fracture. Space between lower cervical vertebrae and trachea should be <1 vertebral body.

Wedge fractures are common, particularly in the osteoporotic. Suspect metastatic secondaries (often seen in the vascular pedicles), especially if adjacent vertebrae are involved. A spondylolysis (usually at L5/S1) is seen as a defect in the pars interarticularis on the lateral, or more easily the oblique film (look for 'Scottie dog'—the 'collar' is the defect); it may be congenital or acquired, the latter being a fatigue fracture, acute trauma or pathological fracture (tumour; TB; Paget's). If L5 slips forward on S1 this is a spondylolisthesis. Prolapsed intervertebral disc cannot be diagnosed from a plain film; MRI gives better definition than CT, and both have largely replaced radiculograms.

Specific conditions: Ankylosing spondylitis, p668 (sacroiliitis; syndesmophytes; bamboo spine); osteomalacia ('rugger-jersey' spine); Scheuermann's disease (apophysitis resulting in fixed smooth kyphosis in adolescents); spina bifida occulta (present in up to 20% of population: may be minor abnormal neurology in legs).

Principal source: American College of Surgeons (1988), Advanced Trauma Life Support (ATLS) course manual, and PA Driscoll 1993 *BMJ* ii 855

Contrast studies

These are performed either with an IV contrast agent or with an oral contrast such as barium or Gastrograffin.

Contrast reactions to IV reagents occur in ~1/1000 patients and death by anaphylaxis in 1/40,000. Reactions include hives, bronchospasm and/or pulmonary oedema. Ask about a history of atopy or allergy to iodine or seafood. Take advice from your radiologists—pre-medication with corticosteroids (eg: prednisolone 40mg PO 12h prior and 2h prior) is effective in reducing the incidence of reactions. The newer non-ionic contrast agents available should be used in this situation. When renal function is impaired, ensure good hydration and diuresis (IV fluids + mannitol) pre- and post-study.

Angiography Used to image aorta, major arteries and branches (atheromatous stenosis and thrombosis, embolism, aneurysms, A–V fistulas and angiomatous malformations), investigation of tumours and may be combined with interventions eg, balloon dilatation, embolization, marking for identification intraoperatively, etc. Digital subtraction angiography (DSA) provides reverse negative views and requires less contrast load for the patient.

- *Cardiac and coronary angiography:* (See p272.)
- *Pulmonary angiography:* Used for visualization of emboli and vascular abnormalities and assessment of right heart pressures. It is the most accurate procedure for diagnosis of PE but only used in cases of massive PE when intervention is considered (p728).
- *Cerebral angiography:* For intra- and extra-cranial vascular disease, atherosclerosis, aneurysms and arterio-venous malformations.
- *Selective visceral angiography:* Used to locate the source of GI bleeding (acute or chronic) when this has not been found by endoscopy. It may be used to selectively infuse drugs or embolic material into the bleeding territory. Coeliac axis angiography is used for intrahepatic pathology (haemangiomas, site and blood supply of hepatic tumours pre-operatively) and is used for selective embolization (especially for carcinoid secondaries in the liver). Late films give good views of the portal vein. Splenic phlebography, performed by percutaneous puncture of the spleen and injection of contrast may be a useful test in portal hypertension for imaging and pressure measurements.
- *Renal angiography:* Originally used to distinguish tumour from cyst (now superseded by ultrasound and CT). Widely used for the embolization of vascular tumours and for the investigations of renal hypertension (2° to atheroma or fibromuscular hyperplasia).
- *Angiography in peripheral vascular disease:* (See p128).

Contrast studies

GI studies with contrast

(Barium swallow, barium meal, small bowel follow through, small bowel or colonic enema).

Commonly done with 'double contrast' technique of air and barium which gives improved demonstration of surface mucosal pattern. Distending the gut wall allows pathology to be more readily seen. Used for demonstration of structural lesions (eg tumour, diverticula, polyps, ulcers, fistulae) as well as abnormal peristalsis, reflux etc. If GI perforation or fistula is suspected, often Gastrograffin is the contrast of choice to prevent contaminating the operative field with barium.

Cleansing the colon is the single most important determinant of quality of barium enemas—régimes vary; contact your radiology department.

Cholangiography may be performed in a variety of ways.
- *Oral cholecystography:* Was the first investigation to be used to visualize the gall bladder. Patient is given oral contrast 12–24h prior to the study. Failure of gall bladder opacification is considered abnormal and indicative of gallstone disease, but may also be due to acute pancreatitis, peritonitis or cholecystitis. Unlikely to work if serum bilirubin >34μmol/l.
- *Intravenous cholecystography:* Is seldom indicated but may be useful in suspected choledocholithiasis and now rarely performed. In order to carry out the investigation, serum bilirubin should be <50μmol/l. Contraindicated in patients with severe hepato-renal disease, allergy to contrast and IgM paraproteinaemia (precipitates out with contrast).
- *Percutaneous transhepatic cholangiography:* May be used to demonstrate dilated ducts in obstructive jaundice and may be combined with external drainage or internal drainage (using an endo-prosthesis across the stricture). Contraindications: bleeding tendency, cholangitis, ascites, allergy.
- *Endoscopic retrograde cholangiopancreatography* (ERCP): (See p498.)

Intravenous urography (See p372.)

GI studies with contrast

Ultrasound

Because imaging by ultrasound entails no ionizing radiation, it has completely taken over conventional radiology in obstetrics (except for pelvimetry, see *OHCS* p118). Also, it is often the first line of investigation of abdominal organs and cardiac lesions. It is also used to guide needle aspiration, or biopsy of masses and collections. However, it is quite operator-dependent and reliability of results may vary. The use of Doppler allows one to assess, qualitatively, patency and direction of flow through vessels, but does not give an accurate quantitative measurement.

Abdominal scan This is often the initial investigation for abdominal masses. It may be used to assess: 1 Liver size and texture (eg fatty liver, shrunken cirrhotic liver), masses within it (cysts vs tumour). 2 The biliary system (dilation in obstructive jaundice, p484; the cause, ie stone vs tumour, is less reliably seen). 3 Doppler of the portal and splanchnic veins is used to rule out thrombosis and assess direction of flow in the portal veins in cirrhosis. 4 The gall bladder (thickening and gallstones within it). 5 The pancreas (pseudocysts, abscesses and sometimes tumour). 6 The aorta and major vessels for aneurysm. 7 The kidneys (size, texture, hydronephrosis in obstruction, polycystic disease, tumour). Often the initial investigation of haematuria and better than IVU for bladder pathology (but see p314). Good for showing fluid and lymphoceles around transplanted kidneys. The adrenals are less reliably imaged and only fairly large masses identified.

Pelvic scan ● Ultrasound is the standard imaging technique for monitoring of normal and abnormal pregnancy, fetal growth and development, localization of the placenta, ectopic pregnancy. ● Ovarian and uterine masses (tumours, cysts and fibroids). ● Testicular masses—hydrocele vs tumour.

Cardiac ultrasound See p274. M-mode scanning allows measurements of chamber size and accurate assessment of valve and wall motion. 2D echocardiography provides real-time images of the anatomy and spatial relationships. Duplex Doppler scanners include colour coding of flow directions and allows direct visualization of shunts and regurgitant flow.

Miscellaneous *Thyroid scan:* To distinguish cyst from tumour. Cannot distinguish benign from malignant nodule.

Orbit and eye: Assessment of retinal and choroidal detachment, localization of foreign bodies and assessment of retro-orbital masses.

Large veins: (eg Femoral and popliteal veins) may be assessed by Doppler ultrasound as an easy, non-invasive initial investigation when thrombosis is suspected (p92).

Other places where an adventurous probe has entered: Oesophagus (mitral valve analysis); vagina (ovarian cancer screening); rectum (prostate).

Radioisotope scanning: 1

Almost any organ of the body may be investigated by using isotopes that are selectively taken up by the organ. Lesions such as tumours may fail to take up the isotope and thus appear as 'cold' areas, or may take up more of the isotope than the surrounding tissue and appear as 'hot' spots. Linear scanners have largely been replaced by the gamma camera which scans a wider field, and the activity can be recorded and processed to provide physiological data about the function of the organ.

Bone scanning ^{99m}Tc (technetium) labelled phosphate complexes are used in the assessment of metastatic disease (especially thyroid, breast, kidney, lung, and prostatic cancer). However, remember that scans are non-specific and will also show up benign lesions, eg: Paget's disease, bone fractures, inflammatory bone lesions, osteoarthritis, rheumatoid arthritis, and fibrous dysplasia.

Gallium-67 scans Used to localize abscesses (5–10 days' old), chronic inflammatory lesions and spread of neoplasms. Most valuable for assessing lymphomas, paticularly in the mediastinum.

Brain scanning (^{99m}Tc) An unreliable way of excluding brain metastases—largely replaced by the more accurate CT and MRI scans.

Cardiac scanning
- ^{99m}Tc *ventriculography:* Labelled RBC or serum albumin demonstrates abnormal wall motion, shunts, size, and function of chambers, cardiac output and ejection fraction. For the MUGA scan (multigated acquisition), data collection is synchronized to the ECG, and selected to create a dynamic picture of cardiac function. May be done at rest or during exercise (but not suitable for fast, irregular rhythms).
- ^{99m}Tc *pyrophosphate scanning:* The isotope is concentrated by recently damaged myocardium. Most sensitive 24–72h after an infarct.
- *Thallium-291 scanning:* Taken up by normal myocardium. An acutely infarcted area (<12h) appears 'hot'; ischaemic or infarcted areas appear 'cold' on exercise, but ischaemic area returns to normal after rest.

Ventilation/Perfusion scan (V/Q scan) ^{99m}Tc macroaggregates or microspheres are used for perfusion lung scanning (Q scan) and areas of hypoperfusion (eg due to pulmonary embolus (PE)) show up as cold areas. However, hypoperfusion due to pneumonia, TB, cysts, or collapse cannot be differentiated from a PE on a perfusion scan alone. Ventilation scans (V scan) may be performed following inhalation of radioactive xenon (^{133}Xe) or krypton (^{81m}Kr). In patients with PE, there will be a defect in the Q scan where the V scan is normal.

Radioisotope scanning: 1

Radioisotope scanning: 2

Hepatobiliary scans ^{99m}Tc colloid is used for liver scanning; masses >2cm diameter appear as 'cold' spots (eg tumour, cysts, abscesses, haematomas). Biliary tree scanning uses ^{99m}Tc-HIDA, a drug which is excreted in bile (hepatic immunodiacetic acid). Serial images show activity in the biliary tree progressing to the gut. The gall bladder will fail to fill in acute cholecystitis or if the cystic duct is obstructed.

Thyroid scan (With ^{99}Tc-pertechnate or ^{125}I.) Used to evaluate nodules (up to 25% of solitary cold nodules are malignant). In conjunction with thyroid function tests and ultrasound, it may help diagnose hyperfunctioning adenomas (single or multiple toxic nodules), and Graves' disease, multinodular goitres, and localize ectopic thyroid tissue (especially after thyroidectomy).

Renal scans
- *^{131}I-Hippuran:* Excreted almost entirely in one passage and useful for evaluation of renal function.
- *^{99m}Tc-DMSA:* Used as a renal cortical imaging agent; uptake reduced and uneven in chronic pyelonephritis, obstructive uropathy, TB; tumours and cysts show up as filling defects.
- *^{99m}Tc-DPTA:* Assessment of renal function; renal blood flow studies; estimation of GFR and evaluation of collecting system.

Adrenal scanning Imaging with ^{79}Se-cholesterol may help localize tumours producing Cushing's or Conn's syndrome, but is not helpful in identifying phaeochromocytoma. The latter, and any adrenal tumour metastases, show up with MIBG (metiodobenzylguanidine).

Testis scans ^{99m}Tc-pertechnate can be used to differentiate between orchitis (where the uptake is increased) and torsion (uptake decreased).

White cell scan Radiolabelled peripheral blood leucocytes (indium or technetium labelled) provide a non-invasive way of localizing and monitoring inflammatory activity: it may be used to find hidden pus, and can be used in patients with colitis (Crohn's or UC) not only to image the extent of disease, but also to monitor progression of disease and response to treatment.

Miscellaneous More recent innovations include using radiolabelled serum amyloid P protein (SAP scans); to diagnose amyloidosis, and radiolabelled octreotide or MIBG to localize the tumour in carcinoid syndrome.

Computerized tomography (CT) scans

Computerized tomography can be performed with or without IV contrast. A dilute oral contrast agent prior to abdominal or pelvic scanning helps delineate the bowel. Density measurements (expressed as Hounsfield Units) can be used to differentiate between masses (eg cysts, lipomas or haematomas). In Hounsfield units, bone is +1000, water is 0, fat is −1000 and the remaining tissues fall between depending on consistency. Vascular lesions 'enhance' with IV contrast while non-vascular lesions do not.

The brain

Most suspected intracerebral lesions are now investigated by CT as a primary diagnostic procedure. CT is used for the diagnosis of hydrocephalus where the ventricles are dilated and sulci flattened against the inner skull. Mass lesions cause midline shift, compressing the lateral ventricles and may have considerable surrounding oedema seen as a low density area. Also used to establish whether the cause of a stroke is intracranial bleeding (haemorrhage is seen as a high density lesion). As the clot absorbs, the lesion becomes isodense, and, later, of low attenuation. Similarly, extradural and subdural haematomas are seen in the acute phase as high density peripheral lesions. Chronic subdural haematomas appear hypo-dense. Cerebral infarction is seen as an area of low attenuation, often enhancing, with no mass effect. Cerebral atrophy is seen as deepening of the sulci and narrowing of the gyri.

The abdomen

Most liver pathology can be assessed by ultrasound but CT is generally more accurate and shows smaller lesions. The pancreas is well demonstrated, but small lesions can be missed. Acute pancreatitis results in generalized swelling and oedema of the gland. CT is particularly useful in delineating renal and adrenal masses. It has largely replaced lymphangiography in evaluation of para-aortic and retroperitoneal lymphadenopathy and is used in the staging of abdominal and haematological malignancy.

The chest

CT is of great value in picking up small lesions such as pleural deposits, which may be missed on plain x-ray: it is able to find 40% more nodules than whole lung tomograms which demonstrate 20% more nodules than plain x-ray. Also useful for assessing mediastinal masses and aneurysms. Modern fast body scanners can even show intracardiac lesions (eg myxomas) although transoesophageal ultrasound and ECG-gated MRI remain superior.

Computerized tomography (CT) scans

Magnetic resonance imaging (MRI)

MRI uses measurements of magnetic movements of atomic nuclei to delineate tissues. Refer to a specialized text for the physics involved. Two types of images are produced:

- *T1 weighted images:* Provide good anatomical planes and better separation of cystic and solid structures due to the wide variance of T1 values among normal tissues. Fat appears as brightest (high signal intensity), the remaining tissues appearing as varying degrees of lower signal intensity (black). Flowing blood appears as black.
- *T2 weighted images:* These provide the best detection of pathology and a decreased visualization of normal anatomy. Tumour surrounded by fat may be lost on T2 imaging. Fat and fluid collections appear brightest.

Advantages of MRI:
- Non-ionizing radiation
- Shows vasculature without contrast
- Images can easily be produced in any plane, eg sagittal and coronal (good for spinal cord, aorta, vena cava)
- Vizualization of posterior fossa and other areas of high CT bony artefact, eg cranio-cervical junction
- High inherent soft tissue contrast

Disadvantages of MRI:
- High cost of equipment
- Claustrophobia (in magnet tunnel for 30–120mins—hypnosis, music, or sedation may be needed)
- Long imaging time causes increased motion artefact
- Unable to scan very ill patients requiring monitoring equipment
- Unsuitable for those with metal foreign bodies (eg pacemakers, CNS vascular clips, cochlear implants, valves, shrapnel)
- Unable to image calcium

Positron emission tomography (PET)

Molecules labelled with positron-emitting radionuclides are injected, and post-annihilation event photon paths are analysed by crystal detectors to produce 3D images of function—eg of blood flow (labelled ammonia) or glucose metabolism (fluordeoxyglucose, FDG).

In cardiology[1,2] it may help answer questions such as 'Is this piece of non-functioning myocardium dead or just hibernating?' If hibernating (low blood flow but glucose metabolism is present), then angioplasty to its supply vessel could improve function.

PET in recurrent tumours PET can help where CT and MRI fail to distinguish cell death following radiotherapy from tumour recurrence. This helps in in planning stereotactic radiotherapy.

PET can spare patients more invasive investigations eg in determining the site of epileptogenic foci, where there is no anatomic lesion which could be demonstrated by MRI.

PET can also diagnose dementia (even before symptoms start), and distinguish Alzheimer's from other dementias (symmetrical hypometabolism in parietal and temporal lobes—not the frontal lobes, which are affected in Pick's dementia, for example).

PET in research PET may help answer questions such as 'Does schizophrenia have a physical basis in the brain?'—and it is interesting to note that it has already shown subtle changes affecting the hippocampus and temporal lobes.

Cost ⩾£350/scan. Cost-effectiveness issues have yet to be resolved.

1 D Eitzman 1992 *J Am Col Card* **20** 559   2 Y Yamamoto 1992 *Circulation* **86** 167–78

Magnetic resonance imaging (MRI)
Positron emission tomography (PET)

Calculating doses of gentamicin

To determine gentamicin dose schedule[1] **1** Join with a straight line the serum creatinine concentration appropriate to the sex on scale A and the age on scale B. Mark the point at which this line cuts line C.
2 Join with a line the mark on line C and the body weight on line D. Mark the points at which this line cuts lines L and M. These will give the loading and maintenance doses respectively.
3 Confirm the appropriateness of this régime at an early stage by measuring serum levels, especially in severe illness and renal impairment.
4 Adjust dose if peak concentrations (1h after IM dose; 30mins after IV dose) outside the range 5–10mg/l. Trough concentrations (just before dose) above 2mg/l indicates the need for a longer dosage interval.

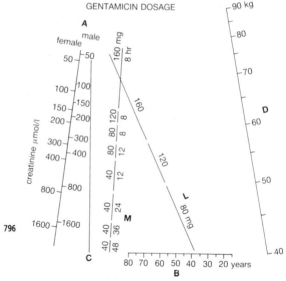

GENTAMICIN DOSAGE

NB: The above allows for **thrice-daily** doses. Many now favour dosing no more frequently than **once daily**, because of fewer adverse effects, and there is quite good evidence from meta-analyses backing up such assertions.[2,3] The big problem is that we do not have good information on how to calculate (and monitor) once-a-day regimes.[4] The *Cooke & Grace régime*: One source provisionally recommends, for feverish neutropenic adults with srum creatinine <300µmol/l, a starting dose of gentamicin 5mg/kg IVI over 30mins, with a serum trough value measured ~24h later. If this proved satisfactory (level <1mg/l) then perform twice-weekly monitoring. If the trough level is 1–2mg/l, halve the dose, and monitor the next trough (ie at 24h). These authors generally stopped gentamicin if this level was >2mg/l, and instituted another antibiotic, such as ciprofloxacin.[5]

1 G Mawer 1974 *Br J Clin Pharm* **1** 45   **2** M Barza 1996 Evidence-Based Medicine (E-BM) **1** 144   **3** R Hatala 1996 *E-BM* **1** 145   **4** PS Millard 1996 *E-BM* **1** 144–5   **5** RP Cooke 1996 *BMJ* **ii** 490

Drug therapeutic ranges

▶ Ranges should only be used as a guide to treatment.
A drug in an apparently too low concentration may still be clinically useful. Some patients require (and tolerate) levels in the 'toxic' range.

Amikacin Peak (½h post IV dose): 20–30mg/l. Trough: <8mg/l.

Carbamazepine Trough: 25–30µmol/l [6–12mg/l].

Clonazepam Trough: 0.08–0.24µmol/l [0.025–0.075mg/l].

Digoxin† (6–12h post dose) 1–2.6nmol/l [0.8–2µg/l]. <1.3nmol/l may be toxic if there is hypokalaemia. Signs of CVS toxicity: arrhythmias, heart block. CNS: confusion, insomnia, agitation, seeing too much yellow (xanthopsia), delirium. GI: nausea.

Ethosuximide Trough: 300–700µmol/l [40–100mg/l].

Gentamicin Peak—½h post IV dose: 9–18µmol/l [5–10mg/l]. Trough (just before dose): ≤2µmol/l. Toxic signs: tinnitus, deafness, nystagmus, vertigo, renal failure.

Lithium† (12h post dose). Guidelines vary: 0.4–0.8mmol/l is reasonable. *Early* signs of toxicity: tremor, agitation, twitching. *Intermediate:* lethargy. *Late:* spasms, coma, fits, arrhythmias, renal failure. See OHCS p354.

Netilmicin† Peak—½h post IV dose: 7–15mg/l. Trough <3mg/l.

Phenobarbitone (& primidone) Trough: 45–130µmol/l [10–30mg/l].

Phenytoin† Trough: 40–80µmol/l [10–20mg/l]. Signs of toxicity: ataxia, nystagmus, sedation, dysarthria, diplopia.

Theophylline† 10–20µg/ml (55–110µmol/l). (▶ see p731) Take sample 4–6h after starting an infusion (which should be stopped for ~15mins just before the specimen is taken[1]). Signs of toxicity: arrhythmias, anxiety, tremor, convulsions.

Tobramycin Peak (½h post IV dose): 11–21µmol/l [5–10mg/l]. Trough: ≤4.3µmol/l.

Valproate Trough: 300-600µmol/l [50–100mg/l].

† Drugs for which *routine* monitoring is indicated. *Trough levels should be taken just before the next dose. The time since the last dose should be specified on the form.

1 B Widdop 1985 *Therapeutic Drug Monitoring*, Churchill Livingstone

Prescribing drugs

▶ Consult the *British National Formulary (BNF)* or your local equivalent before prescribing any drug with which you are not thoroughly familiar.

Remember the dictum *primum non nocere*—first do no harm. The more minor the complaint, the more weight this dictum carries. The more serious the complaint, the more this dictum's antithesis comes into play: *nothing ventured, nothing gained*. If someone is dying in front of you from cerebral malaria, it would be quite inappropriate to delay, Hamlet-like, while weighing up the chances of causing harm through a rare side-effect of the quinine you are about to give. But if you are about to give the quinine to someone with night cramps (for which it is also very effective) then it is your duty to allow each of the ten commandments listed opposite (p799) to filter through your mind. ▶ That this advice is obvious does not prevent it from being regularly ignored—with consequences which may be just as dire as when the real ten commandments are broken.

Prescribing in renal failure Dose modification should be related to creatinine clearance (p632), and the extent to which the drug is renally excreted. This is significant for aminoglycosides (p796), cephalosporins and a few other antibiotics (p177–p181), lithium, opiates and digoxin. ▶ Never prescribe a drug in renal failure before checking how its administration should be altered. Loading doses should not be changed. If the patient is on dialysis (peritoneal or haemodialysis), dose modification depends on how well it is eliminated by dialysis. Consult the drug's data sheet or an expert. Dosing should be timed around dialysis sessions.

Nephrotoxic drugs include: combination analgesics, aminoglycosides, amphotericin B, penicillins (especially methicillin), frusemide, gold, and penicillamine.

The amount by which the dose should be reduced in renal failure (the dose adjustment factor, DAF), depends on the fraction of the drug excreted unchanged in the urine (F). $DAF = 1/(F(kf-1)+1)$, where the kf is the relative kidney function = creatinine clearance/120. The usual dose should be divided by the DAF. In only a few drugs is F large enough to be important. NB: loading doses should not be modified.

Aminoglycosides*	F=0.9	Cephalosporins	F=1.0
Lithium	F=1.0	Sulphamethoxazole	F=0.3–0.5
Digoxin	F=0.75	Procainamide	F=0.6
Ethambutol	F=0.7	Tetracycline	F=0.4–0.6

* eg gentamicin, netilmicin.

Prescribing in liver failure Avoid opiates, diuretics (↑risk of encephalopathy), oral hypoglycaemics and saline-containing infusions. Warfarin effects are enhanced. Hepatotoxic drugs include: paracetamol, methotrexate, phenothiazines, isoniazid, azathioprine, oestrogen, 6-mercaptopurine, salicylates, tetracycline, mitomycin.

The ten commandments

▶ These ten commandments should be written on every tablet.

1 Explore any alternatives to a prescription. Prescriptions lead to doctor-dependency, which in turn frequently leads to bad medicine as described in OHCS p326. (They are also expensive: £2317 million for the UK NHS service in 1992, and increasing at many times the general rate of inflation.) There are 3 places to look for alternatives: the larder (lemon and honey for sore throats, rather than penicillin); the blackboard (eg education about the self-inflicted causes of oesophagitis rather than giving cimetidine); and lastly look to yourself. Giving a piece of *yourself*, some real sympathy, is worth more than all the drugs in your pharmacopoeia to patients who are frightened, bereaved or weary of life.

2 Find out if he wants to take a drug. Perhaps you are prescribing for some minor ailment because you want to solve every problem. But the patient may be happy just to know the ailment *is* minor. If he knows what it is, he may be happy to live with it. Some people do not believe in drugs, and you must find this out.

3 Decide if the patient is responsible. If he now swallows *all* the quinine pills you have so kindly prescribed, death will be swift.

4 Know of other ways your prescription may be misused. Perhaps the patient whose 'insomnia' you so kindly treated is even now grinding up your prescription prior to injecting himself desperate for a fix. Will you be suspicious when he returns to say he has lost your prescription?

5 List the potential benefits of this drug to this patient.

6 List the risks (side-effects, contraindications, interactions). Of any new problem, always ask yourself: 'Is this a side-effect?'

7 Agree with the patient on the risk:benefit ratio's favourability.

8 Record how you will review the patient's need for each drug.

9 Quantify progress (or lack of it) towards a specified, agreed goals. (Eg pulse rate to mark degree of β-blockade; or peak flow monitoring to guide steroid use in asthma.)

10 Make a record of all drugs taken. Offer the patient a copy.

Some important drug interactions

Note: '↑' means the effect of the drug in italics is increased (eg through inhibition of metabolism or renal clearance). '↓' means that its effect is decreased (eg through enzyme induction).

Adenosine ↓ by: Amnophylline. ↑ by dipyridamole.

Aminoglycosides ↑ by: Loop diuretics, cephalosporins.

Antihistamines (eg terfenadine, astemizole) Avoid anything which ↑ concentrations and so risk of fatal arrhythmias—eg erythromycin, other macrolides (eg azithromycin), antifungals, halofantrine, tricyclics, antipsychotics, diuretics and β-blockers and other antiarrhythmics.

Antidiabetic drugs (any) ↑ by: Alcohol, β-blockers, monoamine oxidase inhibitors, bezafibrate. ↓ by: Corticosteroids, diazoxide, diuretics, contraceptive steroids, (maybe lithium).

　Sulphonylureas ↑ by: Azapropazone, chloramphenicol, clofibrate, co-trimoxazole, miconazole, sulphinpyrazone.

　Sulphonylureas ↓ by: Rifampicin, (nifedipine occasionally).

　Metformin ↑ by: Cimetidine. With alcohol: lactic acidosis risk.

Angiotensin-converting-enzyme (ACE)-inhibitors ↓ by: NSAIDs.

Azathioprine ↑ by: Allopurinol.

Beta-blockers—lipophilic (∴ metabolized by liver, eg propranolol but not atenolol) ↑ by: Cimetidine. All ↑ by: verapamil. ↓: NSAIDs.

Carbamazepine ↑ by: Erythromycin, isoniazid, verapamil.

Cimetidine: Theophylline ↑, warfarin ↑, lignocaine ↑, amitriptyline ↑, propranolol ↑, pethidine ↑, phenytoin ↑, metronidazole ↑, quinine ↑.

Contraceptive steroids ↓ by: Antibiotics, barbiturates, carbamazepine, phenytoin, primidone, rifampicin.

Cyclosporin A ↑ by: Erythromycin. ↓ by: Phenytoin.

Digoxin ↑ by: Amiodarone, carbenoxolone and diuretics (as K⁺ levels lowered), quinine, verapamil.

Diuretics ↓ by: NSAIDs—particularly indomethacin.

Ergotamine ↑ by: Erythromycin (ergotism may occur).

Fluconazole: Avoid concurrent asetemizole or terfenadine.

Lithium ↑ by: Thiazide diuretics.

Methotrexate ↑ by: Aspirin, NSAIDs.

Phenytoin ↑ by: Chloramphenicol, cimetidine, disulfiram, isoniazid, sulphonamides. ↓ by: carbamazepine.

Potassium-sparing diuretics with ACE inbitors: (hyperkalaemia).

Theophyllines ↑ by: Cimetidine, ciprofloxacin, erythromycin, contraceptive steroids, propranolol. ↓ by: barbiturates, carbamazepine, phenytoin, rifampicin. See p731.

Valproate ↓ by: Carbamazepine, phenobarbitone, phenytoin, primidone.

Warfarin (and *nicoumalone*) ↑ by: Alcohol, allopurinol, amiodarone, aspirin, chloramphenicol, cimetidine, co-trimoxazole, danazol, dextropropoxyphene, dipyridamole, disulfiram, erythromycin (and broad spectrum antibiotics), gemfibrozil, glucagon, ketoconazole, metronidazole, miconazole, nalidixic acid, neomycin, NSAIDs, phenytoin, quinidine, sulphinpyrazone, sulphonamides, T₄.

Warfarin (and *nicoumalone*) ↓ by: Aminoglutethamide, barbiturates, carbamazepine, contraceptive steroids, dichloralphenazone, griseofulvin, rifampicin, phenytoin, vitamin K.

Zidovudine (AZT) ↑ by: Paracetamol (increased marrow toxicity).

IVI *solutions to avoid Dextrose:* Avoid frusemide, ampicillin, hydralazine, insulin, melphalan, quinine.

0.9% saline: Avoid amphotericin, lignocaine, nitroprusside.

Haematology—reference intervals

B_{12}, folate, Fe, TIBC, see p802–3.

Measurement	Reference interval	Your hospital
White cell count (WCC)	$4.0–11.0 \times 10^9/l$	
Red cell count	♂ $4.5–6.5 \times 10^{12}/l$	
	♀ $3.9–5.6 \times 10^{12}/l$	
Haemoglobin	♂ 13.5–18.0g/dl	
	♀ 11.5–16.0g/dl	
Packed red cell volume (PCV)	♂ 0.4–0.54 l/l	
or haematocrit	♀ 0.37–0.47 l/l	
Mean cell volume (MCV)	76–96fl	
Mean cell haemoglobin (MCH)	27–32pg	
Mean cell haemoglobin concentration (MCHC)	30–36g/dl	
Neutrophils	$2.0–7.5 \times 10^9/l$; 40–75% WCC	
Lymphocytes	$1.3–3.5 \times 10^9/l$; 20–45% WCC	
Eosinophils	$0.04–0.44 \times 10^9/l$; 1–6% WCC	
Basophils	$0.0–0.10 \times 10^9/l$; 0–1% WCC	
Monocytes	$0.2–0.8 \times 10^9/l$; 2–10% WCC	
Platelet count	$150.0–400.0 \times 10^9/l$	
Reticulocyte count	0.8–2.0%* $25–100 \times 10^9/l$	
Erythrocyte sedimentation rate	depends on age (p618)	
Prothrombin time (factors II, VII, X)	10–14 seconds	
Activated partial thromboplastin time (VIII, IX, XI, XII)	35–45 seconds	

*Only use percentages as reference interval if red cell count is normal; otherwise use the absolute value. Express as a ratio versus control.

Proposed therapeutic ranges for prothrombin time (British Society for Haematology guidelines on oral anticoagulants, 1984) See p596 & p290.

British ratio (INR)**	Clinical state
2.0–2.5	Prophylaxis of deep vein thrombosis including high risk surgery (eg for fractured femur).
2.0–3.0	Treatment of deep vein thrombosis, pulmonary embolism, transient ischaemic attacks.
3.0–4.5	Recurrent deep vein thrombosis and pulmonary embolism; arterial disease including myocardial infarction; arterial grafts; cardiac prosthetic valves and grafts.

**British ratios and international normalized ratios (INR) are virtually identical within the therapeutic range. Anticoagulation in AF: see p290.

Cerebrospinal fluid—reference intervals

Opening pressure (mmCSF)	Infants: <80; children: <90; adults: <210	
Substance (see also p442)	Reference interval	Your hospital
Glucose	3.3–4.4mmol/l or ⩾²⁄₃ of plasma glucose	
Chloride	122–128mmol/l	
Lactate	<2.8mmol/l	

Reference intervals—biochemistry

(See p626). See *OHCS* p292 for children

Drugs (and other substances) may interfere with any chemical method; as these effects may be method-dependent, it is difficult for the clinician to be aware of all the possibilities. If in doubt, discuss with the lab.

Substance	Specimen	Reference interval	Your hospital
Acid phosphatase (total)	S	1–5 IU/l	
Acid phosphatase (prostatic)	S	0–1 IU/l	
(ACTH)	P	<80 ng/l	
Alanine aminotransferase (ALT)	P	5–35 IU/l	
Albumin	P*	35–50 g/l	
Aldosterone	P**	100–500 pmol/l	
Alkaline phosphatase	P*	30–300 IU/l (adults)	
α-fetoprotein	S	<10 kU/l	
α-amylase	P	0–180 Somogyi U/dl	
Angiotensin II	P**	5–35 pmol/l	
Antidiuretic hormone (ADH)	P	0.9–4.6 pmol/l	
Aspartate transaminase (AST)	P	5–35 IU/l	
Bicarbonate	P*	24–30 mmol/l	
Bilirubin	P	3–17 μmol/l (0.25–1.5 mg/100 ml)	
Calcitonin	P	<0.1 μg/l	
Calcium (ionized)	P	1.0–1.25 mmol/l	
Calcium (total)	P*	2.12–2.65 mmol/l	
Chloride	P	95–105 mmol/l	
†Cholesterol (see p654)	P	3.9–7.8 mmol/l	
VLDL (see p654)	P	0.128–0.645 mmol/l	
LDL	P	1.55–4.4 mmol/l	
HDL	P	0.9–1.93 mmol/l	
Cortisol	P	a.m. 450–700 nmol/l midnight 80–280 nmol/l	
Creatine kinase (CK)	P	♂ 25–195 IU/l ♀ 25–170 IU/l	
Creatinine (related to lean body mass	P*	70–≤150 μmol/l	
Ferritin	P	12–200 μg/l	
Folate	S	2.1 μg/l)	
Follicle-stimulating hormone (FSH)	P/S	2–8 U/l (luteal) †18	
Gamma-glutamyl transpeptidase (γ-GT, GTT)	P	♂ 11–51 IU/l ♀ 7–33 IU/l	
Glucose (fasting)	P	3.5–5.5 mmol/l	
Glycated (glycosylated) haemoglobin	B	5–8%	
Growth hormone	P	<20 mU/l	
Iron	S	♂ 14–31 μmol/l ♀ 11–30 μmol/l	
Lactate dehydrogenase (LDH)	P	70–250 IU/l	
Lead	B	<1.8 mmol/l	
Luteinizing hormone (LH) (pre-menopausal)	P	3–16 U/l (luteal)	
Magnesium	P	0.75–1.05 mmol/l	
Osmolality	P	278–305 mosmol/kg	
Parathyroid hormone (PTH)	P	<0.8–8.5 pmol/l	
Phosphate (inorganic)	P	0.8–1.45 mmol/l	
Potassium	P	3.5–5.0 mmol/l	
Prolactin	P	♂ <450 U/l; ♀ <600 U/l	
Prostate specific antigen	P	0–4 nanograms/ml	
Protein (total)	P	60–80 g/l	

Substance	Specimen	Reference interval	Your hospital
Red cell folate	B	0.36–1.44 µmol/l (160–640 µg/l)	
Renin (erect/recumbent)	P**	2.8–4.5/1.1–2.7 pmol/ml/h	
Sodium	P*	135–145 mmol/l	
Thyroid-binding globulin (TBG)	P	7–17 mg/l	
Thyroid-stimulating hormone (TSH) NR widens with age, see p538	P	0.5–5.7 mU/l	
Thyroxine (T₄)	P	70–140 nmol/l	
Thyroxine (free)	P	9–22 pmol/l	
Total iron binding capacity	S	54–75 µmol/l	
Triglyceride	P	0.55–1.90 mmol/l	
Tri-iodothyronine (T₃)	P	1.2–3.0 nmol/l	
Urea	P*	2.5–6.7 mmol/l	
Urate	P*	♂ 210–480 µmol/l ♀ 150–390 µmol/l	
Vitamin B₁₂	S	0.13–0.68 nmol/l (>150 ng/l)	

Key: P = plasma (heparin bottle); S = serum (clotted; no anticoagulant); B = whole blood (edetic acid—EDTA—bottle); IU = international unit ♂ = male; ♀ = female.

†Desired upper limit of cholesterol would be ~6 mmol/l.
*See OHCS p81 for reference intervals in pregnancy, as there is significant deviation from these values.
**The sample requires special handling: contact the laboratory.

Arterial blood gases—reference intervals

pH:	7.35–7.45	$PaCO_2$:	4.7–6.0 kPa
PaO_2:	>10.6 kPa	Base excess:	±2 mmol/l
		NB: 7.6 mmHg = 1 kPa (atmospheric pressure ≈100 kPa)	

Reference intervals for urine

Substance	Reference interval	Your hospital
Cortisol (free)	<280 nmol/24h	
Hydroxy-indole acetic acid	16–73 µmol/24h	
Hydroxymethylmandelic acid (HMMA, VMA)	16–48 µmol/24h	
Metanephrines	0.03–0.69 micromol/mmol creatinine	
(or <5.5 micromol/day)		
Osmolality	350–1000 mosmol/kg	
17-Oxogenic steroids	♂ 28–30 µmol/24h ♀ 21–66 µmol/24h	
17-Oxosteroids (neutral)	♂ 17–76 µmol/24h ♀ 14–59 µmol/24h	
Phosphate (inorganic)	15–50 mmol/24h	
Potassium	14–120 mmol/24h	
Protein	<150 mg/24h	
Sodium	100–250 mmol/24h	

For reference intervals for cardiac catheterization, see p272.

Useful addresses and telephone numbers

NB: for Poisons Information Services see p746.

Alzheimers Disease Society 158–160 Balham High Road, London SW12 9BN (081–675 6557)

The British Diabetic Association 10 Queen Ann St, London W1M 0BD (071–323 1531)

British Medical Association (BMA) BMA House, Tavistock Square, London WC1H 9JP (071–387 4499)

Bureau of Hygiene and Tropical Medicine Keppel St, London WC1E 7HT (071–636 8636)

Central Public Health Lab 61 Colindale Avenue, London NW9 5HT (081–200 4400)

Committee on Safety of Medicines CSM, Freepost, London SW8 5BR (071–273 3000)

Communicable Disease Surveillance Centre (for up-to-date advice on overseas travel health requirements) 61 Colindale Avenue, London NW9 5HT (081–200 6868)

Disabled Living Foundation (Gives advice on aids and equipment to help the disabled) 380–384 Harrow Rd, London W9 2HU (071–289 6111)

General Medical Council (GMC) 44 Hallam St, London W1N 6AE (071–580 7642)

Health Information Line (will provide a wide range of health information for both doctors and their patients) 0800 665544

Liverpool School of Tropical Medicine Pembroke Place, Liverpool L3 5QA (051–708 9393)

Malaria Reference Laboratory (for advice on malaria prophylaxis) 071–636 8636 (for advice on treatment ring 071–387 4411)

Medic-Alert Foundation 12 Bridge Wharf, 156 Caledonian Rd, N1 9UD (071–833 3034)

Medical Defence Union 3 Devonshire Place, London W1N 2EA (071–486 6181)

Medical and Dental Defence Union 144 West George St, Glasgow (041–332 6646)

Medical Foundation for the Care of Victims of Torture 96–98 Grafton Rd, Kentish Town, London NW5 3EJ (071–284 4321)

Medical Protection Society 50 Hallam St, London W1N 6DE (071–637 0541)

Multiple Sclerosis Society 25 Effie Road, London SW6 1EE (071–736 6267)

Narcotics Abusers' Register Chief Medical Officer, Drugs Branch, Queen Anne's Gate, London SW1H 9YN (071–273 2213)

National Counselling Centre for Sick Doctors (confidentiality guaranteed) 071 935 5982

The Patients' Association (an advice service for patients) 18 Victoria Park Square, London E2 9PF (081–981 5676)

Transplant service (UK) (Can these organs be used?) 0272–507777

Index

809

813

819

833